W9-BHH-368

Human Genetics

The Jones and Bartlett Series in Biology

Basic Genetics, Second Edition
Daniel L. Hartl

Biochemistry
Robert H. Abeles, Perry A. Frey, William P. Jencks

Biological Bases of Human Aging and Disease
Cary S. Kart, Eileen K. Metress, Seamus P. Metress

Biology in the Laboratory
Jeffrey A. Hughes

The Biology of AIDS, Third Edition
Hung Fan, Ross F. Conner, and Luis P. Villarreal

The Cancer Book
Geoffrey M. Cooper

Cell Biology
David Sadava

Cells: Principles of Molecular Structure and Function
David M. Prescott

Concepts and Problem Solving in Basic Genetics:A Study Guide
Rowland H. Davis, Stephen G. Weller

Early Life
Lynn Margulis

Electron Microscopy
John J. Bozzola, Lonnie D. Russell

Elements of Human Cancer
Geoffrey M. Cooper

Essentials of Molecular Biology Second Edition
David Freifelder, George M. Malacinski

Evolution
Monroe W. Strickberger

Experimental Techniques in Bacterial Genetics
Stanley R. Maloy

From Molecules to Cells: A Lab Manual
Nancy Guild, Karen Bever

General Genetics
Leon A. Snyder, David Freifelder, Daniel L. Hartl

Genetics of Populations
Philip W. Hedrick

The Global Environment
Penelope ReVelle, Charles ReVelle

Handbook of Protoctista
Lynn Margulis, John O. Corliss, Michael Melkonian, and David J. Chapman, editors

Human Anatomy andPhysiology Coloring Workbook and Study Guide
Paul D. Anderson

Human Biology
Donald J. Farish

Human Genetics: A Modern Synthesis
Gordon Edlin

Human Genetics: The Molecular Revolution
Edwin H. McConkey

Living Images
Gene Shih, Richard Kessel

Major Events in the History of Life
J. William Schopf

Medical Biochemistry
N.V. Bhagavan

Methods for Cloning and Analysis of Eukaryotic Genes
Al Bothwell, George D. Yancopoulos, Fredrick W. Alt

Microbial Genetics
David Freifelder

Molecular Biology, Second Edition
David Freifelder

Oncogenes
Geoffrey M. Cooper

100 Years Exploring Life, 1888-1988, The Marine Biological Laboratory at Woods Hole
Jane Maienschein

Origin and Evolution of Humans and Humanness
D. Tab Rasmussen, J. William Schopf

Origins of Life: The Central Concepts
David W. Deamer, Gail Raney Fleischaker

Plant Nutrition: An Introduction to Current Concepts
A. D. M. Glass

Protoctista Glossary
Lynn Margulis

Writing a Successful Grant Application, Second Edition
Liane Reif-Lehrer

Human Genetics

The Molecular Revolution

Edwin H. McConkey

Department of Molecular, Cellular, and Developmental Biology
University of Colorado, Boulder

Jones and Bartlett Publishers

Sudbury, Massachusetts

BOSTON LONDON SINGAPORE

Editorial, Sales, and Customer Service Offices

Jones and Bartlett Publishers
40 Tall Pine Drive
Sudbury, MA 01776
508-443-5000
800-832-0034

Jones and Bartlett Publishers International
7 Melrose Place
London W6 7RL
England

Library of Congress Cataloging-in-Publication Data

McConkey, Edwin H.
Human genetics : the molecular revolution / Edwin H. McConkey.
p. cm.
Includes index.
ISBN 0-86720-854-6
1. Human genetics. 2. Human molecular genetics. I. Title.
[DNLM: 1. Genetics, Medical. QZ 50 M4775h]
QH431.M3298 1993
616'.042—dc20
DNLM/DLC
for Library of Congress 92-39661
CIP

Production Service: Herbert Nolan
Copyediting: Kathy Smith
Design: Deborah Schneck
Typesetting: A & B Typesetting, Inc.
Cover Design: Hannus Design Associates
Printing and Binding: Courier
Cover Art Caption and Credit: A human metaphase chromosome spread, showing a reciprocal translocation between chromosome 7 (pink) and chromosome 12 (green). Photo courtesy of Joe Gray, from an original slide by Marc Vooijs.

Printed in the United States of America
97 96 95 10 9 8 7 6 5 4 3

Contents

Preface

Human Genetics: The Molecular Revolution is a survey of the current status of human genetics with emphasis on the impact of molecular information. It focuses not only on fundamental knowledge about genome organization, function, and variation, but also on understanding, diagnosis, and treatment of genetic disease. I have written it for students and professionals in biology and medicine who want an overview of a dynamic subject that will continue to expand rapidly for many years. It is a subject that is becoming increasingly important in clinical practice, and a subject that will certainly have more and more complex effects on the general public as it makes decisions about reproduction, health, and career issues.

The organization of *Human Genetics: The Molecular Revolution* is unique. It begins with the molecular characterization of the human genome, rather than the conventional descriptions of Mendelian inheritance, pedigree analysis, and chromosome abnormalities. The emphasis on understanding human genetics in molecular terms pervades the entire volume. Dominance and recessiveness are not discussed extensively until the reader has a thorough grasp of the molecular aspects of the human genome and the ways in which it varies. When this is accomplished, those classical concepts are analyzed in terms of biochemical reactions.

I believe this is preferable to a historical approach. Today's scientists and physicians expect molecular explanations. To them, genetics implies the study of DNA—its organization, function, and variation. One cannot ignore classical genetics, of course; but modern students feel that it is more relevant to focus on genes as molecules, and then extend their understanding to molecular interpretations of patterns of inheritance, normal function, and genetic disease.

This book will be appropriate for users at a variety of levels. It can be used as a text for a course in human genetics for students who have had a strong introductory course in biology. They should be familiar with the major properties and functions of DNA, RNA, and proteins. It is also assumed that the reader has at least a general acquaintance with the basic principles of heredity, including the concepts of recessive and dominant inheritance. A third assumption is that students who use this book have heard about recombinant DNA and understand that the ability to insert pieces of foreign DNA into microbial genomes and amplify that DNA to any desired extent has revolutionized our capacity to analyze complex genomes, including our own. Most third-year and fourth-year university students who are majoring in biology or a related field have that much background knowledge.

Human Genetics: The Molecular Revolution will also provide essential fundamental information for medical students. I have attempted to show how genetic diseases should be understood in terms of genetic and biochemical principles, and I explain how powerful new molecular techniques have already brought dramatic improvements in our ability to identify the biochemical basis of genetic diseases, how to diagnose them prenatally and postnatally, and the prospects for molecular cures of some of those diseases. It is my hope that medical students who read this book will acquire a broad perspective on the molecular fundamentals of human genetics that will help them integrate the bewildering clinical details they must master in order to diagnose, treat, and manage specific diseases.

Graduate students working toward degrees in molecular biology, cellular biology, or biochemistry should also find the book to be a useful survey of a field in which the original literature has become too voluminous to master in the time they have available. In addition, professionals whose fields require some familiarity with major advances in human genetics will find this book a good source of information.

Background information on specific aspects of human genetics will be found in the chapters that cover those subjects. For example, determination of the true diploid number of human chromosomes (46) and development of staining methods that permit each type of chromosome to be distinguished from all others were fundamental to human cytogenetics; they are discussed in Chapter 2. Techniques for cell hybridization, which led to a tremendous increase in human gene mapping, are described in Chapter 3. The fundamentals of linkage analysis are given in Chapter 4. Chapter 5 relates the story of positional analysis—the identification of genes responsible for diseases for which the primary biochemical defect is unknown. Chapters 6 and 7 examine the molecular basis of mutation, while Chapter 8 discusses the classical concepts of dominance and recessiveness from a modern point of view.

The rapidly growing power of genetic screening, using molecular assays both before and after birth, is summarized in Chapter 9; and in Chapter 10, the exciting potential of gene therapy is described. The next four chapters deal with more specific topics: cancer (Chapter 11), the immune system (Chapter 12), the sex chromosomes (Chapter 13), and the mitochondrial genome (Chapter 14).

The book concludes with a description of the largest and most ambitious biological research project ever undertaken—the Human Genome Project (Chapter 15), which has the primary goal of identifying all human genes. Eventually, the entire sequence of the human genome and the function of all the genes contained therein will be known. Although the rate at which those goals will be achieved is not yet predictable, the simple fact that the possibility of completely analyzing human genome organization and function is considered to be worthy of extensive government support in several technically advanced nations is symbolic of the excitement and confidence that pervades the field of human genetics today. It is a time of breathtaking progress as we explore the enormously complex universe of our own molecular characteristics and potentialities.

EDWIN H. MCCONKEY
January, 1993

Acknowledgments

I am greatly indebted to the six colleagues who reviewed all or parts of the manuscript of this book: Charles Cantor, Boston University; Michael Goldman, San Francisco State University; Edward McCabe, Baylor College of Medicine; David Prescott, University of Colorado, Boulder; David Sadava, The Claremont Colleges; and William Seltzer, University of Colorado School of Medicine. Their comments and suggestions contributed significantly to the accuracy and readability of the final version. I also wish to thank those who donated photographs: Joe Gray, University of California, San Francisco; Maj Hulten, East Birmingham Hospital, Birmingham, U.K.; Beverly Koller, University of North Carolina; Arthur Robinson, National Jewish Hospital, Denver; William Seltzer, University of Colorado School of Medicine; Grant Sutherland, Adelaide Children's Hospital, North Adelaide, Australia; and Jonathan Van Blerkom, University of Colorado, Boulder. Finally, I would like to express my appreciation to Herb Nolan (production services editor), to Art Bartlett, and to all the staff at Jones and Bartlett for their expert and efficient work and their cheerful cooperation.

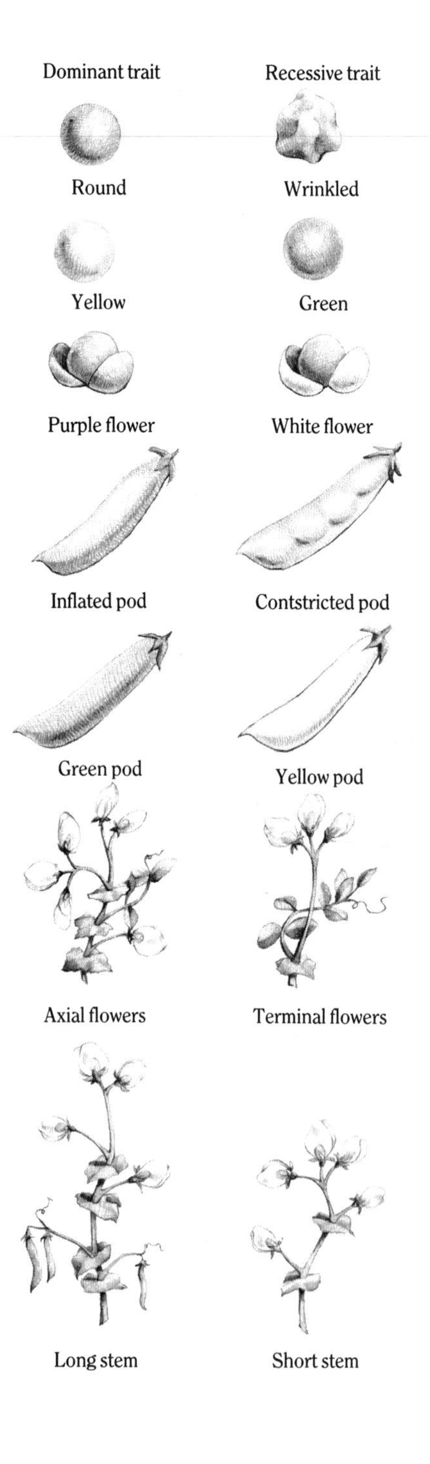

The seven different pairs of traits that Mendel used in his experiments with pea plants. From observing the traits in progeny plants, Mendel deduced the laws of genetics, including the concept of dominant and recessive traits. (Edlin, 1990)

1 Introduction

Some Historical Highlights

The Molecular Era

Sir Archibald Garrod, 1858–1936. The founder of Human Biochemical Genetics. (Harris, 1963)

Human genetics is undergoing a dramatic metamorphosis. A subject that only a few decades ago was characterized principally by the cataloging of genetic diseases and their symptoms, the determination of their patterns of inheritance, and the measurement of their frequencies of occurrence in various populations, human genetics is being revolutionized by molecular biology. The analytic precision and power of human genetics are increasing at a breathtaking rate.

We already know what genes are made of and how their expression is regulated, at least in principle. We know what mutations are, as well as many of their causes, and how they affect the functions of genes. We know how to find genes whose malfunction leads to clinically recognizable disease. We know how to treat some genetic diseases, and we are rapidly devising treatments for others. The largest and most costly biology project ever undertaken—the Human Genome Project—is underway (Chapter 15). The ultimate goals of the project are to identify every gene in the human genome and to sequence as much of the genome as is necessary to achieve that goal. Although the rate at which the Human Genome Program will progress is uncertain for both technical and political reasons, there is little doubt that within the foreseeable future we shall have a molecular description of the human genome that is virtually complete.

Some Historical Highlights

Many authors consider the birth date of molecular biology to be 1953, when Watson and Crick published their model for the double-helical structure of DNA, with its implications for the molecular basis of inheritance. Since then, molecular genetics has built upon the foundation of transmission genetics established by Mendel in 1865 and his successors in the first half of the twentieth century. Prior to Mendel, the prevailing view of inheritance held that the characteristics of parents were somehow blended in the offspring. One of Mendel's greatest insights was the realization that there are *paired units of inheritance* (which we now call *genes*). Sometimes a unit is not expressed in a given individual, if it is a *recessive* allele in the presence of a *dominant* allele, but it retains its identity and may manifest itself in later generations; that is, when two recessive alleles occur in the same genome. (The concepts of dominance and recessiveness, which originated with Mendel, are discussed in detail in Chapter 8.)

Oddly enough, most genetics textbooks do not refer to the concept of inheritable units as one of Mendel's "laws." He is, however, properly given credit for the *Law of Allelic Segregation*, which we now understand to mean that both members of a pair of homologous chromosomes do not normally enter the same gamete; and the *Law of Independent Assortment*, which states that alleles at different genetic loci assort at random during meiosis. Mendel's "Second Law" holds true under two conditions: when the genes under observation are on different chromosomes and when they are on the same chromosome but unlinked (i.e. separated by 50 or more centimorgans).

Mendel did most of his work with the edible pea plant, and as far as is known, he made no effort to extend his conclusions to humans. That being so, when did the study of human genetics begin? A case can be made for the formulation of some basic concepts of human genetics in antiquity. Hippocrates, the father of Western medicine, was aware that individual differences, including illnesses, could be inherited; he explicitly mentioned eye color, bald-headedness, and squinting. In the Near and Middle East in he first few centuries A.D., the authors of the Talmud, a rabbinical commentary on Jewish laws, devoted considerable attention to the condition we now call hemophilia. They were aware that there was a tendency for infant sons in some families to bleed to death after circumcision. Most of the scholars agreed that if two baby boys had suffered that fate, subsequent male children in that family should not be circumcised, at least not until they were older and presumably strong enough to survive the operation.

The year 1865 marked the beginning of human genetics as a science, indirectly because of Mendel's work and directly through the work of Francis Galton. The latter published two articles in 1865 on "Hereditary talent and character," in which he introduced statistical methods for the evaluation of the hypothesis that talent and achievement in the professions were influenced by heredity. Galton went on to found the science of biometric genetics, which brought the power of quantitative analysis into the study of inheritance and laid the groundwork for population genetics.

Charles Darwin apparently never learned of Mendel's work. Darwin wrote extensively on inheritance, carried out breeding experiments on pigeons and plants, and also expressed the belief that many human traits are inherited. He contributed a speculative theory about the mechanism of heredity (pangenesis), but it contained no significant new insight, and it had little impact.

Mendel's work was rediscovered simultaneously by de Vries, Correns, and Tschermak in 1900 and was almost immediately applied to human genetics by Garrod in 1902 in a paper on "The incidence of alkaptonuria: a study in chemical individuality." Alkaptonuria is characterized by urine that darkens quickly upon exposure to air, which is due to the presence of large amounts of homogentisic acid (a fact established in 1891). Garrod presumed that homogentisic acid accumulates in alkaptonuric persons because of an enzyme deficiency, but the state of biochemistry at that time did not permit him to be more specific. We now know that homogentisic acid is a normal product of the catabolism of phenylalanine and tyrosine, and alkaptonuric individuals are homozygous for defects in the enzyme homogentisic acid oxidase (Figure 1–1).

Garrod observed that individuals were either alkaptonuric or not; intermediate states were not found. He also realized that the trait seemed to be

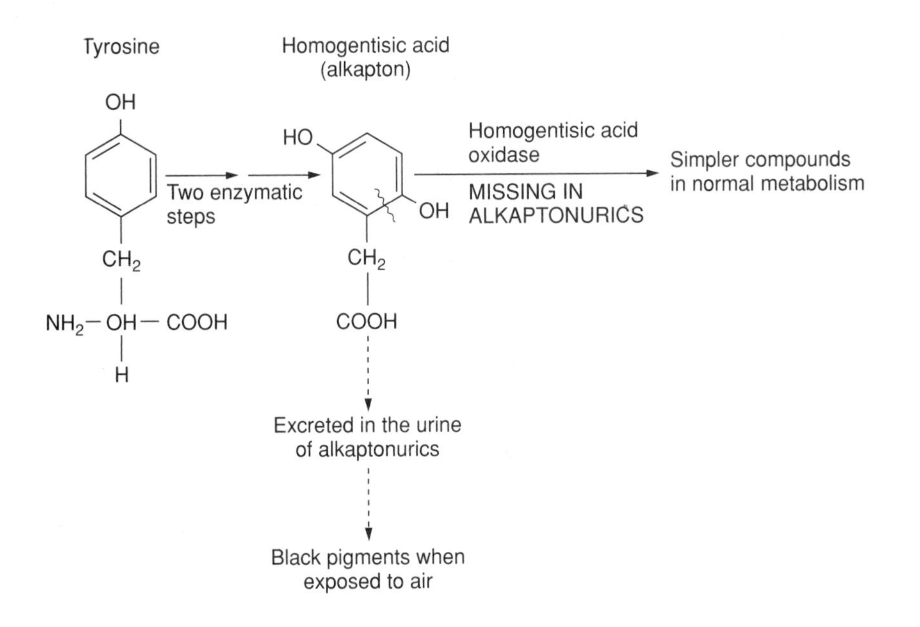

Figure 1–1

The metabolic block in persons with alkaptonuria. Splitting of the six-carbon ring (indicated by the wavy line) by homogentisic acid oxidase during normal metabolism was proposed by Garrod in 1902. His idea that alkaptonurics were missing this enzyme because of a genetic defect was far ahead of its time. (From Mange and Mange, 1990, p. 394; by permission of Sinauer Associates)

inherited, frequently occurring in several siblings from normal parents. Moreover, the parents of alkaptonuric children were often first cousins, and Garrod commented that there must be ". . . an explanation in some peculiarity of the parents, which may remain latent for generations, but which has the best chance of asserting itself in the offspring of a union of two members of a family in which it is transmitted."

Garrod's remarkable insight (stimulated by discussions with Bateson) was to realize that Mendel's concept of recessive inheritable factors provided an explanation for the data on alkaptonuria. Garrod extrapolated this interpretation to other biochemical abnormalities and in 1909 published a classic book, *Inborn Errors of Metabolism*, which clearly presents the paradigm of a chemical basis for genetic disease, resulting from enzyme deficiencies.

He also recognized the significance of biochemical individuality, stating that diseases are ". . . merely extreme examples of variations of chemical behavior which are probably everywhere present in minor degrees and that just as no two individuals of a species are absolutely identical in bodily structure neither are their chemical processes carried out on exactly the same lines."

In the first three decades of the twentieth century there were many fundamental advances in genetics, but these were mostly achieved with experimental organisms, especially maize and *Drosophila*. Among the more important discoveries were the demonstration that chromosomes carry the units of inheritance, that genes are arranged linearly along chromosomes, and that recombination of genes on homologous chromosomes could occur, being associated with the formation of chiasmata (crossovers) at meiosis. In humans, there was the discovery of the ABO blood group in 1900, followed two decades later by the demonstration that the different blood types are caused by multiple alleles at the same locus. And in 1911, E. B. Wilson initiated human gene mapping by deducing that the gene for color blindness is located on the X chromosome (Chapter 13).

Two events in the 1940s foreshadowed the emergence of molecular genetics. First, there was the *one gene-one enzyme hypothesis* of Beadle and Tatum (1941), based largely on work with the bread mold, *Neurospora*. Although the connection between nucleic acids and proteins was not realized immediately, the demonstration that a gene usually specifies the presence of a specific polypeptide was of central importance in the entire field of genetics. Second, there was the proof by Avery, McLeod, and McCarty in 1944 that DNA is the substance that carries inheritable information. Prior to their work on transformation of bacteria, the most popular opinion had been that genes must be proteins.

Another series of events with special significance for human genetics began in the 1940s, involving sickle cell anemia. First, Neel (1949) showed that the pattern of occurrence of the condition had the properties of a genetic disease, with affected individuals being homozygous for a recessive allele. Then Pauling et al. (1949) demonstrated that the disease was associated with a change in the electrophoretic properties of hemoglobin, which implied a chemical difference in the protein. A few years later, Ingram (1957) proved that sickle-cell anemia is caused by a substitution of valine for glutamic acid at amino acid position 6 in beta-globin. This was the first example of an inherited protein variant in which the structural change was precisely identified. More recent studies have proved that the sickle-cell mutation is caused by a change from A to T in the second position of the codon at position 6 (GAG to GTG), as predicted by the genetic code.

The Molecular Era

The transformation of human genetics into a discipline dominated by molecular analysis and molecular explanations of disease has accelerated enormously since the development of recombinant DNA technology in the mid-1970s. The first human gene to be cloned and completely sequenced was the gene for beta-globin (reviewed by Maniatis et al., 1980). Now, there are hundreds of cloned human genes. A panoply of rapid and sensitive techniques for the manipulation of DNA, RNA, and protein stimulated an explosion of activity that touches virtually every aspect of human genetics, ranging from the research lab to the clinic.

From now on, there will be few research projects in human genetics that do not require investigators who are thoroughly trained in molecular biology. Even pure clinicians, who may never enter a research lab themselves, will need to understand the fundamentals of the molecular tests that they order for their patients, so that they neither expect too much nor fail to take advantage of all the information that is potentially available. Assays based on DNA, in particular, will undoubtedly become more numerous and more powerful as time passes. The same will surely be true for immunological assays, which are continually becoming more sophisticated.

Predictions about the potential power of human molecular genetics for the understanding and eventual treatment of disease are easy to make. We can be confident that many optimistic expectations will be fulfilled, although some of the more naive scenarios for gene therapy will surely remain science fiction. Even so, the progress of human genetics is not without its troublesome aspects. The availability of many more sophisticated and expensive assays will add to the nagging problem of how much diagnosis is enough. Gene therapy, if it succeeds at all, will be extremely costly in most cases and will increase the difficulty of deciding who is eligible; this problem has already emerged in the area of organ transplantation.

Social, ethical, and legal problems will proliferate. Should genetic screening programs be mandatory for conditions x, y, and z? Does an employer have a right to test current and potential employees for sensitivity to prevailing unhealthy working conditions and to deny employment on that basis? Do insurance companies have the right to raise rates and/or deny coverage for individuals with certain inherited susceptibilities? How serious must a genetic defect be in order to justify abortion of an affected fetus? What is "normal" anyway?

Despite some obvious potential misuses of genetic information, there is no question that our understanding of the human genome is destined to increase rapidly. The most direct impact of that knowledge will be on the diagnosis and treatment of genetic disease. Although many genetic diseases are rare, the overall burden of genetic disease in the human population is not trivial. Single-gene disorders affect 0.5–1.0% of the general population, and chromosome abnormalities account for about the same frequency of illnesses (Table 1–1).

Common disorders that are not entirely genetic, but that have a significant genetic susceptibility component include heart disease, diabetes, some major cancers, and many autoimmune diseases. Although the genetic basis of susceptibility to these diseases is not yet clear in most cases, one can conservatively attribute at least another 1% of overall morbidity to that category. The actual level is likely to be much higher; for example, recent findings imply that most colon and breast cancers are associated with hereditary susceptibilities (Chapter 11).

Table 1–1 The Total Load of Genetic Disease

TYPE OF GENETIC DISEASE	FREQUENCY/ 1000 POPULATION
Single gene	
Dominant	1.8–9.5
Recessive	2.2–2.5
X-linked	0.5–2.0
Chromosome abnormalities	6.8
Common disorders with a significant genetic component* (The bracketed figures indicate that they are, at best, gross approximations!)	(7–10)
Congenital malformations†	(19–22)
Total (approximate)	(37.3–52.8)

*Genetic contribution, say, one third of such disorders as schizophrenia, diabetes mellitus, and epilepsy.

†Genetic contribution, say, half of malformations like spina bifida, congenital heart disease, cleft lip with or without cleft palate, etc.

Source: Weatherall (1985), p. 33; by permission of Oxford University Press.

The fourth major category in Table 1–1, congenital malformations, is equally difficult to quantitate. It includes conditions like spina bifida, pyloric stenosis, and others where there appears to be a genetic contribution in some cases, but the genetic factors are still obscure. In total, genetic factors probably account for considerably more than 5% of overall disease in developed nations, and this percentage is likely to rise as other sources of morbidity, such as unhealthy diet and environmental pollution, are recognized and controlled.

Another means of estimating the overall impact of inheritance on disease is to use data from pediatric hospitals. Scriver et al. (1973), working in Montreal, found that 11% of pediatric admissions to hospitals were for genetic diseases, and another 18% were for congenital malformations (many of which were presumed to have a genetic basis). Thus, at the pediatric level, almost one-third of admissions were for conditions with a genetic component.

In principle, all genetic diseases can be analyzed in molecular terms, and a great many of them should eventually be treatable and/or curable because of precise molecular descriptions of the underlying defects in DNA and their metabolic consequences. In addition, many non-inherited diseases are certain to be associated with changes in somatic cell DNA. This is already clear for cancers in general (Chapter 11), and there is growing evidence that variations in the DNA rearrangements that accompany differentiation of the immune system (Chapter 12) have an effect on individual susceptibilities to diverse sources of illness, ranging from viral infections to tumors. Somatic variations in mitochondrial DNA (Chapter 14) are another potential source of debility and illness.

The full extent to which detailed knowledge of the human genome and its variations will lead to improvements in human health cannot yet be envisioned, although it is certain that many exciting discoveries await us in the near future. The ethical, legal, and social consequences of that knowledge are even less predictable, but they will surely be profound.

REFERENCES

Avery, O. T., MacLeod, C. M., and McCarty, M. 1944. Studies on the chemical nature of the substance inducing transformation of Pneumococcal types. J. Exp. Med. 79:137–157.

Beadle, G. W. and Tatum, E. L. 1941. Genetic control of biochemical reactions in Neurospora. Proc. Natl. Acad. Sci. USA 27:499–506.

Edlin, G. 1990. Human Genetics: A Modern Synthesis. Jones and Bartlett, Boston.

Galton, F. 1865. Hereditary talent and character. Macmillan's Magazine 12:157–166 and 12:318–327.

Garrod, A. E. 1902. The incidence of alkaptonuria: a study in chemical individuality. Lancet 1902/II:1616–1620.

Garrod, A. E. 1909. Inborn errors of metabolism. Oxford Univ. Press, Oxford, England.

Harris, H. 1963. Garrod's Inborn Errors of Metabolism. Oxford Univ. Press, Oxford, England.

Ingram, V. M. 1957. Gene mutations in human haemoglobin: the chemical difference between normal and sickle cell haemoglobin. Nature 180:326–328.

Mange, A. P. and Mange, E. J. 1990. Genetics: human aspects. 2nd ed. Sinauer Associates, Sunderland, MA.

Maniatis, T., Fritsch, E. F., Lauer, J., and Lawn, R. M. 1980. The molecular genetics of human hemoglobins. Ann. Rev. Genet. 14:145–178.

Mendel, G. J. 1865. Versuche ueber Pflanzenhybriden. Verhandlungen des Naturforschenden Vereins (Bruenn) 4:3–47.

Neel, J. V. 1949. The inheritance of sickle cell anemia. Science 110:64–66.

Pauling, L., Itano, H. A., Singer, S. J., and Wells, I. C. 1949. Sickle cell anemia, a molecular disease. Science 110:543–548.

Scriver, C. R., Neal, J. L., Saginur, R., and Clow, A. 1973. The frequency of genetic disease and congenital malformation among patients in a pediatric hospital. Canad. Med. Assoc. J. 108:1111–1115.

Watson, J. D. and Crick, F. H. C. 1953. Genetical implications of the structure of deoxyribosenucleic acid. Nature 171:964–967.

Weatherall, D. J. 1985. The new genetics and clinical practice. 2nd ed. Oxford Univ. Press, Oxford, England.

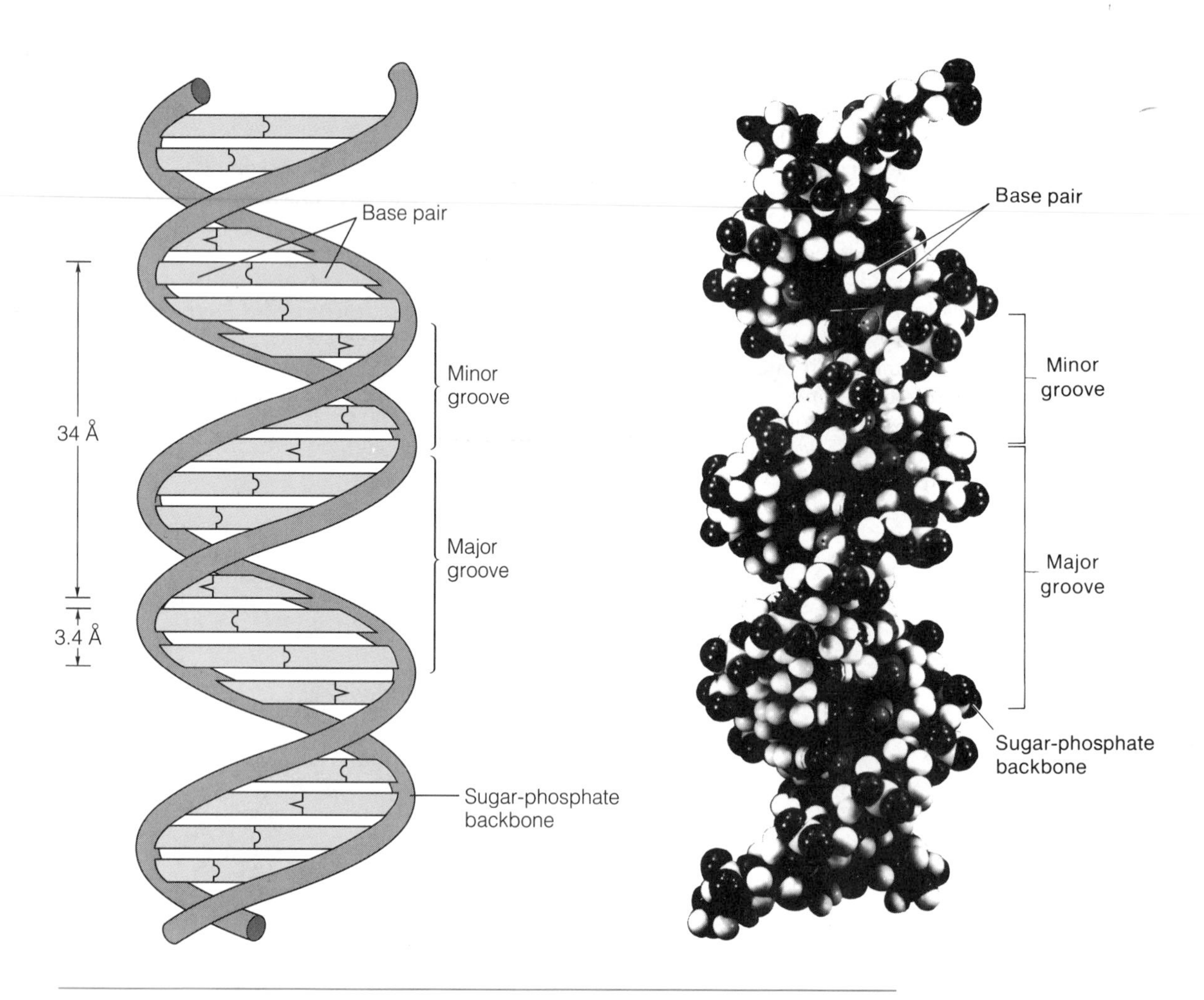

The DNA molecule. On the left is a diagrammatic model of the double helix. At right, a space-filling model of the DNA double helix. The sugar-phosphate backbone winds around the outside of the paired strands, and the base pairs bridge between the strands. (Hartl, 1991)

2 Organization of the Human Genome

Three billion pairs of *nucleotides* in DNA, arranged in the right order, are responsible for the fact that you can read this book, ponder the information and concepts that it contains, and discuss them with other humans. The precise arrangement of those three billion pairs of nucleotides in each person's chromosomes is also the basis for most of the physical details that distinguish each of us from every other human being, and presumably for many of the intellectual and behavioral characteristics that contribute to our individual personalities.

There is nothing quantitatively special about 3×10^9 nucleotide pairs (or *base pairs,* as the subunits of DNA are more frequently, if less accurately, referred to). It is a typical mammalian genome; more precisely, it is a *haploid genome*—the amount of DNA in one complete set of chromosomes. It is the total amount of DNA in a gamete (egg or sperm), but only half the DNA in a somatic cell, which ordinarily contains two sets of nuclear chromosomes. Approximately the same amount of DNA is found in the cells of mice and elephants, aardvarks and armadillos, bats and wombats, pandas and pangolins. There are also some amphibians, some fishes, and even some plants with approximately the same size genomes as humans, as well as some organisms with much larger genomes.

In all living things, from bacteria to blue whales, the genes encoded by DNA are collected together into *chromosomes*. Prior to DNA replication in preparation for cell division, a chromosome consists of a single DNA molecule that contains many thousands, and sometimes many millions, of nucleotide pairs. After DNA replication, but prior to anaphase of mitosis, a eukaryotic chromosome contains two *chromatids*, each of which is a separate DNA molecule, held together by proteins. The presence of two chromatids does not become detectable in the light microscope until prophase of mitosis; actual separation occurs at the beginning of anaphase.

In humans there are 24 types of nuclear chromosomes: 22 *autosomes* (which are represented by two copies in every somatic cell, but only one copy in gametes), an *X chromosome* (present in two copies in female somatic cells, but only once in male somatic cells), and a *Y chromosome* (present only in males). Our nearest evolutionary relatives, the great apes, have 23 pairs of autosomes. Apparently two chromsomes fused to produce human chromosome 2 after the human and chimpanzee lines separated, approximately 5 or 6 million years ago.

There is also a *mitochondrial chromosome* present in the cytoplasm of nearly all cells, but in variable numbers. The mitochondrial chromosome, which is non-nuclear and has a distinctive pattern of inheritance, is not included in the diploid or haploid number of chromosomes (46 and 23, respectively, in humans). Chapter 14 covers mitochondrial DNA and the human diseases that result from mutations therein.

Cells in the late prophase or metaphase stages of mitosis have condensed chromosomes that possess distinctive morphological features visible in the light microscope. In order to make those features detectable, chromosomes from cells arrested in mitosis are spread onto microscope slides, as illustrated in Figure 2–1. The word *karyotype* refers to the collective features of a set of condensed chromosomes. A typical human metaphase karyotype is shown in Figure 2–2. It is not difficult to distinguish human chromosomes from those of other mammals, whatever they may be, using size, shape, and the banding pattern produced by certain cytochemical stains (described in the next section). Even the chromosomes of our nearest relatives, the chimpanzees and the gorillas, have diagnostic differences from the human, although the differences are strikingly few in number (Figure 2–3).

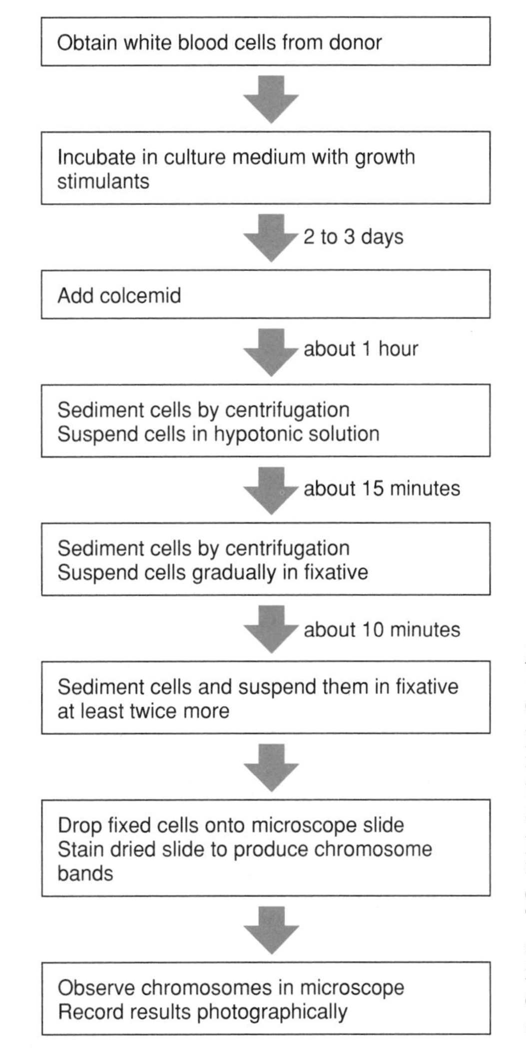

Figure 2–1

Outline of an experimental procedure for observing human metaphase chromosomes. Note the addition of colcemid (or its analog, colchicine), which prevents formation of the mitotic spindle and traps chromosomes in metaphase. The brief soaking in a low-salt solution swells the cells, facilitating spreading of the chromosomes when they are dropped onto a slide.

One cannot look at a human karyotype and see anything qualitatively different from all other mammals; there is no gross evidence of our abilities to make speeches, write poetry, play the violin, or create cosmologies. Presumably these miracles are encoded within the DNA, and, human curiosity being boundless, it is predictable that someday we shall know virtually

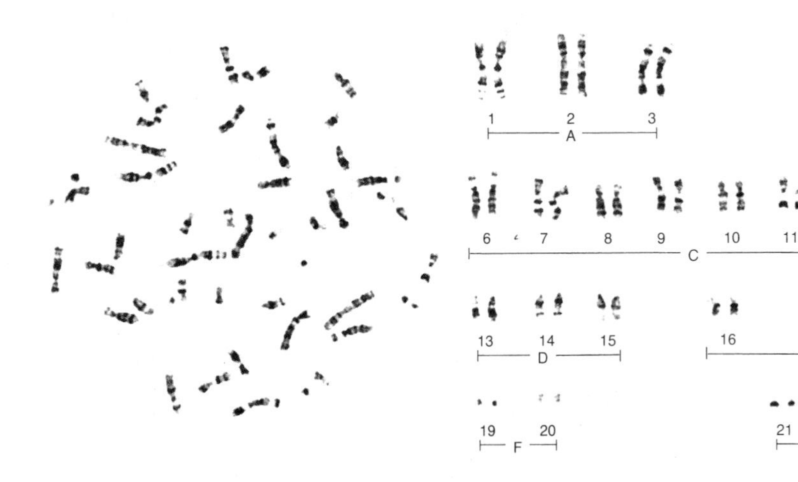

Figure 2–2

A karyotype of a normal human male, showing G bands. Metaphase chromosomes from white blood cells were prepared as shown in Figure 2–1. The left panel shows the chromosomes as they appear on a microscope slide. The right panel shows the chromosomes after they have been cut out of a photograph and arranged in homologous pairs, in numerical order. The letters refer to the seven groups, based on size and centromeric index, which were defined before banding methods made every pair of chromosomes distinguishable from every other pair. (From Snyder et al., 1985, p. 199)

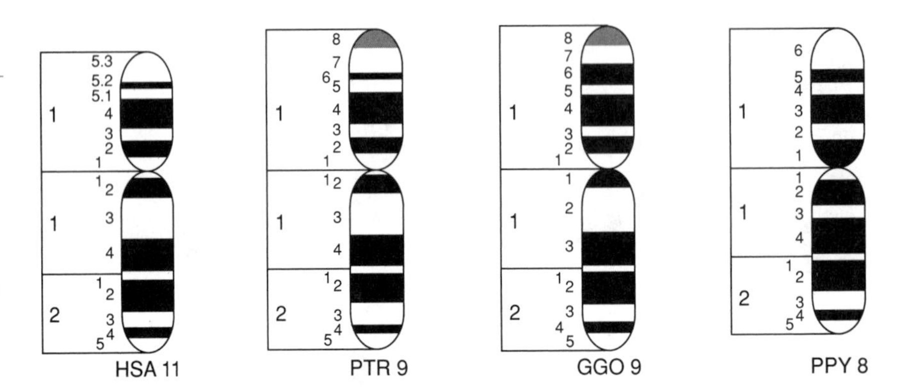

Figure 2–3

Example of similar banding patterns in human and great ape chromosomes. Human (HSA) chromosome 11 is homologous to chromosome 9 from chimpanzees (PTR) and gorillas (GGO) and to chromosome 8 from orangutans (PPY). (From Harnden and Klinger, ISCN 1985, p. 105; by permission of S. Karger AG)

every function and most of the potentialities of the human genome. That day will not come soon, but it is not centuries in the future. The human genome is about to be analyzed with a thoroughness and on a scale that seemed impossible only a few years ago. Before we discuss the future of human genome analysis (Chapter 15), we must survey the general organizational features of the human genome.

General Features of Human Chromosomes

The fact that diploid human cells contain 46 chromosomes was not known until 1956; for many years prior to that, it was thought that the number was 48 because a respected geneticist had come to that conclusion. However, early chromosome preparations were difficult to interpret; metaphases were rare, and there were always overlapping chromosomes. Two technical advances enabled Tjio and Levan (1956) to show unequivocally that the correct number is 46. These were: (1) the use of colchicine to arrest cells in metaphase by preventing formation of the mitotic spindle, and (2) the use of a hypotonic medium to swell cells before fixing them on microscope slides, which results in chromosomes that are well separated from one another (Figure 2–1).

In the early days of human cytogenetics, chromosomes were stained with dyes that bound to DNA indiscriminately and did not reveal any substructure. The only way to distinguish one chromosome from another was by size and the position of the *centromere*, which is either midway between the ends of the chromosome (a *metacentric* chromosome), somewhat closer to one end than the other (a *submetacentric* chromosome), or close to one end (an *acrocentric* chromosome). Truly telocentric chromosomes (centromere at one end) occur in some mammals, but not in humans. Based on size and centromere position, seven groups of human chromosomes were defined in 1960 at a meeting of specialists in Denver, Colorado (see Table 2–1 and Figure 2–2). The individual members within a chromosome group were not distinguishable.

During the next few years, several staining methods were developed that revealed substructures within chromosomes. These substructures had the form of alternating dark and light bands of various widths. One pattern could be produced with quinacrine compounds (Q bands) or with Giemsa stains (G bands, Figure 2–2); another pattern, which reversed the light versus dark pattern, was obtained with Giemsa applied under different conditions (R bands). C bands (Figure 2–4) correspond to constitutive *heterochromatin*, which is found in all centromeres and in the long arm of the Y

Table 2–1 Relative lengths and centromeric indexes of human metaphase chromosomes

CHROMOSOME NUMBER	GROUP[a]	RELATIVE LENGTH[b]	CENTROMERIC INDEX[c]
1	A	8.4	48
2	A	8.0	39
3	A	6.8	47
4	B	6.3	29
5	B	6.1	29
6	C	5.9	39
7	C	5.4	39
8	C	4.9	34
9	C	4.8	35
10	C	4.6	34
11	C	4.6	40
12	C	4.7	30
13	D	3.7	17
14	D	3.6	19
15	D	3.5	20
16	E	3.4	41
17	E	3.3	34
18	E	2.9	31
19	F	2.7	47
20	F	2.6	45
21	G	1.9	31
22	G	2.0	30
X	C	5.1	40
Y	G	2.2	27

[a] The seven groups (A through G) were designated before banding techniques were developed. They are based on relative lengths and centromeric indexes.

[b] Relative length is expressed as a percentage of the total haploid *autosome* length.

[c] The centromeric index is the length of the short arm divided by [total chromosome length × 100].

Source: Data in this table are from Harnden and Klinger (ISCN 1985; by permission of S. Karger AG); they are based on measurements of 95 somatic cells.

chromosome (see next section). C bands are produced by staining after removal of the majority of DNA by acid and alkali treatment.

The chemical reactions that produce these various classes of stainable bands are not entirely clear, but some correlations can be made with base composition and with the distribution of various classes of repetitive DNA (Bickmore and Sumner, 1989). Bands that stain darkly with Giemsa in the standard technique contain DNA that is relatively (A+T)-rich, whereas the Giemsa-light bands contain relatively (G+C)-rich DNA. The DNA in Giemsa-dark bands tends to replicate late in the cell cycle, and relatively few genes are located there. DNA in Giemsa-light bands replicates early and contains about three-fourths of genes that have been localized to specific bands.

The ability to detect band patterns in metaphase chromosomes with cytochemical stains made each chromosome distinguishable from all the others; this was a major advance in cytogenetics and it laid the foundation for gene mapping studies. At a meeting of specialists in Paris in 1971, each pair of autosomes was given a number; the numbers were then arranged in

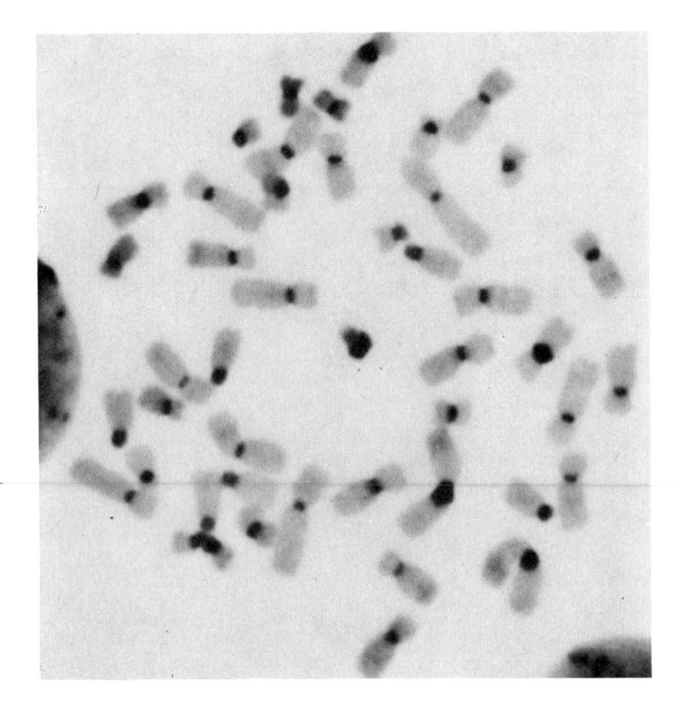

Figure 2–4

Light micrograph of human metaphase chromosomes treated and stained to show C bands, which are located in the centromeric region of every chromosome. (From Prescott, 1988, p. 393)

order of size from largest to smallest, except that number 21 is actually somewhat smaller than number 22.

Table 2–1 gives the relative lengths and *centromeric indexes* (the ratio of the length of the small arm to the total length of the chromosome) of human somatic cell metaphase chromosomes. The sex chromosomes are not numbered. Of course, the two members of a homologous pair of autosomes are not ordinarily distinguishable by cytochemical methods. Another nomenclatorial convention adopted at the Paris Conference was to use the letter *p* for the short arm and the letter *q* for the long arm of each chromosome.

Within each arm of a chromosome, numbers are assigned to large areas called *regions*, and another set of numbers is used for *bands* within regions. As explained in the 1985 Report of the Standing Committee on Human Cytogenetic Nomenclature (Harnden and Klinger, 1985, p. 16), ". . . a *band* is a part of a chromosome clearly distinguishable from adjacent parts by virtue of its lighter or darker staining intensity. The bands are allocated to various regions along the chromosome arms, and the regions are delimited by specific *landmarks*. These are defined as consistent and distinct morphologic features important in identifying chromosomes. Landmarks include the ends of the chromosome arms, the centromere, and certain bands. The bands and the regions are numbered from the centromere outward. A *region* is defined as any area of a chromosome lying between two adjacent landmarks."

Examples of region and band numbering are shown in Figure 2–5. Note that each digit is pronounced separately; for example, band 7q21 is not referred to as "7q twenty-one," it is "7q two-one." A third set of digits and sometimes a fourth set are used for high-resolution banding patterns. The third and fourth digits follow a decimal point. Thus, the terminal band of the short arm of chromosome 1 is 1p36.33 (spoken as "1 p three-six-point-three-three").

The power of chromosomal banding extends far beyond the descriptive goal of assigning numbers to each chromosome. It allows us to identify translocated fragments of chromosomes; it makes it possible to identify

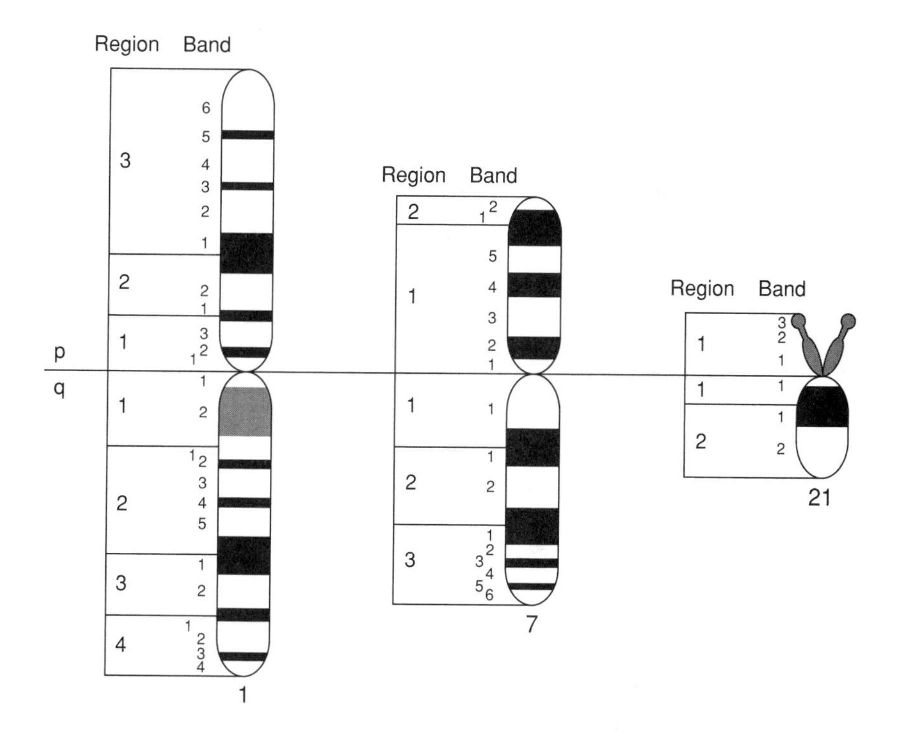

Figure 2–5

The Giemsa banding patterns of several human chromosomes at metaphase. Chromosome 1 is the largest, chromosome 21 is the smallest, and chromosome 7 is of average size. The centromeres have been aligned, with the small (p) arm up and the large (q) arm down. (From Harnden and Klinger, ISCN 1985, pp. 112–113; by permission of S. Karger AG)

small deletions or duplications that may correlate with a particular abnormality; and it enables us to identify specific human chromosomes or portions thereof in human-rodent hybrid cells, which is of great importance in gene mapping (Chapter 3).

The standard Giemsa band map of human chromosomes is diagrammed in Figure 2–6. Approximately 300 bands are detectable when metaphase chromosomes are stained. Considerably more detail can be seen when prometaphase or late prophase chromosomes are stained. This high-resolution (but technically demanding) technique, introduced by Jorge Yunis in 1976, reveals approximately 1200 bands (Figure 2–7), and is proving to be especially useful in the detection of small deletions and other chromosomal rearrangements, as well as in gene mapping by *in situ* hybridization (Chapter 3). It is important to realize that whatever chromosomal bands may represent at the molecular level, they do not correspond to genes; the total number of human genes is unlikely to be less than 30,000 and may be much higher (see last section of this chapter). Therefore, the average metaphase band must contain at least 100 genes. Table 2–2 gives some useful numerical parameters relating chromosomes, base pairs in DNA, and recombination units.

Families of Reiterated Sequences Account for About One-Third of the Human Genome

In all higher eukaryotes, a substantial fraction of the genome consists of *reiterated* (or *repeated*) sequences. In mammals there are typically many degrees of reiteration, and humans are no exception. The presence of reiterated DNA is detected by analysis of the rate at which short pieces of denatured DNA reassociate in solution to form double-stranded molecules under standard experimental conditions; the more copies there are of a given sequence, the faster they reassociate. This technique is explained in Box 2–1.

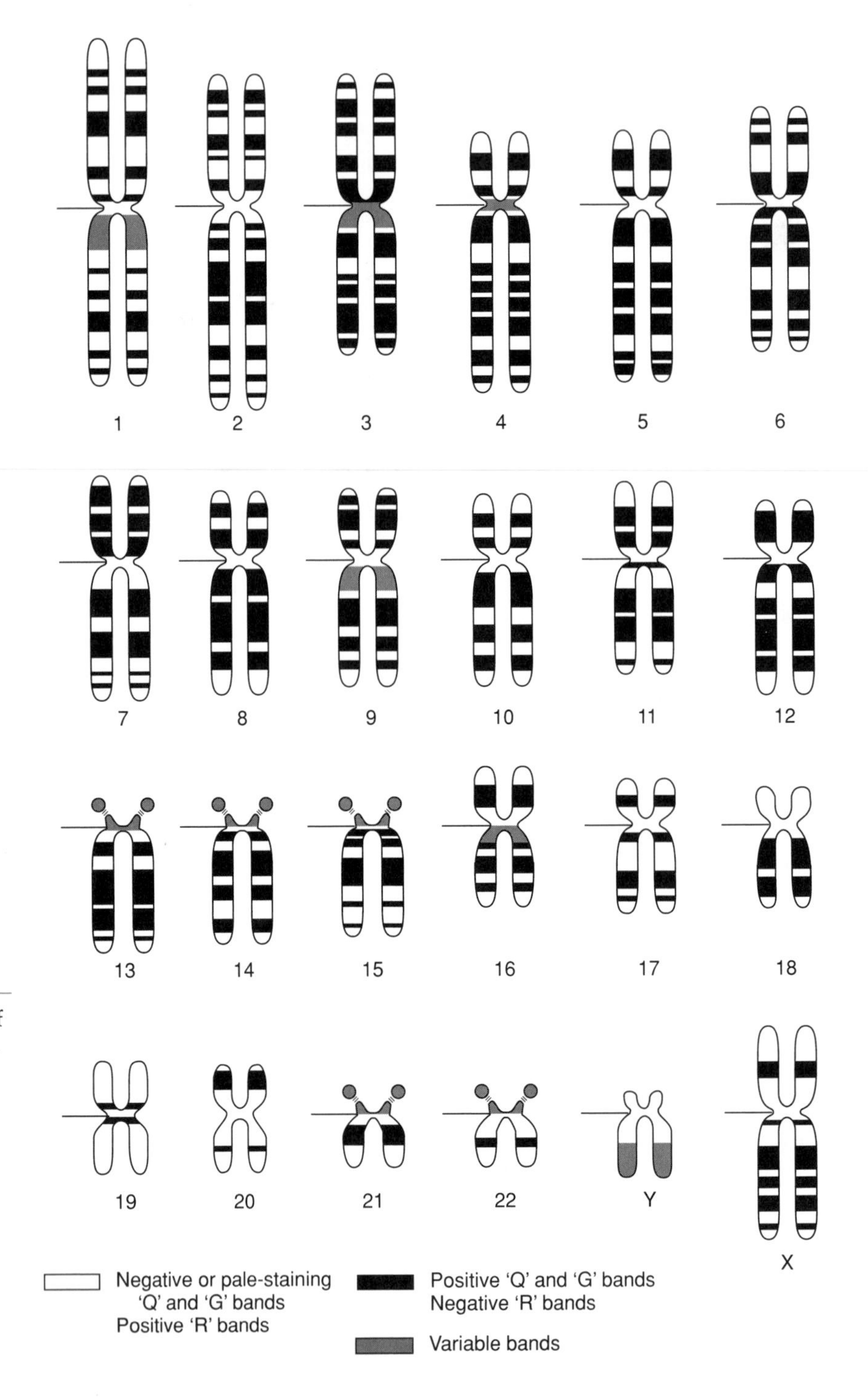

Figure 2–6

Diagram of banding patterns of human chromosomes obtained as a composite of various staining procedures. G bands are obtained with standard Giemsa stain; R bands are obtained with reverse Giemsa stain (R bands are dark where G bands are light and vice versa); Q bands are obtained with quinacrine stain. (From Harnden and Klinger, ISCN 1985, pp. 112–113; by permission of S. Karger AG)

An example of the type of data obtained with human DNA is shown in Figure 2–8. The figure illustrates several important points. First, more than 50% of human DNA renatures at the rate expected for *single-copy sequences* (represented by the lower portion of the curve, extending from approximately C_0t 100 to C_0t 10,000). This is the DNA fraction that contains most of the protein-coding genes. The term "single-copy" should not be interpreted in an absolute sense; genes present in a few copies per genome will undoubtedly also be included in this fraction.

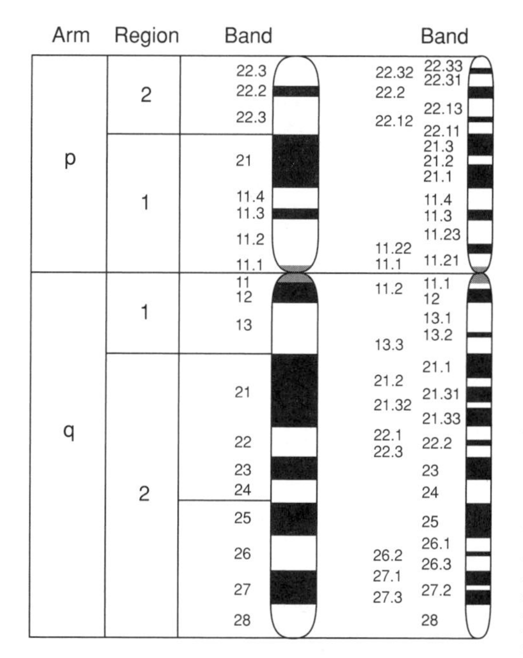

Figure 2–7

Diagrams of G-banding patterns of the human X chromosome at mid-metaphase (left pattern) and at late prophase (right pattern). (From Harnden and Klinger, ISCN 1985, p. 57; by permission of S. Karger AG)

The second kinetic class of DNA implied by the curve in Figure 2–8 extends from about C_0t 10 to C_0t 0.1 ($C_0t_{1/2}$ = 1); it accounts for roughly 20 to 30% of total DNA and represents sequences present about 500X, on average. It is virtually certain that this portion of the reassociation curve is actually the summation of a spectrum of DNA sequences of various degrees of reiteration; no single class is sufficiently abundant to be detected as a separate component by this relatively insensitive technique. Thus, one would expect a sequence repeated 5000X to contribute to the portion of the curve at C_0t 0.1 and a sequence repeated only 50X to contribute to the portion of the curve at C_0t 10. All of these are *middle repetitive DNA*.

The third kinetic class of DNA shown in Figure 2–8 has $C_0t_{1/2}$ = 0.01,

Table 2–2 General characteristics of the haploid human genome

Number of base pairs	3×10^9
Mass	3 picograms = 1.8×10^{12} daltons
Fraction in single-copy DNA	Approximately 2/3
Average size of a gene	20,000 bp
Average exon/intron ratio	1/10
Average intergenic distance	Unknown
Total number of genes	Probably 30,000 to 100,000
Total genetic map distance (recombination units)	About 3000 centimorgans
Base pairs/centimorgan	Approximately 10^6, on average
Chromosomes types	22 autosomes X and Y sex chromosomes 1 mitochondrial chromosome
Chromosome substructure (Giemsa stain)	300 bands (metaphase) 1000–1200 bands (late prophase)
Base pairs/ metaphase band	10^7, on average

BOX 2-1 Detection of Reiterated DNA

The presence of reiterated sequences in the genomes of higher organisms can be detected by analysis of the rate at which complementary single-stranded DNA sequences reassociate with each other to form double-stranded structures. This process, which is also called *renaturation* or *annealing*, is illustrated in Figure 2B1–1.

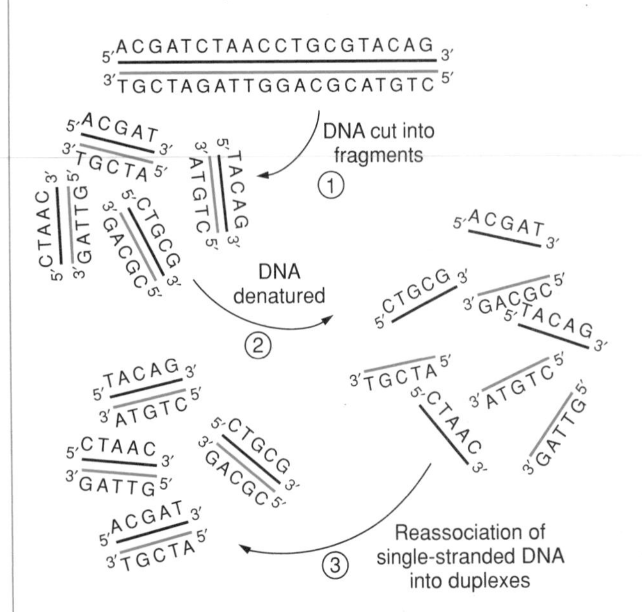

Figure 2B1–1

Denaturation of duplex DNA and reassociation of single strands into duplexes. (1) Duplex DNA cut into short fragments. (2) Duplex fragments denatured into single strands by heat. (3) Reassociation of single strands into duplexes after cooling. In an actual experiment the duplex DNA is cut into fragments of several hundred base pairs. (From Prescott, 1988, p. 330)

The rate at which renaturation takes place depends on several factors, including temperature, ionic strength of the solution, size of the DNA fragments, and the concentration of each sequence in the mixture. *The effect of the concentration of DNA sequences on renaturation rate is the key to the identification of reiterated sequences.* When a simple genome with essentially no repeated sequences, such as bacteriophage T4 or *E. coli*, is cleaved into small pieces of a few hundred nucleotides apiece, the concentration of every sequence in the mixture will be the same as the concentration of every other sequence. For example, if the 160,000 bp genome of phage T4 were cleaved into uniform 500 bp fragments, there would be 320 different pieces (in terms of sequences), but every piece would be at the same molar concentration, regardless of the overall concentration of DNA in the reaction vessel.

The rate-limiting step for reassociation of DNA is the collision of two molecules (the complementary strands) such that hydrogen bond formation is reestablished between them, as shown in the following equation:

$$C/C_0 = 1/(1 + kC_0t),$$

where C_0 is the concentration of DNA in moles of nucleotides per liter at the beginning of the experiment, C is the concentration of single-stranded DNA remaining at time t (measured in seconds), and k is a constant that depends on the physical and chemical parameters of the experiment.

The expression C_0t is pronounced the same as the word for a small, narrow bed. As an example of C_0t, consider a solution of sheared, denatured DNA at 600 micrograms/ml (= 600 mg/L = 2×10^{-3} M, in terms of nucleotides), allowed to renature for 500 seconds; this would generate C_0t = 1. In order for the same solution to reach C_0t = 1000, 139 hours (nearly 6 days) of renaturation would be required.

A plot of C/C_0 versus log C_0t is called a C_0t curve. Another useful quantity is $C_0t_{1/2}$, the C_0t at which 50% of the sequences have renatured.

C_0t curves for genomes without reiterated sequences are simple S-shaped figures that differ only in the rate of renaturation. Several examples are shown in Figure 2B1–2. Notice that $C_0t_{1/2}$ is proportional to the size of the genome. For example, the T4 genome contains 160,000 base pairs and its $C_0t_{1/2}$ is between 0.25 and 0.30; the *E. coli* genome contains 4.6 × 10^6 bp (28–30 times as large as T4), and its $C_0t_{1/2}$ is about 8.

When DNA from a mammal is renatured, a simple S-shaped curve is not obtained; instead, the C_0t curve looks as though several S-curves, each representing a different rate of renaturation, were stacked

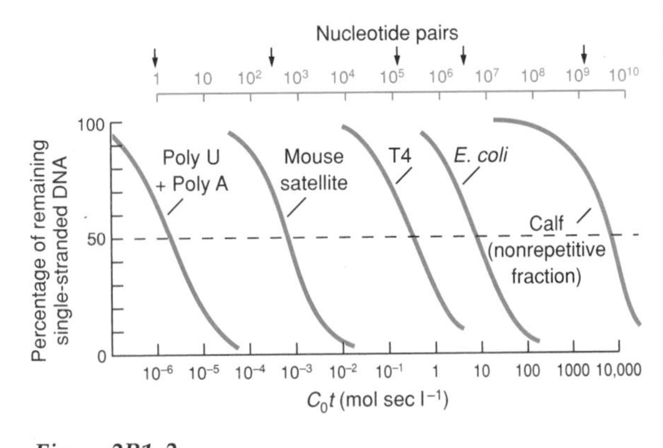

Figure 2B1–2

A set of C_0t curves for various DNA samples. The black arrows on the red scale indicate the number of nucleotide pairs for each sample and also point to the intersection of each curve with the horizontal red line, which is $C_0t_{1/2}$, the point of half renaturation. (From Snyder et al., 1985, p. 173).

on top of one another. A typical mammalian renaturation curve is diagrammed in Figure 2B1-3.

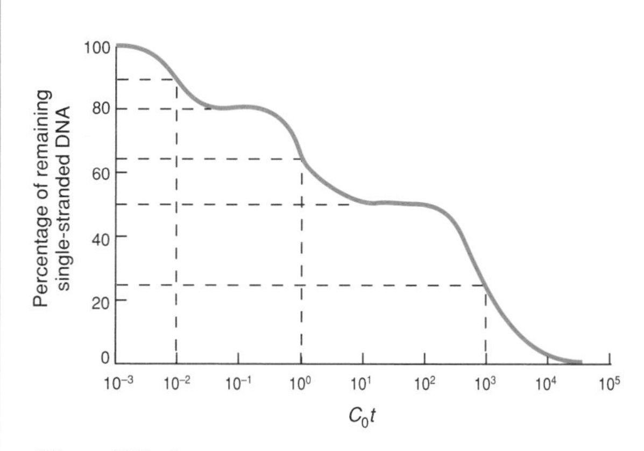

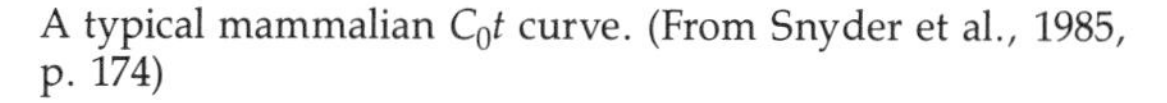

Figure 2B1-3

A typical mammalian C_0t curve. (From Snyder et al., 1985, p. 174)

These complex renaturation curves result from the presence of reiterated DNA. *Highly reiterated DNA* is repeated many thousands of times; it accounts for the upper portion of the curve with the fastest rate of renaturation. *Moderately reiterated DNA* (also called *middle-repetitive DNA*) is the most complex class. It includes sequences that are reiterated from tens to hundreds to about a thousand times; it is represented by the middle portion of the curve, with intermediate rates of renaturation. Finally, there is a large group of unique or nearly unique sequences (*single-copy DNA*), which accounts for the lowest portion of the curve, with the slowest renaturation rate. These are gross classes; the insensitivity of the technique obscures the fact that there are different degrees of repetition within each major group. Nevertheless, it is clear that there is not equal representation of all possible frequencies of reiteration. The tendency for DNA to fall into highly repeated, moderately repeated, and unique sequence classes is indisputable.

REFERENCES

Prescott, D. M. 1988. *Cells*. Jones and Bartlett, Boston.

Snyder, L. A., Freifelder, D., and Hartl, D. L. 1985. *General Genetics*. Jones and Bartlett, Boston.

corresponding to an average repetition frequency of 50,000, but extending to sequences that are repeated at least 500,000 times. All of these sequences are dubbed *highly reiterated DNA*. They constitute roughly 10% of the genome.

Reiterated DNAs may also be classified according to whether they are arranged as *tandem* series of repeats or are *interspersed* among unique sequence DNA. Highly reiterated, tandemly arranged DNA sequences were

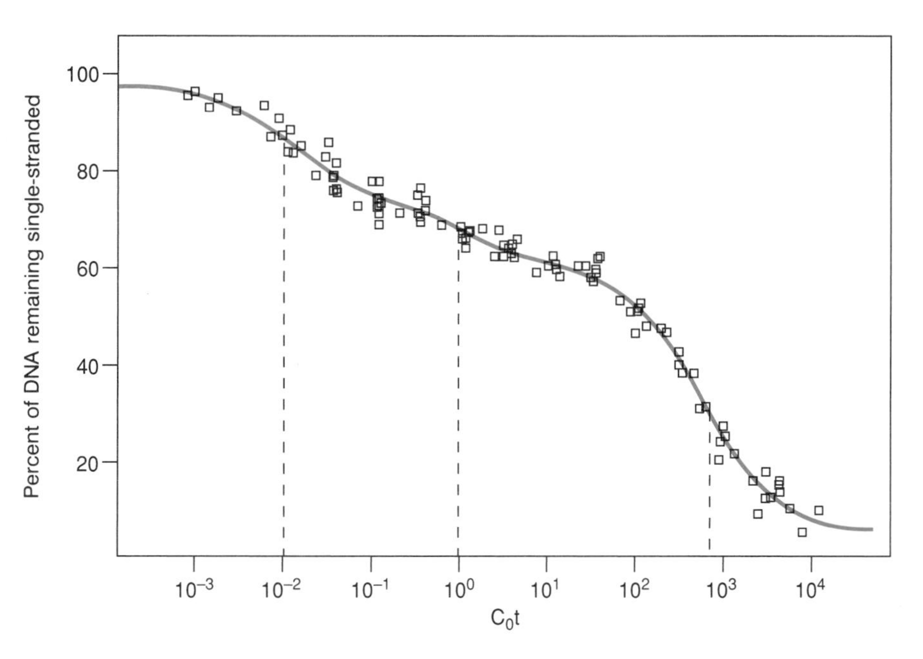

Figure 2–8

Renaturation profile of sheared, denatured human DNA. Vertical dashed lines indicate approximate $C_0t_{1/2}$ values for highly reiterated, moderately reiterated, and single copy DNA classes. (Modified from Schmid and Deininger, 1975; by permission of Cell Press)

first discovered by the technique of CsCl buoyant density gradient centrifugation, because some of them came to equilibrium at slightly different densities than bulk DNA (Figure 2–9). In a CsCl solution, a GC base pair is more dense than an AT base pair. If a tandemly repeated sequence contains a relatively high proportion of G+C, it will come to equilibrium in a CsCl gradient (generated by centrifugal force) at a density that is higher than the density of bulk DNA. Tandemly repeated sequences with relatively high A+T content will form a band in a CsCl gradient that is at lighter density than bulk DNA.

Thus, highly reiterated tandem sequences were seen as separate, minor bands or as bumps on the main band when a plot of density versus DNA concentration in a CsCl gradient was made (Figure 2–9). The term *satellite DNA* originated from this relationship, and it has remained in general use for tandemly arranged, highly reiterated DNAs ever since.

In humans, the best known satellite DNA is the *alpha* family of sequences, which is found in centromeres and has a basic repeat unit of 171 bp. Alphoid sequences tend to be organized in tandem arrays up to several million bp long, within which there may be substantial sequence variation, although the arrays themselves are repeated rather accurately. The size of these alphoid arrays and the number of repeats of the array are chromosome-specific (Willard and Waye, 1987).

Although their association with centromeres raises the possibility that they play a role in centromere function (i.e., the association of sister chromatids and/or the attachment of the mitotic spindle to chromosomes), the function of the alpha satellite sequences has not yet been determined. Recent work indicates that an 80 kiloDalton protein, CENP-B, binds to a 17 bp portion of some members of the alphoid DNA family, but the functional significance of that association is not yet clear (Willard, 1990). Several other centromere-associated proteins have been described. There are also a few minor satellite DNAs found in secondary constrictions of several chromosomes, in the heterochromatic portion of the Y chromosome, and sometimes in centromeres. Satellite DNAs are not transcribed.

Another sequence with a relatively high level of repetition is the hexameric element TTAGGG, which is found at the ends of every human chromosome (the *telomeres*)) in tandem arrays up to 5000–10,000 bp in length (Blackburn, 1991; Moyzis, 1991).

Interspersed repetitive sequences are classified as either short or long. Singer (1982) introduced the term SINEs for short interspersed repetitive sequences, and the term LINEs for long interspersed repetitive sequences. SINE is an acronym for *S*hort *In*terspersed *E*lement and LINE represents *L*ong *In*terspersed *E*lement. Both acronyms have been widely used.

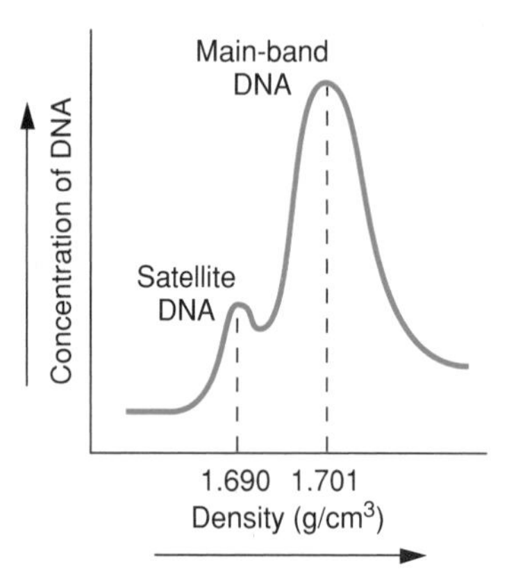

Figure 2–9

Separation of mouse main-band and satellite DNA by equilibrium centrifugation in a CsCl density gradient. (From Snyder et al., 1985, p. 168)

The most abundant and best-studied member of the SINE class is the Alu family, consisting of several hundred thousand copies of a 300 bp sequence, which is flanked by recognition sites for the restriction enzyme,[1] Alu I (Schmid and Jelinek, 1982). Many Alu sequences are transcribed; they are present commonly in nuclear pre-mRNA and to a lesser extent, in mRNA (in noncoding regions). Also, many Alu sequences contain RNA polymerase III initiation sites, and are presumably transcribed separately by that enzyme. Alu sequences may be transposable elements, at least in part, but as yet there is no proof of their origin or function. They may be responsible for many cases of unequal crossing-over, because of misalignment of DNA strands (Chapter 6).

The long interspersed repetitive sequences, which are distinguished from SINEs by being greater than 500 bp long, also contain one predominant member. This sequence, usually called L1, is typically 6400 bp long, but its 5′ end[2] is highly variable, and many members of the L1 family are much shorter. Estimates of reiteration frequency range from as few as 3000 copies to 40,000 copies.

The distribution of SINEs and LINEs along chromosomes has recently been shown (Korenberg and Rykowski, 1988) to correlate with the band pattern revealed by various cytochemical dyes in metaphase chromosomes. Using a new technique for high-resolution *in situ* hybridization (see Chapter 3), Korenberg and Rykowski found that Alu sequences predominate in Giemsa-light bands (also called R-positive bands), and L1 sequences predominate in Giemsa-dark bands (also called R-negative bands or Q-bands). It is not yet clear whether the Alu and L1 sequences cause the band patterns, but the high G+C content of Alu sequences correlates with the G+C content of R bands, and the reverse is true for the A+T content of L1 sequences and G/Q bands. Quite possibly, the staining properties of chromosomal bands are caused, at least in part, by proteins associated with Alu and L1 sequences, but essentially nothing is known about that at the moment.

Perhaps the best known middle repetitive sequences are the genes for 18S and 28S ribosomal RNAs. Each gene copy contains one 18S and one 28S sequence, together with additional nucleotides that are transcribed to form a large pre-rRNA, which is then cleaved by specific nucleases to yield mature ribosomal RNAs.

There are several hundred copies of the genes for 18S and 28S ribosomal RNAs, tandemly arranged in five clusters on the short arms of the acrocentric chromosomes 13, 14, 15, 21, and 22. These are the *nucleolus-organizing regions*. It is here that nucleoli form in interphase cells. Newly synthesized ribosomal RNA precursors bind a large variety of structural proteins, while several types of modifying enzymes act on the RNA-protein complexes to produce mature ribosomal subunits, which are exported to the cytoplasm. There are also numerous copies of the 5S ribosomal RNA gene; these are all on chromosome 1, near the terminus of the short arm.

---

[1] Restriction enzymes are bacterial endonucleases that cleave double-stranded DNA at specific nucleotide sequences. Alu I recognizes the sequence AGCT. Most restriction enzymes have target sequences consisting of four bp or six bp; some examples of the latter are EcoR I (GAATTC) and BamH I (GGATCC). The enzyme names are derived from the names of the organisms from which they are isolated; Alu refers to *Arthrobacter luteus*.

[2] The terms 5′ and 3′ (pronounced "five prime and three prime") refer to the carbon atoms in the ribose portion of nucleotides. When a polynucleotide (RNA or DNA) is synthesized, the first nucleotide is at the 5′ end and the last nucleotide is at the 3′ end. Within a gene, the terms 5′ and 3′ refer to the direction of transcription, which is the same as the direction of translation of a coding sequence.

Many Structural Genes Occur in Small Families of Closely Related Sequences

We have already seen that there are groups of related DNA sequences represented by thousands to hundreds of thousands of copies. So far, none of these has been shown to code for a protein, although some of the middle repetitive sequences code for RNAs with a variety of roles in transcription, RNA processing, and translation. The genes for the large (28S and 18S) ribosomal RNAs and for 5S ribosomal RNA fall into the low end of the frequency spectrum of moderately repeated genes, being present in several hundred copies.

The genes for transfer RNAs (tRNAs) exemplify a borderline situation, being neither single-copy in the strict sense, nor reiterated sufficiently to fall into the conventional definition of "middle repetitive" DNA. There are about 60 different tRNA species in human cells, each encoded by 10–20 gene copies, on average. One of those classes that has been studied is the $tRNA^{Lys}$ group, which is represented by about 14 copies, widely dispersed throughout the genome (Doering et al., 1982). Another is the initiator $tRNA^{Met}$, which has about 12 dispersed copies (Santos and Zasloff, 1981). Further analysis is likely to reveal various types of clusters, either of identifical or different tRNA genes, as have been found in other mammalian genomes.

The first protein-coding genes that were analyzed in detail by recombinant DNA technology were the globin gene clusters. Approximately 98% of adult human hemoglobin is Hemoglobin A, a molecule that consists of two alpha-globin and two beta-globin polypeptides, plus four hemes and associated iron atoms. Somewhat different hemoglobins are present during embryonic and fetal life. Alpha- and beta-globins are related polypeptides, having about 46% identity in amino acid sequence. Although the genes for alpha- and beta-globins presumably had a common ancestor several hundred million years ago, they now occur on separate chromosomes; and each has undergone several rounds of duplication and divergence during human evolution, so it is appropriate to speak of them as separate gene families.

Figure 2–10 shows the arrangement of the human globin genes. The alpha cluster, located on chromosome 16, consists of two copies of alpha-globin, one theta-globin, one zeta-globin, and three pseudogenes. The re-

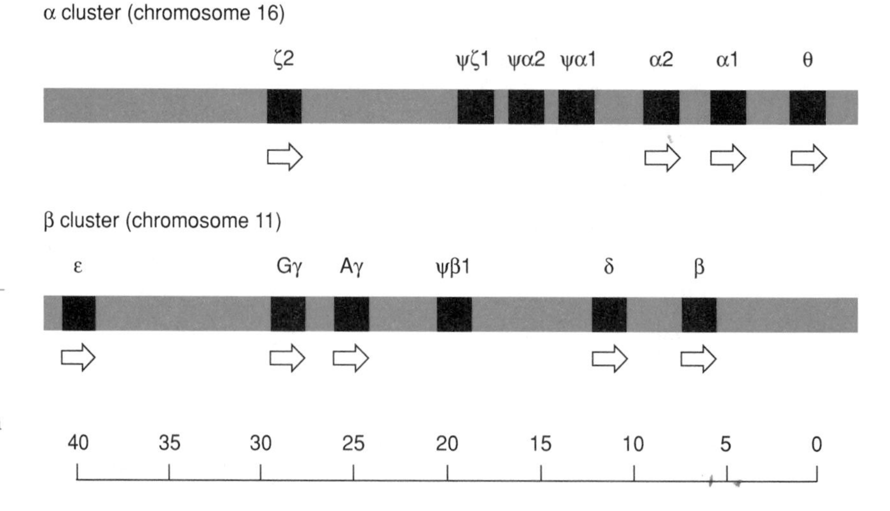

Figure 2–10

The alpha-globin and beta-globin gene clusters. The Greek letter ψ (psi) indicates pseudogenes. Arrows show direction of transcription. (From Lewin, 1990, p. 498; by permission of Cell Press)

cently discovered theta-globin does not appear to be a pseudogene, but whether it is functional is not yet clear. The beta cluster, located on chromosome 11, consists of beta-globin, delta-globin, two gamma-globins, epsilon-globin, and a pseudogene. Pseudogenes cannot code for functional proteins; they are discussed in a later section of this chapter.

The globins illustrate some important principles of gene family organization. First, a gene may exist in more than one copy, as shown here by the two alpha-globin genes, which code for identical proteins. Second, a gene duplication can be followed by evolutionary changes that modify, but do not abolish, the basic function of the protein encoded by that gene. This is well illustrated by fetal hemoglobin (Hb F), where the two gamma-globins confer on Hb F the ability to bind oxygen released by maternal red cells in the placenta. The gamma-globins differ from beta-globin in 39 amino acids out of 146; they also differ from one another by one amino acid (glycine versus alanine at position 136), which is not known to have any functional significance.

Delta-globin is somewhat puzzling; it differs from beta-globin by 10 amino acids, which makes it the closest relative of beta-globin, but its steady-state level in adult red cells is only 2% of the level of beta-globin. It may be an evolutionary failure, a dispensable gene that will ultimately acquire one or more mutations that render it functionless (see a later section of this chapter). The remaining globin genes, epsilon in the beta cluster and zeta in the alpha cluster, are expressed in the embryo (i.e., prior to eight weeks of gestation). Zeta is then replaced by alpha, which functions throughout fetal and adult life; epsilon is replaced by gamma-globin during fetal life, and beta plus delta replace gamma in the postnatal period (Figure 2–11). The control of these developmental switches in expression of the globin gene clusters is not yet well understood, but it is the subject of intense study.

Another small gene family that is of great clinical interest is the set of genes coding for *apolipoproteins*, those proteins that are involved in the transport of cholesterol, phospholipids, and triglycerides in blood. Mutations in apolipoprotein genes may lead to arterial disease and coronary failure. There is one cluster of apolipoprotein genes tandemly arranged on chromosome 11 (apoA-I, apoA-IV, and apoC-III) and another cluster on chromosome 19, involving at least apoE, apoC-I, and apoC-II (Karathanasis, 1985). This organization into two unlinked clusters is reminiscent of the situation for alpha- and beta-globins.

A somewhat larger, partially characterized gene family codes for *actins*, which are major components of muscle filaments and other cytoskeletal structures that are involved in contractile processes such as amoeboid

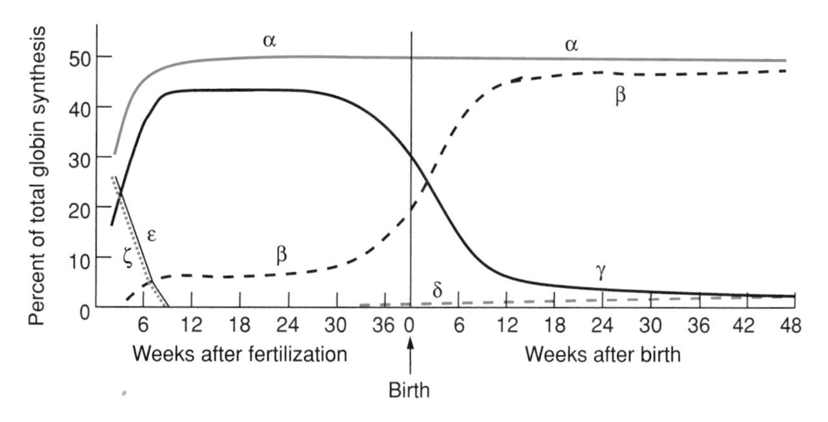

Figure 2–11

Hemoglobin switching during human development. Hemoglobin synthesis begins about three weeks after fertilization. Synthesis of zeta- (ζ) and epsilon-globin (ϵ) ceases by eight weeks. Alpha-globin (α) increases to 50% at 20 weeks and remains at that level throughout life. Gamma-globin (γ) is the predominant product of the beta-globin (β) gene cluster during fetal development, but is replaced by beta-globin post-natally. A small amount of delta-globin (δ) is synthesized, beginning near birth. (From Prescott, 1988, p. 332)

movements and cytoplasmic streaming. There are six classes of actins. The cytoskeletal actins (called beta and gamma) are represented by 15–20 genes apiece, although the actual number of functional copies is not yet certain. Alpha skeletal muscle actin and cardiac muscle actin have single-copy genes (Ponte el al., 1983). The genes for the remaining two classes (smooth muscle actins) have not been described at this time.

Ubiquitin is a protein of only 76 amino acids, and it occurs in all eukaryotic cells examined for its presence. It is one of the most evolutionarily conserved proteins known. Covalent binding of ubiquitin via its C-terminal glycine to lysine residues in other proteins marks those proteins for destruction by the ATP-dependent non-lysosomal protein degradation system. Genes for ubiquitin occur in at least twelve clusters (Wiborg et al., 1985). One of those genes contains an open reading frame 2055 bp long, which codes for nine direct repeats of the ubiquitin sequence, plus one amino acid. Another gene codes only for a ubiquitin monomer, and other variations exist.

In summary, gene families exist in almost all imaginable combinations. As our knowledge of the human genome and related genomes increases, our understanding of genomic evolution will also increase, leading to the recognition of familial relationships among genes that are too distant for us to be aware of today.

Most Protein-Coding Genes Have a Complex Internal Organization

The internal organization of genes in higher eukaryotes is exemplified by the globin genes, each of which consists of three *exons* and two *introns* (Figure 2–12). Exons are nucleotide sequences that are included in mature mRNA, whether they are translated or not; introns are nucleotide sequences that are transcribed, but are spliced out of the pre-mRNA as part of the maturation process that occurs in the nucleus. The remarkable fact that genes can occur "in pieces" was not appreciated until 1977, although it had been known for many years that newly synthesized precursors of mRNA ("hnRNA") were much larger than the mature mRNAs found in cytoplasmic polysomes. Many theories to account for the size discrepancy were put forth, but no one had enough originality to guess the bizarre truth.

The first evidence for the existence of introns (also called *intervening sequences*) came from electron microscope studies on the genomes of

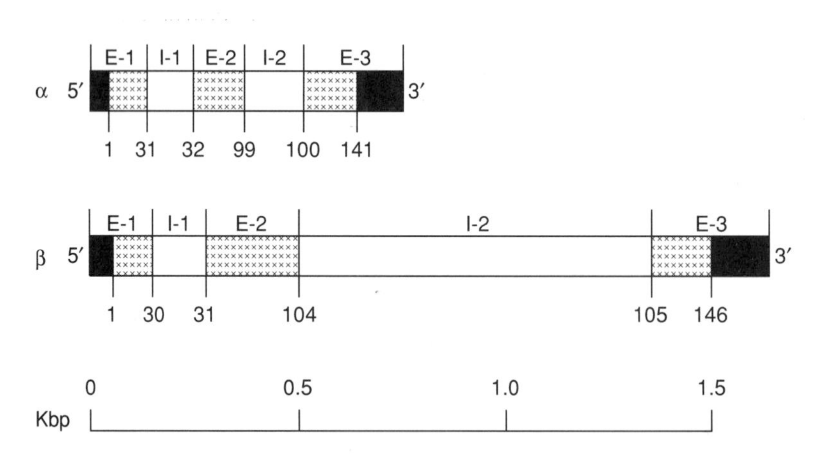

Figure 2–12

The structure of human alpha-globin and beta-globin genes. Transcription begins at the 5′ end and continues to the 3′ end. E-1, E-2, and E-3 are exons (sequences present in mature mRNA); I-1 and I-2 are introns, sequences removed from pre-mRNA during intranuclear processing. The nontranslated leader sequence in E-1 and the nontranslated trailer sequence in E-3 are shown as filled spaces. The remainder of each exon codes for the polypeptide sequence. The numbers below each bar diagram indicate the encoded amino acids.

adenovirus and SV40; comparable studies on the beta-globin gene followed soon thereafter. It was known that a large precursor of beta-globin mRNA (called 15S pre-mRNA) could be obtained from the nuclei of mouse erythroblasts. This 15S RNA could be hybridized to DNA containing the beta-globin gene under conditions that allowed the DNA-DNA duplex to be displaced by the more stable RNA-DNA duplex wherever complementary RNA and DNA sequences existed. Examination of such hybrid structures (called R-loops) in the electron microscope allowed visualization of those parts of the DNA that corresponded to the RNA in sequence.

As shown in Figure 2–13A, the 15S pre-mRNA formed a continuous duplex with the complementary strand of the beta-globin gene, generating one R-loop. However, hybridization of the mature beta-globin mRNA (9S) to an identical DNA sample produced two R-loops, with an intact DNA duplex looped out between them (Figure 2–13B). That DNA duplex, as we now know, represented intron 2; intron 1 was not seen, because of its small size.

Subsequent analysis by DNA cloning, restriction enzyme digestion, and DNA sequencing has shown that every member of the alpha- and beta-globin gene families has three exons and two introns, as diagrammed in Figure 2–12. Nucleotide sequences within introns vary substantially, but the position and size of the introns is rigidly conserved within a family. We do not understand the functional or evolutionary significance of these regularities.

The internal organizations of many other genes are now known and some generalizations can be made, although much remains to be learned. The number and size of both exons and introns vary tremendously among unrelated genes. Most exons are small, averaging less than 200 nucleotides. Exons that are larger than 200 nucleotides tend to occur at the 3′ ends of genes and consist primarily of noncoding sequences. In general, then, only a few dozen amino acids are coded by a given exon, and in some cases the number may be much smaller. The evolutionary significance of such small coding segments in genes is the subject of active debate. Some exons correspond to functional domains of the protein; for example, the central exon of a globin gene encodes the heme-binding segment of the polypeptide. However, attempts to correlate all exons with functional domains within proteins have not been convincing.

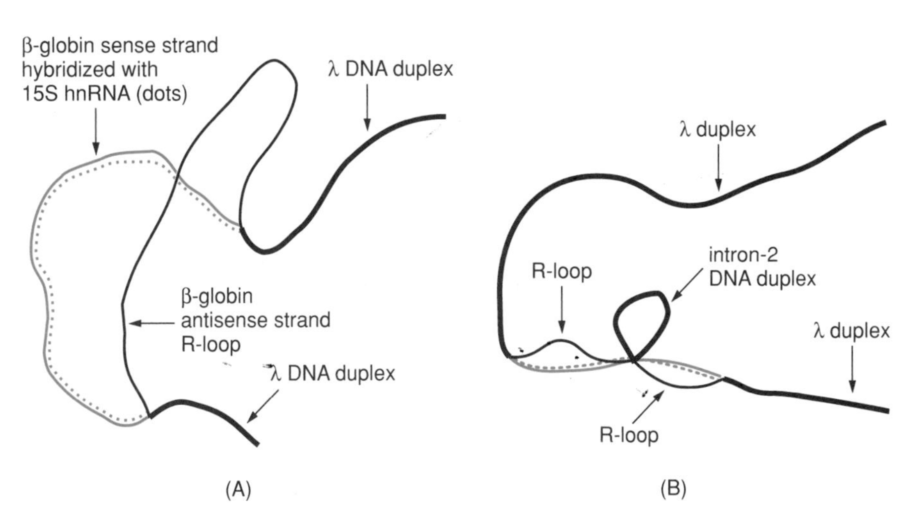

Figure 2–13

R-loop mapping of mouse beta-globin DNA. (A) hybridization of 15S pre-mRNA to the beta-globin gene cloned in a lambda phage vector produces a single R-loop, corresponding to the entire transcribed portion of the gene. (B) hybridization of mature beta-globin mRNA to the cloned beta-globin gene produces two R-loops, separated by a double-stranded DNA region that corresponds to the large intron (the first intron is too small to be detected by this method). (Modified from Tilghman et al., 1978, p. 1312)

The number of introns within a gene varies from zero (e.g., histone genes) to greater than 50 for collagen genes and at least 75 for the dystrophin gene. Of course, the number of exons is always one more than the number of introns in a given gene. Introns also vary over a wide range of sizes, from roughly 100 bp to greater than 100,000 bp. Some examples are given in a later section of this chapter. The ratio of exon DNA to intron DNA is a matter of some interest, because it defines the size of a gene. On the average, there appears to be at least ten times as much intron DNA as exon DNA in mammals, and the ratio may actually be larger.

In the standard pattern of gene expression, transcription begins at a precisely defined point (the first exon) and ends at a less well-defined point somewhat beyond the 3′ end of the last exon. RNA processing enzymes then trim the 3′ end of the primary transcript and remove the sequences corresponding to introns, joining the exons together by a process known as *RNA splicing*, yielding a mature mRNA. However, some genes have alternative promoters, which are used in different tissues. A simple example is that of the mouse amylase gene, which is expressed primarily in the salivary gland and liver. In the salivary gland, transcription begins about 3000 bp upstream of the point where transcription begins in the liver. The mature salivary gland mRNA contains a different first exon than the liver mRNA.

In some other genes, the transcription initiation point is invariant, but the pattern of splicing varies from one tissue to another. As a result, some DNA sequences serve as introns in one tissue, but they are exons in other tissues. Alternative splicing is an important mechanism for the generation of protein diversity in differentiation. It is undoubtedly one factor responsible for the fact that different mutations in a particular gene may have different clinical consequences (see Chapter 7).

One other general property of intron DNA is of major genetic significance. Although most single-base mutations within introns are benign, a few can have catastrophic consequences. For example, mutations that delete normal splice sites or create new splice sites often lead to the production of functionless polypeptides (Chapter 7).

Pseudogenes Are Common

The advent of gene cloning and rapid DNA sequencing has led to some surprising discoveries, not the least of which is the common occurrence of *pseudogenes*—DNA sequences closely related to functional genes, but incapable of coding for the normal gene product because of one or more deletions, insertions, frameshifts, or stop codons.

The first pseudogenes to be recognized were non-functional copies of the 5S ribosomal RNA genes in the frog *Xenopus* (Jacq et al., 1977); subsequently, pseudogenes related to a wide variety of protein-coding genes have been described. Some of the best known examples are found in the globin gene clusters of humans and other vertebrates (see Figure 2-10). The pseudo-beta and pseudo-alpha genes in humans each have several differences from the normal homolog, whereas pseudo-zeta is notable for having the minimal difference—a stop codon has been generated at amino acid 7 by a single base change. Although this does not affect the production of messenger RNA, the polypeptide translated from that message is truncated after only six amino acids have been polymerized.

There are two major classes of pseudogenes. The human globin pseudogenes exemplify the first class, which can be accounted for by gene duplication, followed by mutational inactivation of one of the gene copies in cases where a second normal copy is not selectively beneficial. This type of pseudogene retains the overall organization of the parental gene in terms of exons and introns.

The second (and originally unexpected) class of pseudogenes contains the exon sequences of the functional gene, but not the introns; and a poly(A)[3] sequence is usually present at the 3′ end of the pseudogene. In other words, the pseudogene appears to have been derived from a messenger RNA (Lewin, 1983). This has given rise to the term *processed genes*, referring to the processing of RNAs, whereby introns are excised from the original transcript, leading to mature mRNAs. A widely held hypothesis is that a messenger RNA is occasionally copied into a cDNA in a germ cell precursor in vivo, and the cDNA is subsequently inserted more or less at random into the genome (Figure 2–14). The reverse transcriptase that makes the cDNA and the integrase that inserts the cDNA into a chromosome could be supplied by a retrovirus (see Chapter 11) or they could be enzymes encoded by transposable elements that are a normal part of the human genome (see Chapter 7).

The fact that many pseudogenes are not adjacent to their normal counterparts is consistent with the cDNA synthesis and random insertion hy-

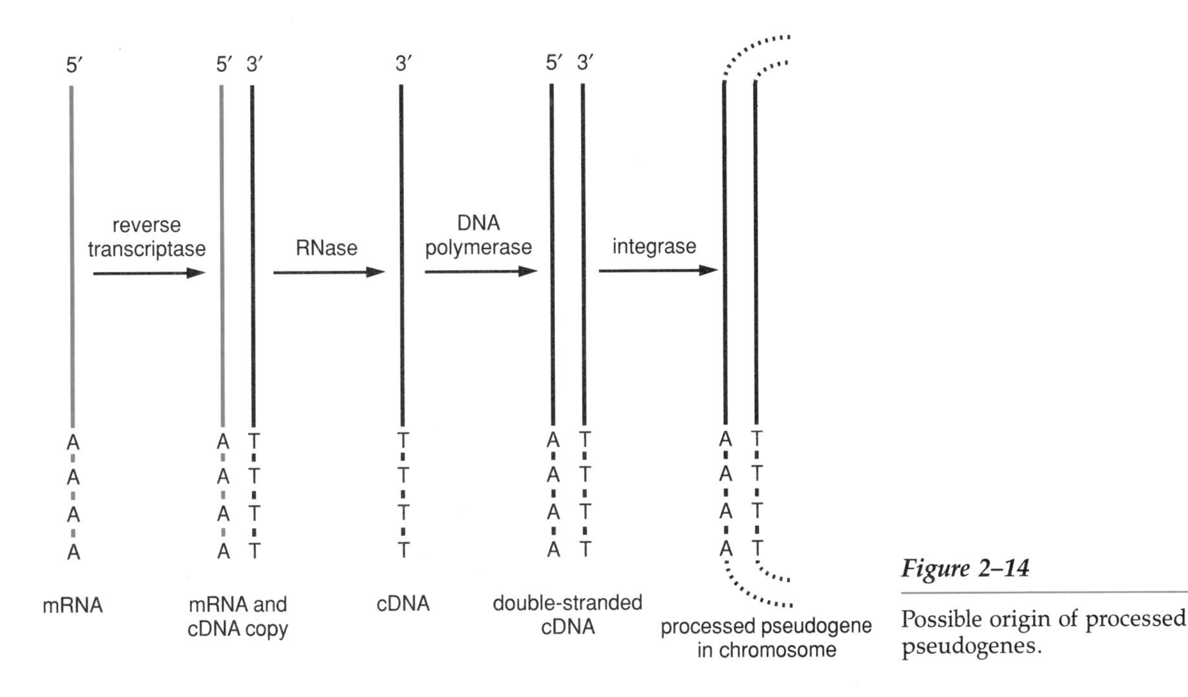

Figure 2–14

Possible origin of processed pseudogenes.

[3] Nearly all mammalian mRNAs acquire a sequence of roughly 50–150 adenylic acid residues at the 3′ end, which are added post-transcriptionally by the enzyme poly(A) polymerase, before the mRNAs are exported from nucleus to cytoplasm. The function of the poly(A) sequence is not well understood, although it appears to increase mRNA stability. In any case, the poly(A) sequence has been very useful to molecular biologists, because it allows us to separate mRNAs from the much more numerous rRNAs and tRNAs, using oligo(dT) columns, to which the poly(A) binds.

pothesis. Some examples from humans are pseudogenes for the lambda immunoglobulin chain (Hollis et al., 1982) and for beta-tublin (Wilde et al., 1982), neither of which occurs on the chromosome that includes the normal gene.

The Structure and Variation of Specific Genes

What is the average size of a gene? How many introns does an average gene contain and how are they dispersed between the first and last transcribed nucleotide? It is too soon to compute answers to those questions, but there is no doubt that the organization of specific genes varies tremendously. A few examples follow.

We have already described the organization of the genes for alpha- and beta-globins and their close relatives. They are all relatively small genes, coding for relatively small polypeptides. The coding nucleotides make up somewhat more than one-fourth of the total gene. A first cousin of the hemoglobins is *myoglobin*, the protein that functions as an oxygen carrier in muscle. Myoglobin contains 155 amino acids, only 9 more than beta-globin, but the myoglobin gene is over 10 kbp long (Weller et al., 1984). There are two introns, as in the alpha- and beta-globin gene families, but the introns cover 5.8 and 3.6 kbp in myoglobin. Those long introns are not unique to humans; they exist in other mammals in which the myoglobin gene has been examined, so they must have some evolutionary significance, although what that might be is beyond our understanding at the moment.

Figure 2–15

Nucleotide sequence of the human interleukin-2 gene and flanking regions. The sequence, excluding the introns, was numbered from the putative mRNA start site (indicated by an asterisk in the second line). Note the ATG translation initiation codon at the beginning of the third line. The arrow in the third line indicates the presumed cleavage site for the polypeptide signal sequence. A TATAAATT-sequence, present 32 nucleotides upstream from the putative start site, as well as the poly(A) addition signal (AATAAA), are indicated by red boxes. The three introns are also boxed (only the first intron sequence is given in total). (From Degrave et al., 1983; by permission of Oxford University Press)

As another example of a gene coding for a small protein, we will consider the *interleukin-2* gene. Interleukin-2, also known as *T-cell growth factor*, came to the public's attention recently when large amounts of the protein, produced by recombinant bacteria, were successfully used to treat a few cases of human cancer. Although most cancers do not respond to IL-2, the protein still holds promise as an agent to be used in combination with other drugs and/or immunostimulants. Newly synthesized IL-2 contains 153 amino acids. The first 20 amino acids constitute the signal sequence; they are removed and several sugars are added to the polypeptide before it is secreted. The gene contains four exons and three introns (Figure 2–15). The first 47 nucleotides of the mRNA are a noncoding leader and the last 284 are a noncoding trailer; the coding sequences make up almost 9% of the gene (Degrave et al., 1983).

The enzyme *hypoxanthine-guanine phosphoribosyl-transferase* (abbreviated HPRT or HGPRT) catalyzes one of the steps on the salvage pathway for the purines hypoxanthine and guanine, returning those products of nucleic acid degradation into the pathway for synthesis of more nucleic acids. HPRT is of considerable clinical interest, because partial deficiency of the enzyme leads to *gout* and total deficiency leads to *Lesch-Nyhan syndrome*, which is characterized by severe physical and neurological abnormalities. Both diseases occur predominantly in males, as expected for an X-linked gene. The human gene consists of nine exons, dispersed over 42–44 kbp (Patel et al., 1986; Figure 2–16A).

However, the protein encoded by the HPRT gene contains only 218 amino acids when first synthesized. The sizes of the exons, in bp, are: 145, 107, 184, 66, 18, 83, 47, 77, and 637. As is usually the case, the last exon contains a long untranslated trailer sequence following the termination co-

```
.....CATCAGAAGAGGAAAAATGAAGGTAATGTTTTTCAGACAGGTAAAGTCTTTGAAAATATGTGTAATATGTAAAACATTTTGACACCCCATAA
                                                 -100
                                                   *
TATTTTTCCAGAATTAACAG[TATAAATT]GCATCTCTTGTTCAAGAGTTCCCTATCACTCTCTTTAATCACTACTCACAGTAACCTCAACTCCTGCCACA
 -50                                                1

Met Tyr Arg Met Gln Leu Leu Ser Lys Ile Ala Leu Ser Leu Ala Leu Val Thr Asn Ser↓Ala Pro Thr Ser Ser
ATG TAC AGG ATG CAA CTC CTG TCT TGC ATT GCA CTA AGT CTT GCA CTT GTC ACA AAC AGT GCA CCT ACT TCA AGT
 50                                                    -100

Ser Thr Lys Lys Thr Gln Leu Gln Leu Glu His Leu Leu Leu Asp Leu Gln Met Ile Leu Asn Gly Ile Asn
TCT ACA AAG AAA ACA CAG CTA CAA CTG GAG CAT TTA CTT CTG GAT TTA CAG ATG ATT TTG AAT GGA ATT AAT | GTA
                              150
                       1° intron (90bp)                                                          Asn Tyr Lys
AGTATATTTCCTTTCTTACTAAAATTATTACATTTAGTAATCTAGCTGGAGATCATTTCTTAATAACAATGCATTATACTTTCTTAG | AAT TAC AAG
                                                                                                     200

Asn Pro Lys Leu Thr Arg Met Leu Thr Phe Lys Phe Tyr Met Pro Lys Lys
AAT CCC AAA CTC ACC AGG ATG CTC ACA TTT AAG TTT TAC ATG CCC AAG AAG | GTAAGTACAATATTTTATGTTCAATTTCTGT
                                                                250

TTTAATAAAATTCAAAGTAATATGAAAATTTGCACAGATGGGACTAATAGCAGCTCATCTGAGGTAAAGAGTAACTTTAATTTGTTTTTTGAAAACCC

AAGTTTGATAATGAAGCCTCTATTAAAACAGTTTTACCTATATTTTTAATATATATTTGTGTGTTGGTGGGGGTGGGAAGAA-----------------

------2° intron (±2400bp)----------------------TGCAGAAAGTCTAACATTTTGCAAAGCCAAATTAAGCTAAAACCAGTGAG

TCAACTATCACTTAACGCTAGTCATAGGTACTTGAGCCCTAGTTTTTCCAGTTTTATAATGTAAACTCTACTGGTCCATCTTTACAGTGACATTGAGAA

CAGAGAGAATGGTAAAAACTACATACTGCTACTCCAAATAAAATAAATTGGAAATTAATTTCTGATTCTGACCTCTATGTAAACTGAGCTGATGATAAT

           Ala Thr Glu Leu Lys His Leu Gln Cys Leu Glu Glu Glu Leu Lys Pro Leu Glu Glu Val Leu Asn
TATTATTCTAG | GCC ACA GAA CTG AAA CAT CTT CAG TGT CTA GAA GAA GAA CTC AAA CCT CTG GAG GAA GTG CTA AAT
                                                                       300

Leu Ala Gln Ser Lys Asn Phe His Leu Arg Pro Arg Asp Leu Ile Ser Asn Ile Asn Val Ile Val Leu Glu Leu
TTA GCT CAA AGC AAA AAC TTT CAC TTA AGA CCC AGG GAC TTA ATC AGC AAT ATC AAC GTA ATA GTT CTG GAA CTA
                                    350

Lys
AAG | GTAAGGCATTACTTTATTTGCTCTCCTGGAAATAAAAAAAAAAAAGTAGGGGGAAAAGT------3° intron (±1900bp)------CTT

GAAAATAAAGGCAACAGGCCTATAAGACTTCAATTGGGAATAACTGTATATAAGGTAAACTACTCTGTACTTTAAAAAATTAACATTTTTCTTTTATAG

Gly Ser Glu Thr Thr Phe Met Cys Glu Tyr Ala Asp Glu Thr Ala Thr Ile Val Glu Phe Leu Asn Arg Trp Ile
GGA TCT GAA ACA ACA TTC ATG TGT GAA TAT GCT GAT GAG ACA GCA ACC ATT GTA GAA TTT CTG AAC AGA TGG ATT
400                                                                450

Thr Phe Cys Gln Ser Ile Ile Ser Thr Leu Thr
ACC TTT TGT CAA AGC ATC ATC TCA ACA CTG ACT TGA TAATTAAGTGCTTCCCACTTAAAACATATCAGGCCTTCTATTTATTTAAAT
                                 500                                                  550

ATTTAAATTTTATATTTATTGTTGAATGTATGGTTTGCTACCTATTGTAACTATTATTCTTAATCTTAAAACTATAAATATGGATCTTTTATGATTCTT
                                      600                                                  650

TTTGTAAGCCCTAGGGGCTCTAAAATGGTTTCACTTATTTATCCCAAAATATTTATTATTATGTTGAATGTTAAATATAGTATCTATGTAGATTGGTTA
                               poly A     700                                                750
                                 ↓
GTAAAACTATTT[AATAAA]TTTGATAAATATAAACAAGCCTGGATATTTGTTATTTTGGAAACAGCACAGAGTAAGCATTTAAATATTTCTTAGTTACTT
                                        800                                                 850

GTGTGAACTGTAGGATGGTTAAAATGCTTACAAAAGTCACTCTTTCTCTGAAGAAATATGTAGAACAGAGATGTAGACTTCTCAAAAGCCCTTGCTTT.
                                         900                                                 950
```

don and there is a short leader sequence between the beginning of the mRNA and the initiation codon for protein synthesis in the first exon. These two regions plus the coding sequence result in an mRNA 1345 nucleotides long. The average intron size in this gene is greater than 5000 bp, while the coding sequences amount to less than 2% of the gene.

Phenylalanine hydroxylase (PAH) is a liver enzyme whose function is to convert phenylalanine into tyrosine. Its absence leads to the well-known recessive genetic disease, *phenylketonuria* (PKU), which occurs in one birth in 10–15,000 Caucasians. The gene structure is quite similar to that of the HPRT gene, but on a larger scale. The protein is encoded by 1353 bp, which are contained in 13 exons dispersed over 87 kbp of DNA (DiLella et al., 1986). Again, the coding sequences account for less than 2% of the gene.

The *serum albumin* gene is much more compact than the preceding two examples. Serum albumin is one of the most abundant proteins in human blood, reaching about 40 mg/ml in adults. All of this protein is the product of a single-copy gene located on chromosome 4. The protein contains 623 amino acids when first synthesized. There are 15 exons distributed over 16,961 bp of gene (Minghetti et al., 1986; Figure 2–16B); the ratio of coding sequences to gene size is about 11%.

The structures of two X-linked genes for blood-clotting disorders make an interesting comparison. Defects in *factor IX* lead to *hemophilia B*, a relatively mild disease that occurs about once per 30,000 males. Factor IX is a protein of average size, containing 461 amino acids, which are encoded by 1383 bp. The mRNA is 2800 nuceotides long; the gene contains 8 exons spread over 34 kbp of DNA (Anson et al., 1984). The coding sequences amount to slightly more than 4% of the gene.

Factor VIII, whose absence or malfunction leads to the more serious *hemophilia A* in one male per 10,000 births, is an exceptionally large protein with an exceptionally large gene. There are 7053 coding nucleotides in an mRNA approximately 9000 nucleotides long; the 26 exons are dispersed over 186 kpb of DNA (Gitschier et al., 1984). But, as in the case of the factor IX gene, the ratio of coding sequence to total gene size is about 4%.

There are genes even larger than the factor VIII gene. One of them is the gene for *thyroglobulin*. This protein consists of two identical polypeptide chains, each about 2800 amino acids long. It is the source of the hormone *thyroxine*, the synthesis of which is paradoxically inefficient. Thyroglobulin is synthesized by the thyroid gland, iodinated at several tyrosine residues, secreted, taken up again into thyroid cells, and degraded within lysosomes to yield at most four thyroxine molecules from the initial 5600 amino acids in each thyroglobulin dimer. The polypeptides are encoded by an mRNA of approximately 8600 nucleotides. The gene spans at least 300 kpb, divded into 37 exons (Baas et al., 1986). The approximately 8400 coding bases amount to less than 3% of the gene.

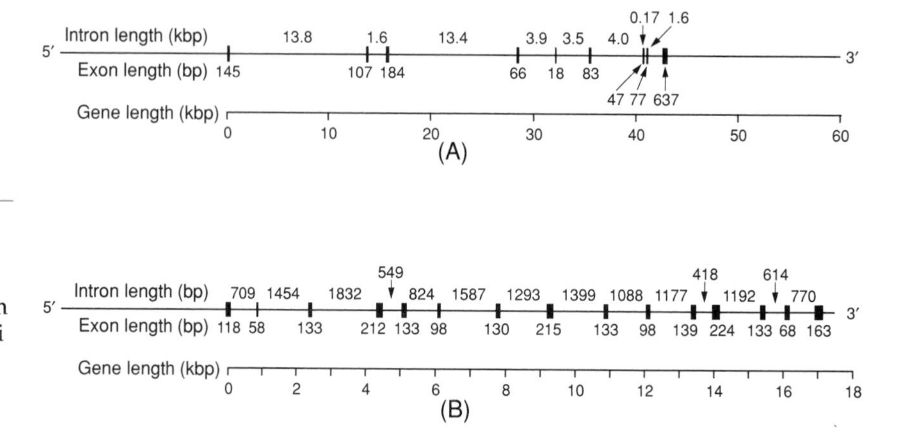

Figure 2–16

Organization of some representative human genes. (A) The HPRT gene. (From Patel et al., 1986) (B) The serum albumin gene. (From Minghetti et al., 1986) In both diagrams, exons are indicated by vertical bars.

But the grand champion gene, in terms of known size, is the one for a recently described protein whose abnormality causes *Duchenne muscular dystrophy* (Chapter 5), the most common lethal sex-linked recessive disease in humans (Monaco and Kunkel, 1987). This protein, named *dystrophin*, contains about 4000 amino acids making it roughly 10X as large as an average protein. The corresponding 12,000 coding bases are distributed over no less than 75 exons covering at least 2000 kbp—a stretch of DNA nearly half as large as the entire genome of *E. coli*. The coding sequences represent less than 1% of the gene. If this monster gene is transcribed in the usual fasion from one end to the other as a continuous pre-mRNA molecule, it would take RNA polymerase more than 18 hours to make a complete copy!

The preceding set of examples is heavily weighted with genes responsible for common genetic diseases, and it is not possible to say which of them are "typical" genes in terms of internal organization. It would not be surprising if genes with numerous and/or very large introns are particularly prone to deletion mutations, and thus come to the attention of molecular geneticists more readily than genes with typical structure. Whatever the average size of human genes in general may be, that number is of more than academic interest. It will set an upper limit on the total number of genes the genome may contain, and it will have important implications for analysis of the mapping and sequencing data that the Human Genome Project will generate (Chapter 15).

The Total Number of Human Genes Is Unknown

Human genetics is a rapidly expanding field of knowledge, and in a basic sense, the limits of that field will be defined by the total number of genes in the human genome. There is no direct way to count genes, unless the entire genome has been sequenced, and even then, some small genes may be missed. Various viral genomes have been completely sequenced; these range from about 5000 bp of SV40 to the 48,000 bp of coliphage lambda, to the 170,000 bp of Epstein-Barr virus. Several animal mitochondrial genomes have been sequenced, including the 16,519 bp of human mitochondrial DNA. In the near future, we may expect the full sequence of *E. coli* (about 4.6×10^6 bp). In due time, there will be complete or nearly complete sequences of a yeast (*S. cerevisiae*), a nematode (*C. elegans*), a fruitfly (*D. melanogaster*), a mouse (*M. musculus*), and *Homo sapiens*.

There are two general indirect approaches to the determination of total gene number. The genetic approach is exemplified by the work of Burke Judd on *Drosophila* (Judd et al., 1972), in which the total number of genes in a limited region of one chromosome was estimated, following production of a large series of recessive lethals in response to mutagens. Obviously, it is not an approach that can be intentionally applied to humans. Moreover, it is likely to overlook genes whose absence is not lethal.

The second indirect approach is to determine the total number of different messenger RNAs present in an organism, or in selected cells, tissues, or organs. This is done by *nucleic acid hybridization*. For example, one might prepare total mRNA from rat brain, mix a large quantity of the mRNA with a small quantity of radioactive, denatured DNA from rats, carry out hybridization to completion, and then measure the fraction of the DNA that has been hybridized to mRNA. This is the technique of saturation hybridization; that is, all sequences in the DNA that are complementary to an mRNA sequence, no matter whether the mRNA is abundant or rare, are hybridized. Many studies of this type have been done, and a typical finding is that 5% of the single-copy DNA can be hybridized by brain mRNA. From this, an estimate of the total number of different polypeptides that might be encoded by the mRNAs from the tissue of origin is obtained as follows.

Sequence complexity of single-copy DNA = 2×10^9 bp.

Percent of DNA hybridized = 5%. Multiply by two, because transcription is usually asymmetric, to get 10%, the fraction of DNA that codes for mRNAs expressed in this particular organ.

$10\% \times (2 \times 10^9) = 2 \times 10^8$ nucleotides, the sequence complexity of the mRNA.

Average size of an mRNA may be 2000 nucleotides.

$(2 \times 10^8)/2000 = 10^5$, implying that 100,000 different proteins of average size might be synthesized in rat brain.

Many such studies were carried out in the 1960s and 1970s, with the following general conclusions: (1) individual cell lines, such as fibroblasts, have enough nucleotide diversity in their mRNAs to code for 10,000–20,000 proteins; (2) mammalian liver mRNAs have enough diversity to code for 20,000–30,000 proteins; and (3) the mammalian brain, which is anatomically and functionally the most complex organ, also has the most complex mRNA populations, with enough sequence diversity to code for 100,000–200,000 proteins, according to some investigators.

Such calculations give us nothing more than an approximation of the gene number problem, rather than an answer to the problem. The calculations involve too many assumptions to be accepted as fact, as many have done. In the first place, nucleic acid hybridization is a complex procedure, and it is easy for two equally competent investigators using the same experimental system to get results that differ by a factor of two or three. There is also considerable uncertainty about the average size of mRNAs. One study found that most brain mRNAs are quite large; overall, their data implied that at most 30,000 proteins might be coded by rat brain (Milner and Sutcliffe, 1983).

Finally, the extent to which mRNA preparations, however carefully they may have been prepared from polysomes, may be contaminated by pre-mRNAs is impossible to estimate. The fact that a large majority of the sequence diversity in so-called mRNA preparations is found in the least abundant species raises the possibility that transcripts of introns, which we know to have at least 10X the sequence diversity of exon transcripts, account for a significant fraction of apparent mRNA complexity.

Another point worth considering is that if an average gene spans at least 20,000 bp, as discussed earlier in this chapter, then 100,000 genes would occupy a total of 2×10^9 bp, which is equivalent to the entire single-copy fraction of a mammalian genome. This would leave no room for intergenic DNA. Thus, the hypothesis that there are tens of thousands of brain-specific genes virtually requires that most of those genes be organized in a special way; for example, they might contain few, if any, introns, or there might be very little space between genes. There is no information that would allow us to confirm or refute those possibilities at present. The answer will emerge when we have extensive sequence data on genes that encode brain-specific proteins.

Unfortunately, there is no direct way to find out whether 10,000 or 100,000 proteins are present in a mammalian organ. The most powerful protein-resolving technique available (two-dimensional gel electrophoresis) is not adequate for mixtures containing more than 1000–2000 different polypeptides. However, it is worth noting that all the reactions of intermediary metabolism, all of the anabolic and catabolic pathways, plus all of the structural and regulatory proteins that are presently known, would not add up to more than 10,000 gene products.

Clearly, it is important to know how many different protein-coding genes occur in the human genome. If there are 30,000—let alone 150,000—then, not only is our knowledge of human biochemistry grossly incomplete, but the full extent of the possibilities for genetic disease cannot even be estimated. One of the early benefits of the Human Genome Project should be an answer to this question. Certainly by the time that 10% of the genome has been sequenced, we should have a reliable estimate of the average density of genes per million base pairs, and from that information, the gene content of the entire genome can be extrapolated.

SUMMARY

The human haploid nuclear genome contains three billion base pairs of DNA, organized into 24 types of chromosomes: 22 pairs of autosomes plus the X and Y chromosomes.

Late prophase and metaphase chromosomes can be distinguished from one another by patterns of bands that are revealed by Giemsa stain.

Many DNA sequences are reiterated; that is, they are present in more than one copy per haploid genome. The amount of reiteration varies from a few copies to hundreds of thousands of copies.

Some sequences are tandemly reiterated; examples include centromeric DNA and telomeric DNA. Other reiterated sequences are interspersed with unique (single-copy) sequences. The most abundant short interspersed sequences are members of the Alu family, which are about 300 bp long and have about 300,000 members per haploid genome. The long interspersed repetitive sequences include the L1 family, represented by several thousand members with a typical length of a few thousand bp.

Many structural genes occur in small families of related sequences. The best-studied examples are the alpha-globin and beta-globin gene families.

Most protein-coding genes are organized into alternating exons and introns. Although the entire gene is transcribed, RNA processing enzymes in the cell nucleus excise the introns and splice the exons together to produce mature mRNA. Most exons are smaller than 200 bp; the size of introns varies widely, ranging up to tens of thousands of bp. The number of introns varies from zero to at least 75 in different genes.

Pseudogenes are DNA sequences that closely resemble known genes, but cannot code for a functional polypeptide. Pseudogenes may arise by gene duplication, in which case they are located close to their relatives, or by the formation of a DNA copy of mRNA and insertion of that DNA into the genome at a random location. In both cases, subsequent mutations inactivate the function of the gene copy.

The organization of specific genes varies widely. In general, there appears to be about 10X as much DNA in introns as in exons. Gene size varies from a few hundred bp to more than two million bp.

The total number of human genes is unknown. Estimates range from 30,000 to more than 100,000.

REFERENCES

Anson, D. S., Choo, K. H., Rees, D. J. G. et al. 1984. The gene structure of human antihaemophilic factor IX. EMBO J. 3:1053–1060.

Baas, F., Van Ommen, G-J. B., Bikker, H., Arnberg, A. C., and Vijlder, J. J. M. 1986. The human thyroglobulin gene is over 300 kb long and contains introns of up to 64 kb. Nuc. Acids Res. 14:5171–5186.

Bickmore, W. A. and Sumner, A. T. 1989. Mammalian chromosome banding—an expression of genome organization. Trends in Genetics 5:144–148.

Blackburn, E. H. 1991. Structure and function of telomeres. Nature 350:569–573.

Degrave, W., Tavernier, J., Duerinck, F. et al. 1983. Cloning and structure of the human interleukin-2 chromosomal gene. EMBO J. 2:2349–2353.

Denver Conference. 1960. A proposed standard system of nomenclature of human mitotic chromosomes. Am. J. Hum. Genet. 12:384–388.

DiLella, A. G., Kwok, S. C. M., Ledley, F. D., Marvit, J., and Woo, S. L. C. 1986. Molecular structure and polymorphic map of the human phenylalanine hydroxylase gene. Biochemistry 25:743–749.

Doering, J. L., Jelachich, M. L., and Hanlon, K. M. 1982. Identification and genomic organization of human $tRNA^{Lys}$ genes. FEBS Letters 146:47–51.

Forget, B. G., Benz, E. L., Jr., and Weissman, S. M. 1983. Normal human globin gene structure and mutations causing the beta-thalassemia syndromes. In "Recombinant DNA Applications to Human Disease," C. T. Caskey and R. L. White, eds., pp. 3–17. Cold Spring Harbor Laboratory, Cold Spring Harbor, NY.

Gitschier, J., Wood, W. I., Goralka, T. M. et al. 1984. Characterization of the human factor VIII gene. Nature 312:326–330.

Harnden, D. G. and Klinger, H. P., eds. 1985. ISCN (1985): An international system for human cytogenetic nomenclature. S. Karger AG, Basel, Switzerland.

Hartl, D. L. 1991. Basic Genetics. Jones and Bartlett, Boston.

Hartl, D. L. 1983. Human genetics. Harper and Row, New York.

Hollis, G. F., Hieter, P. A., McBride, O. W., Swan, D., and Leder, P. 1982. Processed genes: a dispersed human immunoglobulin gene bearing evidence of RNA-type processing. Nature 296:321–325.

Jacq, C., Miller, J. R., and Brownlee, G. G. 1977. A pseudogene structure in 5S DNA of *Xenopus laevis*. Cell 12:109–120.

Judd, B., Shen, M. W., and Kaufman, T. C. 1972. The anatomy and function of a segment of the X chromosome of *Drosophila melanogaster*. Genetics 71:139–156.

Karathanasis, S. K. 1985. Apolipoprotein multigene family: tandem organization of human apolipoprotein AI,CIII, and AIV genes. Proc. Natl. Acad. Sci. USA 82:6374–6378.

Korenberg, J. R. and Rykowski, M. C. 1988. Human genome organization: Alu, LINES and the molecular structure of metaphase chromosome bands. Cell 53:391–400.

Lewin, B. 1990. Genes IV. Oxford University Press, Oxford, England.

Lewin, R. 1983. How mammalian RNA returns to its genome. Science 219:1052–1054.

Milner, R. J. and Sutcliffe, J. G. 1983. Gene expression in rat brain. Nuc. Acids Res. 11:5497–5520.

Minghetti, P. P., Ruffner, D. E., Kuang, W-J. et al. 1986. Molecular structure of the human albumin gene is revealed by nucleotide sequence within q11-22 of chromosome 4. J. Biol. Chem. 261:6747–6757.

Monaco, A. P. and Kunkel, L. M. 1987. A giant locus for the Duchenne and Becker muscular dystrophy gene. Trends in Genetics 3:33–37.

Moyzis, R. K. 1991. The human telomere. Scientific American, Aug.:48–55.

Paris Conference. 1971. Standardization in human cytogenetics. Birth Defects: Original Article Series 8(No. 7). The National Foundation—March of Dimes, New York.

Patel, P. I., Framson, P. E., Caskey, C. T., and Chinault, A. C. 1986. Fine structure of the human hypoxanthine phosphoribosyltransferase gene. Mol. Cell. Biol. 6:393–403.

Ponte, P., Gunning, P., Blau, H., and Kedes, L. 1983. Human actin genes are single copy for alpha-skeletal and alpha-cardiac actin but multicopy for beta- and gamma-cytoskeletal genes. Mol. Cell. Biol. 3:1783–1791.

Prescott, D. M. 1988. Cells. Jones and Bartlett, Boston.

Santos, T. and Zasloff, M. 1981. Comparative analysis of human chromosomal segments bearing nonallelic dispersed $tRNA^{Met}$ genes. Cell 23:699–709.

Schmid, C. W. and Deininger, P. L. 1975. Sequence organization of the human genome. Cell 6:345–358.

Schmid. C. W. and Jelinek, W. R. 1982. The Alu family of dispersed repetitive sequences. Science 216:1065–1070.

Singer, M. F. 1982. SINEs and LINEs: highly repeated short and long interspersed sequences in mammalian genomes. Cell 28:433–434.

Snyder, L. A., Freifelder, D., and Hartl, D. L. 1985. General genetics. Jones and Bartlett, Boston.

Tjio, J. H. and Levan, A. 1956. The chromosome number of man. Hereditas 42:1–6.

Tilghman, S. M., Curtis, P. J., Tiemeier, D. C. et al. 1978. The intervening sequence of a mouse beta-globin gene is transcribed within the 15S beta-globin mRNA precursor. Proc. Natl. Acad. Sci. USA 75:1309–1313.

Weller, P., Jeffreys, A. J., Wilson, V., and Blanchetot, A. 1984. Organization of the human myoglobin gene. EMBO J. 3:439–436.

Wiborg, W., Pedersen, M. S., Wind, A. et al. 1985. The human ubiquitin multigene family. EMBO J. 4:755–759.

Wilde, C. D., Crowther, C. E., and Cowan, N. J. 1982. Diverse mechanisms in the generation of human beta-tubulin pseudogenes. Science 217:549–552.

Willard, H. F. 1990. Centromeres of mammalian chromosomes. Trends in Genetics 6:410–416.

Willard, H. F. and Waye, J. S. 1987. Hierarchical order in chromosome-specific human alpha satellite DNA. Trends in Genetics 3:192–198.

Yunis, J. J., Sawyer, J. R. and Ball, D. W. 1978. The characterization of high-resolution G-banded chromosomes of man. Chromosoma 67:293–307.

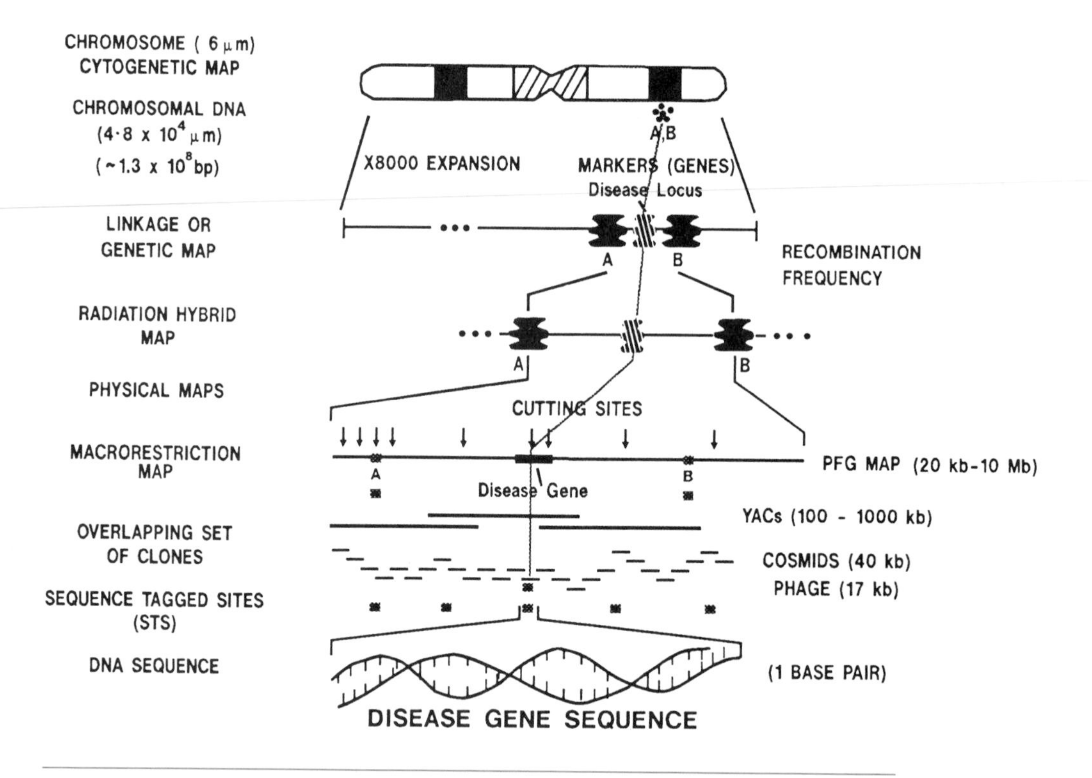

Multiple levels of human chromosome mapping. The line running vertically through the diagram represents the tracking of markers A and B through progressively more precise levels of mapping. In this way, investigators can follow a candidate disease gene from the coarsest to the finest map resolution, which is the DNA sequence. The cytogenetic map provides the lowest level of resolution, measuring the distance between chromosomal features (i.e., bands or breakpoints) visible under the light microscope. Chromosome banding can resolve features to about 5 Mb. The linkage or genetic map (Chapter 4) measures the recombination frequency between two linked markers, which can be genes or polymorphisms (A and B in this diagram). Radiation hybrid maps are produced by breaking chromosomes with radiation and then identifying the fragment carrying the marker (the breakpoint); the resolution of these maps is comparable to that of linkage maps (Box 4–3). At the next resolution level, macrorestriction fragments of 1 to 2 Mb are separated and the markers localized and mapped. Finer mapping resolution is provided by ordered libraries of yeast artificial chromosomes (YACs), which have insert sizes from 100 to 1000 kb. Ordered libraries of cosmids have smaller insert sizes, usually about 40 kb, and produce higher-resolution maps. The DNA base sequence is the highest-resolution map, with sequence tagged sites (STSs) (Chapter 15) used as unique reference points. (Figure provided by C. E. Hildebrand, LANL. Reproduced from Human Genome: 1991–1992 Program Report)

3 Gene Mapping, I

PHYSICAL MAPS

The ultimate goal of human gene mapping is to know the chromosomal location of every human gene and the order in which the genes are arranged. We expect to know the complete nucleotide sequence of the human genome eventually, as discussed in Chapter 15, but whether the location of most genes will be deduced from the DNA sequence, or whether the majority of genes will be identified and mapped before the sequence of the genome has been completely analyzed, is not yet clear. The least that can be said is that excellent progress in gene mapping has already been made and the future is bright.

The first human genes to be assigned to a specific chromosome were X-linked genes. Their distinctive pattern of inheritance (see Chapter 13) was noticed in the early part of this century, and Wilson concluded in 1911 that the genes for color vision must be carried on the X chromosome. The first autosomal assignment did not take place until 1968. This was achieved by showing linkage (see Chapter 4) between the Duffy blood group locus and a morphological variant of chromosome I (Donahue et al., 1968). The advent of somatic cell hybridization gave a major impetus to human gene mapping, as will be described presently; and the more recent molecular techniques using restriction enzymes, cloned DNA probes, and high-resolution *in situ* hybridization have greatly accelerated the process.

Gene Mapping by Somatic Cell Hybridization

Mapping at the Chromosome Level

The development of methods for fusing cells of the same or different species has had a major impact on mammalian gene mapping. *Cell fusion* (also known as *cell hybridization*) can be induced to occur when cultured cells are treated with polyethylene glycol or inactivated Sendai virus, both of which mediate fusion of plasma membranes between cells. The formation of a common cytoplasm is followed somewhat later by mitosis, when the chromosomes from both parental cells attach to a common spindle. Following mitosis, cytokinesis produces two daughter cells, each with a single nucleus containing a complete set of chromosomes from each parental cell type (Figure 3–1).

In the case of human-rodent hybrid cells, human chromosomes are gradually lost during subsequent cell divisions, while the complete rodent genome is retained. The basis for the preferential loss of human chromosomes is unknown, but the fact that hybrid cells containing only one or a few human chromosomes can be produced has been a great boon to gene mappers.

Even with the best techniques, hybrid cells are not initially present in large numbers, being greatly outnumbered by both parental cell types, fused and not fused. Therefore, the second step in the production of a hybrid cell line is to eliminate the parental cells. The classical selection system is HAT (hypoxanthine-aminopterin-thymidine) medium, first applied to hybrid cells by Littlefield (1964). If one parental cell type is thymidine kinase deficient (TK^-) and the other parental cell type is deficient in hypoxanthine phosphoribosyl transferase ($HPRT^-$), neither will grow in HAT medium. Aminopterin blocks *de novo* synthesis of both purines and thymidylate (normally the major pathway) thus forcing cells to rely upon the "salvage pathways" (Figure 3–2A).

The result is that any cell that can use both hypoxanthine (via HPRT)

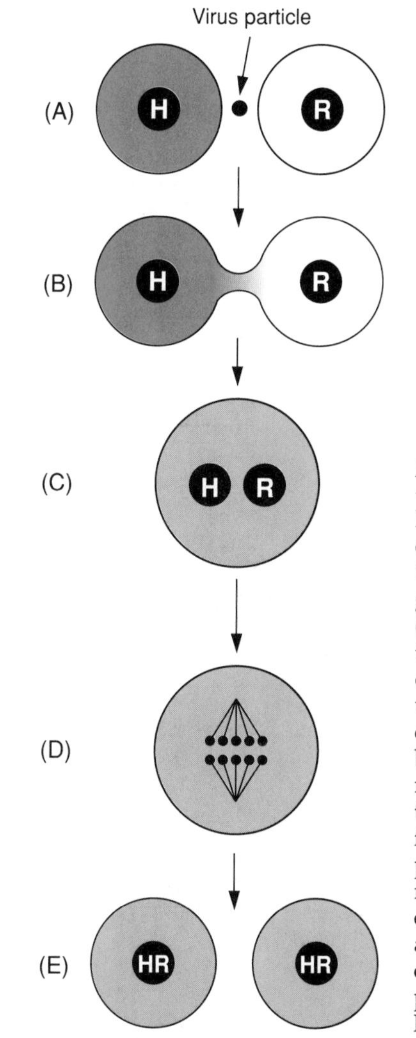

Figure 3–1

Hybridization of mammalian cells induced by inactivated Sendai virus. H = human genome; R = rodent genome. (A) Binding of a virus particle to two adjacent cells. (B) Fusion of the membrane of the virus with the membranes of both cells, creating a cytoplasmic bridge. (C) Binucleated cell resulting from fusion. (D) At the first mitosis following cell fusion, chromosomes from both parent cells attach to a single mitotic spindle. (E) Daughter cells contain one nucleus apiece, each of which has chromosomes from both parents. (Modified from Prescott, 1988, p.155)

and thymidine (via TK) will survive in HAT medium, which supplies both compounds. Neither parental cell type meets this test, but hybrid cells do, and the hybrids will therefore multiply while the parental cells either die or remain quiescent (Figure 3–2B). Colonies of hybrid cells can be isolated after they are large enough to be visible (at least 1000 cells), and the hybrid clones may thereafter be multiplied to any desired extent.

If a rodent cell line lacking TK is fused with a human cell line lacking HPRT and the resulting hybrid cells are grown in HAT medium, human chromosomes will gradually be lost from the hybrids, until the only one that remains will be the chromosome that contains the gene for TK, an enzyme necessary for the synthesis of DNA in HAT medium. This strategy led to the demonstration that the TK gene is on human chromosome 17—the first assignment of a human gene by somatic cell genetics (Migeon and Miller, 1968; Miller et al., 1971). The reciprocal protocol ($HPRT^-$ rodent cells and TK^- human cells) showed that the human X chromosome contains the HPRT gene.

Various other selective systems exist; for example, Chinese hamster cells that are glycine auxotrophs and thus will not grow in glycine-deficient medium have been extensively used by Puck and his colleagues (e.g., Kao

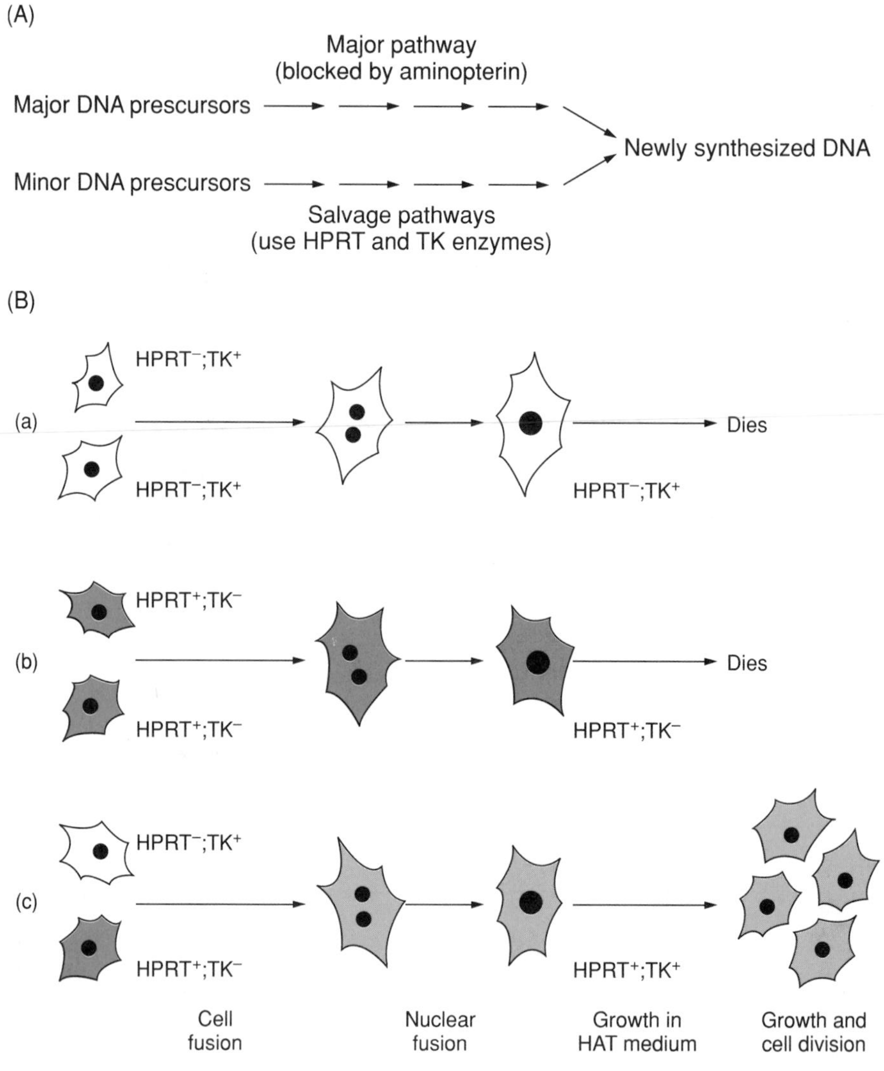

Figure 3–2

Selection of hybrid cells in HAT medium, in which the only cells that can grow are the HPRT+, TK+ hybrids. (A) HAT medium blocks the major DNA synthesis pathways, but supplies precursors for the salvage pathway. (B) Parental cells and fused cells containing only one parental cell type die; fused cells containing both parental types ("hybrids") survive and multiply. (Modified from Snyder et al., 1985, p. 471)

et al., 1976). Another convenient experimental fact is that human lymphocytes will not grow in ordinary culture conditions; they can therefore be used as the source of human chromosomes in a cell hybridization experiment without the need for a selective system to eliminate the human parental cells that do not participate in hybrid cell formation.

However, if every gene to be mapped required the use of a rodent cell line lacking any functional copies of that gene, progress would have been slow. The development of the *clone panel* method by Ruddle and his colleagues has made possible much more extensive use of somatic cell hybrids for human gene mapping (Creagan and Ruddle, 1975). The essential concept behind the clone panel strategy is as follows. If several hybrid cell clones are isolated from the products of a cell fusion experiment, there will be a time when each clone still contains a few human chromosomes, but not the entire human genome. Because loss of human chromosomes from the hybrids is random except for the chromosome bearing the selected marker, each clone will have a different subset of human chromosomes. (In fact, every cell in a clone will not be identical, but if a clone is established after several human chromosomes have been lost, every descendant of that cell will be alike insofar as the absence of those chromosomes is concerned.

The remaining chromosomes may vary somewhat from cell to cell within the clone, but as long as one can show that most of the cells contain a specific subset of human chromosomes, the clone will be useful for gene mapping.)

If one chooses several different hybrid cell clones with suitably different subsets of human chromosomes, and if one can assay for a particular gene product or the gene itself, an unambiguous decision can be made about the human chromosome that contains the gene in question, even though there is no selection for that gene. The way in which the clone panel method works is illustrated in Figure 3–3 for a hypothetical genome of 8 chromosomes, for simplicity. Note that only 3 hybrid cell clones are required, each containing 4 of the original 8 chromosomes.

The minimum number of clones (m) needed to assign a gene to a chromosome is $C = 2^m$, where C is the number of different chromosomes in the genome. For humans, C = 24 (22 autosomes plus X and Y sex chromosomes), which lies between 2^4 and 2^5; so 5 clones would be required, in principle. For the human genome to be mapped with optimal efficiency, each of the five ideal clones would have to have 16 chromosomes.

In practice, ideal clone panels are virtually impossible to obtain and are not necessary. What usually happens is that a panel of more than five clones is obtained which allows an unambiguous decision to be made, despite the presence of considerable redundancy for several chromosomes. One of the early examples of this method led to the assignment of the gene for UMP kinase to human chromosome 1, as shown in Table 3–1.

Mapping a gene via its protein product requires that the human protein be separable from the corresponding rodent protein, which will also be expressed in the hybrid cells. The most common separation method is electrophoresis. In the example in Table 3–1, UMP kinase was detected via its enzyme activity after electrophoresis of cell extracts in starch gels. Some-

Clones of hybrid cells	Chromosome number 1	2	3	4	5	6	7	8
A	+	+	+	+	−	−	−	−
B	+	+	−	−	+	+	−	−
C	+	−	+	−	+	−	+	−

Figure 3–3

Three-clone panel for mapping a genome with eight chromosomes. Presence of a specific chromosome in a clonal cell line is indicated by +; absence is indicated by −. Each chromosome has a unique binary signature, in terms of presence or absence in the three clones. Thus, if a gene or gene product were present in clones A and C, but absent from clone B, the gene could be assigned to chromosome 3 (shaded column). (Reproduced with permission from Ruddle and Creagan, 1975. © Annual Reviews, Inc.)

Table 3–1 Assignments of UMP kinase gene to chromosome 1 by the clone planel method

CELL LINE	UMPK ACTIVITY	CHROMOSOME* 1	2	3	4	5	6	7	8	9	10	11	12	13	14	15	16	17	18	19	20	21	22	X
WA-IIa	−	−	+	+	+	−	−	−	−	−	−	+	+	−	+	−	+	−	−	−	−	−	−	+
JFA-14b	+	+	+	−	−	+	−	−	+	−	−	−	−	+	−	−	+	+	+	−	−	−	−	+
WA-Ia	+	+	+	−	+	−	−	+	+	−	+	−	+	+	−	+	+	−	+	+	+	+	+	+
J-10-H-12	+	+	−	+	−	−	+	+	−	−	+	+	+	−	+	+	+	+	+	−	−	+	−	+
AIM-3a	−	−	+	−	+	+	−	+	+	−	+	+	+	+	+	−	−	+	+	−	+	−	−	+
AIM-8a	−	−	−	−	−	−	+	+	−	+	−	+	+	+	+	+	+	+	+	+	+	+	+	+
AIM-11a	−	−	+	+	−	−	−	+	−	−	+	−	+	−	−	−	−	−	−	−	−	−	−	+
AIM-23a	+	+	+	+	+	+	−	+	+	−	+	−	+	−	+	+	+	−	+	−	−	+	+	+

*All cell lines retained the chromosomes indicated with a frequency of 5% or greater.
Source: Satlin et al., 1975.

times, immunological methods can be used to distinguish the human protein from the rodent protein.

For nearly a decade, the somatic cell genetics approach to gene mapping was limited to genes that were expressed in tissue culture cells; and because the rodent parent cells used for cell hybrids were almost always transformed fibroblasts, which could be grown reliably and indefinitely, all of the genes that characterize differentiated tissues such as blood, muscle, and nervous system were inaccessible. This problem was overcome when methods for detecting genes themselves, rather than their products, were developed.

A transient technical stage was exemplified by the mapping of human alpha- and beta-globin genes, loci that are not expressed in fibroblasts, of course. This was done with DNA-DNA hybridization in solution. Purified human globin mRNAs were obtained from bone marrow cells and radioactive cDNA copies of the mRNAs were made with reverse transcriptase. Each cDNA was then hybridized to total DNA from several mouse-human hybrid cells containing various subsets of human chromosomes, using conditions where the cDNAs would pair with human globin genes, but not mouse globin genes, which differ sufficiently from the human genes for the reaction to be species-specific.

In this way, it was found that cDNA from human alpha-globin mRNA would form a molecular hybrid only with DNA from cells that contained human chromosome 16 (Deisseroth et al., 1977), and beta-globin cDNA would hybridize only to DNA from cells that contained human chromosome 11 (Deisseroth et al., 1978). Both of these results have been fully confirmed by recombinant DNA techniques, which rapidly rendered solution hybridization obsolete.

The two most important tools for gene mapping by recombinant DNA technology are restriction enzymes and *cloned DNA* segments. (Readers beware! Do not confuse cloned DNA with hybrid cell clones.) The way in which these tools are used is exemplified by the mapping of the insulin gene by Owerbach et al. (1980), as outlined in Figure 3–4. The first step is to obtain a probe for the gene of interest. In this case, mRNA was purified from Islets of Langerhans from rat pancreas, and then separated by size on a sucrose gradient. The most abundant size class of mRNAs was the right size to encode insulin.

Those mRNAs were copied with reverse transcriptase to yield cDNA, which was then cloned in a bacterial plasmid. The cloned cDNA is more desirable than the original cDNA because it can be obtained in unlimited quantities, although cDNA is also usable as a probe. A clone containing insulin sequences was identified by the size of the insert, by its sensitivity to certain restriction enzymes, and finally, by the nucleotide sequence. The cloned insulin cDNA was then made radioactive by enzymic labeling with radioactive nucleotides *in vitro*.

The cloned rat insulin cDNA could be used to detect genomic DNA sequences that code for rat insulin, and because the amino acid sequence of insulin does not vary greatly from one mammal to another, the rat probe could also be used to detect mouse and human insulin genes. This was essential, because the somatic cell hybrids that were to be used for the mapping were mouse-human hybrids.

The restriction enzyme EcoR I was used to digest both mouse DNA and human DNA, and the DNA digests were subjected to electrophoresis in order to separate the various pieces according to size. Following electropho-

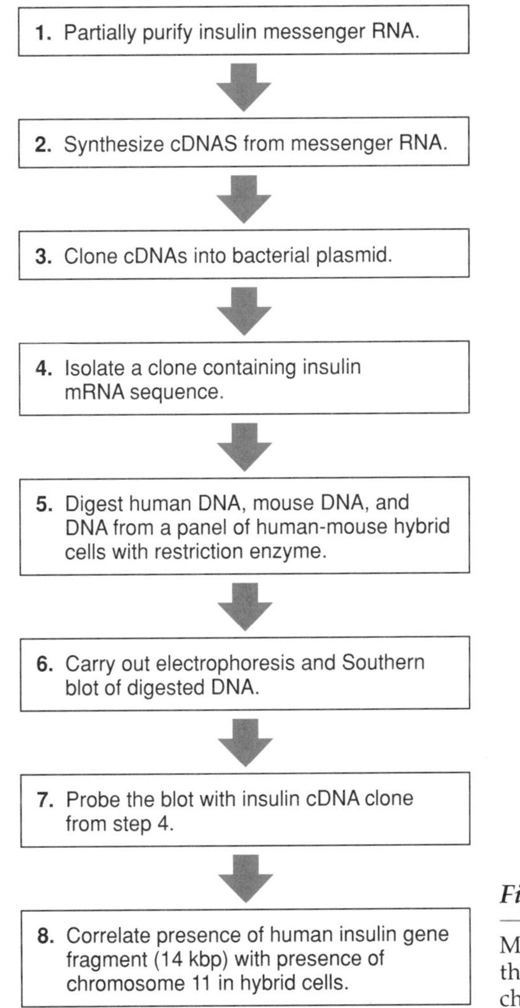

Figure 3–4

Major steps in the mapping of the human insulin gene to chromosome 11.

resis, the DNA fragments were denatured and transferred to nitrocellulose by the process known as *Southern blotting,* and the radioactive cDNA insulin probe described previously was hybridized to the genomic DNA attached to the nitrocellulose membrane. Detection of DNA-DNA hybrids by autoradiography showed that the two mouse insulin genes were located in 9 kbp, 1.8 kbp, and 1.4 kbp fragments, and the single human insulin gene was in a 14 kbp fragment (Figure 3–5). This separation gave assurance that the human gene could be detected in mouse-human hybrids and distinguished from the mouse insulin genes, which would be expected to be present in all of the hybrids.

DNAs from a clone panel of hybrid cells were then examined by the preceding technique, and it was found that the presence of the human insulin gene correlated perfectly with the presence of chromosome 11 (Table 3-2). Several hundred human genes have now been assigned to specific chromosomes by this general method. In principle, all that is needed is a gene-specific probe, a restriction enzyme that makes different-sized pieces from mouse DNA and human DNA, and a panel of hybrid cells with the right subsets of human chromosomes.

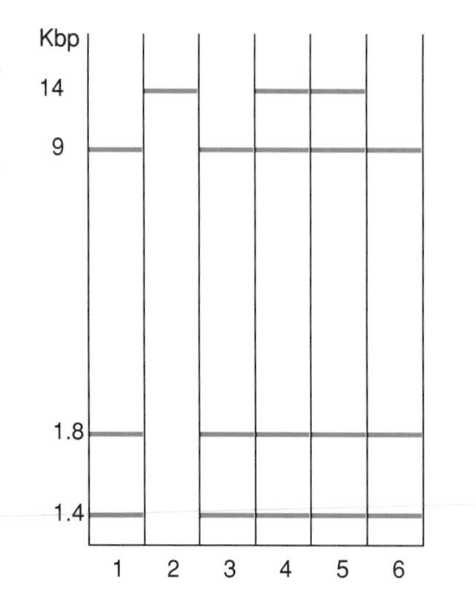

Figure 3–5

Restriction fragments detected by cloned insulin cDNA probe after EcoR I digestion of mouse, human, and mouse-human hybrid cell DNAs. Lane 1 = mouse, lane 2 = human, lanes 3 to 6 = different hybrids (TSL2, SIR8, DUA5, and JSR6C respectively; refer to Table 3-2 for chromosome content of the hybrids). The hybrid cells represented by lanes 4 and 5 contain the human insulin gene; the hybrid cells represented by lanes 3 and 6 do not. (Based on data in Owerbach et al., 1980)

Subchromosomal Mapping

Somatic cell methodology can also be used to assign genes to specific regions of chromosomes, provided that cell lines with suitable chromosomal modifications are available. The first subchromosomal assignments of human genes were made with the KOP translocation (Figure 3–6), which involves transfer of most of the long arm of the X chromosome to the long arm of chromosome 14 and the reciprocal transfer of a small piece of 14q to the X chromosome (Riciutti and Ruddle, 1973). The initials KOP refer to the donor of the cells and the two scientists who first studied the translocation.

Table 3–2 The distribution of human insulin, LDH A and human chromosomes in human-mouse cell hybrids

			CHROMOSOMES																							
HYBRID	INSULIN[a]	LDHA[b]	1	2	3	4	5	6	7	8	9	10	11	12	13	14	15	16	17	18	19	20	21	22	X	Y
WIL-2	−	−	−	−	−	−	−	−	−	+	−	+	−	+	−	−	+	−	+	−	−	−	+	−	+	−
WIL-8Y	+	+	−	−	−	−	−	+	+	−	−	+	+	−	−	+	−	−	+	+	−	+	+	−	+	−
TSL-2	−	−	−	+	−	−	+	+	−	−	−	+	−	+	−	−	−	−	−	+	−	+	+	−	+	−
TSL-5	+	+	−	−	−	+	+	−	−	−	−	+	+	+	−	−	−	−	+	+	−	−	+	−	−	−
TSL-6	−	−	−	−	−	−	+	−	−	−	−	+	−	−	−	−	−	−	−	−	+	−	−	−	−	−
TSL-8	−	−	−	−	−	−	−	−	−	−	−	−	−	−	−	−	−	−	+	+	−	+	−	−	−	−
RAS-8	+	+	+	+	+	+	+	+	+	+	−	+	+	+	+	+	+	+	+	+	+	+	+	+	+	−
RAS-9DT	+	+	+	+	+	+	−	−	+	+	+	+	+	+	−	+	+	+	+	+	+	+	+	+	+	−
SIR-8	+	+	+	+	+	+	+	−	+	+	+	+	+	+	+	+	+	+	+	+	+	−	+	+	+	+
XTR-3BSAgA	−	−	−	−	−	+	+	+	+	+	+	+	−	−	−	+	−	−	+	+	+	−	+	−	−	−
DUA-5	+	+	−	−	+	−	+	−	−	−	−	−	+	−	−	+	+	−	+	+	−	−	+	−	−	−
DUM-13	+	+	+	+	+	+	−	+	+	+	−	+	+	−	+	−	+	+	+	+	+	+	+	+	+	−
ALR-2	+	+	+	−	+	+	+	+	+	−	+	+	+	+	+	+	+	+	+	−	+	+	+	−	−	−
JSR-6C	−	−	−	+	+	−	+	−	−	−	−	+	−	−	−	+	+	+	−	+	−	−	−	−	+	−
JSR-24D	+	+	+	−	+	−	+	+	−	+	−	+	+	−	+	+	+	+	+	+	+	+	+	−	+	−

[a]The insulin gene was measured as the 14 kbp fragment.
[b]The LDH A gene has been assigned to chromosome 11 by several laboratories, and co-segregates with the insulin gene.
Source: Owerbach et al., 1980, p. 83

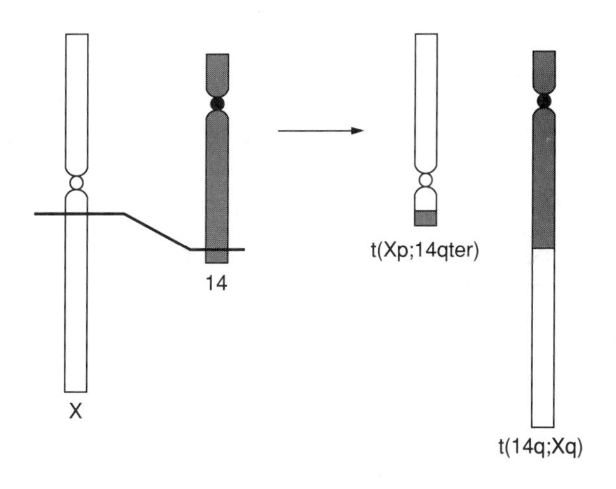

Figure 3–6

The KOP (14qXq) translocation. Normal chromosomes X and 14 are shown on the left, with the breakpoints indicated by a line. The translocation products are shown on the right. (Reproduced with permission from Riciutti and Ruddle, 1973. © Annual Reviews, Inc.)

When human cells containing the KOP translocation were fused with mouse cells, some of the hybrids that survived in HAT medium had the t(14qXq) translocated chromosome, but not the normal X, the normal 14, or the t(Xp14qter) translocation. This immediately showed that the HPRT gene was on the long arm of the X chromosome, because HPRT was the selected marker in this case. Two other human X-linked genes, PGK and G6PD, were also expressed in the hybrid cells, and were therefore also assignable to the translocated portion of Xq. Fortuitously, one other gene assignment resulted from the study of the hybrids with the KOP translocation because they always expressed human nucleotide phosphorylase (NP). At that time, it was known only that NP was coded by an autosomal gene; the new information placed the NP gene on chromosome 14.

Translocated chromosomes have been used many times for subchromosomal mapping after a gene has been assigned to a specific chromosome by use of a panel of somatic cell hybrids. The Human Genetic Mutant Cell Repository lists about 500 fibroblast lines with translocations in its catalog. Included there are translocations involving every human chromosome.

Another approach to subchromosomal mapping employs cell lines in which different parts of a given chromosome are missing. For example, in Theodore Puck's laboratory, a series of Chinese hamster-human hybrid cell lines carrying various fragments of human chromosome 11 was created, using X-rays to induce chromosome breakage (Figure 3–7). Gusella et al. (1979) prepared DNA from each of these cell lines, and then used a cloned human beta-globin DNA probe to find out which cells contained the corresponding gene. It was already known that beta-globin was encoded by chromosome 11. Lines J1-23, J1-10, and J1-11 were positive, whereas lines J1-9 and J1-7 were negative. These results implied that the gene for beta-globin is approximately in the middle of band 11p12, not far from LDH-A (in cytological terms).

Gene Mapping by In Situ *Hybridization*

It is a surprising fact that DNA does not need to be extracted from cells in order to be available for hybridization with exogenous RNA or DNA. Gall and Pardue (1969) showed that after metaphase-arrested cells had been

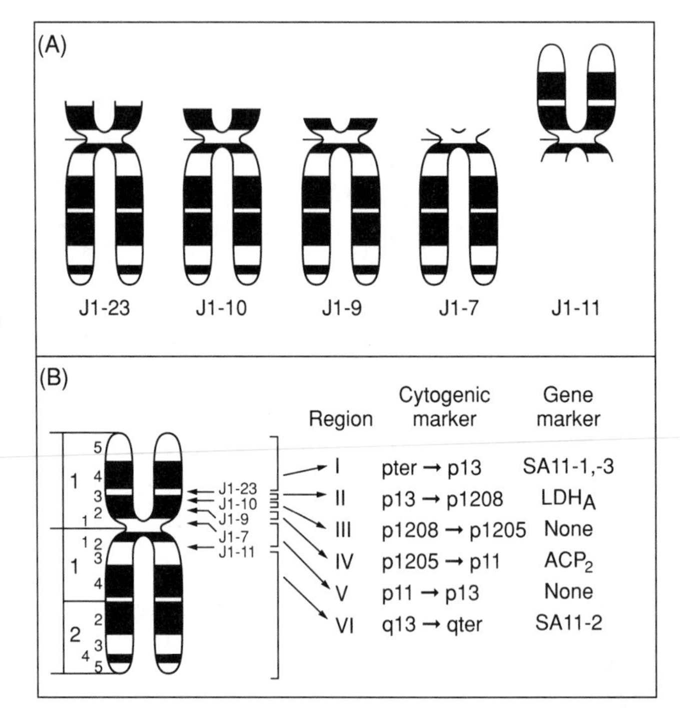

Figure 3–7

(A) Diagram showing the various terminal deletions of human chromosome 11 in five hybrid cell clones. (B) Scheme of chromosome 11, with arrows indicating the breakpoints. The five breakpoints divide chromosome 11 into six regions, which are characterized by cytogenetic and gene markers as indicated to the right of the drawing. (From Gusella et al., 1979, p. 5240)

fixed onto a microscope slide and chromosomal DNA had been denatured by dipping the slides in alkali briefly, RNA-DNA hybrids could be formed with radioactive ribosomal RNA in a solution applied to the fixed cells. Following removal of unhybridized RNA, the slides were dipped in a photographic emulsion, which formed a layer of highly sensitive film in close contact with the cells. After a suitable exposure time, during which the emission of beta particles from the labeled, hybridized RNA created silver grains in the emulsion, the film was developed and the location of the grains relative to the underlying chromosomes was determined microscopically (Figure 3–8).

The best isotope for autoradiographic localization is tritium, because the weak beta particle that it emits produces silver grains only in the immediate vicinity of its origin, thus providing the best resolution available. However, it is not possible to label nucleic acids with tritium to a sufficiently high specific activity to detect single-copy genes, when only one molecule of probe (radioactive RNA or DNA) hybridizes to a complementary sequence in a metaphase chromosome. For this reason, *in situ* hybridization was at first limited to mapping genes that were reiterated about one hundred times or more. The technique was excellent for *Drosophila* salivary gland chromosomes, where most genes are represented by 1000 side-by-side copies; but in humans, only the genes for ribosomal RNAs were initially mappable.

More recently, the sensitivity of *in situ* hybridization with radioactive probes has been substantially increased by the use of conditions that promote probe network formation, so that many probe molecules become associated with one target sequence in the chromosome. Another increase in sensitivity resulted from the realization that reliable conclusions can be obtained from statistical analysis of autoradiographic grains over a large num-

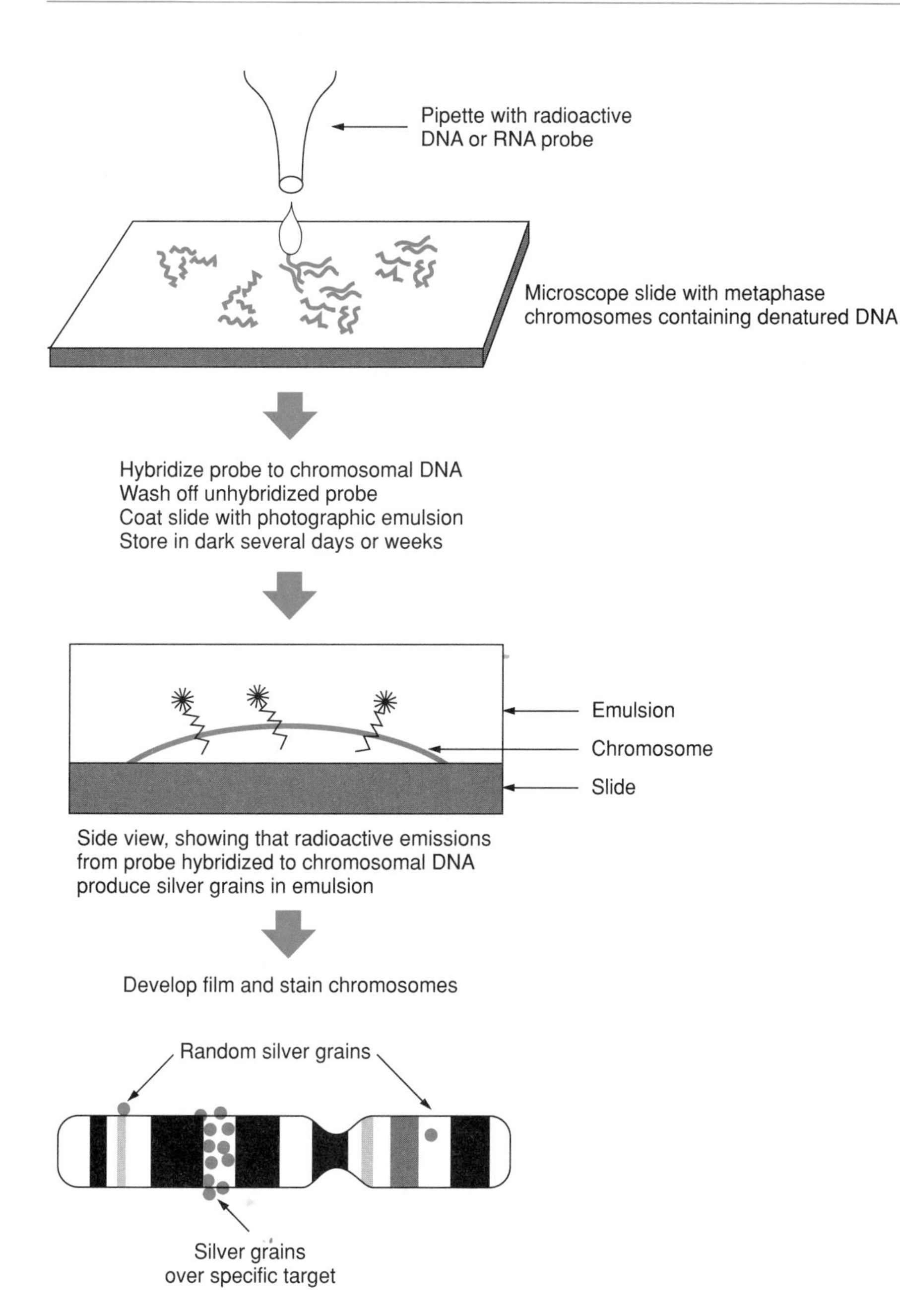

Figure 3–8

Major steps for *in situ* hybridization with radioactive probes.

ber of metaphases. Even though there may be only one silver grain over the chromosomal region where a single-copy gene actually resides and several silver grains elsewhere in the same metaphase spread (because of non-specific sticking of the probe), experience has shown that analysis of about 50 good metaphases will reveal that the total number of silver grains over the genuine target area will be significantly greater than the grain distribution elsewhere. An example of *in situ* hybridization mapping data is shown in Figure 3–9.

Evidently, mapping by *in situ* hybridization has significant advantages over the somatic cell hybrid method; no panel of cloned hybrid cells needs to be grown or karyotyped, and the *in situ* result immediately assigns a

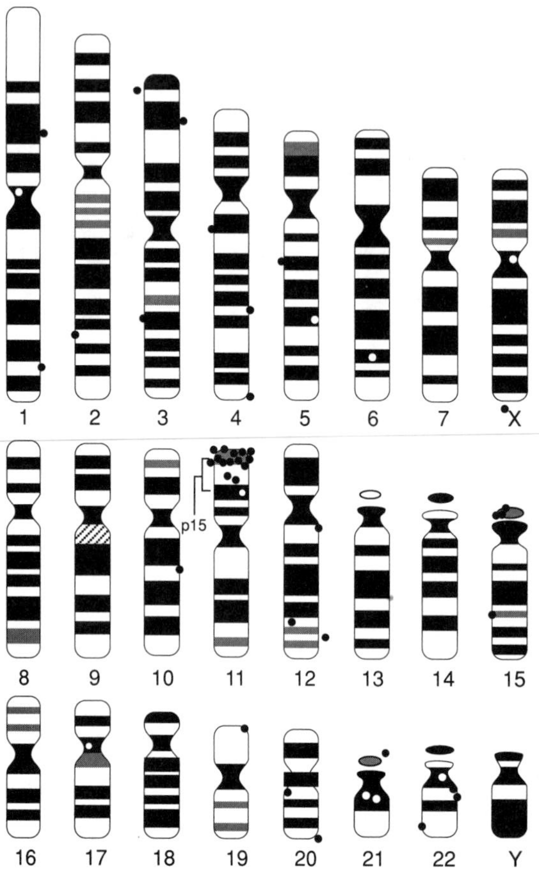

Figure 3–9

Diagrams of G-banded human metaphase chromosomes, with combined autoradiographic data from 35 metaphases, hybridized with a radioactive DNA probe for the insulin gene. Twelve of the 35 metaphases were labeled on band 11p15; those grains represented 26% of total labeled sites. (From Harper et al., 1981, p. 4459)

gene, not only to a chromosome, but to a subregion of a chromosome containing only one or two Giemsa bands. The major disadvantage of mapping single-copy genes by *in situ* hybridization of tritium-labeled probes is the length of time (usually several weeks) that one must wait for an adequate autoradiogram to form. A secondary disadvantage is that even with tritium, the grain size gives only moderately accurate information about the subchromosomal location of the target gene.

Both of these difficulties have been overcome by the development of nonradioactive probes and microscopic techniques that permit substantially increased resolution as well as detection of weak signals. The nonradioactive probes are based on the extraordinarily high affinity of the proteins avidin or streptavidin for the small organic molecule biotin (a vitamin in mammals). Biotin can be attached via linker molecules to the purine or pyrimidine portions of nucleotides in a variety of ways, so that it does not interfere with the hydrogen-bonding capacity of the nucleotide (Figure 3-10 and Langer et al., 1981). There are now several commercial sources of biotinylated dATP or dUTP. These molecules can be used in the synthesis of oligonucleotides of known sequence, which can then be used as probes for the corresponding DNA in chromosomes.

In situ hybridization with a biotinylated probe is carried out in the same manner as with a radioactive probe. Detection of the hybrids formed between the probe and its target in a chromosome is achieved by adding some form of labeled avidin or streptavidin after the hybridization reaction is complete. These molecules will bind to the hybridized biotinylated probe

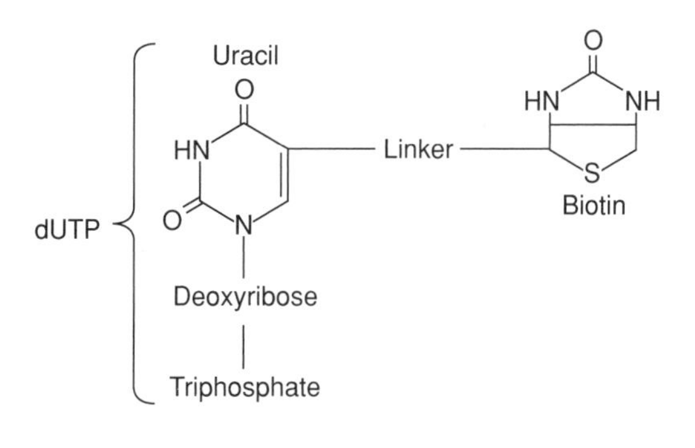

Figure 3–10

Biotinylated dUTP. Linkers usually contain 11 to 21 carbon and nitrogen atoms. The pyrimidine used is uracil instead of thymidine, because the linker must be attached to carbon 5 of the pyrimidine ring, which is occupied by a methyl group in thymine. DNA polymerase readily accepts dUTP in place of TTP. A biotinylated RNA can also be made, using biotin-UTP and RNA polymerase.

with great efficiency, but are not readily detectable by themselves. In some detection systems, avidin or streptavidin is coupled with an enzyme, usually alkaline phosphatase, which can be used to produce a microscopically visible deposit of colored reaction product at the site where the probe has been bound (Figure 3–11A).

More often nowadays, the avidin is coupled, directly or indirectly, to a fluorescent molecule (Figure 3–11B). This leads to an amusing acronym, FISH (fluorescent *in situ* hybridization). When the preparation is examined in a fluorescence microscope, a colored spot corresponding to the location of the target gene can be seen. The availability of laser-scanning confocal microscopes and a variety of other technical improvements now makes it possible to detect single-copy genes with impressive resolution. Some workers now claim to be able to pinpoint a gene to within one megabase by this technique—an amount of DNA corresponding to about 1% of an average human chromosome (Lichter et al., 1990). As the examples in Figure 3–12 show, the two copies of a gene on sister chromatids are clearly resolved, and the simultaneous use of multiple probes in a single hybridization allows one to establish gene order efficiently. It seems likely that *in situ* hybridization with

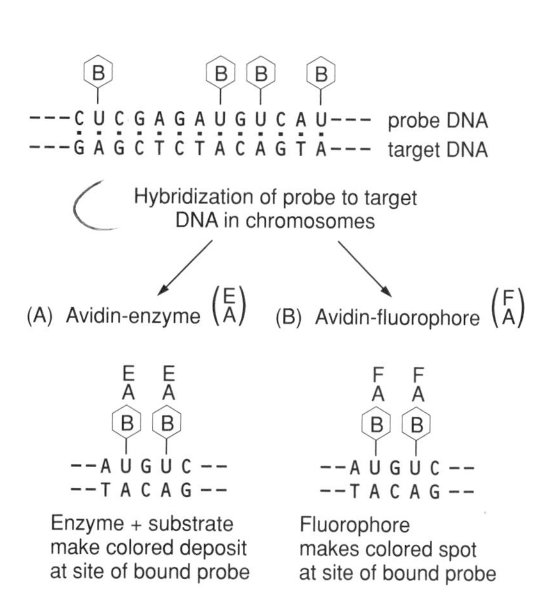

Figure 3–11

Detection of biotinylated probes hybridized to target DNA.

After the probe has hybridized to target DNA in chromosomes, avidin or streptavidin coupled to some type of reporter molecule is bound to the biotin. In (A), the avidin is bound to an enzyme, such as alkaline phosphatase or beta-galactosidase. When a suitable substrate is supplied, the enzyme will form a deposit of colored product that may be detected by conventional microscopy. In (B), the avidin is coupled to a fluorescent molecule, such as fluorescein or Texas Red, which will form a colored spot when the fluorophore is excited by light of the appropriate wavelength.

Sometimes a "sandwich" technique is used to generate a stronger signal. For example, after avidin-fluorescein has bound to the probe (which is bound to target DNA), a biotinylated anti-avidin antibody can be added, followed by more avidin-fluorescein.

Another nonradioactive detection system uses nucleotides coupled to a steroid, digoxigenin, to make the labeled probe. Following binding to target DNA, the probe is detected with anti-digoxigenin antibodies coupled to an enzyme or to a fluorophore. The principle is the same as for the biotin-avidin system.

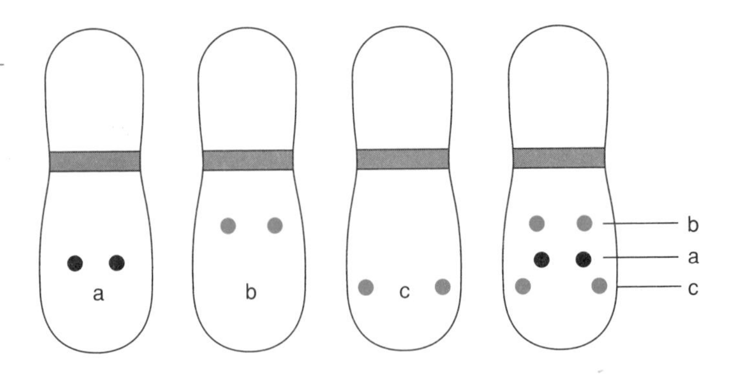

Figure 3–12

Mapping of three different cloned DNAs by FISH. The order of the corresponding chromosomal loci can be approximately determined by single probe hybridizations and by measuring the distance of each fluorescent spot from the nearest telomere. For closely spaced loci, the order can often be ascertained by simultaneous hybridization of three probes, two of them labeled with one color (such as red for probes b and c in the diagram on the right) and the other with a different color (such as green for probe a in the right-hand diagram).

biotinylated probes, followed by detection of hybrids with fluorescent secondary probes and high resolution microscopy, will rapidly become the method of choice for gene mapping at the chromosome level.

Isolation of Individual Chromosomes

In order to progress to a finer level of resolution than is possible with light microscopy, molecular mapping approaches must be employed. These always involve cloning a genome or fragments of a genome by recombinant DNA methods, followed by ordering of the clones to create a map. For small genomes, such as those of bacteria and viruses, a clone library containing the entire genome is suitable for mapping. Complete genomic clone libraries are easy to prepare, but for complex genomes such as our own, the number of components is so large that it is not practical to use the entire library to make an ordered map of an entire chromosome or even a large fraction of a chromosome.

The obvious first step in fractionating the human genome is to isolate individual types of chromosomes. Metaphase chromosomes from synchronized cells can be isolated in small quantities by *flow cytometry,* using a fluorescence activated cell sorter (FACS). Chromosomes in suspension, prepared from metaphase-arrested cells, are first stained with one or two fluorescent dyes, the binding of which primarily reflects the amount of DNA in each chromosome, but is also affected by other properties such as clusters of reiterated sequences, the extent of supercoiling, and the (A+T)/(G+C) ratio. The stained chromosomes are forced to flow at high speed through an orifice under conditions that create microdroplets containing at most one chromosome apiece. As the droplets pass through a laser beam that excites the fluorescent stain, the amount of fluorescence is measured photometrically and interpreted electronically. The machine can be programmed so that chromosomes exhibiting different amounts of fluorescence are deposited in different collection pools; this is done by electronic deflection of microdroplets containing chromosomes of the desired type (Figure 3–13).

Efforts to purify individual types of human chromosomes by flow cytometry began in the late 1970s and met with partial success. A few chromosomes had sufficiently distinctive staining properties to permit their isolation in almost pure form, but most chromosomes fell into groups containing 2 or more of the 24 different types. Recent technical advances have drastically improved the situation. Lebo et al. (1986) used a dual-laser cy-

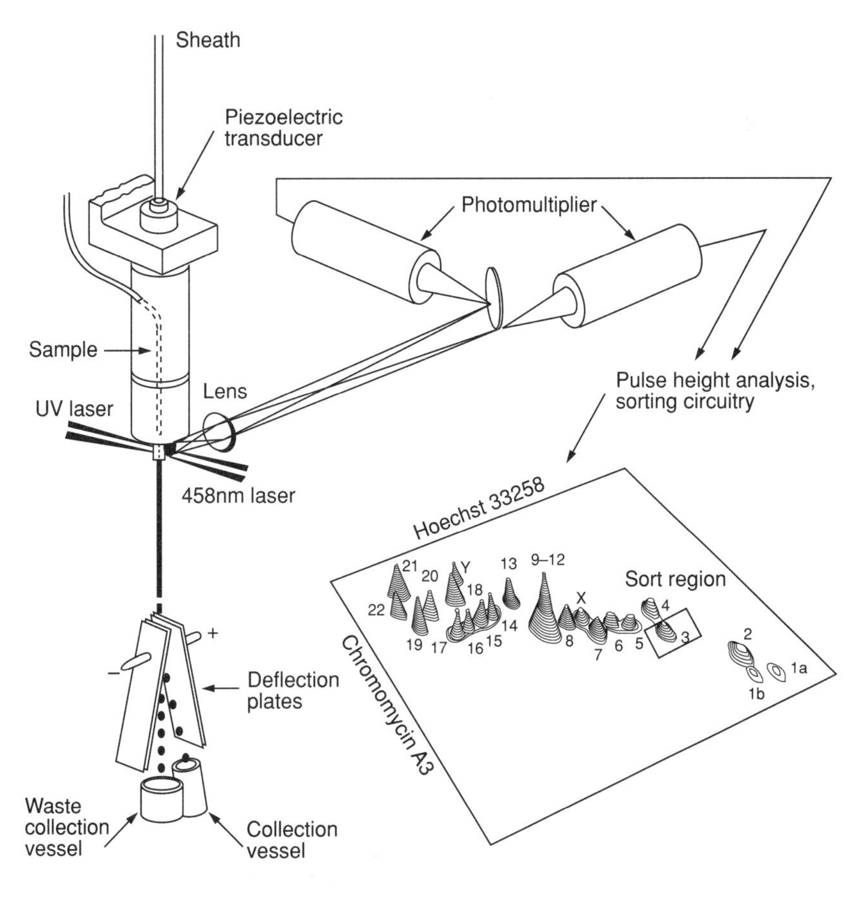

Figure 3–13

Diagram showing the principles of dual-beam flow cytometry and sorting of chromosomes that have been stained with two fluorescent dyes. Chromosomes enter the tube marked "sample" and mix with a sheath fluid. As each chromosome passes through the laser beams, each type of dye is excited to emit light of a characteristic wavelength, which is measured by a photomultiplier tube.

As the liquid exits from the orifice of the flow tube, it breaks up into microdroplets, each containing at most one chromosome. Droplets carrying chromosomes of a desired type can be given an electric charge and diverted into a separate collection vessel. The inset shows a bivariate distribution for fluorescence of two commonly used dyes, measured for chromosomes from karyotypically normal cells. (From Gray et al., 1986, p. 142)

tometer and two dyes with different fluroescence emission spectra to separate 21 different human chromosomes; only chromosomes 10, 11, and 12 appeared in the same fraction.

The instrument described by Lebo et al. (1986) can sort chromosomes directly onto nitrocellulose filters so that the DNA can be used for "dot-blot" hybridization. About 30,000 chromosomes of a given type, containing a total of 3 to 15 ng of DNA, are required for an analysis. The approach is quite similar to Southern blotting. First, DNA is denatured on the filters and exposed to a radioactive probe in solution; after a time, the unhybridized probe is washed off and hybrids are detected by autoradiography. If the probe hybridizes to a single-copy gene, the autoradiogram of the filter will reveal the chromosome that contains the gene.

However, the principal use for flow-sorted chromosomes is in the preparation of chromosomal clone libraries in various types of vectors. For several years, the U.S. Department of Energy (DOE) laboratories at Los Alamos and Livermore have been working on a National Laboratory Gene Library Project, with the goal of preparing clone libraries of every human chromosome, for the use of investigators throughout the world. Both DOE labs have been active in the development of FACS machines for this specific purpose. Some human chromosomes have been isolated directly from normal human cells, as indicated previously; others have been isolated from human-hamster hybrid cell lines containing only one human chromosome.

It is often possible to separate all of the hamster chromosomes from the single human chromosome more efficiently than it is to separate certain human chromosomes from one another.

The first clone libraries prepared by the DOE labs from flow-sorted human chromosomes were made from complete restriction enzyme digests; these were small-fragment, nonoverlapping libraries that are of limited use for gene mapping. These libraries now exist for every human chromosome; they are available from the American Type Culture Collection. The second stage in the National Laboratory Gene Library Project is the preparation of large-insert, overlapping clone libraries from individual types of human chromosomes. This project is currently underway; it uses lambda and cosmid vectors for the most part.

The Top-Down Approach to Molecular Mapping

The ultimate goal of gene mapping is the complete nucleotide sequence of the human genome, but at present, there is no way to sequence DNA in intact chromosomes. Instead, it is necessary to fragment chromosomes (or whole genomes) and prepare recombinant DNA clones of a few hundred nucleotides each, which can be sequenced by the methods outlined later. However, in order to assemble the sequence of an entire chromosome from a set of sequences of 500–1000 base pairs apiece, it is necessary to know the order in which the multitude of sequenceable fragments were arranged in the chromosome originally. This cannot be achieved in one step; a hierarchy of maps of different levels of resolution must be created. That is, chromosomes must first be fragmented into large pieces, which are subsequently broken down into medium-sized pieces, which are finally cleaved into pieces small enough to sequence. This is referred to as the *top-down mapping strategy*.

Let us first consider the problem of breaking an average human chromosome (120 Mbp) into 100–200 fragments, that is, into pieces a few hundred kilobases long, on average. For many years, this level of resolution was a serious obstacle to molecular gene mapping. There were two difficulties: first, there was no technique available for analyzing DNA molecules in that size range; second, there were no cloning vectors that could accept inserts greater than roughly 40 kilobases. Both problems have been solved recently.

Size Fractionation of Very Large DNA Molecules

The separation of DNA molecules according to size by electrophoresis in agarose or acrylamide gels is a basic procedure in molecular biology. Within the range of a few nucleotides to 50 kbp, one could easily determine the sizes of a set of DNA molecules simply by choosing the concentration of agarose (for large molecules) or acrylamide (for small molecules) that was appropriate for the size range in the sample. However, there was no fractionation of DNA molecules larger than 50 kilobase pairs. The problem was not that the big DNAs refused to move through the agarose; they did move, but they all moved at the same rate.

Charles Cantor and his colleagues (Schwartz et al., 1983; Schwartz and Cantor, 1984) reasoned that DNA molecules greater than 50 kbp must move through dilute agarose gels in a serpentine fashion, rather than as random coils, which would be too large to penetrate the agarose mesh (Figure 3–14).

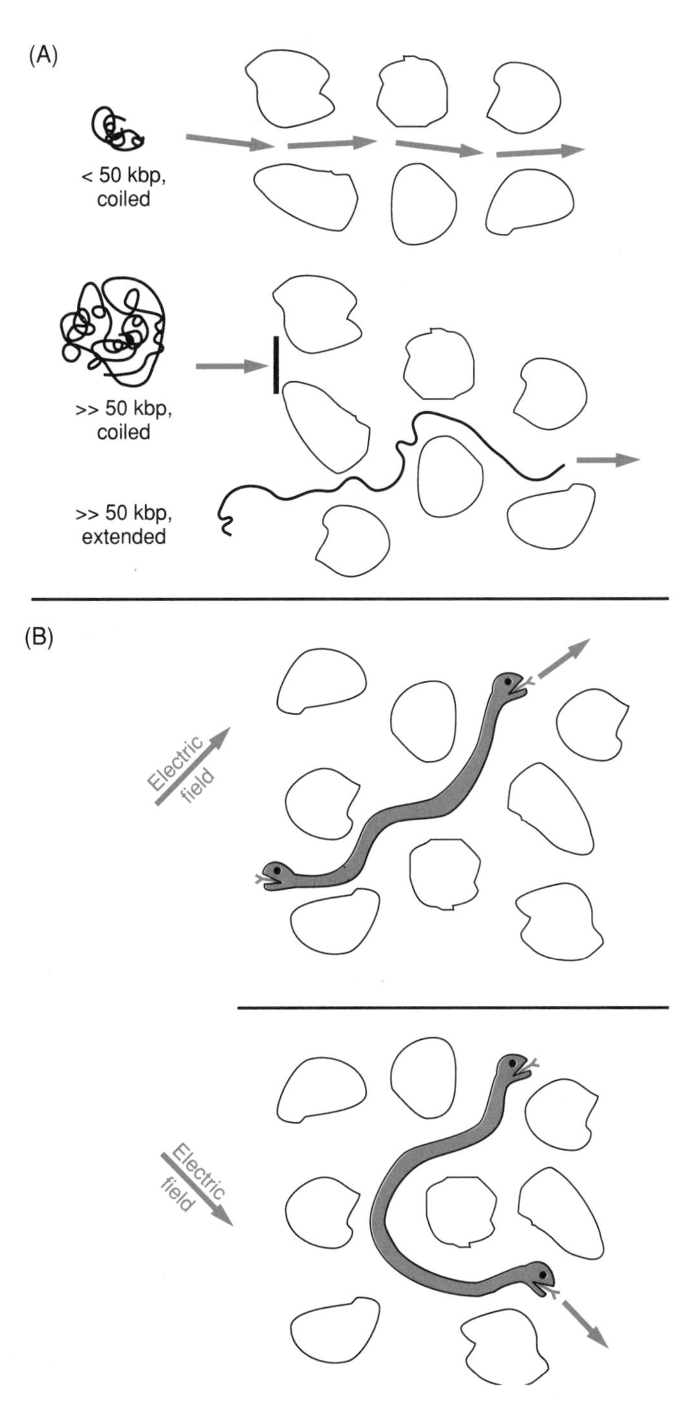

Figure 3–14

Separation of DNA by ordinary and by pulsed-field gel (PFG) electrophoresis. (A) During standard, constant-field electrophoresis, DNA molecules less than 50 kbp can pass through the pores in the agarose matrix as random coils (top drawing), while coiled molecules substantially greater than 50 kbp are blocked (middle drawing). One end of a large DNA molecule may enter the matrix when it becomes extended, because of random thermal motion. The rest of the molecule gradually follows, until it is entirely in the serpentine form. (B) DNAs greater than 50 kbp traveling in a gel with alternating electric field directions (PFG electrophoresis) are forced to spend much of their time reorienting, relative to their long axes. Either end may lead when the field direction is changed. Larger molecules spend more time reorienting than smaller molecules; hence, size fractionation of molecules greater than 50 kbp can occur.

They suggested that one of the ends of a coiled DNA molecule would find its way into a pore in response to an electric field, and that the entire molecule would eventually follow, rather like a snake entering a burrow. They coined the term *reptation* for this process. This provided a theoretical explanation for the observation that DNAs of all sizes greater than 50 kbp would move at the same rate in an electric field. All DNAs have the same diameter, regardless of length; once the serpentine configuration was achieved, all DNAs would have nearly the same mobility. DNAs smaller than 50 kbp would be able to enter the agarose pores as random coils, at least part of the time; the smaller the coil, the less time it would spend being retarded by the agarose, and therefore, fractionation according to size would occur.

Pursuing this idea further, Cantor and co-workers realized that a variable in the process of reptation would be the length of time it took for a DNA molecule to become extended into the serpentine form. Thus, large DNA molecules should take longer than small DNA molecules. In a continuous electric field, the fraction of total experimental time needed for reptation to begin is trivial; virtually all of the time is spent moving endwise through the agarose. However, if the direction of the electric field were varied frequently, then each time the field changed, every DNA-snake would have to pull back into a random coil and find a new hole in the agarose to slither through in response to the new direction of the electric field. By choosing a suitable pulse time, such that the DNA molecules spent a large percentage of time rearranging their configurations, molecules greater than 50 kbp could be fractionated according to size. In this manner, the technique known as *pulsed-field gel electrophoresis* (abbreviated PFGE) was devised.

Southern et al. (1987) argued that extended DNA molecules probably do not reorient into a coiled form and then re-extend themselves in response to each change in the field gradient. They felt that their observations were better explained by a model in which a DNA molecule that had reached a stretched-out configuration in response to one electrical pulse would remain extended when the field direction was changed, and would begin to move in response to the new field when the opposite end of the molecule found a pore that it could enter. In effect, their model treats a DNA molecule as a serpentine pushmi-pullyu; and the reason that large DNAs are fractionated by PFGE is that the longer the distance between the two heads, the longer time it takes that molecule to make net progress each time the field changes.

Whatever the physical reality may be, in the few years of its existence PFGE has had a major impact on molecular genetics, because it filled a resolution gap between cytogenetic analysis, which is accurate to a few Mbp, and cloning vectors that could accept 20–40 kbp of DNA. PFGE is very effective in separating DNA molecules from about 40 kbp to at least 1 Mbp; and some work has been done with molecules that appear to be 5 to 10 Mbp in size. PFGE gives excellent resolution of intact yeast (*S. cerevisiae*) chromosomes, which range from about 150 kbp to 2 Mbp; indeed, yeast chromosomes are often used as size markers on PFGE gels (Figure 3–15).

Large-Insert Cloning Vectors

Although pulsed-field gel electrophoresis made it possible to construct long-range restriction maps, the maps were no more than intangible representations of genetic reality, because there was at first no way to clone comparably large pieces of DNA. Conventional cloning vectors based on lambda phage are limited to roughly 20 kbp of insert DNA; and cosmids, the most capacious vectors available until recently, could accept at most 40–45 kbp of insert DNA. An average human chromosome of 120 Mbp represents 3000 40 kbp pieces laid end-to-end, and inasmuch as it is necessary to analyze overlapping clones in order to assemble a linear array, it would be a gigantic task to establish an ordered series of cosmids for an entire chromosome.

The need for a cloning vector that can accept inserts in the size range of hundreds of kilobases has been met within the past few years in a novel way by the creation of artificial chromosomes in yeast, now universally

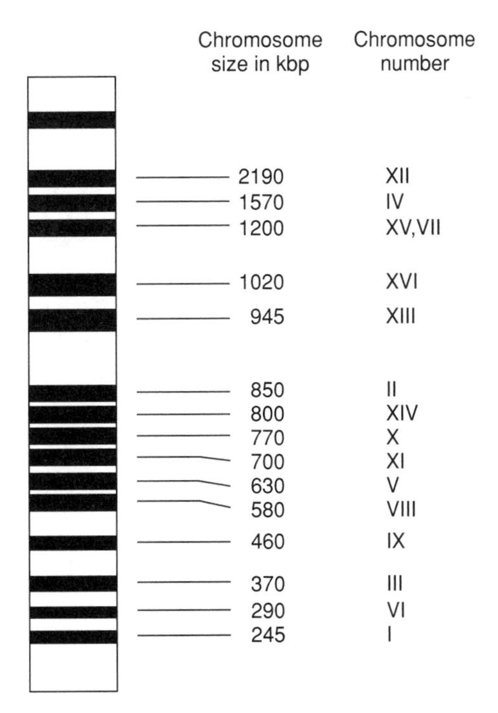

Figure 3–15

Chromosomes of *Saccharomyces cerevisiae*, strain YNN295, separated by pulsed-field gel electrophoresis. (Based on data in Vollrath and Davis, 1987)

called YACs (Burke et al., 1987). The fundamental requirements for a chromosome are diagrammed in Figure 3–16. They are: (1) a centromere, the structure to which mitotic spindle fibers attach; (2) telomeres, which are specialized DNA sequences at the ends of chromosomes; and (3) at least one site for the origin of DNA replication (called an ARS, or autonomously replicating sequence). Large chromosomes have many ARSs; replication proceeds bi-directionally from each of them. All three classes of yeast sequences were available as recombinant bacterial plasmids.

However, in order for a chromosome to be faithfully replicated and accurately distributed to daughter cells during mitosis, there is a need for a minimum size. Although the minimum size is not clearly defined, the need for DNA in addition to the centromere, telomeres, and the ARS implied that a stable artificial chromosome could be created from those three sequences plus an insert consisting of foreign DNA. That is exactly what Burke et al. accomplished, using for inserts human DNA that had been partially digested with restriction enzymes that cut at widely separated sites (Figure 3–17).

Libraries of human DNA cloned in YACs have now been created and are being used by a rapidly increasing number of laboratories. A variety of technical problems exist; such as, it is not as easy to work with large numbers of yeast clones as it is with bacteria, and artificial chromosomes are present in only one copy per cell, in contrast to multicopy bacterial plasmids or bacteriophage. However, it is possible to cope with these problems, and there is no doubt that YACs will play an important role in mapping projects involving complex genomes.

A cloning vector based on the phage P1 has been described by Sternberg (1990). This vector accepts inserts in the range of 80–100 kbp, twice the

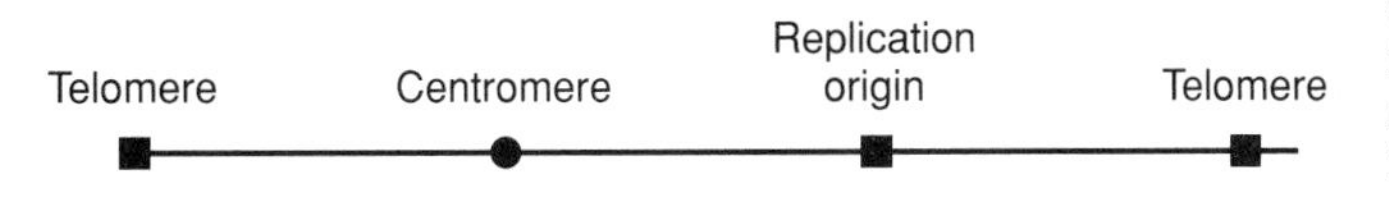

Figure 3–16

Essential sequences in a eukaryotic chromosome are those present in the two telomeres, the centromere, and an origin of replication (an ARS). Multiple replication origins are usually present, as well as genes and intergenic regions that give the chromosome a length required for reliable segregation at mitosis. (From Prescott, 1988, p. 375)

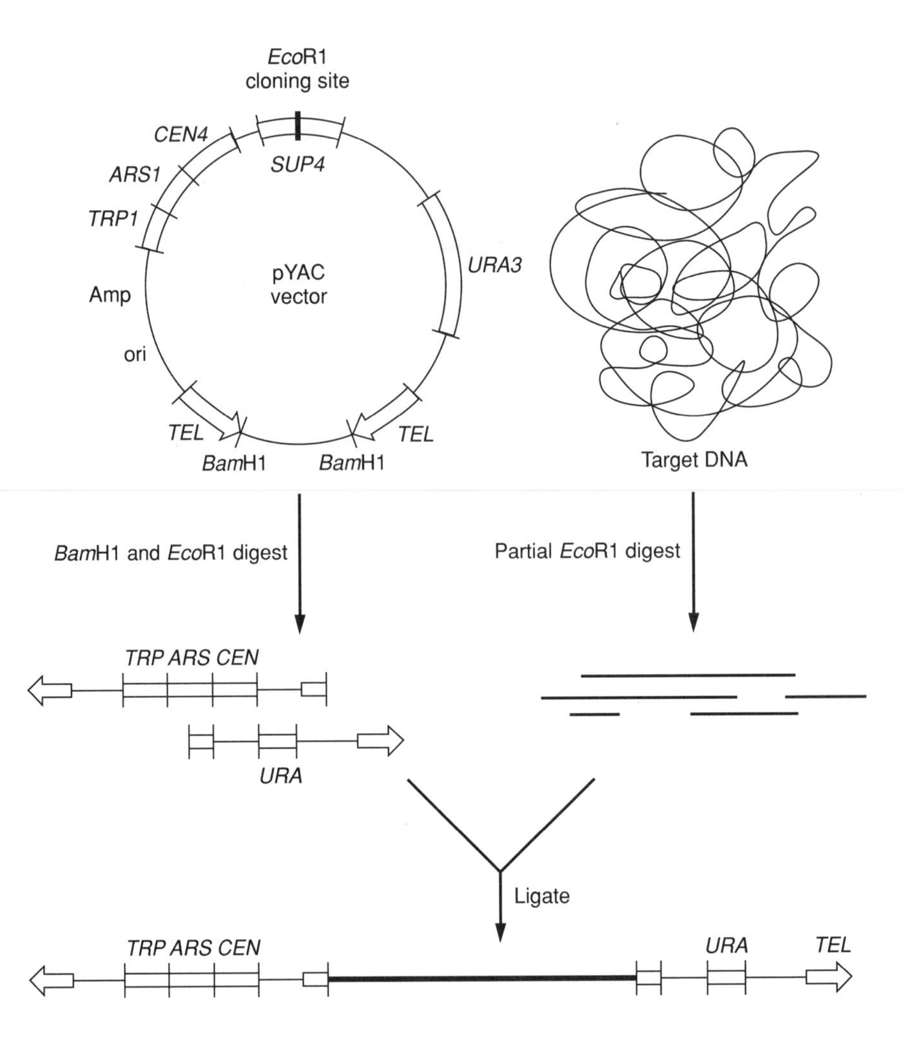

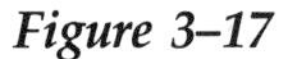

Figure 3–17

A yeast artificial chromosome (YAC) cloning system. A plasmid containing inverted repeats of telomeric (TEL) sequences, a centromere sequence (CEN4), and selectable markers (TRP1 and URA3), provides the two vector arms after cleavage with restriction enzymes in the SUP4 gene and at the BamH1 sites (middle drawing on left side).

The vector arms are dephosphorylated and ligated to large DNA fragments from a target DNA (e.g., human). The recombinant molecules are introduced into yeast, where they are maintained as extra chromosomes. Only one of many possible recombinant molecules is shown, and the actual size of the human DNA insert would be much larger, relative to the vector arms, than is shown in the diagram. (From Cooke, 1987, p. 173)

size that can be cloned in cosmids. After the insert DNA has been ligated into the vector and recombinant molecules of the desired size range have been isolated, they are packaged into P1 phage heads in a cell-free extract, in much the same fashion that recombinant lambda phage are prepared. The virions bind to suitable bacterial hosts and inject their DNA, where it replicates as a plasmid. It is too soon to predict whether this cloning system will replace cosmids; it clearly cannot substitute for YACs, which can accept foreign DNA inserts of several hundred kbp. Additional large-insert vectors are being developed in several laboratories.

Restriction Maps and Contig Construction: the Bottom-Up Approach

As we have learned from the preceding sections, the process of obtaining a complete sequence of a genome requires that the genome be fragmented into a moderate number of large pieces, each of which can be broken into smaller pieces, which can be fragmented into still smaller pieces, and so on until pieces small enough to sequence are obtained. This is the top-down approach. However, at every level prior to the final sequence determination, it is necessary to create an ordered array of the fragments, so that their positions relative to one another in the original DNA will be known. When the original order of fragments obtained from a larger piece of the genome has been established, we say that the *bottom-up mapping strategy* has been used.

In order to understand the bottom-up strategy, let us imagine that we are going to prepare a YAC library from the DNA of chromosome 21, the smallest human chromosome, which contains about 45 Mbp of DNA. We shall use a restriction enzyme that cuts, on the average, about once every 225 kbp. If we did a complete digest of the DNA, we would obtain 200 different pieces, and there would be no direct way to deduce their order in the intact chromosome. On the other hand, if we did a *partial digest* with that restriction nuclease, we would generate a collection of many more than 200 fragments of the chromosome, and every piece would have some regions of overlap with other pieces in the collection (Figure 3–18). Now we could find out which pieces were adjacent to a given piece by identifying the regions of overlap. This process is called *contig* construction, that is, the identification of cloned fragments of DNA that represent contiguous regions in the original chromosome.

The basic strategy for contig construction is to cleave each cloned DNA with a restriction enzyme that will produce 20–40 fragments, separate the fragments according to size by electrophoresis in a gel, and compare the fragment patterns obtained from various clones until a region of partial identity has been discerned. Figure 3–19 presents a hypothetical example. In a real situation involving hundreds of patterns to be compared, the initial identification of potentially overlapping fragments would be done by computer, with subsequent confirmation by a researcher.

Contig construction has been extensively employed in the analysis of the genomes of *Escherichia coli* (Kohara et al., 1987), brewers' yeast, *Saccharomyces cerevisiae* (Olson et al., 1986), and the nematode worm, *Caenorhabditis elegans* (Coulson et al., 1986). The worm project, which is being done at the MRC Laboratory of Molecular Biology in Cambridge, England, has been the most extensive of those three, and it has the greatest relevance to human genetics, because the genome of *C. elegans* at 80 Mbp is approximately the same size as one human chromosome. The Cambridge group cloned the entire worm genome into cosmids and then attempted to create contigs by the method outlined previously.

Of particular importance for the Human Genome Project is the fact that they reached a point of diminishing returns after 90–95% of the genome had been assembled into 700 contigs of greatly different sizes. Apparently, there are numerous sequences that clone inefficiently, and/or sequences that are unstable, and/or sequences that are toxic to the bacterial host in some fashion. Whatever the causes of the gaps may have been, the experience of the worm researchers showed that it is not feasible to assemble a complete map of a DNA molecule the size of a human chromosome from a set of fragments in the 30–40 kbp range. This emphasizes the need for an

Figure 3–18

Partial restriction endonuclease digestion of a defined DNA segment. The top line represents intact DNA, with four cleavage sites shown by arrowheads. The bottom four lines show the cleavage products expected when the DNA has been cut at one (and only one) of the four possible sites. Every fragment has regions of overlap with several other fragments.

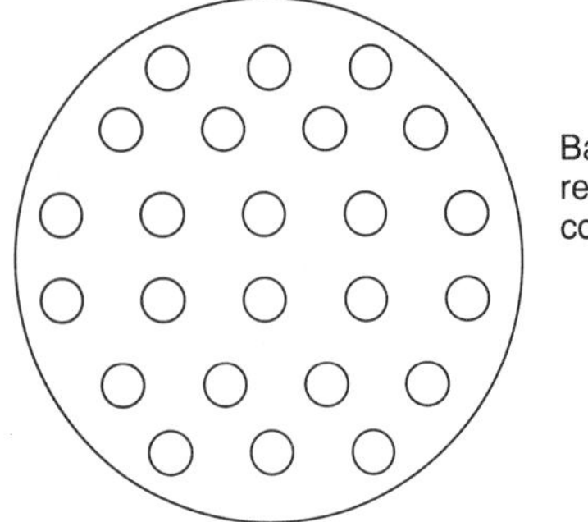

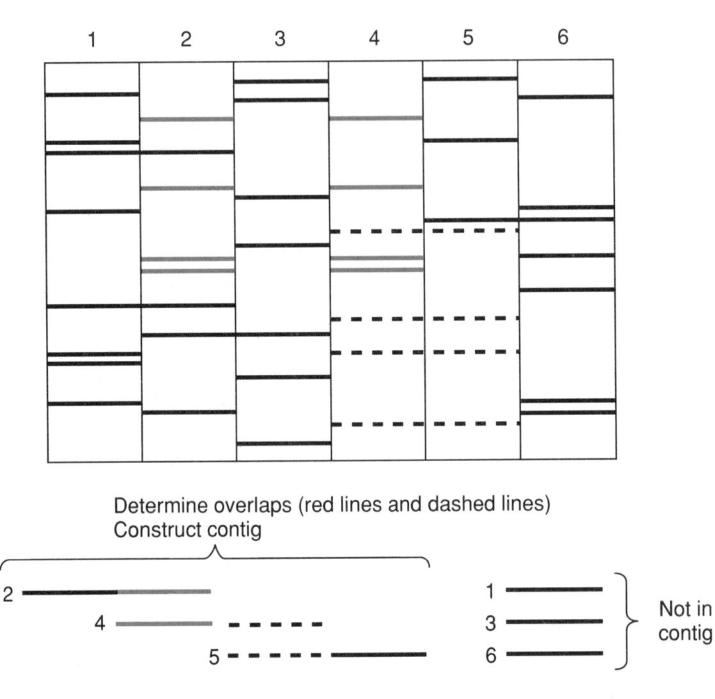

Figure 3–19

Contig construction. Individual cosmids are isolated and digested with one or more restriction enzymes. The fragments from each cosmid are labeled with a fluorochrome and separated by electrophoresis. The fragment sizes are determined and compared among cosmids. Those cosmids that share a substantial subset of fragments (represented by red lines in lanes 2 and 4 and by dotted lines in lanes 4 and 5) are considered to overlap and form a contig. Note that single bands may be shared by one or more cosmids (e.g., the third band from the top in lanes 1 and 2), but these are presumed to be mostly coincidences, rather than evidence of overlap.

intermediate stage of analysis, involving fragments several hundred kbp long; and indeed, the Cambridge group has been able to close many of the gaps in the *C. elegans* map by using YAC clones, a technique that became available long after their project began (Coulson et al., 1988).

DNA Sequencing

At the finest level of resolution, a physical gene map contains the nucleotide sequence. Prior to 1977, direct sequencing of DNA was not possible, and the prospects for detailed molecular analysis of mammalian genomes seemed bleak. But in that year, two simple and effective procedures were described that have revolutionized molecular genetics, especially in con-

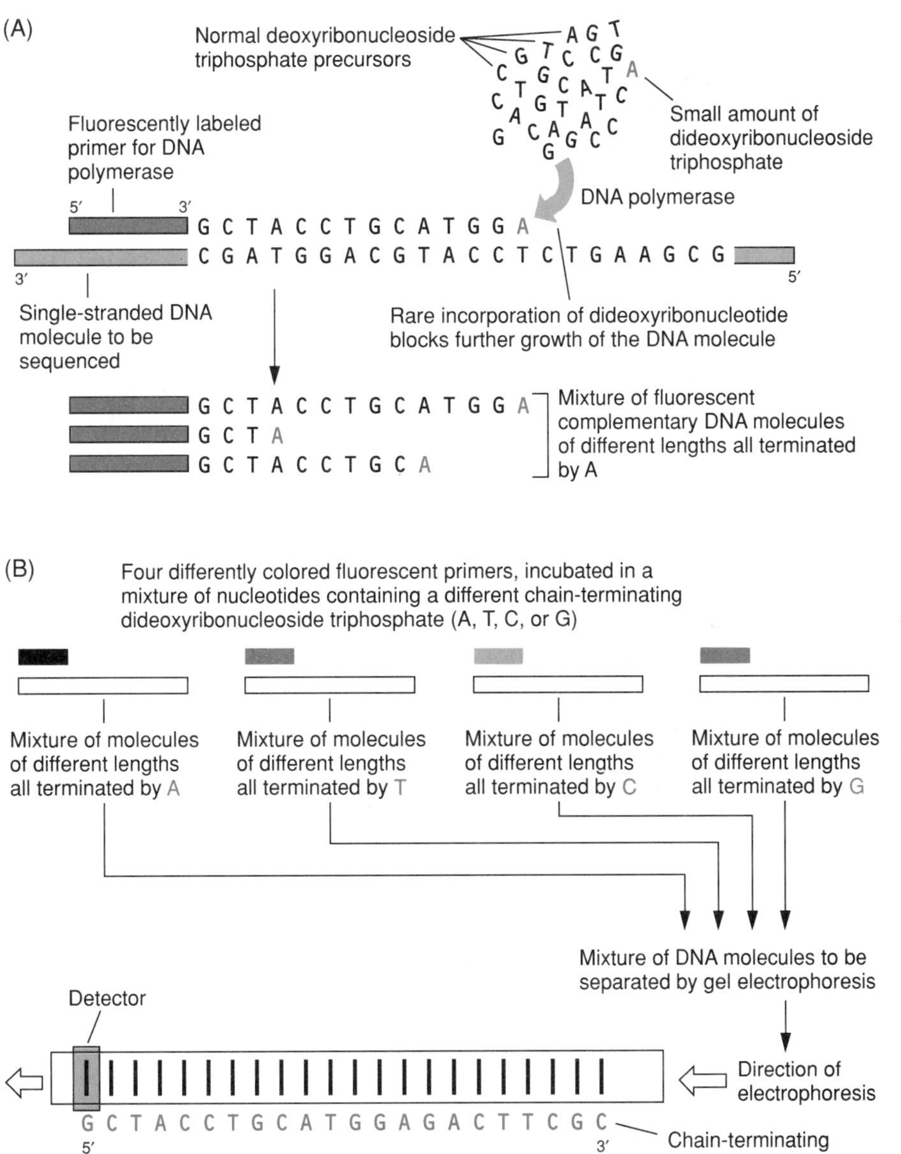

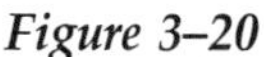

Figure 3–20

Major steps in sequencing DNA. (A) The Sanger method, which produces a set of partial copies of a template strand, terminated at various points by random incorporation of a dideoxynucleotide. (B) Four separate reactions of the type diagrammed in panel A are carried out, each with a different chain-terminating nucleotide and a different fluorochrome attached to the primer. The products of the four reactions are combined and electrophoresed together. As each partial copy of the template strand passes through a laser beam, the fluorescence is read and the nucleotide that terminated fragments of that size is recorded. (From Alberts et al., 1989, p. 187; by permission of Garland Publishing)

junction with recombinant DNA procedures. Both the Maxam and Gilbert procedure (1977) and the Sanger, Nicklen and Coulson procedure (1977) are used extensively, but only the latter will be described here, because it appears to be more adaptable to the automated, rapid sequencing techniques that are now being developed for the Human Genome Project.

Technical aspects of the Sanger procedure are illustrated in Figure 3–20. The strategic question that will be of interest to readers of this volume is, "Can the procedure be automated, so that it becomes technically feasible to undertake sequencing the human genome (3 billion bp) and all the other genomes that will be needed as comparative materials for the Human

Genome Project?" In their original form, both the Gilbert and Sanger procedures for DNA sequencing are labor intensive. Labeled DNA fragments must be subjected to electrophoresis on long acrylamide gels, the dried gel must be autoradiographed, and the developed film must be interpreted by an experienced technician. Several analyses must be done on every piece of DNA being sequenced, in order to ensure reproducibility and to resolve ambiguities. Interpretation of the gels is boring and fatiguing; technician "burn-out" is a common problem.

Intensive efforts to automate the process are underway. There are already several machines on the market that offer significant improvements over the original methods. All of these machines employ some form of fluorescent labeling system, instead of the radioisotopes required for the autoradiographic methods. The basic idea is to use four different fluorescent labels, so that reaction products that terminate in A will be distinguishable by their fluorescence emission spectrum from those that terminate in T, G, or C, and so forth. In some cases the primer used in the Sanger method is labeled in separate reactions with four different fluorescent molecules, and in another system, each of the dideoxynucleotides that are used for chain termination carries a different fluorescent adduct.

In some machines, the fluorescence is detected as the DNA fragments of various sizes undergo electrophoresis; the fragments move through a laser beam or pair of laser beams that excite the fluorescent molecules, and a photomultiplier tube and associated electronics determine which of the four types of fluorescence is present in that particular DNA fragment (Figure 3–20). In other systems, the gel is run for a fixed length of time and dried, as though it were to be autoradiographed, but the fluorescent labels are detected by a device that scans the gel. The makers of these various machines claim that they can produce sequences of about 10,000–20,000 bp/day, and it is generally expected that efficiencies will improve with time.

The preparative stages immediately preceding DNA sequencing are more complex. Consider a typical cloned DNA about 20 kbp long. In order to be sequenced, it must be broken down into pieces about 400 bp long; that is, 50 subfragments must be generated. It would be difficult to determine the order of so many fragments in one step, so at least a two-step procedure must be followed. One would probably first cleave the 20 kbp DNA with two or three restriction enzymes that recognize 6-base sequences (6-cutters), both singly and in combination, and deduce the restriction map by one of several available methods. Inasmuch as cleavage sites for 6-cutters occur every 4096 bp, on the average, 5 fragments would be expected from an average 20 kbp molecule. (Of course, cleavage sites are not spaced uniformly, and the results for a given 20 kbp DNA may vary widely from the average.)

Several techniques for sequencing a piece of DNA that is a few kbp long are available. These are described in technical manuals, and no details will be given here. In all cases, the DNA must be broken down into fragments about 400 kbp long, which can sequenced in one operation. Those short pieces must be cloned into a vector, such as phage M13, that produces a single-stranded DNA product (the virion DNA in the case of M13), because the sequencing reaction involves copying a single-stranded DNA. It is the partial copies of the template strand, which have been terminated at various points by incorporation of dideoxynucleotides, from which the sequence is read (Figure 3–20).

The procedures for preparing DNA fragments to be sequenced are easily as time-consuming as the sequencing procedure itself, and they have not yet benefited from automation to the same extent that sequencing has. However, many possibilities are being explored, and major technical advances are certain to occur. One potentially powerful new method is scanning tunneling microscopy (STM), which is capable of producing atomic level resolution, but is not yet able to distinguish the four DNA bases from one another. It is hoped that this technique may eventually make it possible to read the sequence of DNA molecules without fragmenting the DNA and separating the fragments. Other techniques, such as atomic force microscopy and mass spectroscopy, are also under investigation for their potential in DNA sequencing.

In summary, it is clear that methods for DNA sequencing are becoming faster, cheaper, and more automated. Although the very high level of efficiency that will ultimately be required by the Human Genome Project has not yet been achieved, there is good reason to be optimistic. It is quite likely that sequencing—the final stage of gene mapping—will not long remain as a major obstacle to large-scale genome projects; and for the average laboratory, where only a few dozen kilobases may be of interest, sequencing will be a trivial matter.

SUMMARY

Mapping genes to specific human chromosomes benefited greatly from the exploitation of somatic cell hybrids. Key factors were the tendency of human-rodent hybrids to lose human chromosomes rapidly, and the development of selective media that favored survival of hybrids retaining a specific human gene.

In the early 1970s, mapping of human genes with somatic cell hybrids depended upon identification of human gene products (enzymes and other proteins) in the hybrid cells. Development of recombinant DNA methods increased the power of somatic cell genetics tremendously by providing DNA probes that could be used to identify the gene itself, regardless of whether it was expressed in hybrid cells.

Use of translocations and other chromosomal rearrangements in somatic cell hybrids provides a way to map a gene to a limited subregion of a chromosome.

Hybridization of radioactive DNA probes to metaphase chromosomes on microscope slides provided a method of gene mapping that was not dependent on hybrid cell lines. Mapping by *in situ* hybridization also gave information on subchromosomal location of the target gene. More recently, the use of fluorescent probes has simplified the technique of *in situ* hybridization and increased resolution.

Individual human chromosomes can be isolated in small quantities with the use of a Fluorescence Activated Cell Sorter. DNA clone libraries have been prepared from flow-sorted chromosomes at the Los Alamos and Livermore National Laboratories.

Recombinant DNA techniques offer the potential of cloning any given gene in microorganisms. In order to analyze entire human chromosomes, it is necessary to break the DNA down in stepwise fashion. Digestion with restriction enzymes that cut rarely (about once every Mbp), followed by separation of fragments by pulsed-field gel electrophoresis, is one useful technique. DNA fragments that are several hundred kbp long can be cloned in Yeast Artificial Chromosomes.

A series of overlapping cloned DNA segments is called a *contig*. Construction of contigs is an important step in analyzing any substantial region of a genome, whether the goal is characterization of a specific gene or merely sequencing of anonymous DNA. Sequencing is a multistep procedure that was initially labor intensive, but is undergoing rapid technological innovation. Current machines with an output of about 15,000 bp/day are likely to be replaced by much more powerful machines and/or entirely new techniques in the near future.

REFERENCES

Alberts, B., Bray, D., Lewis, J., Raff, M., Roberts, K., and Watson, J. D. 1989. Molecular biology of the cell. 2nd ed. Garland Publishing, New York.

Beebe, T. P., Jr., Wilson, T. E., Ogletree, D. F. et al. 1989. Direct observation of native DNA structures with the scanning tunneling microscope. Science 243:370–372.

Burke, D. T., Carle, G. F., and Olson, M. V. 1987. Cloning of large segments of exogenous DNA into yeast by means of artificial chromosome vectors. Science 236:806–812.

Cooke, H. 1987. Cloning in yeast: an appropriate scale for mammalian genomes. Trends in Genetics 3: 173–174.

Coulson, A., Sulston, J., Brenner, S., and Kam, J. 1986. Toward a physical map of the genome of the nematode *Caenorhabditis elegans*. Proc. Natl. Acad. Sci. USA 83:7821–7825.

Coulson, A., Waterston, R., Kiff, J., Sulston, J., and Kohara, Y. 1988. Genome linking with yeast artificial chromosomes. Nature 335:184–186.

Creagan, R.P. and Ruddle, F.H. 1975. The clone panel; a systematic approach to gene mapping using interspecific somatic cell hybrids. Cytogenet. Cell Genet. 14:282–286.

Deisseroth, A., Nienhuis, A., Lawrence, J. et al. 1978. Chromosomal localization of human beta-globin gene on human chromosome 11 in somatic cell hybrids. Proc. Natl. Acad. Sci. USA 75:1456–1460.

Deisseroth, A., Nienhuis, A., Turner, P. et al. 1977. Localization of the human alpha-globin structural gene to chromosome 16 in somatic cell hybrids by molecular hybridization assay. Cell 12:205–218.

Donahue, R. P., Bias, W. B., Renwick, J., and McKusick, V. A. 1968. Probable assignment of the Duffy blood group locus to chromosome 1 in man. Proc. Natl. Acad. Sci. USA 61:949–955.

Gall, J. G. and Pardue, M. L. 1969. Formation and detection of RNA-DNA hybrid molecules in cytological preparations. Proc. Natl. Acad. Sci. USA 63:378–383.

Gray, J. W., Lucas, J., Peters, D. et al. 1986. Flow karyotyping and sorting of human chromosomes. Cold Spring Harbor Symposia on Quantitative Biology 51:141–149.

Gusella, J., Varsanyi-Breiner, A., Kao, F-T. et al. 1979. Precise localization of human beta-globin gene complex on chromosome 11. Proc. Natl. Acad. Sci. USA 76:5239–5243.

Harper, M. E., Ullrich, A., and Saunders, G. F. 1981. Localization of the human insulin gene to the distal end of the short arm of chromosome 11. Proc. Natl. Acad. Sci. USA 78:4458–4460.

Kao, F-T., Jones, C., and Puck, T. T. 1976. Genetics of somatic mammalian cells: genetic, immunologic, and biochemical analysis with Chinese hamster cell hybrids containing selected human chromosomes. Proc. Natl. Acad. Sci. USA 73:193–197.

Kohara, Y., Akiyama, K., and Isono, K. 1987. The physical map of the whole *Escherichia coli* chromosome: application of a new strategy for rapid analysis and sorting of a large genomic library. Cell 50:495–508.

Langer, P. R., Waldrop, A. A., and Ward, D. C. 1981. Enzymatic synthesis of biotin-labeled polynucleotides: novel nucleic acid affinity probes. Proc. Natl. Acad. Sci. USA 78:6633–6637.

Lebo, R. V., Anderson, L. A., Lau, Y. F-C., Flandermeyer, R., and Kan, Y. W. 1986. Flow-sorting analysis of normal and abnormal human genomes. Cold Spring Harbor Symp. Quant. Biol. 51:169–176.

Lichter, P., Tang, C-J. C., Call, K. et al. 1990. High-resolution mapping of human chromosome 11 by *in situ* hybridization with cosmid clones. Science 247:64–69.

Littlefield, J. W. 1964. Selection of hybrids from matings of fibroblasts in vitro and their presumed recombinants. Science 145:709–710.

Maxam, A. M. and Gilbert, W. 1977. A new method of sequencing DNA. Proc. Natl. Acad. Sci. USA 74:560–564.

Migeon, B. R. and Miller, C. S. 1968. Human-mouse somatic cell hybrids with single human chromosome (group E): link with thymidine kinase activity. Science 162:1005–1006.

Miller, O. J., Allerdice, P. W., and, Miller, D. A. 1971. Human thymidine kinase locus: assignment to chromosome 17 in a hybrid of man and mouse cells. Science 173:244–245.

Olson, M. V., Dutchik. J. E., Graham, M. Y. et al. 1986. Random-clone strategy for genomic restriction mapping in yeast. Proc. Natl. Acad. Sci. USA 83:7826–7830.

Owerbach, D., Bell, G., Rutter, W., and Shows, T. 1980. The insulin gene is located on chromosome 11 in humans. Nature 286:82–84.

Prescott, D. M. 1988. Cells. Jones and Bartlett, Boston.

Riciutti, F. C. and Ruddle, F. H. 1973. Assignment of three gene loci (PGK, HGPRT, G6PD) to the long arm of the human X chromosome by somatic cell genetics. Genetics 74:661–678.

Ruddle, F. H. and Creagan, R. P. 1975. Parasexual approaches to the genetics of man. Annu. Rev. Genet. 9:407–486.

Sanger, F., Nicklen, S., and Coulson, A. R. 1977. DNA sequencing with chain-terminating inhibitors. Proc. Natl. Acad. Sci. USA 74:5463–5467.

Satlin, A., Kucherlapati, R., and Ruddle, F. H. 1975. Assignment of the gene for human uridine monophosphate kinase to chromosome 1 using somatic cell hybrid clone panels. Cytogenet. Cell Genet. 15:146–152.

Schwartz, D. C. and Cantor, C. R. 1984. Separation of yeast chromosome-sized DNAs by pulsed-field gradient gel electrophoresis. Cell 37:67–75.

Schwartz, D. C., Gaffran, W., Welsh, J. et al. 1983. New techniques for purifying large DNAs and studying their properties and packaging. Cold Spring Harbor Symp. Quant. Biol. 47:189–195.

Smith, C. L. and Cantor, C. R. 1987. Purification, specific fragmentation, and separation of large DNA molecules. Methods in Enzymology 155:449–467.

Snyder, L. A. Freifelder, D., and Hartl, D. L. 1985. General genetics. Jones and Bartlett, Boston.

Southern, E. M., Anand, R., Brown, W. R. A., and Fletcher, D. S. 1987. A model for the separation of large DNA molecules by crossed field gel electrophoresis. Nuc. Acids Research 15:5925–5943.

Sternberg, N. 1990. Bacteriophage P1 cloning system for the isolation, amplification, and recovery of DNA fragments as large as 100 kilobase pairs. Proc. Natl. Acad. Sci. USA 87:103–107.

U.S. Department of Energy. 1992. Human Genome: 1991–92 Program Report. U.S. Department of Energy, Office of Energy Research, Office of Health and Environmental Research, Washington, D.C.

Vollrath, D. and Davis, R. W. 1987. Resolution of DNA molecules greater than 5 megabases by contour-clamped homogeneous electric fields. Nucleic Acids Research 15:7865–7876.

Wilson, E. B. 1911. The sex chromosomes. Arch. Mikrosk. Anat. Entwicklungsmech. 77:249–271.

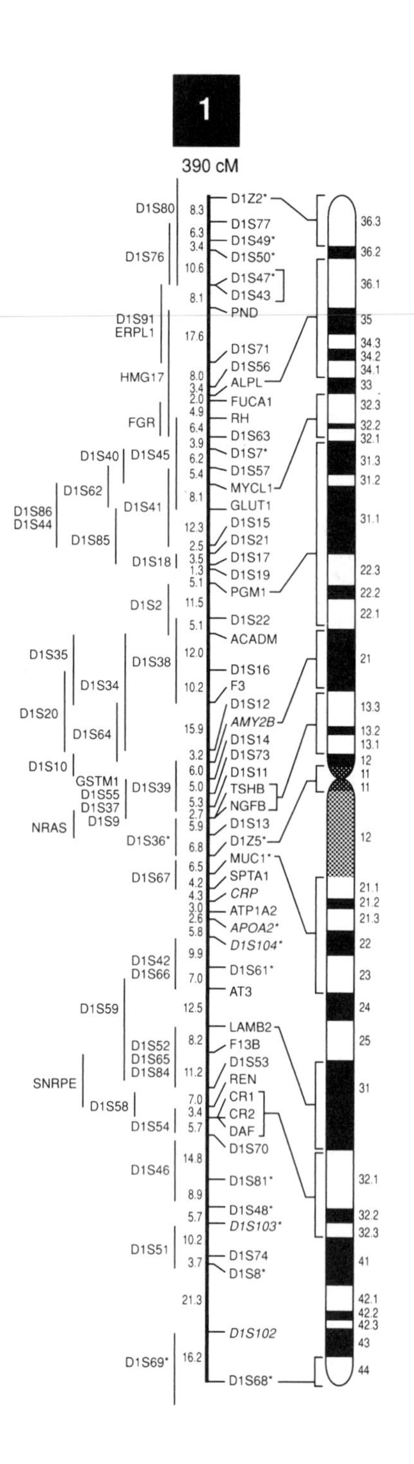

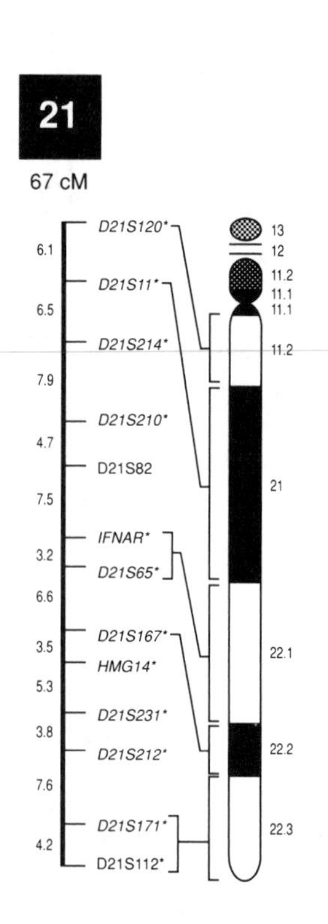

Recent linkage maps of the largest and smallest human chromosomes. (Compiled by the NIH/CEPH Collaborative Mapping Group, 1992)

4 Gene Mapping, II

LINKAGE ANALYSIS AND GENETIC MAPS

Genes that are located on the same chromosome are physically connected; that is, they are part of the same DNA molecule. In almost all organisms, each chromosome contains only one continuous DNA molecule, even though it may consist of hundreds of millions of base pairs. Genes on the same chromosome are said to be *syntenic* (which means, "on the same thread"). In humans, on average, a chromosome must contain at least 1000 genes, and possibly several thousand (see Chapter 2).

During germ cell formation, specifically during prophase of the first meiotic cell division, homologous members of a pair of chromosomes undergo *synapsis*; that is, they associate side-by-side in a very precise manner. At this time, each chromosome consists of two fully formed chromatids, joined only at their centromeres by special proteins; thus there are four chromatids in each pair of chromosomes (Figure 4–1A). Each chromatid represents a separate DNA molecule.

While the chromosomes are synapsed, *crossing-over* may occur between any two non-sister chromatids. If there are allelic differences at two or more locations in the homologous chromosomes, then the genetic result of crossing-over between two heterozygous loci will be *recombination* (Figure 4–1, B–E). Recombination generates *new groups of alleles* on each of the participating chromatids, but it does not ordinarily produce new alleles. Normally, crossing-over is perfectly reciprocal; there is neither loss nor gain of nucleotides in either chromatid. Unequal crossing-over sometimes occurs, particularly in areas where there are tandemly repeated sequences, but we will not concern ourselves with that until Chapter 6.

A single crossover involves only two of four chromatids in a synapsed pair of homologous chromosomes. Therefore, two recombinant chromatids and two non-recombinant chromatids are produced by one crossover (Figure 4–1). If we consider the results of many meioses, either in one individual or in a population, the overall frequency of recombinant chromosomes is called the *recombination fraction* (designated RF or theta).

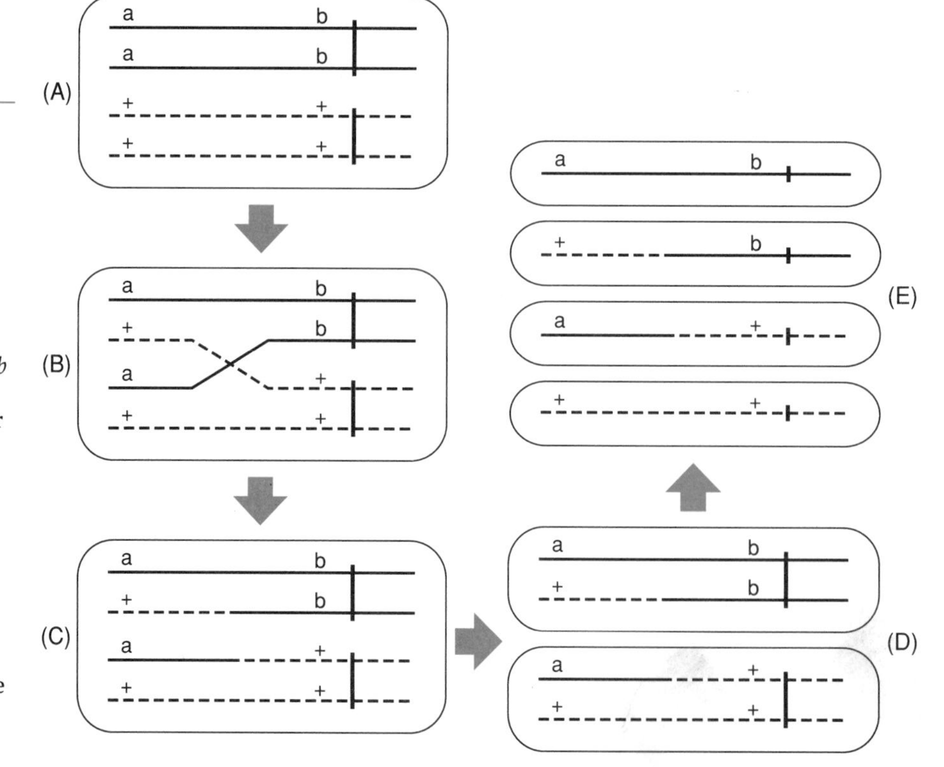

Figure 4–1

Crossing-over and recombination. (A) Two homologous chromosomes, each containing two chromatids, are precisely aligned side-by-side in prophase of the first meiotic division. Dark bars represent centromeres. (B) A crossover takes place between the *a* and *b* loci, involving the two central chromatids. After crossing-over is complete (panel C), the first meiotic division takes place (panel D). During the second meiotic division, the centromeres that held the chromatids together are separated, and each chromatid (now a mature chromosome) enters a separate daughter cell. Two of the meiotic products are recombinant; the other two are non-recombinant.

If two genes on the same chromosome are close enough so that the recombination fraction is less than 50%, the two genes are *genetically linked* (that is, alleles at the two loci do not assort independently). If two loci are syntenic, but so far apart on a chromosome that crossovers between them produce a recombination fraction of 50%, such genes are genetically unlinked, even though they are both physically part of the same DNA molecule that constitutes a given chromosome. In other words, when RF = 50%, we cannot distinguish genes that assort independently because they are on different chromosomes from genes that assort independently because they are far apart on the same chromosome.

The recombination fraction cannot exceed 50%, no matter how many crossovers occur between two loci. In order to understand this point, consider all the possibilities for double crossovers. If two crossovers involve the same two chromatids, none of the resulting chromosomes will be recombinant for the flanking markers (loci *a* and *b* in Figure 4–2), although there will actually be a section between the marker loci where a reciprocal exchange of DNA has occurred (Figure 4–2A). However, the second crossover does not have to involve the same two chromatids as the first crossover: it may be a three-strand double crossover (of which there are two types; Figure 4–2B and 4–2C), which results in two recombinant and two non-recombinant chromosomes; or it may be a four-strand double crossover, which results in four recombinant chromosomes (Figure 4–2D).

As Figure 4–2 indicates, *if these double crossover possibilities occur randomly*, which seems to be the case, the net result is the same as an average of one crossover between the flanking markers; 50% of the resulting chromosomes are recombinant and 50% are not. The same result would be obtained for triple, quadruple, and higher numbers of crossovers.

Genetic linkage was not recognized until nearly half a century after

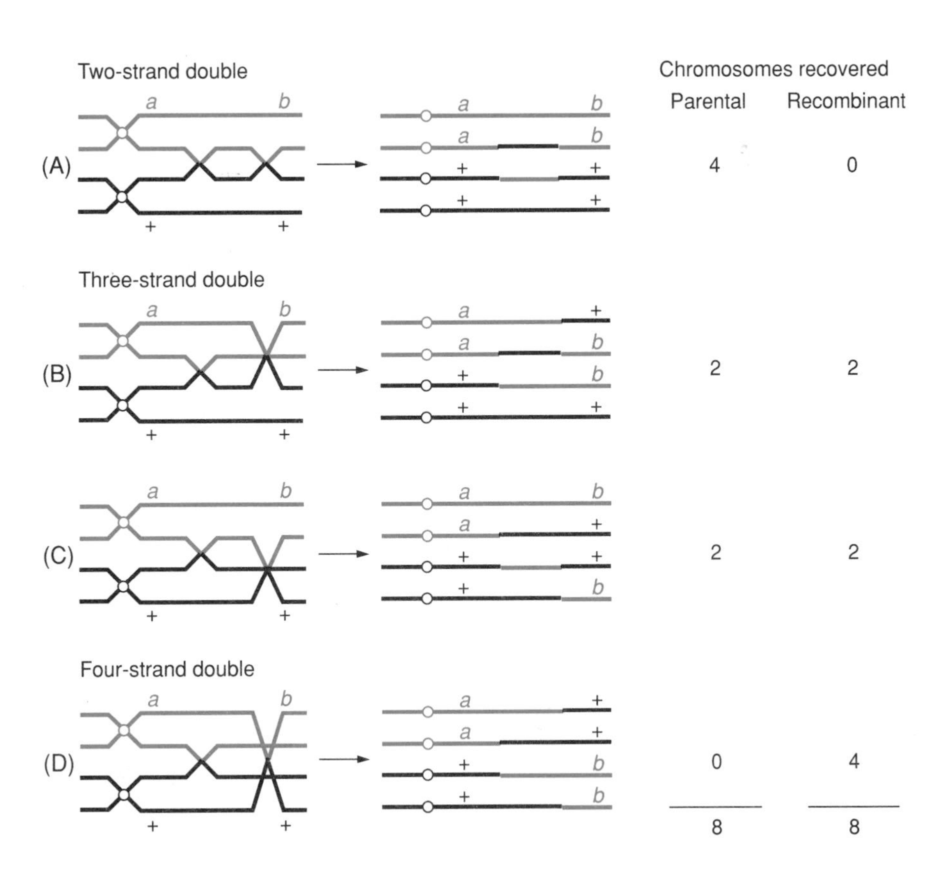

Figure 4–2

A diagram of the possible types of double crossovers and the results of each, in terms of recombination between flanking markers. (From Snyder et al., 1985, p. 77)

Mendel's pioneering work on the inheritance of seven characteristics in pea plants. Inasmuch as *Pisum sativum* has only seven pairs of chromosomes, more than one modern author has commented that Mendel was either very lucky in his choice of traits to study, or he chose to ignore results that did not fit his concept of independent assortment. Bateson et al. (1905) were the first to publish data on the inheritance of two characteristics that did not assort independently during gamete formation, but it was Thomas Hunt Morgan who interpreted such phenomena in terms of exchanges between homologous chromosomes, for which he used the term "crossing over" (Morgan and Cattell, 1912). A year later, Sturtevant (1913) produced the first genetic map, using data on recombination frequencies in *Drosophila*.

The cytological counterparts of genetic recombination events are the *chiasmata* that may be seen in meiotic cells (Figure 4–3). Janssens (1909) interpreted chiasmata as points of exchange between chromatids from homologous chromosomes, while they are paired during meiosis. The essence of that concept is still accepted today, although it has been modified and greatly extended by molecular models of crossing-over; the most widely accepted of those models is the one created by Robin Holliday (1964), with the modification suggested by Meselson and Radding (1975).

The overall frequency of crossovers in humans has been estimated in several ways. Morton et al. (1982) used the simple process of counting them in cytological preparations of spermatocytes; they found an average of 52 crossovers per male meiosis (slightly more than two per chromosome pair). It is convenient to express this number in terms of *map units* or *centiMorgans (cM)*. One genetic map unit or one cM is equal to a recombination fraction of 1% over small distances. One crossover implies a genetic map length of 50 cM; that is, if two loci are separated by a distance such that an average of one crossover occurs between them in every meiotic cell, then those loci are 50 cM apart.

Extrapolating to the whole genome, we see that 52 crossovers implies a total genetic map length of 2600 cM in humans. *Terminalization*, a process by which chiasmata supposedly move toward the ends of chromosomes, could lead to an underestimate of chiasma frequency, but recent work indicates that terminalization does not occur (Hulten, 1990). A different estimate was obtained by Renwick (1971), who extrapolated from the limited linkage map data then available on certain regions of the genome to predict a total, sex-averaged haploid autosomal map length of 3300 cM.

It is important to realize that map lengths are not additive over large distances, because of the occurrence of multiple crossovers. As we saw in Figure 4–2, multiple crossovers produce the same ratio of recombinant gametes to non-recombinant gametes as single crossovers. Therefore, the occurrence of multiple crossovers within a given interval is not detectable

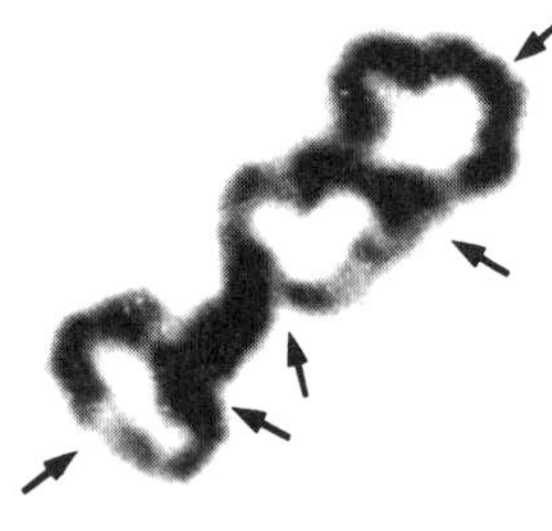

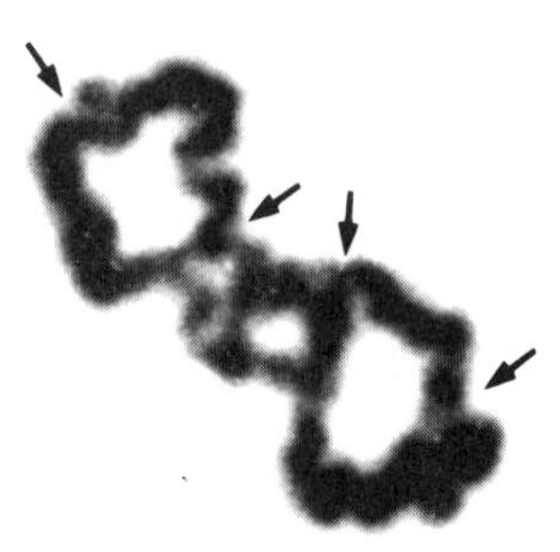

Figure 4–3

Photomicrographs of two human meiotic bivalents at diakinesis/metaphase I. Arrows indicate four chiasmata in one bivalent and five chiasmata in the other. (Courtesy of Maj Hulten)

by analysis of only two marker loci. For example, if loci *A*, *B*, *C*, *D*, and *E* are located on a chromosome in that order, and if they are separated from each other by 5 cM apiece, it is probable that the frequency of recombination between *A* and *E* would be significantly less than 25 cM if it were measured directly on those two loci, because some multiple crossovers would be likely to occur in that interval.

The quantitative relationship between frequency of recombination and the distance between genetic loci is not simple. The existence of a crossover at one point often appears to suppress the formation of other crossovers nearby; this is called *positive interference*. However, there are regions that are recombinational "hot-spots." The occurrence of an exceptionally high frequency of recombination in a limited portion of the genome is called *negative interference*. Several geneticists have written formulas that attempt to describe the non-linear relationship between genetic distance and recombination frequency; these are called *mapping functions*. More information may be found in the reviews by Ott (1985) and by White and Lalouel (1987).

A further complexity is that recombination rates are not the same in both sexes; they tend to be higher in females, although the sex difference is not constant from one region of the genome to another. For example, Renwick (1971) estimated the haploid autosomal map length as 2750 cM in males and 3850 cM in females. This paradoxical phenomenon is not understood. Unless otherwise noted, we shall ignore the difference in the rest of this text, in order to keep calculations from becoming excessively complex. Considering the uncertainties involved in the chiasmata count of Morton, the extrapolations by Renwick, and the sex difference, it seems reasonable to use a "round number"—3000 cM—as the total genetic map length in humans, which corresponds to 1 cM per million base pairs of DNA (Table 2-2).

Genetic maps play an important role in the analysis of any genome. A genetic map gives the order in which genes occur in a chromosome and the approximate location of a gene of interest. Genetic maps do not require knowledge of gene function; they can be constructed for genes that are known only from their effects on phenotypes. This is particularly important in humans, where many diseases are known to be genetic in origin, but little or nothing is known about their biochemical basis. Although we expect to have the complete sequence of the human and several other genomes within two decades, those complete physical maps will not eliminate the usefulness of information on genetic linkage. In particular, knowledge of the frequency of recombination in selected regions of the genome where important disease-causing loci occur will always be important for genetic screening and counseling. And, as we study the genetic basis of disease for the next century or more, linkage analysis will often be crucial for determining whether a given clinical syndrome is caused by defects at more than one locus. The principles of linkage analysis are outlined in the next sections.

Restriction Fragment Length Polymorphisms and Linkage Analysis in Humans

The Need for Polymorphic Marker Loci

Linkage analysis requires the existence of individuals who have distinguishable alleles at both loci whose linkage or non-linkage is to be determined; in other words, *individuals must be double heterozygotes in order to be informative for linkage analysis*. In the human context, where the goal often is to test for linkage between a disease and a set of marker loci, heterozygosity at the disease locus is usually present. A marker locus is any gene or DNA segment whose location in the genome is already known. In order to map an unknown disease-causing gene, marker loci from all regions of the genome are needed; these loci must have a high level of heterozygosity in order to be useful for linkage analysis.

For autosomal dominant diseases, informative matings almost always involve one affected parent who has only one copy of a mutant allele; homozygotes for dominant diseases are quite rare (see Chapter 8). For autosomal recessive diseases, the parents of affected offspring are necessarily heterozyous at the disease locus. For X-linked diseases, affected male parents are hemizygous; affected female parents may be either heterozygous or homozygous at the disease locus. Determination of the mode of inheritance is a necessary first step in linkage analysis. The characteristics of autosomal dominant and autosomal recessive patterns are reviewed in Chapter 8; sex-linked inheritance is discussed in Chapter 13.

In the early days of human linkage analysis, there were few useful marker loci. A useful marker locus must be highly polymorphic so that there is a reasonably good chance of a person with a disease gene of interest also being heterozygous at the marker locus. The word *polymorphic,* which literally means "many forms," has received a number of special definitions by geneticists. Usually, a locus is defined to be polymorphic if there are two or more alleles, each of which occurs with a frequency of 1% or more. Thus, a locus with two alleles, one of which had a frequency of 99% and the other a frequency of 1%, would qualify as a polymorphic locus by that definition. However, it would be of little use for linkage analysis, because only 2% of the population would be heterozygous. For linkage analysis, a locus must be polymorphic enough to generate at least 20% heterozygotes, and preferably much more.

The first human autosomal linkage groups were identified by pedigree analysis in the 1950s, and all of them involved one or another of the blood groups because they were the most polymorphic loci known at that time. For example, the Rh blood group locus was shown to be linked to the gene for elliptocytosis[1] in 1954, and linkage between the ABO blood group locus and the gene for nail-patella syndrome[2] was demonstrated in 1955. Progress was slow, and by 1968, when human genes began to be mapped by somatic cell genetics, only nine autosomal linkage groups had been established by analysis of family pedigrees. In no case was the specific autosome identified (McKusick, 1986).

The availability of highly polymorphic marker loci was dramatically increased when restriction enzyme digestion of DNA, followed by separation of fragments and identification of specific fragments by Southern blotting, became a routine procedure in biology laboratories (Figure 4–4). It was not long before some investigators noticed that different individuals sometimes had different-sized restriction fragments at certain sites in the genome (Figure 4–5), reflecting presence or absence of restriction enzyme cleavage sites. The fact that the different fragments could be easily assayed by electropho-

[1] An autosomal recessive defect that produces abnormally shaped red blood cells and causes anemia.

[2] An autosomal dominant condition characterized by malformation of the nails and reduced or absent patellae; abnormalities of the elbows and the ileum are often present.

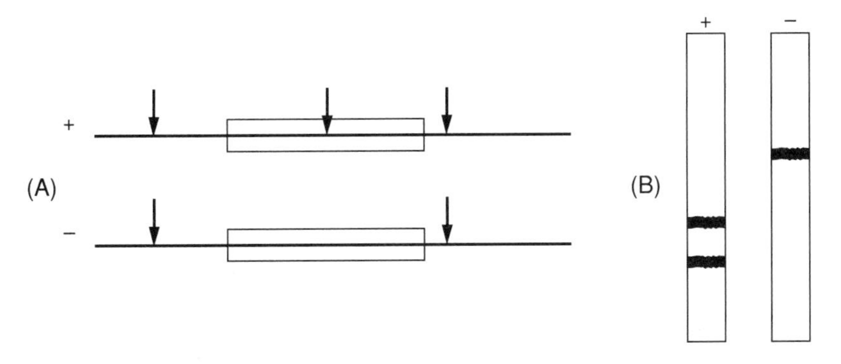

Figure 4–4

Detection of restriction fragment length polymorphisms (RFLPs). (A) Diagram of two homologous chromosomes with two invariant cleavage sites for an unspecified restriction endonuclease, and a polymorphic site for the same nuclease between the invariant sites. Points of cleavage by the enzyme are indicated by arrows. The boxed area is the chromosome region detected by the probe. (B) Diagram of a Southern blot, showing the relative sizes of the fragments produced by the restriction endonuclease from the chromosomes in panel A. Direction of electrophoresis would be from top toward bottom. In both panels, "+" indicates that a cleavage site is present at the polymorphic locus, and "−" indicates that it is absent.

resis meant that polymorphisms in a restriction enzyme cleavage site generated *codominant alleles*; that is, the presence or absence of either allele could be recognized in any individual, regardless of pedigree information. Codominant alleles are a very useful property of marker loci, because heterozygotes can be distinguished from homozygotes.

An early compilation of available data on the beta-globin gene cluster (Jeffreys, 1979) revealed the fact that restriction fragment length polymorphisms are quite common in the human population. Jeffreys suggested that as many as 1% of the base pairs may vary polymorphically. Even if that estimate should prove to be too high by a factor of ten, and remembering that restriction enzymes will only detect a small fraction of nucleotide sequence variants, it is clear that a virtually inexhaustable supply of polymorphisms must exist, and these can be used as marker loci. Hindsight makes this high level of nucleotide sequence variation plausible. After all, we know that at least 90% of the DNA in the overall genome does not code for proteins; therefore, small variations in nucleotide sequence in most of that noncoding DNA should be selectively neutral, and neutral mutations should accumulate over time.

In a seminal paper in 1980, Botstein et al. pointed out that inter-individual variations in nucleotide sequence, some of which could be detected as restriction fragment length polymorphisms (RFLPs), were potentially a treasure house of marker loci for linkage analysis. The polymorphic sites were there in large numbers; with enough time and effort, a set of marker loci covering the entire genome could be identified.

One must realize, however, that there is no general method for the detection of multiple RFLPs in a single assay; each RFLP must be found with a specific probe. Fortunately, any cloned segment of unique-sequence DNA

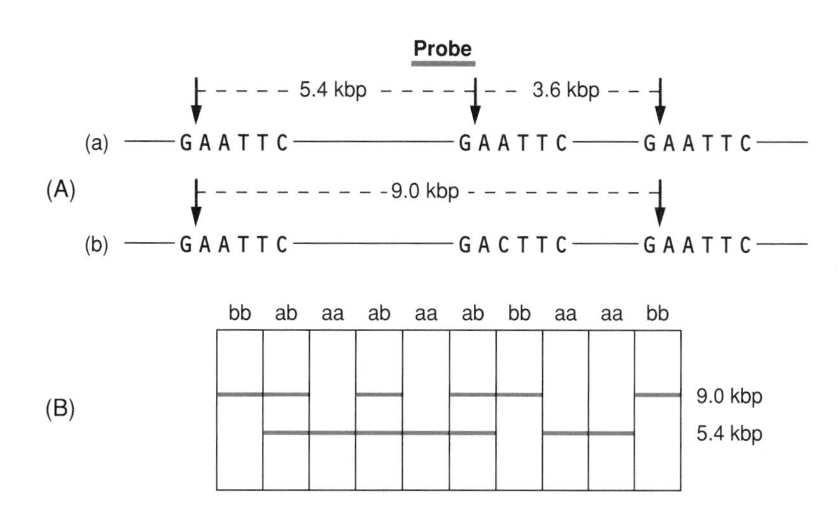

Figure 4–5

Detection of a polymorphism, resulting From variable presence of an EcoR I cleavage site (GAATTC). In (A), an arrow indicates that the enzyme will cleave the DNA at that point; absence of an arrow indicates no cleavage. In chromosomes of type (a), EcoR I will produce two DNA fragments from the diagrammed area, but the probe will only hybridize to the 5.4 kbp piece. In chromosomes of type (b), only a 9.0 kbp fragment of DNA will be produced. (B) Gel patterns of 10 unrelated individuals, showing how genotypes may be deduced from the bands detected by the probe after Southern blotting. In this hypothetical example, the frequency of heterozygotes is 30%.

can be used as an RFLP probe, whether it represents a known gene or not. Any probe can be used to detect complementary DNA in restriction fragments from a series of people, and if there is a polymorphic restriction enzyme cleavage site in or near the region recognized by the probe, then at least one of the DNA fragments that can be detected with the probe will vary in size in some individuals (Figure 4–5).

Consider the total genetic map distance of the human genome, which is approximately 3000 centiMorgans. Botstein et al. (1980) argued that if RFLPs were located at regular intervals of 20 cM, then any unknown gene would be at most 10 cM from an RFLP. A distance of 10 cM implies that only 10% of the progeny of matings in which a genetic disease and an RFLP are segregating will be recombinants; that is, tight linkage can be demonstrated with a few families of moderate size. It only requires 150 RFLPs spaced 20 cM apart to cover the human genome, but since the process of finding RFLPs involves random choice of probes, it seemed likely that at least 1000 RFLPs would have to be identified before the entire genome would be covered with a density of at least one RFLP site per 20 cM.

Following the 1980 paper by Botstein et al., an explosion of activity in RFLP identification occurred. Although many groups participated, the two largest efforts were led by Ray White at the University of Utah and Helen Donis-Keller at Collaborative Research, Inc. The latter group published a summary of their work in 1987 (Donis-Keller and 32 others, 1987), in which they described more than 400 RFLPs distributed over the entire human genome, with an average distance of 10 cM between markers. Some markers were as close as 3 cM from one another, while other gaps as large as 30 cM existed. The White group are publishing their data one chromosome at a time (e.g., a linkage map for chromosome 1 is in O'Connell et al., 1989a; chromosome 2 is covered in O'Connell et al., 1989b, etc.). Together, the two sets of data provide a human linkage map with approximately 5 cM resolution—far more detailed than that envisioned in the original proposal of Botstein et al. (1980). Completion of the linkage map of polymorphic marker loci has been accepted as a major goal of the Human Genome Project. Within the first 5 years of the project (that is, by 1996) a map with 2 cM resolution is targeted; eventually, a 1 cM map will be achieved (Chapter 15).

When a highly polymorphic locus has been identified, it must be placed into the overall human linkage map. Approximate physical location can be achieved by somatic cell hybrid studies and by *in situ* hybridization, but the new marker locus must also be mapped relative to nearby loci whose genetic linkage relationships have already been determined. In an admirable example of international collaboration, Jean Dausset and others at the Centre d'Etude du Polymorphisme Humain (CEPH) in Paris have established a panel of immortalized cell lines for this purpose (Dausset et al., 1990). DNAs from more that 40 three-generation families (4 grandparents, 2 parents, and 6 to 16 children) are available to qualified investigators worldwide. Many of these cell lines were provided by Ray White from his extensive collection of multigeneration material from Mormon families in Utah.

A scientist who receives the CEPH DNAs first carries out a Southern blot analysis with his/her probe on the parental DNAs, to find out which of the families will be informative for that polymorphic locus. Then the pattern of inheritance of the alleles at that locus is determined for all members of the informative families. The results are sent to Paris, where they are compared with the data on all other relevant loci that have been analyzed so far. The pattern of inheritance at the new locus will correspond to the pattern shown by one or more already mapped loci. From this analysis, a location for the new polymorphic locus on the human linkage map is obtained. The CEPH families have been enormously useful in establishing the current linkage map, with its overall resolution of roughly 5 cM. As the map is extended to finer levels of resolution, additional families will be needed, but efforts are already underway to obtain the required material.

Maximizing the Information Content of Marker Loci

The more polymorphic a marker locus is, the higher will be the frequency of heterozygotes at that locus, and therefore, the more useful the locus will be for linkage analysis and for genetic screening. Botstein et al. (1980) introduced the idea of *polymorphism information content* (PIC) and derived a formula to express it. For autosomal dominant diseases, they defined informativeness as ". . . the probability that a given offspring of a parent carrying the rare allele at the index locus will allow deduction of the parental genotype at the marker locus." PIC is a function of the probability that an individual will be heterozygous at that locus, because homozygous persons are not informative.

An ordinary RFLP can have only two forms; either a cleavage site is present or it is absent (Figure 4–4). The maximum possible frequency of heterozygotes in a two-allele system is 0.5, which would occur in a randomly mating population where each of the two alleles had a frequency of 0.5 (an unlikely situation). If we designate the presence of a restriction site as "+" and its absence as "–", then the probability of a +/– individual would be 0.5 × 0.5, and the probability of a –/+ individual would also be 0.5 × 0.5; in sum, the probability of a heterozygote would be 0.25 + 0.25 = 0.5.

Inasmuch as two-allele loci cannot produce more than 50% heterozygotes (without special assumptions), it is important to find marker loci that have more than two alleles. Occasionally, it happens that two or more RFLP sites occur so close together that they can be detected with a single probe. Then the variants at both loci can be treated as haplotypes. A *haplotype* is a set of alleles at two or more closely linked loci on a single DNA molecule; that is, a portion of a chromosome. If one or more of those loci are heterozygous in a given person, then there will be two different haplotypes in the diploid genome of that person. In general, the concept of a haplotype is only applied to small regions of a chromosome that will not be frequently disrupted by recombination.

For a two-locus system, where each locus has two alleles, there are four possible haplotypes, as diagrammed in Figure 4-6. These haplotypes can be treated as a single complex locus with four alleles. The maximum hetero-

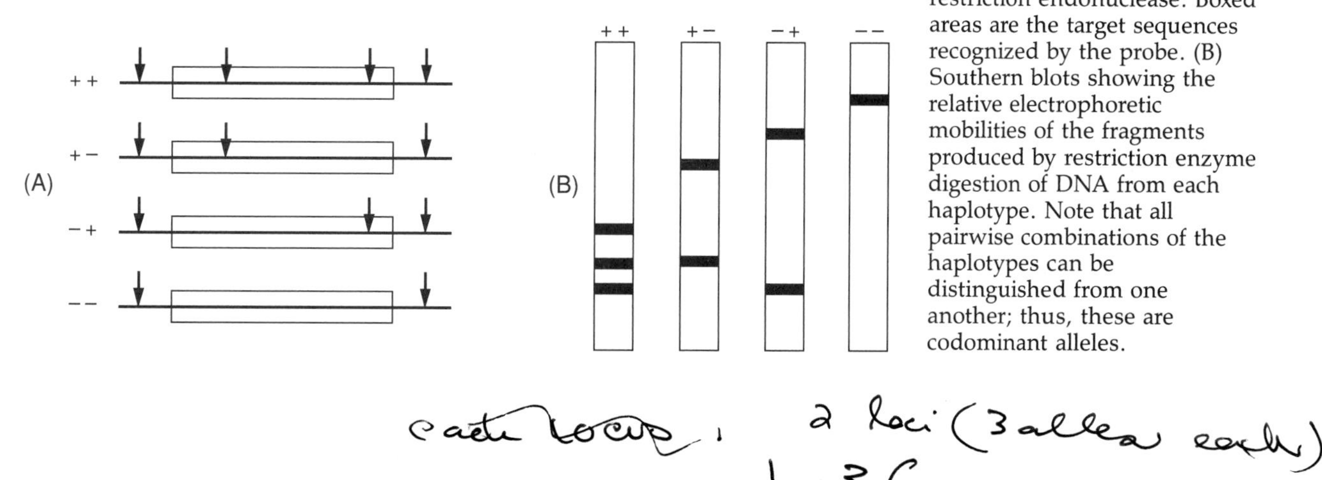

Figure 4–6

A complex polymorphic locus consisting of two adjacent RFLP sites. (A) The four possible haplotypes. Arrows indicate presence of a cleavage site for a restriction endonuclease. Boxed areas are the target sequences recognized by the probe. (B) Southern blots showing the relative electrophoretic mobilities of the fragments produced by restriction enzyme digestion of DNA from each haplotype. Note that all pairwise combinations of the haplotypes can be distinguished from one another; thus, these are codominant alleles.

zygosity for a two-locus RFLP haplotype is 0.75, which would occur if all four alleles had the same frequency.[3] Of course, a locus in which every allele has the same frequency is extremely unlikely. Nevertheless, it is clear that complex loci consisting of at least two RFLP sites offer the possibility of greater heterozygosity than simple RFLP loci.

Much higher levels of heterozygosity are associated with a different type of polymorphism that occurs in repeated sequences known as *minisatellites* (Jeffreys et al., 1985). These polymorphisms are described by the term, *Variable Number of Tandem Repeats*, or VNTRs (Nakamura et al., 1987). These hypervariable loci consist of tandem repeats of oligonucleotide sequences that are usually 10–60 bp long; the number of repeats varies from less than ten to several dozen. As shown in Figure 4–7, the variable restriction fragment length does not arise from presence versus absence of a cleavage site for a restriction endonuclease; it is caused by the number of repetitions of the unit sequence between two invariant cleavage sites.

Some of the VNTR loci that have been studied have more than 10 alleles, and the frequency of heterozygotes in the general population may exceed 90%. Fortunately, there appear to be many VNTR loci scattered throughout the human genome, although they are not as common as simple RFLP sites. A continuing search for VNTR loci is underway, because they represent ideal marker loci for many genetic analyses.

There is another class of tandem repeats involving dinucleotides, such as CA or TG, where the dinculeotide may be repeated roughly 15–60 times. Another name for tandem dinucleotide repeats is *microsatellites*. The number of repeats in a specific location is indicated by a subscript; for example, $(CA)_{23}$ means that the dinucleotide CA is repeated 23 times. It has been estimated that there are 50,000 to 100,000 $(CA)_n$ sequences in the human genome, or about one every 30–60 kbp. A similar number of $(GT)_n$ repeats has been estimated, and various other simple sequence repeats, some of them involving three or four nucleotides, are also numerous.

Recent work has shown that dinucleotide repeats are often polymorphic. For example, Weber and May (1989) studied 10 $(CA)_n$ repeats in the human genome and found all of them were polymorphic, with the number of alleles varying from 4 to 11. It appears that many of these microsatellites will be highly informative for linkage analysis, and because microsatellites are abundant, they are certain to provide an important source of marker loci.

Polymorphisms in dinucleotide repeat loci cannot be detected by conventional restriction enzyme digestion followed by Southern blotting, be-

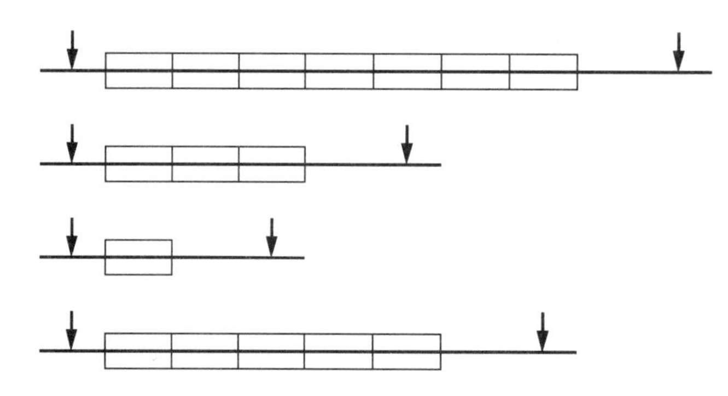

Figure 4–7

Diagram of a hypothetical VNTR locus. Invariant restriction enzyme cleavage sites flanking the variable repeat region are shown by arrows. The tandem repeats are indicated by boxes. The examples shown, from top to bottom, have 7, 3, 1, and 5 copies of the repetitive unit. After restriction enzyme digestion, these would yield DNA fragments of different sizes, identifiable by electrophoresis and Southern blotting.

[3] To reach that conclusion, make use of the fact that the frequency of heterozygotes at any locus, regardless of the number of alleles, must be (1 – homozygotes). Now let the frequency of each allele at a four-allele locus be 0.25; then the frequency of each class of homozygotes will be 0.0625. It follows that $1 - (4 \times 0.0625) = 0.75$.

cause that technique is not sufficiently sensitive to reveal differences as small as two bp (one unit of a dinucleotide repeat). Instead, differences in the number of repeat units in different individuals are found by using the Polymerase Chain Reaction (PCR) to amplify small segments of DNA containing the tandemly repeated units, followed by electrophoresis in a gel suitable for resolving single nucleotide differences in size. The PCR is an extremely powerful and versatile new technique; it is described in Box 4–1. Another method for detecting polymorphisms with PCR is described in Box 4–2.

Linkage Analysis with Autosomal Dominant Diseases

General Aspects

In experimental organisms, the classical test for linkage involves a two-factor testcross, where double heterozygotes are mated with homozygous recessives, and the genotypes of a large number of progeny are scored.

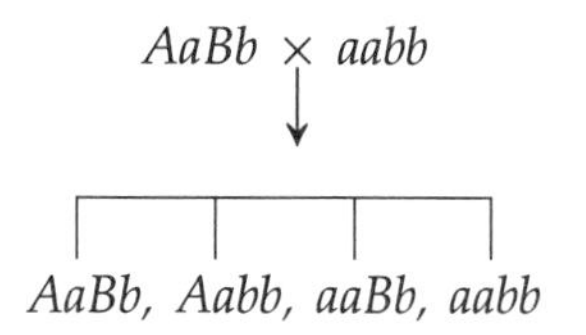

If locus *A* and locus *B* are on different chromosomes, or far apart on the same chromosome, alleles at the two loci will obey Mendel's Law of Independent Assortment, and the four progeny genotypes will occur with equal frequency. However, if the two loci are genetically linked, two of the four possible genotypes will be more frequent in the progeny than the other two genotypes. In the case of very closely linked genes, there may be no recombinants (depending on the sample size) and all the progeny will fall into only two genotype classes.

In theory, any two markers can be separated by recombination, but in practice, sample size limitations in humans will make it impossible to detect recombination between markers that are separated by less than 1 cM, except for a few exceptionally common diseases. Readers who like to play with numbers may be amused by considering the most extreme imaginable case, which would be two marker loci defined by variants at adjacent nucleotides. Inasmuch as 1 cM is about one million bp, one nucleotide corresponds to 10^{-6} cM, or a recombination fraction of 10^{-8}. In other words, one would have to analyze approximately one hundred million progeny of informative matings in order to find one recombinant for markers at adjacent nucleotides!

Human linkage analysis is made difficult by the fact that human matings are arranged with total disregard for the preferences of human geneticists. One must study what happens to be available, rather than experimentally designed crosses. Fortunately, the vast size of the human population means that informative families are often available. Another drawback is that human families are usually small; therefore, it is necessary to study several or many families in which alleles are segregating at the two loci in question, and special statistical methods must be used to interpret the data.[4]

[4] A technique that combines features of linkage analysis and physical mapping in somatic cell hybrids is described in Box 4–3.

BOX 4–1 PCR and Polymorphisms

Detection of RFLPs and VNTRs has traditionally depended on the use of probes for specific regions of the genome. Many restriction enzyme digestions are done for each probe, followed by electrophoresis of the DNA, Southern blotting, hybridization of the probe to the blot and detection of hybridized probe by autoradiography. All of this must be done on DNA from 10–20 different persons, in the hope of finding a polymorphic site. The process is laborious. By contrast, the detection of polymorphisms in microsatellites (e.g., tandem dinucleotide repeats) is readily achieved with the use of the *Polymerase Chain Reaction* (PCR), a powerful technique for the rapid amplification of specific DNA sequences, which can then be characterized by other methods.

The Polymerase Chain Reaction uses DNA polymerases to amplify a nucleic acid segment that is flanked by two short regions of known sequence, for which complementary oligonucleotides can be synthesized, to be used as primers by the polymerase. The method is explained and diagrammed in Figure 4B1-1 and reviewed in White et al. (1989).

The main feature of the PCR is its ability to amplify very small amounts of starting material at least a million-fold, thereby producing enough product to analyze by non-radioactive techniques. The DNA or RNA sequences to be amplified do not need to be extensively purified, because the use of specific primers restricts the activity of the polymerase to the target sequence. Amplification is rapid and relatively cheap, thereby making it feasible to apply the PCR to large series of samples. In a sense, the Polymerase Chain Reaction is a form of cell-free cloning; it often bypasses the need for cloning in microorganisms with recombinant DNA technology.

Many applications of the PCR have been described. These include evolutionary biology, developmental biology, and forensic analysis. In human genetics, there are extensive applications in genetic screening (Chapter 9) and in the detection of mutations in certain circumstances (Chapter 6). We will become acquainted with the technique in the present context, the search for polymorphisms in human DNA.

As mentioned on page 74, polymorphisms in microsatellites are best detected by amplifying the short segments of DNA that contain the tandem repeats, and then using electrophoresis of the amplified DNA through a gel that is capable of distinguishing size differences involving only one or two nucleotides. The way in which the amplication is accomplished with the PCR is diagrammed in Figure 4B1–2.

Another application of the PCR to the search for polymorphisms is explained in Box 4–2.

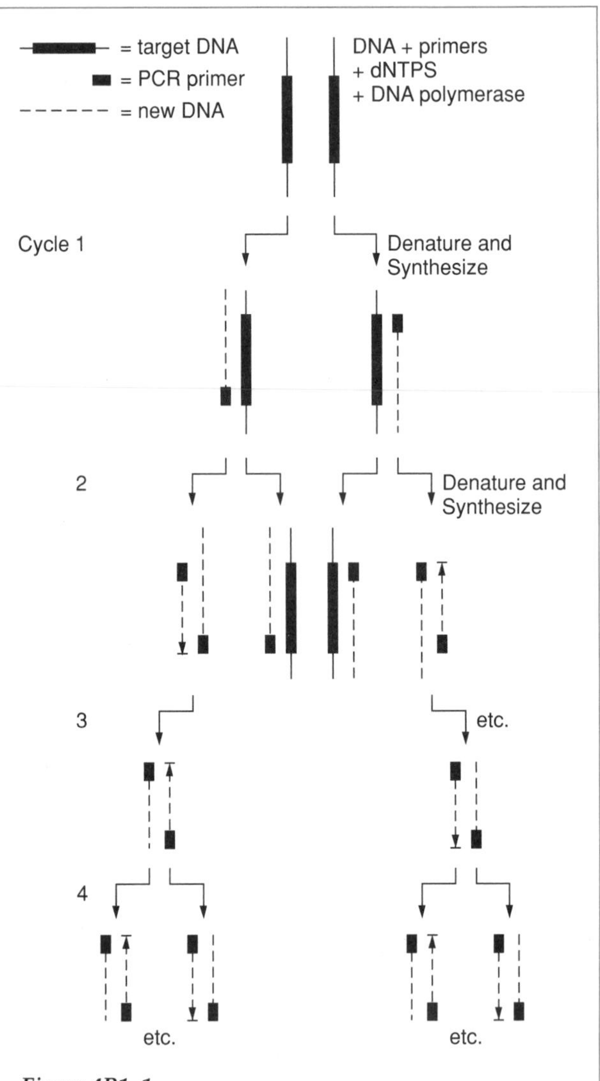

Figure 4B1–1

The Polymerase Chain Reaction. PCR is based on the enzymatic amplification of a DNA fragment that is flanked by two oligonucleotide primers that hybridize to opposite strands of the target sequence. The primers are oriented with their 3′ ends pointing toward each other. Repeated cycles of heat denaturation of the template, annealing of the primers to their complementary sequences, and extension of the annealed primers with a DNA polymerase result in the amplification of the segment defined by the 5′ ends of the PCR primers.

Only the first two cycles are shown completely. Beginning with the third cycle, the diagram does not show the fate of the original DNA and the extension products made from it. Note that the long primer extensions can be made only from the original template, and that they increase additively with each cycle (e.g. only 20 copies of each strand would be synthesized in 20 cycles). In contrast, the short, discrete, primer-terminated copies, which first appear in the second cycle, proceed to double with each subsequent cycle and rapidly become the predominant form of amplification product (e.g., 10^6 copies of each target strand would be made in 20 cycles). The use of the thermostable *Taq* polymerase means that fresh polymerase does not have to be added after each denaturation step. (From White et al., 1989)

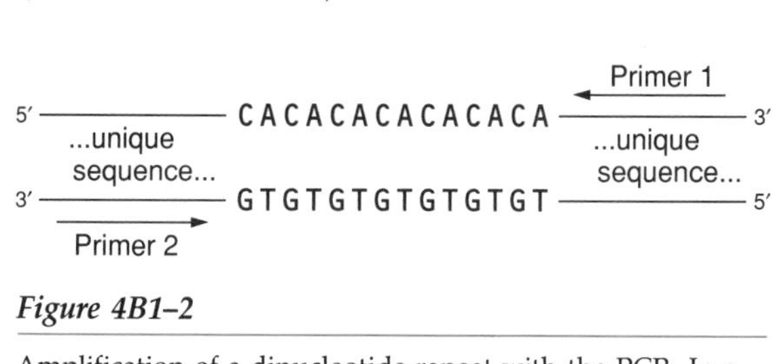

Figure 4B1–2

Amplification of a dinucleotide repeat with the PCR. In a real situation, there would usually be a larger number of dinucleotide repeat units, and the unique sequences might not abut the dinucleotide repeats directly (sometimes other repeated sequences, which would not be suitable for primers, lie between the dinucleotide repeat region and truly unique sequences). When the amplification products from a series of individuals are analyzed by electrophoresis, variability in the amount of DNA between the primers can be detected.

REFERENCE

White, T. J., Arnheim, N., and Ehrlich, H. A. 1989. The polymerase chain reaction. Trends in Genetics 5:185–189.

BOX 4–2 *RAPD: A New Way to Find Polymorphisms*

Williams et al. (1990) described a clever new strategy for finding polymorphisms without using cloned DNA as probes for RFLPs or VNTRs. The technique scans many sites in the genome simultaneously; it has the potential to significantly increase the number of known polymorphisms in the human genome. This method depends upon *Random Amplification of Polymorphic DNA* (abbreviated RAPD), and in order to understand it, readers should first become familiar with the Polymerase Chain Reaction (PCR) by studying Box 4–1.

The RAPD technique is based upon two principal facts: (1) a PCR primer containing 10 nucleotides (a 10-mer) will bind to many complementary sequences in a mammalian genome, from which amplification can be initiated; and (2) if a given binding site is polymorphic in the population (i.e., if some individuals have one or more base substitutions at that specific site), then the 10-mer will not serve as a primer in some individuals. The result is that certain regions of the genome will be amplified in some individuals, but not in others, and those differences can be easily detected by electrophoresis.

First, consider the requirements for amplification of a portion of a genome using *one* PCR primer (remember that in most other PCR applications, two different primers are used, one at each end of the sequence to be amplified). In order for both strands of DNA to be amplified, sequences complementary to that single primer sequence must be present on both strands of DNA within a few kbp of each other (Figure 4B2–1). We shall use 2 kbp as the upper limit for amplification of a DNA segment by PCR, although some workers claim to be able to amplify as much as 5 kbp. Then, regions of the DNA that do not contain

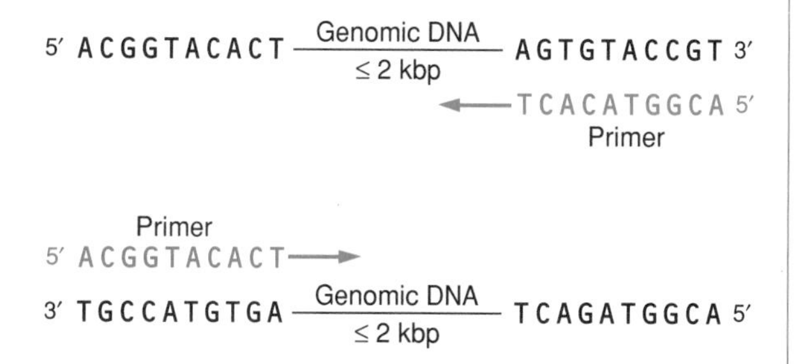

Figure 4B2–1

Diagram of the sequence relationships that must exist in order for a segment of DNA to be amplified by PCR, when only one primer is used. Note that the primer sequence (and its complement, of course) must be present on both strands of DNA. From the reader's point of view, the primer sequence has opposite orientation on the two DNA strands; however, it actually has the same polarity (when read 5′ to 3′) on both strands.

two copies of the primer sequence in reverse orientation within about 2 kbp of each other will not be amplified with a single primer.

The probability of finding any 10-base sequence in a genome is 4^{-10} (assuming 50% G + C content and random distribution of nucleotides), and there are approximately 3×10^9 10-base sequences in a human genome. Thus, we expect about 2800 copies of a given 10-base sequence ($3 \times 10^9 \times 4^{-10}$) in the entire genome.

However, PCR with a single primer requires an inverse copy to occur within 2 kbp, and the probability of that occurring is about 0.002 (2000×4^{-10}). Therefore, the total number of amplifiable segments of DNA in an average human genome, using one 10-mer as primer, should be about 5 or 6 (2800×0.002).

Some of these segments will be too small to be detected on a standard gel, and various other factors, including statistical fluctuations, may affect the results; however, to a first approximation, we can predict 5 to 10 segments of DNA will be amplified from a human genome using one random 10-mer as a PCR primer. Gel analyses presented by Williams et al. (1990) are in agreement with that prediction (Figure 4B2-2).

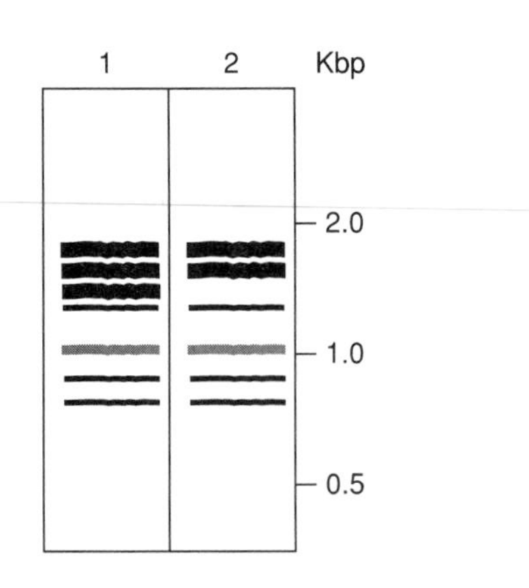

Figure 4B2–2

Amplified DNA segments from two unrelated human genomes, using the primer shown in Figure 4B2-1. Note that the third band from the top reveals a polymorphism. (Based on data in Williams et al., 1990, p. 6532)

Notice that the primers used for RAPD are decidedly shorter than primers used for most other PCR applications, which generally are focused on a unique genomic site. A primer that will bind to only one site in a human genome needs to be at least 16 nucleotides long; otherwise, it may occur more than once, strictly on the basis of chance ($3 \times 10^9 \times 4^{-15} = 2.8$; $3 \times 10^9 \times 4^{-16} = 0.7$). To be safe, it is advisable to use primers about 20 nucleotides long when a unique site is targeted. In contrast, the goal of a search for random polymorphisms is to find as many candidate sites as can be conveniently assayed; for that reason, primers of 9 or 10 nucleotides are optimal. Smaller primers would bind to more sites, but too many amplified bands would be produced.

The types of polymorphisms that are detectable by RAPD include single or multiple base substitutions, deletions of a primer binding site, and insertions that move the primers so far apart that PCR amplification fails. All of these changes should eliminate amplification of a given segment from the variant chromosome. Insertions or deletions between primer sites that change the size of the amplified region by a detectable amount should lead to production of a new band on a gel pattern.

One disadvantage of the technique for genetic purposes must be mentioned: the polymorphic sites identified by RAPD will usually not be codominant markers. The reason is that if a change in one chromosome abolishes the production of an amplified segment from that chromosome, it will not affect amplification of the corresponding segment on the homologous chromosome. Thus, heterozygotes will have the same number of bands as homozygotes for the presence of a band, and although one band will be 50% lighter in heterozygotes than in homozygotes, that difference cannot be reliably detected. Codominant markers will be produced when a polymorphism creates a new band of different size, as will usually be the case at a VNTR locus.

Finally, it is worth noting that the name, RAPD, is a partial misnomer, although a good mnemonic. All sites that meet the criteria for primer binding are amplified, whether they are polymorphic or not. The utility of the technique arises from the number of sites that can be surveyed with a single primer, and the ease with which the PCR can be applied. Large numbers of 10-mers for RAPD are now commercially available; the technique is likely to be widely used.

REFERENCE

Williams, J. G. K., Kubelik, A. R., Livak, K. J., Rafalski, J. A., and Tingey, S. V. 1990. DNA polymorphisms amplified by arbitrary primers are useful as genetic markers. Nuc. Acids Res. 18:6531–6535.

For each informative family in a linkage study, it is necessary to calculate, (1) the likelihood that the observed progeny genotypes could have arisen from linked genes, with a given frequency of recombinants (RF); and (2) the likelihood that the observed progeny genotypes could have arisen from unlinked genes (i.e., by independent assortment). We shall call the first likelihood L(RF) and the second likelihood L(0.5). RF can have any value less than 0.5; if it is not substantially less than 0.5, demonstration of linkage will be difficult.

If the ratio L(RF)/L(0.5) is greater than 1, it indicates that linkage is more likely than non-linkage to explain the family's genotypes; if it is less than 1, then non-linkage is more likely. This *odds ratio* can be calculated for every available informative family, and the overall odds can be obtained by multiplying all the individual odds together. This is logically valid, because each family can be regarded as an independent event. Note, however, the assumption that we are dealing with the same disease gene in every family (i.e., the same genetic locus, but not necessarily the same alleles).

Human geneticists have devised a system of linkage analysis called *the method of lods,* where *lod* stands for the common logarithm of the odds that linkage is a better explanation than non-linkage for a set of data. The use of logarithms, rather than the raw odds ratio, has some advantages; one is the ability to combine data from a series of families by addition of lod scores, rather than multiplication of odds ratios. The theoretical justification for this method of handling data in linkage studies was developed in a classic paper by Morton (1955). Lod scores are usually designated by the letter Z, thus

$$Z(\mathrm{RF}) = \log[L(\mathrm{RF})/L(0.5)]$$

By convention, a lod score of 3 (odds = 1000:1) is considered to be "proof" of linkage, and a lod score of 2 (odds = 100:1) is "strong evidence." A score of -2 is taken as proof that two loci are not linked at the RF value that leads to that score (they may be more loosely linked). The calculation and use of lod scores will become clear as we examine some simple examples. These examples are presented so that the reader may understand the principles of human genetic linkage analysis. Anyone wishing to undertake a real linkage analysis project should consult the general references (e.g., Conneally and Kivas 1980; Ott, 1985; White and Lalouel, 1987) and/or an authority on the subject.

Basic Calculations in Human Linkage Analysis

Consider the family in Figure 4–8, which is equivalent to a testcross in experimental organisms. Note that the affected parent is a double heterozygote and the unaffected parent is a double homozygote. The following steps are involved.

1. *If* there is linkage between the disease locus and the marker locus, the linkage phase in individual I-1 is unknown. *Linkage phase* refers to the different sets of alleles on each member of a pair of homologous chromosomes; the concept is nearly identical to that of haplotypes. When dealing with an individual who is a heterozygote at two loci, each of which involves a dominant allele and a recessive allele, it is customary to refer to the genotype *AB*/*ab* as

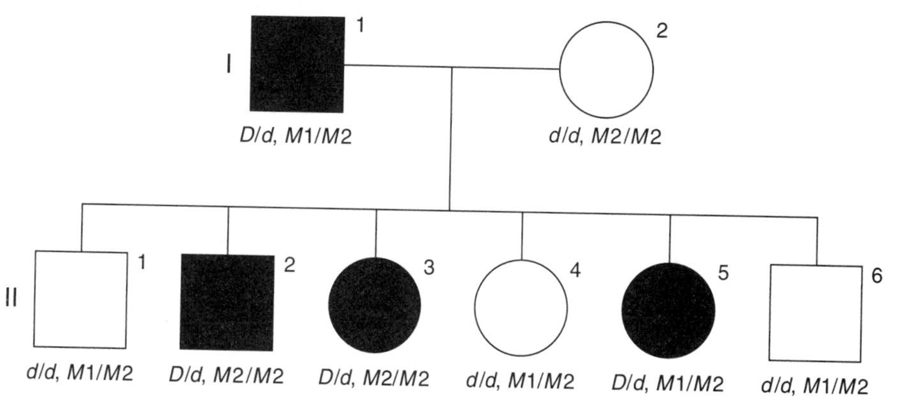

Figure 4–8

Hypothetical pedigree of a family in which an autosomal dominant disease (locus *D*) and codominant alleles at a marker locus (*M*) can be tested for linkage. Filled symbols indicate affected individuals.

having alleles *in coupling*, and to the genotype *Ab/aB* as having alleles *in repulsion*.

The terms *coupling* and *repulsion* lose their significance when codominant alleles are present, as is the case for RFLP sites. Nevertheless, they are a standard form of genetic nomenclature, so we shall arbitrarily define coupling in the present case to be *DM1/dM2* and repulsion to be *dM1/DM2*.

We do not know the linkage phase in individual I-1 in Figure 4–8. The *prior probability* of each genotype is 0.5, provided we know nothing about his parents' genotypes.

2. We next calculate the *conditional probabilities* of I-1 producing each possible type of gamete. The two conditions to be evaluated are coupling and repulsion. Note that we leave RF numerically unspecified for the moment.

PHASE	GAMETE FREQUENCIES			
	DM1	*DM2*	*dM1*	*dM2*
coupling	(1 − RF)/2	RF/2	RF/2	(1 − RF)/2
repulsion	RF/2	(1 − RF)/2	(1 − RF)/2	RF/2

Also note that every value in the table is divided by 2. This is because there are two classes of recombinant gametes and two classes of non-recombinant gametes, whereas RF is defined as the total frequency of recombinant gametes.

3. Now we determine the number of each class of paternal gametes observed in the progeny. Because we are dealing with an autosomal dominant disease in a family with only one affected parent, the genotype of each child at the disease locus will be evident from the child's phenotype; the genotype at the marker locus will be determined by restriction enzyme digestion of a sample of DNA and measurement of the size of the resulting fragments.

To state the general case, let $n1$ equal the number of *DM1* gametes, $n2$ the number of *DM2* gametes, $n3$ the number of *dM1* gametes, and $n4$ the number of *dM2* gametes. Also let $n1 + n2 + n3 + n4 = n$, the total number of progeny in the family. In the example in Figure 4–8, there are six children, and $n1 = 1$, $n2 = 2$, $n3 = 3$, and $n4 = 0$.

4. The probability of obtaining the observed family, given any specified value of RF, can be obtained by combining information in steps 1, 2, and 3 above. For example, the probability of a child receiving a gamete of class *DM2* is RF/2 if the father's alleles are in coupling, and the probability of having $n2$ such children in one family is $(\text{RF}/2)^{n2}$ Similar considerations lead to the general formula:

$$L(\text{RF}) = L(\text{coupling})L(\text{RF}) + L(\text{repulsion})L(\text{RF}).$$

Because $L(\text{coupling}) = L(\text{repulsion}) = 0.5$, and because $\text{RF}/2 = 0.5\,(\text{RF})$ and $(1 - \text{RF})/2 = 0.5\,(1 - \text{RF})$, we get:

$$\begin{aligned} L(\text{RF}) &= 0.5[0.5^n(1 - \text{RF})^{n1+n4}(\text{RF})^{n2+n3}] \\ &\quad + 0.5[0.5^n(1 - \text{RF})^{n2+n3}(\text{RF})^{n1+n4}] \\ &= 0.5^{n+1}[(1 - \text{RF})^{n1+n4}(\text{RF})^{n2+n3} + (1 - \text{RF})^{n2+n3}(\text{RF})^{n1+n4}] \end{aligned}$$

For the example given in Figure 4–8, it is logical to begin evaluating this formula by choosing RF = 0.167, because there is one probable recombinant (II-5) in the family, *if* the father's alleles are in repulsion. However, we have no information on the linkage phase in the father as yet, so we must also consider the possibility that his alleles are in coupling and there are five recombinants among the progeny. Evaluating those possibilities, we get

$$L(0.167) = (0.5)^7[(0.833)(0.167)^5 + (0.833)^5(0.167)] = 0.000524$$

5. Now we must calculate the probability of obtaining the observed progeny by independent assortment. If the disease locus and the marker locus are unlinked, the probability of each of the four possible gamete genotypes is 0.25. In the general case, therefore, $L(0.5) = 0.25^n$, where n is again the total number of progeny.

For the example in Figure 4–8

$$L(0.5) = (0.25)^6 = 0.000244.$$

6. Finally, we divide L(RF) by $L(0.5)$ to obtain the relative likelihood of obtaining the observed progeny by linkage versus non-linkage. For the example in Figure 4–8, we have 0.000524/0.000244 = 2.147. In other words, linkage is about twice as likely as non-linkage for this six-child family.

The lod score, designated by Z, is 0.332. Evidently, the data from this family would have to be combined with data from many similar families—or preferably, from larger families—before a decision about the linkage or non-linkage of locus D and locus M could be made.

Before we go on to other aspects of linkage analysis, let's look back at the example just presented. We calculated a lod score for a family with six children, using 0.167 as the value of RF, because the pedigree suggested that there was one recombinant genotype among the offspring. What would have happened to the lod score if we had used a different value of RF? In pedigrees like this, where the putative number of recombinants (r) can be counted directly, it can be shown that the *maximum likelihood estimate* (MLE) of the lod score will be obtained when RF is chosen to be r/n, where n is again the total number of offspring. In our example, r/n was 1/6, so we have already determined the maximum lod score for this family. Unconvinced readers should repeat the calculations, using slightly higher or lower values for RF; the lod scores obtained will be lower than the lod score given in the example.

In practice, one would compute Z (the lod score) for a variety of RFs for each family under study. Inasmuch as the frequency of recombinant progeny will surely vary from family to family because of chance, the MLEs of RF will also vary from family to family. However, in linkage analysis we ultimately want a maximum likelihood estimate of RF for the combined data set, which will be that value of RF which gives the largest total lod score when all the family data are summed.

More Examples

In order to more fully illustrate the factors that affect human linkage analysis, we shall now consider two variations on the preceding theme. First, let's ask, "How would the lod score for the family in Figure 4–8 be affected if the linkage phase of the father were known?" Figure 4–9 presents a hypothetical pedigree for the father in Figure 4–8.

Now it is clear that the father's alleles must be in repulsion, $DM2/dM1$, if the two loci are linked. The prior probability of coupling is now zero, and the calculation of the $L(\text{RF})$ becomes

$$L(\text{RF}) = 0.5^{n}(1 - \text{RF})^{n2+n3}(\text{RF})^{n1+n4}$$

$$L(0.167) = (0.5)^{6}(0.833)^{5}(0.167) = 0.001046$$

This is almost twice the value obtained when we did not know the linkage phase of the father's alleles. Clearly then, an approximate doubling of the odds favoring linkage over non-linkage will occur (the lod score will increase by roughly 0.3, the common logarithm of 2). This emphasizes the usefulness of multigeneration pedigrees for linkage analysis.

Next, we ask whether it is better to have one family with 12 children than two identical families with 6 children each. According to the principle that the lod scores for different families may be added to give an overall estimate of linkage probability, we see that if we had analyzed two families identical to the one in Figure 4–8, the total lod score would be 0.332 + 0.332 = 0.664. For comparison, let us now suppose that the parents in Figure 4–8 had 12 children, with each of the progeny genotypes being represented twice as often as in that figure, so that the best estimate of RF is again 0.167. What would be the lod score for that family?

We now have

$$n = 12,\ n1 = 2,\ n2 = 4,\ n3 = 6, \text{ and } n4 = 0.$$

$$L(0.167) = (0.5)^{13}[(0.833)^{10}(0.167)^{2} + (0.833)^{2}(0.167)^{10}]$$

We also have

$$L(0.5) = (0.25)^{12} = (0.5)^{24}$$

Then

$$L(0.167)/L(0.5) = 9.22 \text{ and } Z = 0.965$$

We conclude that the odds favoring linkage over non-linkage are twice as high for the family of 12 as for two identical families of 6 children each, which in this case would be $(2.147)^2 = 4.610$. This emphasizes the value of large families for linkage analysis. However, note that if we knew the linkage phase of all the affected parents, then two families with 6 offspring apiece would yield the same lod score as one family with 12 offspring.

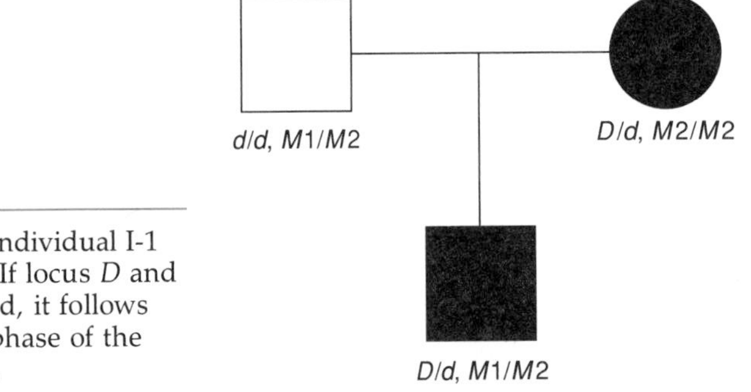

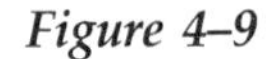

Figure 4–9

The pedigree of individual I-1 from Figure 4–8. If locus D and locus M are linked, it follows that the linkage phase of the son is $dM1/DM2$.

Multilocus Linkage Mapping

The examples given previously dealt with only two loci at a time; they were presented in the context of finding linkage between a locus for a biochemically unknown disease and a marker locus, such as the site of an RFLP. As the human linkage map becomes filled in, it is often necessary to do multipoint analysis, because the uncertainties in the locations of loci that have been studied two at a time may not allow us to specify a definite order for a series of closely linked loci. Suppose that loci *S*, *T*, and *U* have been analyzed pairwise and the maximum likelihood estimates of RF are: 0.06 for *ST*, 0.08 for *TU*, and 0.12 for *SU*. Although those numbers suggest that the true order of the loci actually is *S-T-U*, that is by no means certain, because each of the RFs is likely to be characterized by a large statistical uncertainty, unless the sample is extraordinarily large by human genetic standards.

Methods for three-point analysis of linkage data have been available for a number of years; they are discussed in the reviews by White and Lalouel (1987) and by Ott (1985). Three-point analysis can help to determine the order of loci and can he used to measure interference in crossing over. Multipoint analysis can also be extended to more than three loci, but the computations rapidly become complex.

Linkage Analysis with Autosomal Recessive Diseases

The principles that were described in the preceding section on autosomal dominant diseases also apply to linkage analysis with autosomal recessive diseases, but the calculations are more complicated. We will not go into the mathematics here. However, it is important to have a conceptual understanding of the special problems that apply to linkage analysis of autosomal recessive genetic diseases.

In our discussion of linkage analysis with dominant diseases, we saw that normal and affected children born to a potentially informative couple were equally useful in testing the hypothesis that the disease gene was linked to a given marker locus. However, when dealing with an autosomal recessive disease, we do not know the genotype of normal children at the disease locus; they are either homozygous for the normal allele (*RR*) or heterozygous (*Rr*). This means that normal offspring, by themselves, cannot be used to deduce the linkage phase of a doubly heterozygous parent. The presence of normal children in a pedigree involving a recessive genetic disease is not without value, but for our purposes, it suffices to note that there is less information available from normal children in recessive pedigrees than in dominant disease pedigrees. The effect of these considerations is that more families are required to demonstrate linkage between a marker locus and an autosomal recessive disease than for an autosomal dominant disease. Consequently, some of the rarer recessive diseases do not provide enough material for standard linkage analysis.

A very clever method for demonstrating linkage with relatively small numbers of affected individuals has been described by Lander and Botstein (1987). The method, known as *homozygosity mapping*, depends upon the fact that the children of marriages between blood relatives (consanguineous marriages, usually between first cousins) represent a significant fraction of affected individuals for any recessive disease. Such children have a relatively high probability of being homozygous for any given allele carried by

BOX 4–3 Gene Mapping by X-ray Zapping

There is a relatively new technique that combines features of physical mapping and genetic mapping. It is called *Radiation Hybrid (RH) mapping* (Cox et al., 1990) and is based upon a strategy devised in the 1970s (Goss and Harris, 1975).

In RH mapping, chromosomes are broken by X-irradiation, and then the frequency with which pairs of markers remain on the same fragment is determined. In general, the farther apart two markers are on a chromosome, the more likely they are to be separated by X-irradiation. By analyzing a series of markers known to occur on a single chromosome, it is possible to determine the order of the markers. The result is analogous to a linkage map, but it is independent of meiosis.

Practical aspects of RH mapping are diagrammed in Figure 4B3–1. A culture of human-hamster hybrid cells, preferably containing only one human chromosome per cell, is exposed to a dose of X-rays sufficient to break every chromosome into 5 to 10 fragments. Broken chromosome ends are rapidly ligated together by cellular enzymes; the result is that most fragments of the human chromosome become attached to fragments of hamster chromosomes. Those with centromeres can undergo normal mitotic segregation.

However, the irradiated cells have suffered too much damage to survive independently. Therefore, irradiated cells are fused with untreated hamster cells that are HPRT$^-$. In HAT medium (Chapter 3) the only cells that can grow and multiply will be those hybrids that have acquired the HPRT gene from the irradiated hamster genome. Coincidentally, most of them will also retain one or more fragments of human chromosome.

If many cells from the preceding protocol are cloned, and if a set of markers representing loci on the original human chromosome can be assayed in each clone, then the frequency with which each pair of markers is retained together after irradiation can be determined. For example, Cox et al. (1990) assayed approximately 100 radiation hybrid clones for 14 markers known to occur within the proximal 20 Mbp on the long arm of human chromosome 21. The data were subjected to statistical analysis, and when some results from pulsed field gel electrophoresis were added, the order of all 14 markers was determined, with an average resolution of 500 kbp.

By analogy with meiotic recombination, Cox et al. defined the centiRay (cR). At a given radiation dose, a distance of one cR between markers implies that a break occurs between them in 1% of treated cells. Initial studies on chromosome 21 indicate that the relation between centiRays and physical distance is relatively constant; 1 cR is about 50 to 55 kbp.

Based on the data of Cox et al. (1990), the resolving power of RH mapping (500 kbp) is somewhat better than that of *in situ* hybridization on metaphase chromosomes (1 or 2 Mbp). Clearly, the resolution of RH mapping can be increased simply by analyzing more cell clones. However, fluorescent *in situ* hybridization on interphase cells can determine the order of markers separated by no more than 50 to 100 kbp (Trask et al., 1991). It remains to be seen whether radiation hybrid mapping will prove to be the method of choice for many analyses.

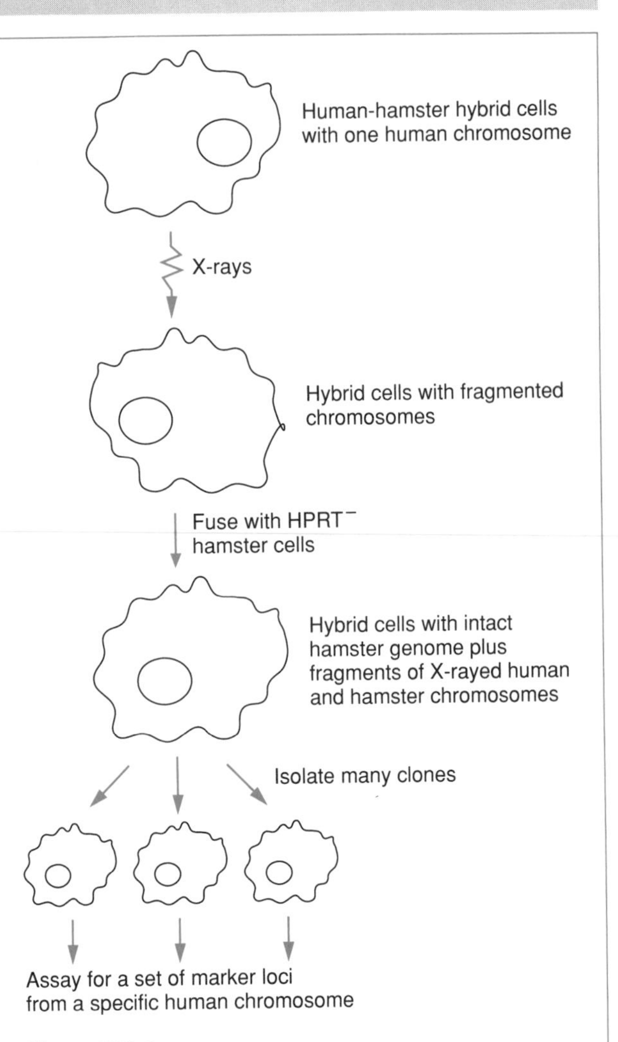

Figure 4B3–1

Major steps involved in radiation hybrid mapping. See text for details.

REFERENCES

Cox, D. R., Burmeister, M., Price, E. R., Kim, S., and Myers, R. M. 1990. Radiation hybrid mapping: a somatic cell genetic method for constructing high-resolution maps of mammalian chromosomes. Science 250:245–250.

Goss, S. J. and Harris, H. 1975. New method for mapping genes in human chromosomes. Nature 255:680–684.

Trask, B. J., Massa, H., Kenwrick, S., and Gitschier, J. 1991. Mapping of human chromosome Xq28 by two-color fluorescence in situ hybridization of DNA sequences in interphase cell nuclei. Am. J. Hum. Genet. 48:1–15.

one of their ancestors; that is, at a given genetic locus, they may be *homozygous by descent*. The rarer the genetic disease, the higher the proportion of affected individuals who are homozygous by descent.

This phenomenon is illustrated in Figure 4–10. A substantial portion of the chromosome carrying the recessive disease-causing allele will usually be homozygous in affected offspring of consanguineous matings. If enough polymorphic marker loci can be examined, there will be a set of closely linked marker loci that are homozygous in the child, but heterozygous in the parents. Of course, this can happen anywhere in the genome, so one must examine several affected children in order to find a consistent correlation between homozygosity for a given region and presence of the disease.

Lander and Botstein (1987) present a theoretical justification for their claim that conventional lod scores can be obtained from such data. Computer-generated graphs are given, showing the relationship between the RFLP map and the number of children needed to map a gene. Their conclusion is that 5 to 10 affected children of first cousin matings will suffice to locate many recessive disease genes by homozygosity mapping, provided a dense map of marker loci (RFLPs) is available. This is substantially less than the total number of informative offspring required for conventional linkage mapping, and families with only one affected child can be informative.

Linkage Analysis Can Provide Evidence of Genetic Heterogeneity

Although the most frequent use of linkage analysis in human genetics at the present time is the mapping of genes responsible for diseases by reference to previously mapped RFLPs, another important use is the detection of heterogeneity resulting from a clinical phenotype being caused by defects at two or more genetic loci. The underlying concept is quite simple. If a disease is caused by abnormalities in two or more unlinked or very

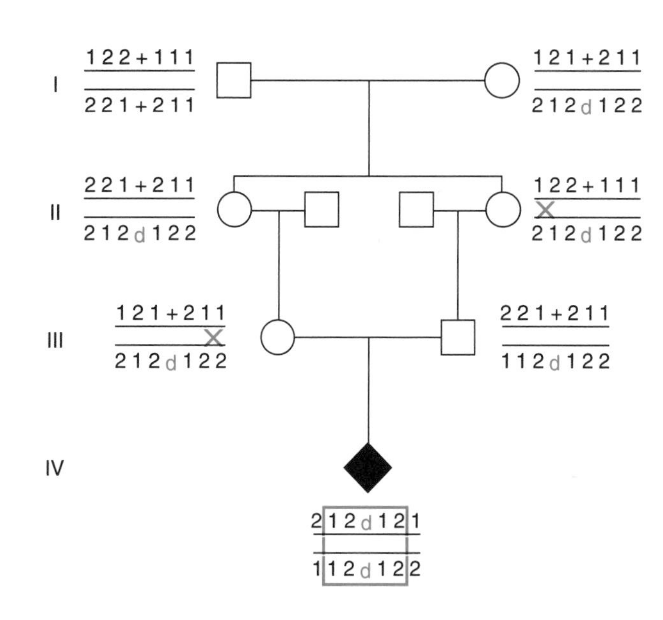

Figure 4–10

Homozygosity mapping. Hypothetical pedigree showing a first cousin marriage (generation III) that produced a child affected with a recessive disease (generation IV), because of homozygosity by descent for a disease-causing allele (*d*). 1s and 2s represent different alleles at closely linked marker loci. A hypothetical set of crossovers is indicated in generations II and III, which reduces the number of loci that become homozygous by descent. For the six RFLPs illustrated, a contiguous region with four homozygous marker loci would be found in the affected child.

loosely linked genes, then the observed RF in different families will sometimes be very different from the overall RF calculated for the rest of the families that have been analyzed.

A classic example comes from the work of Morton (1956) on linkage between elliptocytosis (a recessive disorder that changes the shape of red blood cells and produces anemia) and the rhesus (Rh) blood type. Morton studied seven large families that were informative at both loci, and as shown in Figure 4–11, he found that there was close linkage in four of them; but in the other three families, Rh and elliptocytosis assorted independently. Thus, there appear to be two unlinked loci capable of causing elliptocytosis.

The problem in detecting genetic heterogeneity lies in deciding how big a difference in RF values is a significant deviation from the maximum likelihood estimate of RF for the entire data set. Consider the following situation. Suppose that we have calculated RFs for six large families; the values are 0.09, 0.07, 0.16, 0.13, 0.38, and 0.12. Does the 0.38 number imply that a second locus may be responsible for the disease? Statistical methods for evaluating that possibility are discussed in Conneally and Rivas (1980), in Ott (1985), and elsewhere. For our purposes, it is sufficient to be aware that genetic heterogeneity can be revealed by linkage analysis, if enough information is available.

Recognition of heterogeneity can be important in elucidating the molecular basis of a disease, and it can be essential in genetic counseling. The latter point has been repeatedly emphasized by Gusella and others working with Huntington's Disease (see Chapter 5), where it was vitally important to be sure that only the locus on chromosome 4p was responsible for the disease, before closely linked RFLP markers were used to tell persons at risk whether they were likely to develop the disease later in life.

Linkage Equilibrium and Disequilibrium

The preceding sections of this chapter have contained the tacit assumption that the loci under consideration are in *linkage equilibrium*, which means that all possible combinations of alleles on a single chromosome (i.e., all

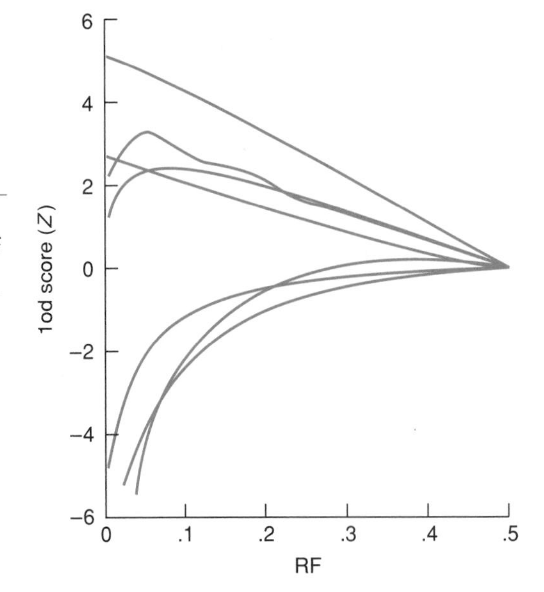

Figure 4–11

The relationship between lod scores (Z) and various values of RF for seven families in which elliptocytosis and Rh blood types were segregating. The upper four lines show maximum Z values at or near RF = 0, which implies close linkage. The lower three lines reach a maximum at RF = 0.5, indicating independent assortment. (From Morton, 1956; by permission of University of Chicago Press)

possible haplotypes or all possible gamete genotypes) occur as frequently as would be predicted from the random association of individual allele frequencies.

Consider an example where we have the following allele frequencies at two linked loci:

$$A = 0.2 \qquad M = 0.6$$
$$a = 0.8 \qquad m = 0.4$$

Four haplotypes are possible, and if there is linkage equilibrium, their frequencies in the population will be:

$$\begin{aligned} AM &= 0.2 \times 0.6 = 0.12 \\ Am &= 0.2 \times 0.4 = 0.08 \\ aM &= 0.8 \times 0.6 = 0.48 \\ am &= 0.8 \times 0.4 = \underline{0.32} \\ & \qquad\qquad\quad 1.00 \end{aligned}$$

However, for closely linked loci, *linkage disequilibrium* may exist; that is, the actual haplotype frequencies may be different from those predicted by the hypothesis of random association. One potential explanation for linkage disequilibrium is that a mutant allele may have arisen fairly recently (in generational terms), and there has not been sufficient time for recombination to establish equilibrium with all alleles at neighboring loci. Another possibility is that there is selection pressure for or against a particular combination of alleles.

Suppose we were able to measure the haplotype frequencies in the preceding example and we found the following:

$$\begin{aligned} AM &= 0.04 \\ Am &= 0.16 \\ aM &= 0.56 \\ am &= 0.24 \end{aligned}$$

These observed frequencies differ from the expected frequencies, and if the sample were large enough, the differences would be statistically significant. We can express linkage disequilibrium quantitatively by defining a disequilibrium coefficient, such that

$$\text{D} = \text{observed frequency} - \text{expected frequency}$$

For the preceding example:

$$\begin{aligned} AM\!: D &= 0.04 - 0.12 = -0.08 \\ Am\!: D &= 0.16 - 0.08 = +0.08 \\ aM\!: D &= 0.56 - 0.48 = +0.08 \\ am\!: D &= 0.24 - 0.32 = -0.08 \end{aligned}$$

Notice that D may be either positive or negative, but its absolute value will be the same for every haplotype.

The maximum value that D may have is determined by setting one of the haplotypes involving the least common allele at a frequency of zero. For the current example, $D_{max} = 0.12$, which would occur if the frequency of AM were zero in the population under study. The absolute value that D can attain for any two-locus system with two alleles at each locus is 0.25, which could occur if the frequency of each of the four alleles were 0.25.

Now we ask, what is the effect of linkage disequilibrium on linkage analysis in humans? For pedigrees where no assumptions need to be made about any of the genotypes, linkage disequilibrium is irrelevant. However, there are situations where linkage analysis requires that a guess be made about one or more individuals' genotypes. If linkage disequilibrium exists and is unrecognized, the guesses may be wrong and the resulting total lod score will be correspondingly inaccurate.

Linkage disequilibrium may also affect the accuracy of genotype prediction for fetuses and thus is of interest in genetic counseling. In addition, linkage disequilibrium is sometimes helpful in finding a gene responsible for a disease, because the closer a polymorphic marker locus is to the target gene, the more likely it is to show linkage disequilibrium with that gene. The identification of genes that cause diseases for which the primary biochemical defect is unknown is described in the next chapter.

SUMMARY

Crossing-over, the physical exchange of DNA that often occurs between homologous chromosomes during meiosis, leads to recombination of alleles at genetic loci flanking the crossover region. Linkage analysis uses the frequency of recombination between two or more loci to determine whether the loci are near one another on the same chromosome, and if so, the order of the loci relative to one another.

Independent assortment of alleles at two loci is indicated when the recombination frequency is approximately 50%. This may occur because the loci are on different chromosomes or far apart on the same chromosome.

A genetic map unit is called a centiMorgan (cM); it represents a recombination fraction of 1%, which corresponds to about 1 Mbp in humans, on average.

Genes of unknown function can be mapped by linkage analysis, using only the phenotypes produced by variants at that locus. This is especially important in humans, where many genes that cause disease have not been characterized in terms of primary biochemical effects.

In order to map a gene responsible for a specific disease, a set of marker loci with known map locations must be available. Each marker locus must be polymorphic, so that there is a high probability that individuals who are heterozygotes for a disease-causing allele and a normal allele at the locus to be mapped will also be heterozygotes at the marker locus. Restriction Fragment Length Polymorphisms (RFLPs) provide an important source of marker loci. The most highly polymorphic marker loci are associated with Variable Numbers of Tandem Repeats (VNTRs), including dinucleotide repeats.

Linkage analysis in human families requires a mathematical approach distinct from that used for experimental organisms because of the small size of the families. The standard "method of lods" involves a comparison of the probability that a given family could have arisen because the loci under study are linked, compared to the probability that the same family could occur if the loci are not linked. Data from many families usually have to be combined in order to determine whether a disease locus and a marker locus are actually linked.

A haplotype is a set of alleles at two or more loci on the same DNA molecule (the same chromosome or chromatid). Recombination occurring over many generations tends to randomize the possible combinations, generating haplotypes in proportion to allele frequencies. This is called *linkage equilibrium*. Linkage disequilibrium can be present if two loci are very close to one another and one or more alleles have arisen recently, so that not enough time has elapsed for equilibrium to arise. Linkage disequilibrium can be helpful in identifying a gene that causes a disease.

REFERENCES

Bateson, W., Saunders, E. R., and Punnett, R. C. 1905. Experimental studies in the physiology of heredity. Rep. Evol. Comm. R. Soc. 2:1–55, 80–99.

Botstein, D., White, R. L., Skolnick, M., and Davis, R. W. 1980. Construction of a genetic linkage map in man using restriction fragment length polymorphisms. Am. J. Hum. Genet. 32:314–331.

Conneally, P. M. and Rivas, M. L. 1980. Linkage analysis in man. Adv. Hum. Genet. 10:209–266.

Dausset, J., Cann, H., Cohen, D. et al. 1990. Centre

d'Etude du Polymorphisme Humain (CEPH): collaborative genetic mapping of the human genome. Genomics 6:575–577.

Donis-Keller, H., Green, P., Helms, C. et al. 1987. A genetic linkage map of the human genome. Cell 51:319–337.

Hartl, D. L. 1983. Human genetics. Harper and Row, New York.

Holliday, R. 1964. A mechanism for gene conversion in fungi. Genet. Res. 5:282–304.

Hulten, M. A. 1990. The topology of meiotic chiasmata prevents terminalization. Ann. Hum. Genet. 54:307–314.

Janssens, F. A. 1909. Spermatogenese dans les Batraciens. V. La theorie de la chiasmatypie. Cellule 25:387–411.

Jeffreys, A. J. 1979. DNA sequence variants in the G-gamma, A-gamma, delta and beta globin genes of man. Cell 18:1–10.

Jeffreys, A. J., Wilson, V., and Thein, S. L. 1985. Hypervariable "minisatellite" regions in human DNA. Nature 314:67–73.

Lander, E. S. and Botstein, D. 1987. Homozygosity mapping: a way to map human recessive traits with the DNA of inbred children. Science 236:1567–1570.

McKusick, V. A. 1986. The gene map of *Homo sapiens*: status and prospects. Cold Spring Harbor Symp. Quant. Biol. 51:15–27.

Meselson, M. S. and Radding, C. M. 1975. A general model for genetic recombination. Proc. Natl. Acad. Sci. USA 72:358–361.

Morgan, T. H. and Cattell, E. 1912. Data for the study of sex-linked inheritance in *Drosophila*. J. Exp. Zool. 13:79–101.

Morton, N. E. 1955. Sequential tests for the detection of linkage. Am. J. Hum. Genet. 7:277–318.

Morton, N. E. 1956. The detection and estimation of linkage between the genes for elliptocytosis and the Rh blood type. Am. J. Hum. Genet. 8:80–96.

Morton, N. E., Lindsten, J., Iselius, L., and Yee, S. 1982. Data and theory for a revised chiasma map of man. Hum. Genet. 62:266–270.

Nakamura, Y. et al. 1987. Variable number of tandem repeat (VNTR) markers for human gene mapping. Science 235:1616–1622.

NIH/CEPH Collaborative Mapping Group. 1992. A comprehensive genetic linkage map of the human genome. Science 258:67–86.

O'Connell, P., Lathrop, G. M., Nakamura, Y., et al. 1989a. Twenty-eight loci form a continuous linkage map of markers for human chromosome 1. Genomics 4:12–20.

O'Connell, P., Lathrop, G. M., Nakamura, Y. et al. 1989b. Twenty loci form a continuous linkage map of markers for human chromosome 2. Genomics 5:738–745.

Ott, J. 1985. Analysis of human genetic linkage. Johns Hopkins University Press, Baltimore, MD.

Renwick, J. H. 1971. The mapping of human chromosomes. Ann. Rev. Genet. 5:81–120.

Snyder, L. A., Freifelder, D., and Hartl, D. L. 1985. General genetics. Jones and Bartlett, Boston.

Sturtevant, A. H. 1913. The linear arrangement of six sex-linked factors in *Drosophila*, as shown by their mode of association. J. Exp. Zool. 14:43–59.

Weber, J. L. and May, P. E. 1989. Abundant class of human DNA polymorphisms which can be typed using the polymerase chain reaction. Am. J. Hum. Genet. 44:388–396.

White, R. and Lalouel, J-M. 1987. Investigation of genetic linkage in human families. Adv. Human Genetics 16:121–228.

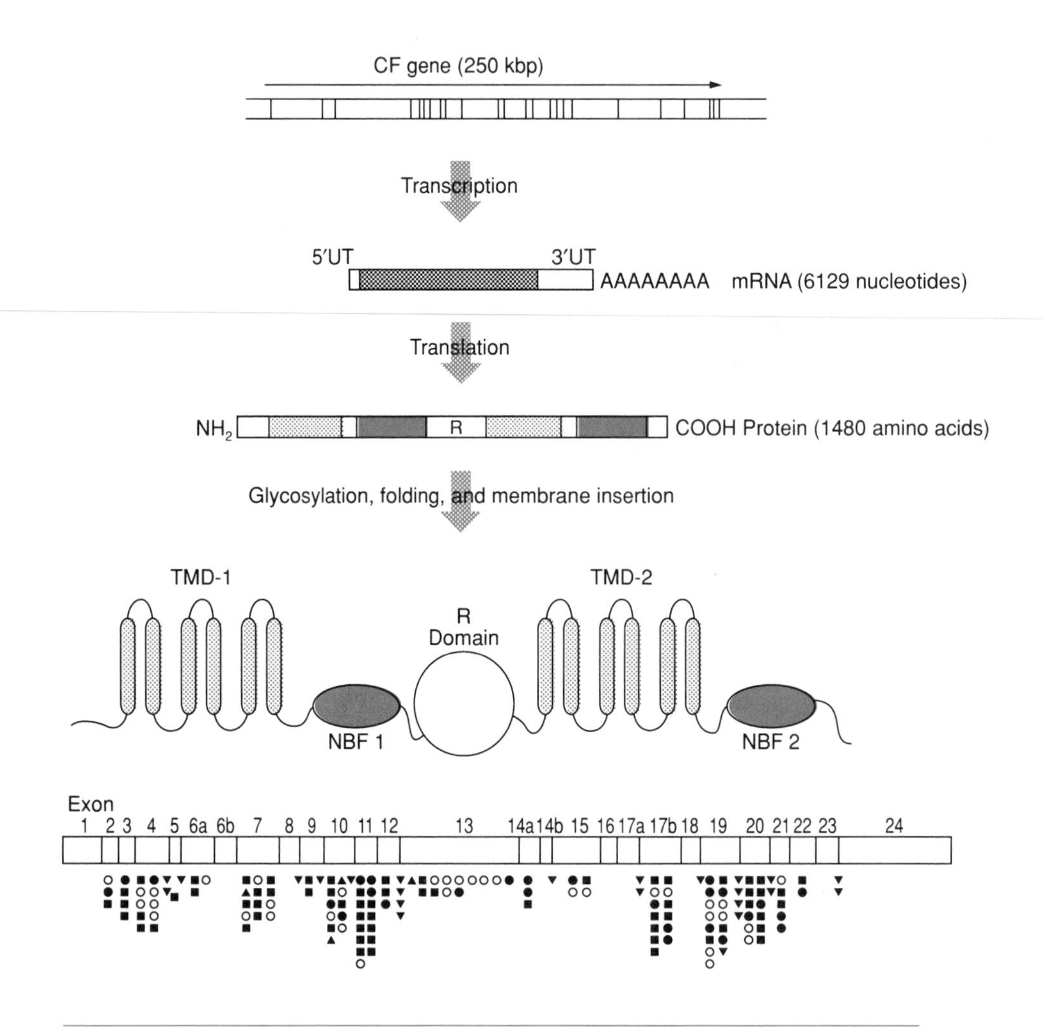

Molecular biology of cystic fibrosis. CFTR, the protein product of the *CF* gene, forms a chloride channel with complex regulation. 5′UT and 3′UT are the untranslated leader and trailer of the mRNA, TMD-1 and TMD-2 are transmembrane domains, NBF-1 and NBF-2 are nucleotide-binding folds, and R, the regulatory domain that apparently opens and closes the channel, is a substrate for protein kinase. Some of the many mutations that produce clinical symptoms are shown below the diagram: (▲) in-frame deletion; (■) missense mutation; (●) nonsense mutation; (O) frame-shift mutation; and (▼) splicing mutation. (Modified from Collins, 1992)

5 Identifying the Genetic Basis of Disease

Several thousand inherited abnormalities have been described and, doubtless, thousands more will be recognized as our understanding of human genetics matures. However, very few of the genes whose malfunction causes specific clinical syndromes have been identified. The reason is that the primary metabolic defect that causes a given genetic disease is rarely obvious. We are all familiar with some of the spectacular success stories; for example, some mutations in the genes for alpha- or beta-globins cause thalassemias, mutations in the gene for phenylalanine hydroxylase cause phenylketonuria (PKU), and mutations in the gene for HPRT cause Lesch-Nyhan syndrome. In all such cases, the patients' symptoms suggested a defect in a specific aspect of metabolism, and the investigators were able to identify the abnormal protein, using a combination of deductive reasoning and intuition. With the arrival of recombinant DNA technology, it has been possible to confirm that the genes that code for the abnormal or missing proteins actually do contain mutations. Thus, the full path from clinical manifestations to altered metabolism to abnormal protein to mutant gene has been described.

Such cases are rare, however. Most genetic diseases cause symptoms whose biochemical basis is obscure. As recombinant DNA technology developed in the late 1970s, some geneticists realized that it should soon be feasible to identify a gene responsible for a specific disease *without knowing the function of that gene*. The process by which such genes are found is the subject of this chapter. It has been called *reverse genetics* and *positional cloning;* both terms have shortcomings (see Box 5–1), so the term used here will be *positional analysis.*

There are two main stages in the experimental strategy. First, the chromosomal location of a specific disease-causing gene can be determined by the use of linkage analysis (Chapter 4); this reduces the number of gene candidates significantly, sometimes eliminating 99.9% of the genome from consideration. Second, the precise gene responsible for the disease in question can then be identified by a variety of nucleic acid techniques which are used to analyze the DNA at that chromosomal position. We shall now survey the main steps in a typical positional analysis investigation, and then we shall see how the general strategy has been applied to four specific diseases.

A Typical Positional Analysis Experimental Strategy

Map the Gene by Linkage Analysis

If a disease is not extremely rare, it is usually possible to map the locus of the corresponding gene within a few million base pairs of DNA, using methods described in Chapter 4. In that manner, approximately 99.9% of the genome may be ruled out. Nevertheless, dozens of genes may lie within the region identified by mapping, so more detailed analysis by other techniques is necessary.

Isolate DNA from the Mapped Region

This is necessary so that recombinant DNA methodologies can be used to characterize the genes in the region of interest. The precise method to be used at this stage varies from disease to disease. Sometimes, a cytologically

BOX 5-1 Forward and Reverse Genetics: Which Direction Is Which?

Since the mid-1980s, it has been fashionable to use the phrase *reverse genetics* to refer to the process by which a gene responsible for a disease is identified, when nothing is known about the specific function of the gene or the protein that it encodes (Ruddle, 1984; Orkin, 1986). After identifying the gene and showing that its mRNA is absent or abnormal in affected individuals, the protein sequence is deduced, and eventually the protein's function is determined. The process is exemplified by the four disease gene searches that are described in this chapter. It differs from the older approach (which is rarely, if ever, called "forward genetics") where investigators proceed from clinical symptoms to specific biochemical steps to showing that a protein involved in one of those steps is mutant in affected individuals. The final step is to go from information on the protein to the gene, usually via mRNA, using cDNA as a probe to identify genomic clones that contain the corresponding gene.

This comparison shows that the word "reverse" refers to the fact that the newer gene identification strategy goes from gene to protein, in contrast to the older strategy, which goes from protein to gene. But from a larger perspective, both strategies go in the same direction: *from phenotype to genotype*. Paul Berg (1990) argues that "reverse genetics" should be reserved for processes that go *from genotype to phenotype*. As an example, he refers to experiments where directed mutations are introduced into a gene, the gene is transferred to an experimental organism such as the mouse, and the effects of mutant gene expression on the organism's phenotype are determined.

It has been suggested that *positional cloning* should be used in place of "reverse genetics" (Collins, 1991). However, "positional cloning" suggests that a gene of unknown function is found simply by cloning DNA from the region to which the gene has been mapped by linkage analysis. In fact, the process is much more complicated, as explained in the text of this chapter. For that reason, the phrase *positional analysis* will be used in this text. The word "analysis" is broad enough to cover the entire series of steps that ordinarily occur between mapping a gene and proving that a specific gene in that region is actually responsible for the disease in question.

Another term describing a gene identification strategy is the *candidate gene* approach. This refers to situations where a gene that is known to have a role in the overall process that is abnormal in individuals with a specific genetic disease is examined for the presence of mutations in affected persons. Evidently, the protein encoded by the candidate gene must be known. The mutations can be identified directly at the DNA level with modern techniques; or they can be initially identified at the protein level, as was the case for many mutations in alpha- and beta-globins, which produce various anemias and thalassemias. The candidate gene strategy can be considered to be a subdivision of positional analysis.

A recent successful example of the candidate gene approach (reviewed by McKusick, 1991) is the identification of a gene responsible for most cases of *Marfan syndrome*, an autosomal dominant disorder characterized by many and variable defects in the skeleton, eye, and cardiovascular system (see Table 8-1). The gene was first mapped by linkage analysis to a region of chromosome 15q. Then attention was focused on fibrillin, a large connective tissue protein encoded by a gene in that area. Fibrillin is an important component of extracellular microfibrils, which participate in elastic fiber formation. A defect in elastic fibers would be consistent with the symptoms of Marfan syndrome.

Polymorphic marker loci were found that showed zero recombination with the Marfan disease locus. Then, single base changes that caused a missense mutation (proline replaces arginine) in the fibrillin protein were found in two unrelated patients. No doubt other mutations in the same gene will soon be found to be associated with clinical manifestations. The full story will be complicated, because fibrillin appears to be encoded by a small family of genes, at least some of which are located on other chromosomes. Mutations in the other fibrillin genes may be responsible for several other diseases that affect connective tissue, producing symptoms that partially overlap those of Marfan syndrome.

Each of the names for gene identification strategies described here has its partisans, but whether any of them will be universally adopted remains to be seen. Whatever the techniques may be called, the identification of genes whose abnormal function underlies human genetic diseases has become a powerful and productive area of research.

REFERENCES

Berg, P. 1991. Reverse genetics: its origins and prospects. Bio/Technology 9:342–344.

Collins, F. S. 1991. Identification of disease genes: recent successes. Hospital Practice 26, No. 10:93–98.

McKusick, V. A. 1991. The defect in Marfan syndrome. Nature 352:279–281.

Orkin, S. N. 1986. Reverse genetics and human disease. Cell 47:845–850.

Ruddle, F. H. 1984. The William Allen memorial award address: reverse genetics and beyond. Am. J. Hum. Genet. 36:944–953.

recognizable deletion is associated with the disease in at least some patients. In such cases, it is possible to isolate DNA from normal persons which corresponds to the deletion in affected persons, as will be explained later for Duchenne muscular dystrophy.

If there is not an obvious disease-associated deletion, then it may be beneficial to create a rodent-human hybrid cell line containing as the sole human component only a portion of the chromosome where the gene is known to reside. Alternatively, it may be possible to isolate a region of DNA extending over no more than a few megabases, using pulsed-field gel electrophoresis. In any case, there comes a time when a certain amount of "brute-force" work is necessary to isolate a set of overlapping clones, each containing 20–40 kb of DNA, that collectively span the region where the disease-causing gene is known to occur.

Identify Conserved Sequences

As is true for mammals in general, the vast majority of human DNA does not code for proteins; rather, it is either intron DNA or intergenic DNA (Chapter 2). Evolutionary constraints have been much more severe on protein-coding DNA than on noncoding DNA. Accordingly, it is possible to identify protein-coding sequences by doing an evolutionary comparison. One treats a variety of non-human DNAs with a restriction enzyme, separates the fragments on a gel by electrophoresis, transfers the fragments to a membrane suitable for DNA-DNA hybridization, and probes the membrane with a radioactive cloned human DNA from the region identified in the mapping and isolation steps described above. This is often referred to as a "zoo blot."

If the probe contains coding DNA, it will usually react with several of the non-human DNAs on the zoo blot (an example is shown in Figure 5–9); if it does not contain any coding sequences, it will usually hybridize only to human DNA. By the sequential examination of all the clones covering the region of DNA in question, it is possible to identify all or most of the clones that contain protein-coding sequences. This further reduces the amount of DNA that needs to be studied more extensively.

Identify mRNAs Transcribed from Each Candidate Gene

Typically, a genetic disease will primarily affect a limited number of tissues in the patient; for example, skeletal muscles in Duchenne muscular dystrophy and exocrine secretory cells in cystic fibrosis. In a positional analysis investigation, therefore, it is important to show that a candidate gene is expressed in normal persons in those tissues that are primarily affected in patients with the disease. If a protein-coding sequence identified with a zoo blot is not expressed as mRNA in appropriate tissues, that gene may be deleted from the list of candidates.

Detection of mRNAs is ordinarily accomplished with "Northern blots." RNA from appropriate normal tissues is separated according to size by electrophoresis on an agarose gel, transferred to a membrane, and hybridized to a radioactive DNA probe prepared from a cloned candidate gene (an example is shown in Figure 5–7). When a probe detects an mRNA on a Northern blot, one learns (a) that the gene is expressed in tissues that can be affected by the disease and (b) the size of the normal mRNA. The latter information is important in the next step.

Correlate Abnormal mRNAs with Presence of the Disease

Northern blots are then made from mRNAs prepared from tissues of persons affected with the disease, together with controls consisting of mRNAs from normal persons. When these blots are probed with a candidate gene clone, the absence of mRNA or the presence of mRNA of abnormal size in some patients is strong evidence that the gene responsible for the disease is represented by the probe (Figure 5-7). It is not necessary that all affected persons have abnormal mRNAs; many mutations that lead to functionless proteins do not affect the synthesis or size of the mRNA. However, it is a general rule that a given gene can be inactivated in many ways (i.e., multiple abnormal alleles exist; see Chapter 7). Therefore, it is to be expected that if a particular probe really represents the gene responsible for the disease, it will reveal some abnormality in the mRNA from some fraction of patients with the disease.

Correlate DNA Deletions with Presence of the Disease

An alternative approach to the identification of a gene responsible for a genetic disease is to correlate the occurrence of submicroscopic DNA deletions with the presence of the disease. Deletions that are too small to be detected cytologically are a common cause of mutant alleles; they have been extensively documented for the thalassemias, for Lesch-Nyhan syndrome, for Duchenne muscular dystrophy, and for several other diseases that have been studied at the DNA level.

In order to detect DNA deletions, one uses the same set of candidate genes identified by the previous steps to probe Southern blots of DNA from affected and normal persons. A deletion will be indicated by a change in the size of one or more restriction fragments in the DNA from an affected person, compared to normal controls. Most patients will not have deletions, but if DNAs from several dozen patients are available, it would not be unusual to find that 5–10% of them contain deletions in the DNA of the gene responsible for the disease. In some diseases, notably Duchenne muscular dystrophy, the fraction of deletion-bearing patients may be much higher (greater than 50%); in others, such as cystic fibrosis, deletions large enough to be detected on Southern blots are virtually nonexistent.

Evidently, a DNA deletion may also alter the size of the RNA transcribed from the mutant gene, or it may abolish the production of mRNA altogether. But this is not an obligatory effect, and demonstrating that a candidate gene contains deletions in at least some patients, but not in normal controls, can be an important aspect of a positional analysis investigation.

Identify Mutations in the Candidate Gene from Affected Persons

When a gene has been identified by the preceding steps, it is necessary to show that there are mutations in that gene from persons with the disease. The first step is to obtain the sequence of the normal mRNA, by cloning cDNA from normal donors and sequencing it. Then, one compares the sequence of cloned cDNA from affected persons, if it is available. If the disease-causing mutations abolish the synthesis of detectable mRNA, it may be necessary to clone the entire gene and look for differences between normal and affected persons by a variety of methods.

One of the difficulties that arises at this stage is the inevitable occurrence of polymorphisms. The nucleotide sequence of any two individuals may vary at one or more places within a specific gene, but those differences may not have functional consequences. Proving that a difference will lead to loss of protein function is not always easy, but it is a necessary step in definitively identifying the nucleotide changes responsible for a specific disease.

Positional Analysis and Specific Diseases

We shall now consider some outstanding examples of the success of the positional analysis strategy. Other examples exist (see the section on retinoblastoma in Chapter 11), and it is a safe assumption that the genes responsible for many other inherited diseases will be identified in the near future, but these examples will illustrate most of the principles involved.

Huntington Disease

The first major success in mapping an autosomal gene of unknown function involved Huntington disease. Recognized by its clinical symptoms since the late 19th century, Huntington disease is a classical autosomal dominant condition; that is, in matings between a person with one mutant allele and a normal person, half of the children, on the average, will be affected with the disease. A remarkable aspect of this particular disease is that symptoms do not usually develop until the fourth or fifth decade of life, although cases of onset as early as age 6 and as late as age 80 have been reported (Figure 5–1).

The earliest symptoms are involuntary movements, usually twitching of the limbs or head, or facial tics. With the passage of time, these movements become more vigorous and more frequent, so that the affected person is more-or-less always moving; this accounts for the alternate name, Huntington's Chorea (a dance). Over the course of one or two decades neurological degeneration worsens progressively; the loss of muscular control leads to confinement in bed and ultimately to total helplessness. The patient's personality also deteriorates, passing through irritability to depression to dementia. Eventually, the weakened body can no longer support life. There is no cure and no treatment. The biochemistry of the disease is not understood, although the symptoms are evidently related to premature neuronal death, especially in the basal ganglia.

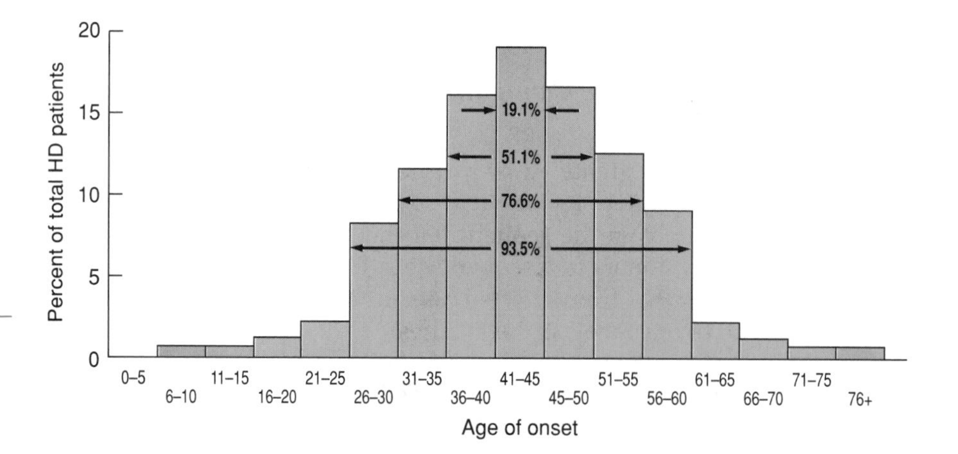

Figure 5–1

Distribution of ages at onset in 802 cases of Huntington disease. (From Vogel and Motulsky, 1979, p. 85)

Huntington disease (HD) was an excellent candidate for a reverse genetics study because of the large number of informative families available. Geneticists have been fascinated by the disease for many years, no doubt because of its being a dominant condition with 100% penetrance, its puzzling late onset, and its inexorable course. Families of affected persons have often been cooperative with geneticists, and several large pedigrees are known. In the United States, a roster of families with HD has been kept for some years by the Hereditary Disease Foundation. Some of these families include several dozen informative individuals.

However, the most extraordinary genetic resource is a Venezuelan kindred living in the vicinity of Lake Maracaibo. There are now about 3000 people descended from a woman who developed HD in the early 1800s. In all likelihood, she received the mutant gene from a father of European origin, possibly a British sailor. Among the living descendants of that woman are at least 100 HD patients and 1100 young people who are at risk of developing the disease. Families with 15 children are common; the situation is ideal for genetic studies. A small portion of the Venezuelan pedigree is shown in Figure 5-2.

James Gusella and his colleagues at Harvard Medical School used linkage analysis (Chapter 4) on DNA from these families to study the relationship between randomly isolated RFLPs and the presence or absence of HD. Other investigators had done similar studies on more than 30 polymorphic enzymes and other proteins, but had failed to find evidence of linkage to HD. Those studies, however, only excluded about 20% of the genome as the locus of the HD gene. In principle, one might have to test 100–200 marker loci, with an average spacing of 20 cM, before finding a clear case of linkage to any disease gene, but Gusella and his group were amazingly fortunate. After examining only 12 probes, they found that one of them, called G8, was unambiguously linked to HD (Gusella et al., 1983). The lod score reported was approximately 8; that is, the odds favoring linkage over non-linkage were 100,000,000 to one! Subsequent analyses of a much larger number of individuals raised the odds favoring linkage to a truly astronomical level (Figure 5-3).

Initially, the chromosomal locus detected by probe G8 was not known, but somatic cell genetic analysis, using a panel of rodent-human hybrid cells as described in Chapter 3, quickly assigned it to chromosome 4. That locus, now designated formally as D4S10, was subsequently localized to the tip of the short arm of chromosome 4 (Figure 5-4), both by *in situ* hybridization and by the discovery that it is missing from the DNA of patients with Wolf-Hirschorn syndrome, who have a visible deletion of the tip of 4p.

D4S10 is a complex RFLP locus; it consists of two closely linked polymorphic restriction enzyme cleavage sites. Thus, there are four haplotypes, which can be considered as alleles of a single locus, as explained in Chapter 4 (Figure 5-5). Recombinants between D4S10 and the HD gene occur in about 4% of meioses, suggesting that the marker locus is roughly 4×10^6 bp (Chapter 4) distant from the disease gene. Intensive efforts to find closer markers have had some success, in that RFLPs between the HD locus and D4S10 have been found; but so far, no polymorphic marker on the telomeric side of the HD locus has been identified. It is possible that the HD locus is the closest gene to the 4p telomere. This lack of flanking markers has annoying consequences when predictive genetic screening is attempted (Chapter 9).

At this writing (May, 1992), the HD gene has not yet been identified. One of the complicating factors surely must be the difficulty of obtaining

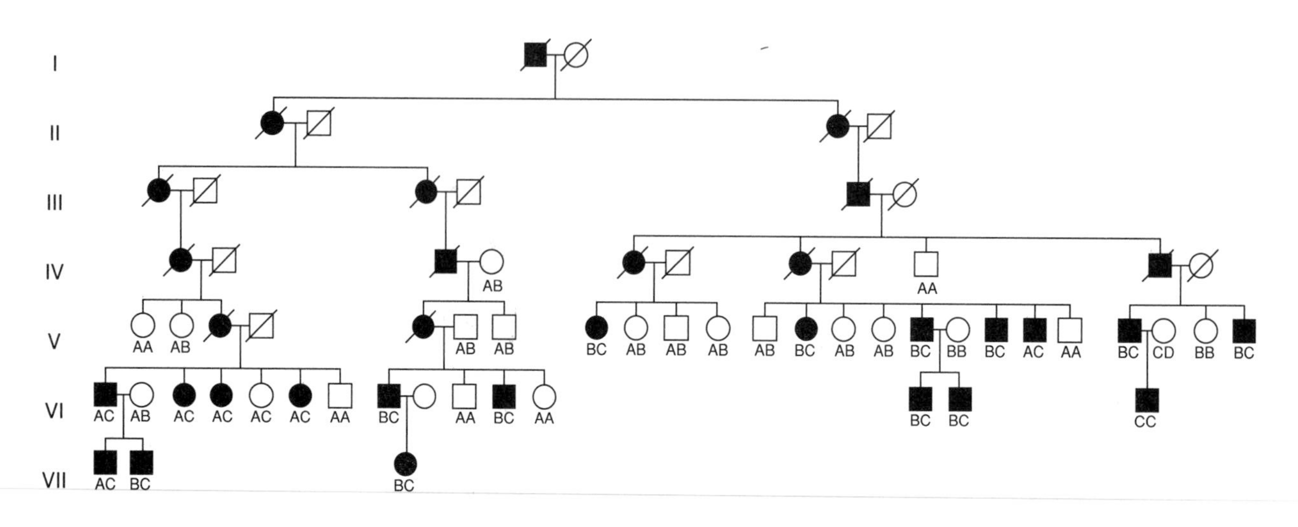

Figure 5–2

Partial pedigree of the Venezuelan Huntington disease family. Lymphoblastoid cell lines were established from blood samples from these individuals. The letters below the circles and squares are the haplotypes determined at the D4S10 locus by analysis of restriction enzyme-digested DNA with the G8 probe. It is clear that haplotype C segregates with the disease in this family (affected individuals are indicated by the filled symbols). Deceased individuals are indicated by a symbol with a slash through it. (Reprinted from Gusella et al., 1983, p. 236; © Macmillan Magazines Ltd.)

mRNAs from either normal or affected persons; brain biopsies, after all, are out of the question. Even autopsy material would be of doubtful value, because the long course of the disease would probably lead to numerous secondary effects on the metabolism of basal ganglia neurons. However, the region in which the HD gene must lie has been narrowed to at most 2.5 Mbp, and several candidate genes that are expressed in appropriate neurons are under investigation.

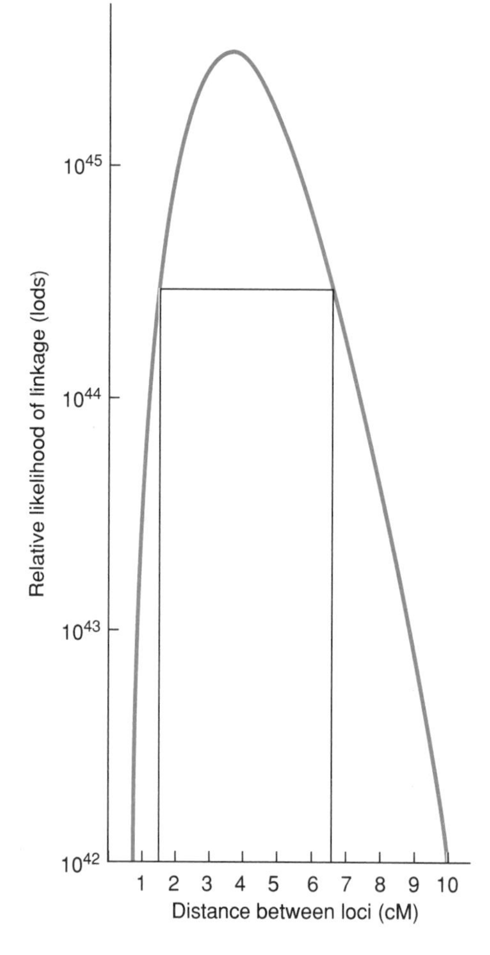

Figure 5–3

Linkage of locus D4S10 to Huntington disease. Computer analysis was performed on linkage data from a large number of individuals in families with HD, and lod scores were plotted as a function of distance between the disease locus and the marker locus. The best estimate for the genetic distance between the two loci is given by the peak of the curve. (From Gusella et al., 1986, p. 361)

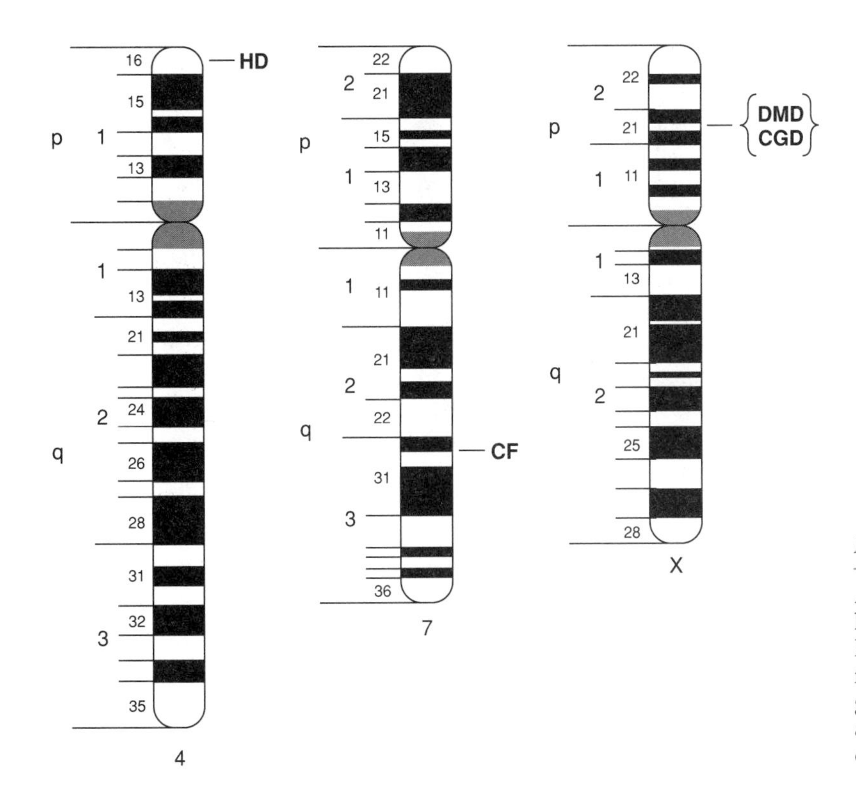

Figure 5–4

Approximate chromosomal location of the genes for Huntington disease (HD), cystic fibrosis (CF), X-linked chronic granulomatous disease (CGD), and Duchenne muscular dystrophy (DMD).

Chronic Granulomatous Disease (CGD)

One of the body's chief lines of defense against invading microorganisms is provided by phagocytic cells (neutrophils, macrophages, and eosinophils) that engulf and then inactivate the microbes. The ability to carry out this essential function can be abolished by several single-gene disorders, which are collectively known as *chronic granulomatous disease (CGD)*. In two-thirds of affected persons the disorder is caused by an X-linked recessive gene; in the remaining one-third the disorder is caused by autosomal recessive mutations. Patients suffer from recurrent, severe bacterial or fungal infections; sites of chronic inflammation produce the granulomas to which the name of the disease refers.

Phagocytes from patients with X-linked CGD (X-CGD) are defective in the production of superoxide, which is produced by an oxidase associated with the plasma membrane, via the reaction: $NADPH + 2O_2 \rightarrow NADP^+ + 2O_2^- + H^+$.

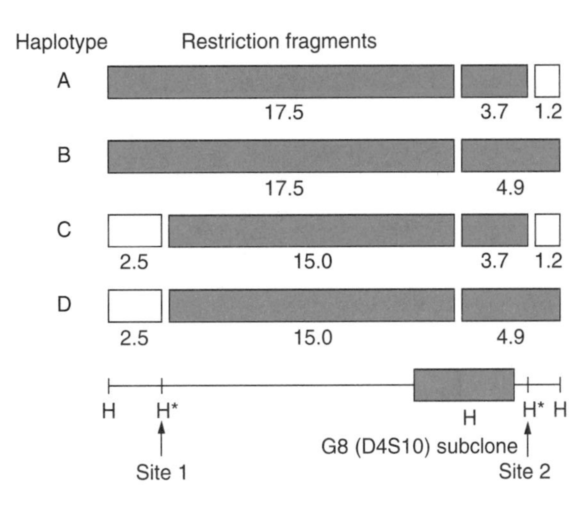

Figure 5–5

Haplotypes at the D4S10 locus, as defined by the G8 probe. There are two polymorphic HindIII sites at this locus, with two alleles at each site. The combination of alleles at these sites constitutes a haplotype; there are four haplotypes, indicated by the letters A, B, C, and D. The part of the locus that corresponds to the G8 probe is shown in the bottom line, and the HindIII fragments that are detected by the probe are shown as shaded in the upper four bar diagrams. The unshaded fragments are not detected by the probe. H = a HindIII cleavage site. Sizes of fragments are given in kbp. (From Jenkins and Conneally, 1989, p. 171; by permission of University of Chicago Press)

Apparently the superoxide radical is necessary for the killing action of the phagocytes.

Prior to the investigation of Orkin and his colleagues (Royer-Pokora et al., 1986; Orkin, 1987), the gene inactivated by the X-linked CGD mutations was not known, but from the work of others, it was suspected that a b-cytochrome was involved, because the absorption spectrum of this cytochrome is absent from the phagocytes of patients with that disease. The story of how Orkin and co-workers began with this information and progressed to the isolation of the X-CGD gene, and then to the identification of the encoded protein, represents the first complete success of the positional analysis strategy and illutrates the full power of the method.

Linkage analysis of the relationship between the X-CGD gene and a variety of RFLPs from the X chromosome implied that the gene is located in Xp21 (Figure 5-4). This conclusion was strengthened by the existence of a boy with a cytologically visible deletion of Xp21; the boy was simultaneously affected with X-CGD, Duchenne muscular dystrophy (DMD), and two other genetic defects. Inasmuch as a collection of cloned segments of DNA from Xp21 had already been produced in the search for the DMD gene (as will be detailed presently), the task of finding the X-CGD gene was enormously simplified. In essence, the question to be answered was whether any of the available Xp21 clones represented the X-CGD gene.

The first step was to identify mRNAs from Xp21 which are made in appropriate cells. To do this, Orkin's group carried out a "subtractive hybridization" (Figure 5-6). First, they prepared total mRNA from HL60 cells—a human leukocyte line that has been transformed so that it grows indefinitely in culture and thus can provide as much material as the investigation may require. HL60 cells can be induced to express the phagocyte-specific oxidase system that is faulty in X-CGD patients. Next, they prepared mRNA from cultured white blood cells that had originated from the patient with the Xp21 deletion.

Now, they were able to determine which mRNAs were present in HL60 cells and were absent from the Xp21-deletion patient. This was achieved by making cDNAs from the HL60 mRNA and hybridizing them to an excess of mRNA from the deletion patient. At that point, all the hybridized cDNAs were removed from the preparation, thereby subtracting most of the mRNA sequences that were present in both preparations. The remaining, unhybridized cDNAs were significantly enriched for sequences that originated from Xp21 genes.

The next question was, did any of the Xp21 genomic clones (obtained by the Kunkel lab as a byproduct of the search for the DMD gene) contain sequences that were present in the enriched cDNA preparation described here? To answer this question, it was first necessary to make a radioactive probe from the enriched cDNAs by copying them with DNA polymerase in the presence of radioactive DNA precursors. Then the probe was hybridized to Southern blots of restriction enzyme-fragmented DNA from each of the Xp21 clones. If the subsequent autoradiogram revealed a radioactive band, then the clone from which that DNA fragment originated would be a candidate for the X-CGD gene.

Two overlapping subclones from pERT 379 (a plasmid containing an insert from Xp21) gave positive results. The Orkin group then asked, what size mRNA from white blood cells is produced by the gene from which the pERT 379 clone originated? The mRNA from HL60 cells was fractionated according to size by electrophoresis, a Northern blot was prepared, and a radioactive probe prepared from clone pERT 379 was used to scan the blot.

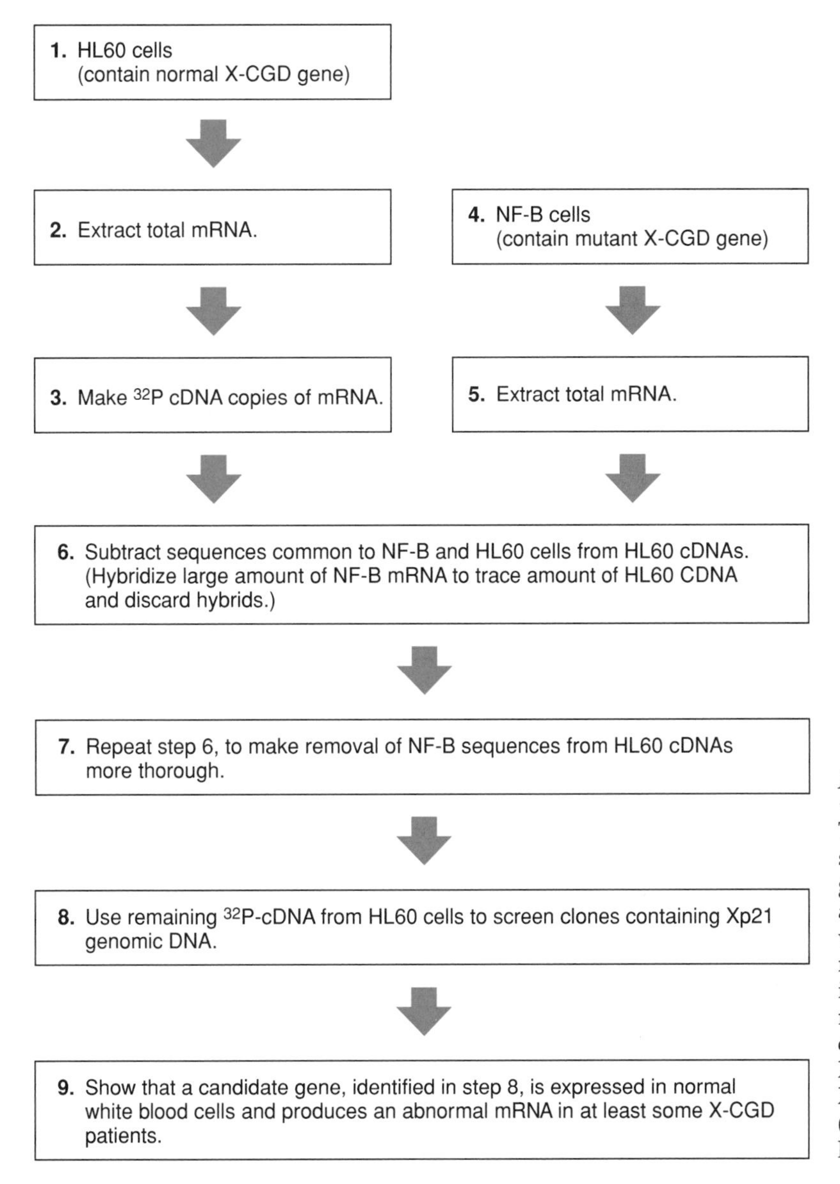

Figure 5–6

The subtractive hybridization scheme used to identify the gene for X-linked CGD. HL60 is a human leukemic cell line; it was a convenient source of mRNA that was expected to include the transcript of the normal X-CGD gene. NF-B cells came from a patient with X-CGD and a deletion of region Xp21. See text for explanation. (Based on data from Royer-Pokora et al., 1986)

A 5kb RNA was detected in HL60 cell RNA, but not in mRNA from fibroblasts, kidney, or liver. It is present in normal granulocytes and monocytes, where it amounts to about 0.1% of total mRNA. All of these properties suggested that the 5kb mRNA would be a good candidate for the product of the X-CGD gene.

Northern blots of mRNA from four X-CGD patients' phagocytes were probed with pERT 379; three were completely negative for the 5kb RNA that is present in normal persons (Figure 5–7). The fourth patient had a 5kb RNA that hybridized to the probe, but more detailed analysis showed that this RNA had a small internal deletion. These results almost guaranteed that the X-CGD gene had been found.

Subsequent studies led to isolation of complete cDNA clones, which, when sequenced, gave a predicted protein of 486 amino acids. The sequence also had five potential glycosylation sites, which was consistent with the expectation that the protein is associated with cell membranes. The sequence of this protein was not homologous to any known protein.

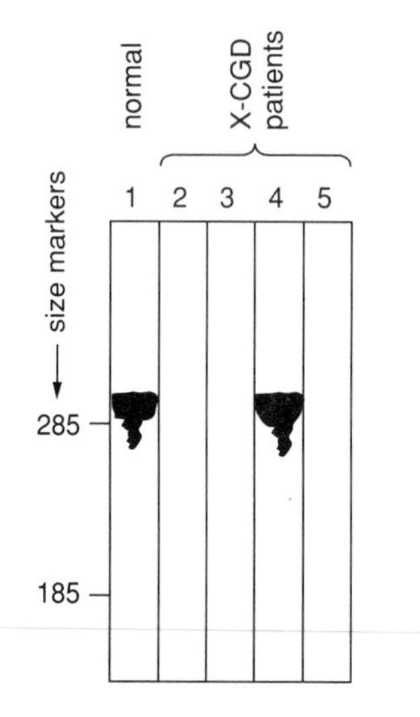

Figure 5–7

Diagrammatic representation of Northern blots of total cell RNA from white blood cells of a normal person (lane 1) and four CGD patients (lanes 2–5), probed with clone 379. Three patients had no RNA detectable with the probe (lanes 2,3,5); one patient (lane 4) had an RNA approximately the same size as in normal persons, but it was later shown to contain a small deletion. Size markers were ribosomal RNAs, indicated at the left edge of the diagram: 28S rRNA contains about 4700 nucleotides; 18S rRNA has nearly 1900 nucleotides. (Based on data from Royer-Pokora et al., 1986)

Concurrently, the b-cytochrome complex of white blood cells was isolated in another laboratory. It contains a 22kD polypeptide (presumably the apo-cytochrome) and a 90kD polypeptide that is probably the product of the gene found by the Orkin group. This polypeptide is approximately 50kD in size when de-glycosylated, which is consistent with the 486 amino acids predicted by the cDNA sequence. Moreover, immunological assays show that the 90kD polypeptide is absent from neutrophils of all X-CGD patients thus far examined. The 22kD polypeptide is also absent; presumably it is unstable when the 90kD component of the cytochrome complex is unavailable.

The genes that are defective in autosomal CGD have not yet been identified, but two laboratories have recently described a cytosolic GTP-binding complex that has NADPH oxidase-activating capacity and is abnormal in autosomal-CGD patients. There are at least two polypeptide components; most autosomal-CGD patients lack the 47kD polypeptide, but some lack a 67kD polypeptide. A third component is needed for oxidase activity in cell-free preparations, but patients deficient in this component have not yet been found.

Duchenne Muscular Dystrophy (DMD)

Duchenne Muscular Dystrophy is our most serious X-linked recessive genetic disease. It occurs in approximately one of every 3500 male children. Patients are essentially symptom-free for the first two to four years of life, but then they begin to exhibit weakness of the skeletal muscles. Their condition progressively worsens and by age 10 to 12, they are confined to wheelchairs. Death from respiratory failure usually occurs in the late teens or early twenties. There is no cure and there is virtually no treatment, other than making them as comfortable as possible.

At the cellular level, Duchenne muscular dystrophy is characterized by continual breakdown of skeletal muscle cells. One of the diagnostic criteria is the release of muscle-specific enzymes, especially creatine kinase, into the blood. In normal muscles, there is also a small amount of myocyte degeneration, sometimes caused by extreme exertion, but doubtless also caused by the stresses and strains of ordinary body use. There is a population of undifferentiated cells, known as "satellite cells" dispersed throughout normal muscles. In response to local myocyte death, these satellite cells differentiate to form myoblasts, which fuse as usual to form myotubes, thereby repairing the damage. In normal people the supply of satellite cells is sufficient for an average lifetime, but in DMD patients the population of satellite cells is exhausted within the first few years of life; after that, the myocytes are no longer replenished and the child begins to show symptoms.

Identifying the DMD Gene by Deletion Cloning In a *tour de force* of positional analysis, Kunkel (1985) and his colleagues at Harvard Medical School demonstrated how one could employ deletions to clone DNA that is not there (Monaco et al., 1986; Koenig et al., 1987). The key to their research was a boy with a cytologically visible deletion in the short arm of the X chromosome at Xp21. This boy was simultaneously affected with DMD, retinitis pigmentosa, and X-CGD. Linkage studies had also implicated Xp21 as the locus of the DMD gene.

It may sound strange, but it is possible to clone the DNA that is missing from a person with a deletion. The basic concept is to take normal DNA, remove everything that it has in common with DNA from the deletion-carrying person's DNA, and whatever is left represents DNA that is missing from the patient. This is a straightforward extrapolation of an idea that was earlier applied to the isolation of DNA specific to the Y chromosome; that is, if you take male DNA and remove all the DNA that it shares with females of the same species, you are left with Y chromosome DNA. The technique is analogous to subtractive hybridization of RNAs, described previously.

Kunkel and coworkers prepared DNA from the boy with the Xp21 deletion and DNA from a normal male. They mixed a 500-fold excess of the deletion patient's DNA with some of the normal DNA, denatured them, and let them renature. In such a case, most of the normal DNA fragments will reassociate with DNA from the deletion patient, simply because the latter are more abundant. However, those sequences that are absent from the deletion patient but present in normal DNA will only have each other with which to hybridize.

The next problem is: how do you distinguish one category of renatured DNA from the others? The trick is as follows: the deletion patient's DNA is broken by shearing, so that it will have random ends; the normal person's DNA is cut with a restriction enzyme (Mbo I in this case), so that it has uniform, defined ends. After denaturation and renaturation, three kinds of double-stranded molecules will exist, as shown in Figure 5–8.

Only the renatured DNAs that have Mbo I ends on both strands can be ligated into a plasmid with compatible ends; the DNAs with one or two sheared-end strands will not fit, because base pairing at the ends of insert DNAs and vector DNAs is required in order for DNA ligase to link the insert to the vector. Kunkel et al. actually used a vector cleaved with Bam H1, which cuts at G'GATCC, whereas Mbo I cuts at 'GATC. Mbo I was used on the normal human DNA because they wanted to cut it into small pieces; Mbo I was not used on the plasmid vector because it would cut the plasmid in too many places.

One other special technical feature of this procedure must be explained. In Kunkel's experiment, the ratio of deletion DNA to normal DNA was 500:1, and this meant that the concentration of the normal DNA fragments would be so low that complementary strands had little chance of finding each other within an acceptable time. In order to increase the rate of

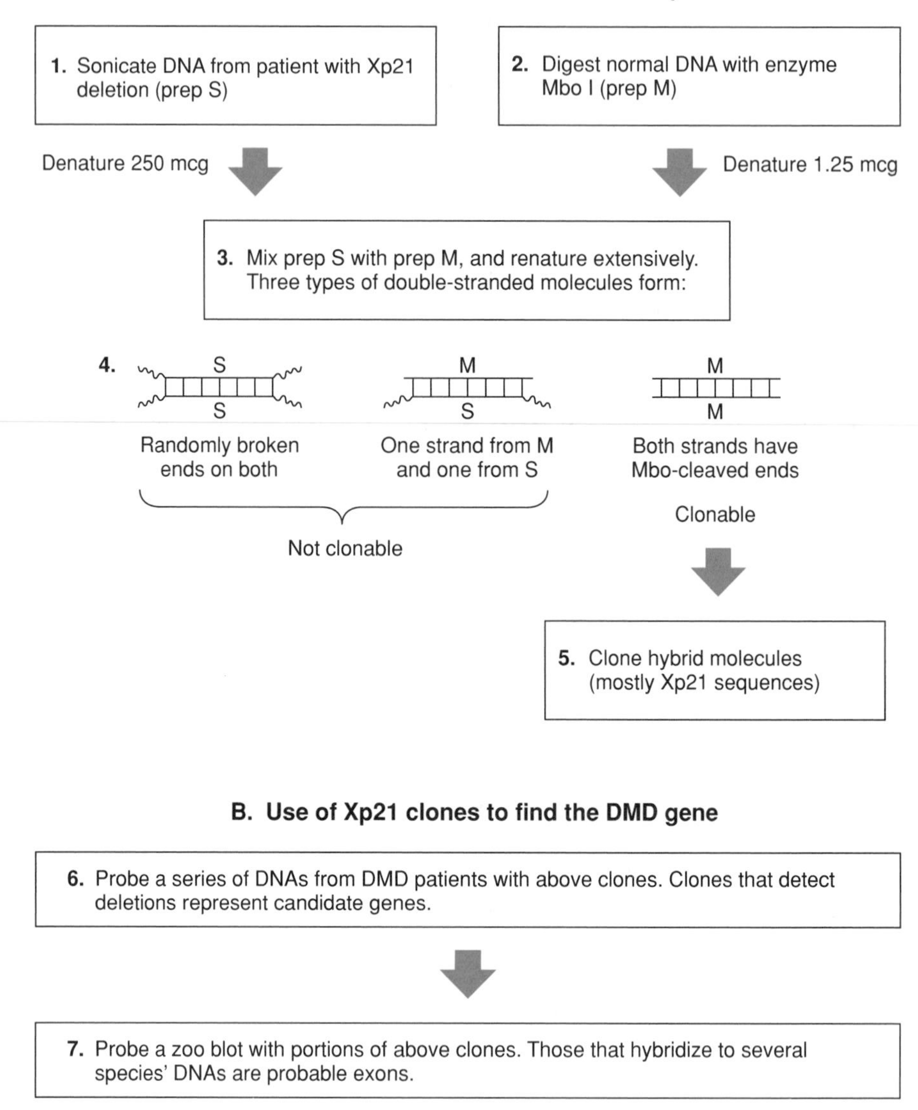

Figure 5–8

Highlights of the positional analysis strategy employed by Kunkel and co-workers to identify the gene responsible for Duchenne muscular dystrophy. (Based on data from Kunkel et al. (1985) and Monaco et al. (1986))

reassociation, they used a phenol-enhanced reassociation technique (PERT). By adding phenol to the reassociation mixture and making an emulsion, the volume of aqueous solution available to the DNA molecules is reduced drastically, and the rate of reassociation of complementary strands is increased several thousand-fold. This allows complementary strands to find one another before they are degraded by the high temperature necessary for the renaturation reaction.

After the renaturation process was complete, the phenol was removed, the DNA was concentrated and ligated into a Bam-cleaved plasmid, bacteria were transformed, and recombinant clones were isolated. DNAs from a series of these clones were then hybridized to Southern blots of DNA from a variety of sources, including normal males, normal females, triple-X females, and of course, the boy with the Xp21 deletion. Those probes that hybridized to normal X chromosomes and did not hybridize to DNA from the Xp21 deletion boy were putative representatives of Xp21 DNA and were candidates for pieces of the DMD gene. Four such clones were isolated.

To further characterize their pERT clones, the Kunkel group reasoned that many DMD patients may have deletions in the DMD gene, even though most of these would be too small to detect cytologically. Therefore, they used each of the four clones identified previously to screen a panel of DNAs from 57 DMD boys. They found one (pERT 87) that failed to hybridize to five of them, implying that those five boys suffered from DMD as a result of a deletion involving the region corresponding to pERT 87. This probe was then used for chromosomal walks in both directions, expanding the locus to about 200 kb. The locus recognized by the pERT 87 probe was given the name DXS164.

The next step was to find the messenger RNA transcribed from the DXS164 region. Inasmuch as most of the DNA in any gene is likely to be from introns, the investigators made a series of subclones from DXS164 (from a normal person, of course), and used them to screen zoo blots, as described in the introduction to this chapter (Figure 5-9). Two were found that hybridized to all tested mammalian DNAs at high stringency, suggesting that these probes contained exon sequences that had been strongly conserved during evolution. The use of these probes with Northern blots of mRNA from various human tissues identified a messenger RNA about 14 kb in size, which subsequently was shown to code for the protein that is defective in DMD.

Finding the DMD Gene with a Translocation Junction The fascinating story of the identification of the gene responsible for Duchenne muscular dystrophy would be incomplete without a description of an independent

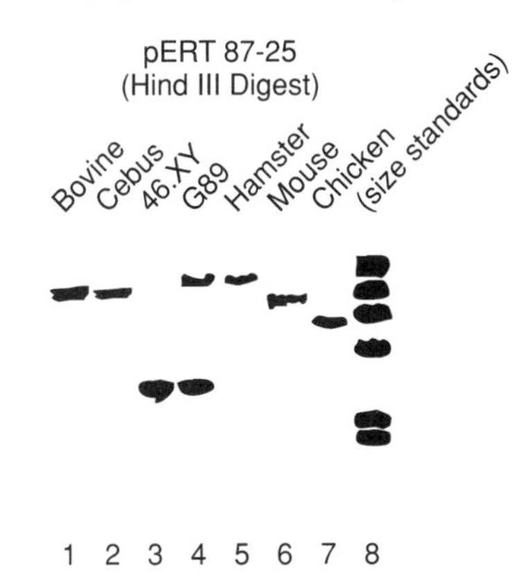

Figure 5–9

A zoo blot of DNAs digested with HindIII, electrophoresed in an agarose gel, transferred to a membrane, and probed with pERT 87-25, a clone from human chromosome Xp21. The lane labeled Cebus is from a monkey, 46,XY indicates a normal human male, and G89 is a hamster-human hybrid cell line whose sole human component is the X chromosome. The standards in the far right lane are size markers derived from bacteriophage lambda. The fact that pERT 87-25 hybridizes to all tested DNAs implies that it may be from a portion of a gene that encodes a conserved amino acid sequence. The variable sizes of the fragments detected by the probe presumably result from evolutionary drift of nonessential sequences. (Reprinted from Monaco et al., 1986, p. 647; by permission; © Macmillan Magazines Ltd.)

approach that was successfully pursued by Worton (1984) and his colleagues (Ray et al., 1985; Bodrug et al., 1987). They made use of DNA from one of seven known girls affected with DMD. It would be natural to assume that girls affected with an X-linked recessive disease are homozygous for the mutant gene, but in fact, none of these girls had an affected father. Each of them had a translocation involving the short arm of the X chromosome and an autosome; the breaks in the autosomes were all different, but the breaks in the X chromosomes all involved band Xp21. In the affected girls, who also had one normal X chromosome, the normal X was inactivated in all of their cells. (The reason why this occurs, instead of the usual random X inactivation described in Chapter 13, is not known.) The translocated X was the only active X chromosome, so if the translocation interrupted the DMD gene, the fact that the girls expressed the disease phenotype would be understandable.

One of the DMD girls had an X;21 translocation, with the break in chromosome 21 in the ribosomal DNA cluster on the short arm (Figure 5–10). In theory, if one could isolate a cloned DNA that included the X/21 junction, a probe for the DMD gene could be obtained. (There are two 21s in this story; do not confuse chromosome 21 with band p21 in the X chromosome).

Worton and his group began by making human-mouse hybrids with cells from the X;21 DMD girl as the source of human chromosomes. The hybrid cells were grown until some clones with no human component except the der(X) were obtained (i.e., the chromosome with the long arm, centromere, and proximal short arm of the X chromosome attached to the distal part of the short arm of chromosome 21). This was necessary to remove other human chromosomes that carry ribosomal DNA (numbers 13, 14, 15, and 22), which would interfere with subsequent assays for the junction fragment.

How could the X/21 junction be recognized? It should contain a restriction fragment that is not found in the normal cluster of genes for rRNA. Worton's group eventually found such a fragment, using a probe derived from the 3′ end of the 28S rRNA gene and DNA cleaved with the restriction enzyme BamH I (Figure 5–11). The next step was to obtain a clone of the junction fragment. A library was made from total genomic DNA from the hybrid cells that contained the der(X) chromosome, and this library was screened with the probe from the 3′ end of the 28S rRNA gene. Every clone that reacted with the probe was then subjected to Southern blot analysis after the DNA was cleaved with BamH I, and a clone that had the diagnos-

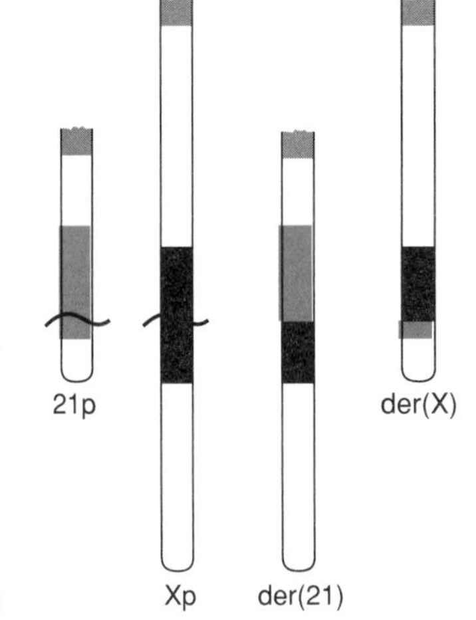

Figure 5–10

Schematic of the short arms of chromosome 21, the X chromosome, and the two translocation products, der(21) and der(X). Centromeres are at the top of each diagram. (Based on data from Worton et al., 1984)

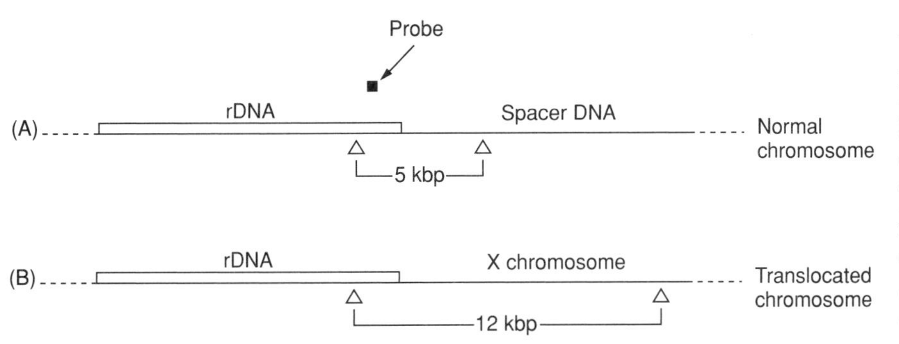

Figure 5–11

Identification of the X;21 junction fragment. (A) Portion of a ribosomal RNA gene cluster (rDNA) from a normal chromosome 21. (B) The junction fragment from a girl with an X;21 translocation. Triangles show the points of cleavage with the restriction enzyme BamH I. A fragment about 5 kbp long is produced from the normal chromosome 21, whereas a fragment 12 kbp long comes from the der(X) chromosome. These fragments were detected with the probe 100-3, which detects a small portion of the rDNA repeat unit near its 3′ end (indicated by the small, solid box above panel (A)). (Diagram based on data from Ray et al., 1985)

tic 12 kb fragment was identified. Presumably, this represented the junction fragment.

The junction fragment clone (now called Xj) was studied further. When used as a probe, it detected a 15 kb fragment from BamH I-digested DNA from normal cells, a 12 kb fragment from the hybrid cells with the der(X) chromosome, and both 15 kb and 12 kb fragments from the DNA of the DMD girl with the X;21 translocation. Linkage analysis showed that the locus defined by Xj was either identical to, or very close to, the DMD locus. Finally, the presumed junction fragment clone failed to hybridize to DNA from six out of 107 DMD boys, suggesting that the corresponding region was deleted in those boys. More detailed analyses carried out later have thoroughly confirmed that the Xj clone came from the DMD gene.

The DMD Gene and its Product, Dystrophin Identification of DNA segments from the DMD gene in Kunkel's and Worton's laboratories quickly led to the mRNA, which led to the protein encoded by the gene. Both the protein and its gene are remarkable. The protein, now called *dystrophin*, is exceptionally large, consisting of nearly 4000 amino acids (Hoffman et al., 1987). There are at least four domains within dystrophin, as diagrammed in Figure 5–12: (1) a 240-amino-acid N-terminal domain with strong similarity to the actin-binding domains of alpha-actinins (a group of cytoskeletal proteins), (2) 24 repeats of a roughly 109-amino-acid sequence with some resemblance to spectrin (a large protein that forms part of a meshwork on the inner surface of red blood cell membranes), (3) a cysteine-rich domain of 150 amino acids that is similar to the C-terminal domain of some alpha-actinins, and (4) a 420-amino-acid C-terminal domain that has no known relatives.

Immunological studies show that dystrophin is located adjacent to the sarcolemma, the plasma membrane of myotubes; it may be part of the cytoskeletal system in those cells, or at least participate in connecting the cytoskeletal system to the plasma membrane. In particular, its absence or abnormal composition may make the sarcolemma abnormally sensitive to

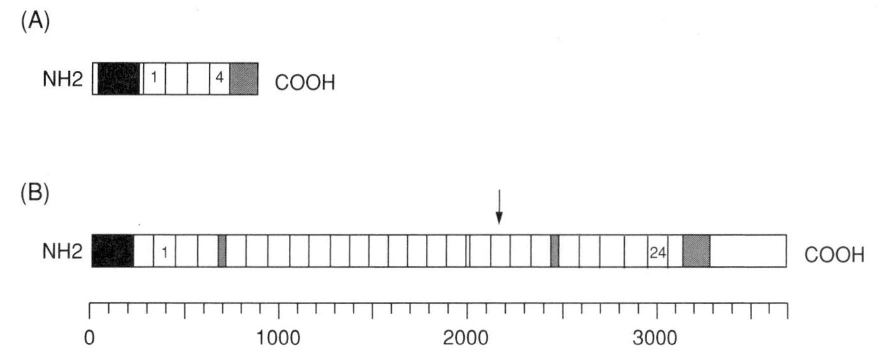

Figure 5–12

Domain organization of alpha-actinin (A) and dystrophin (B). The domains showing similarity between actinin and dystrophin are shown in black (amino terminal domain) or obliquely hatched (cysteine-rich domain). There are 24 spectrin-like repeats in dystrophin (only the first and last are numbered), which are presumed to form an "extended-rod" domain. Two putative hinge regions are indicated by horizontal hatches. An area where deletions frequently occur, between exons 44 and 45 in the gene, is indicated by an arrow. The scale at the bottom gives the number of amino acids. (Reprinted by permission from Mandel, 1989; © Macmillan Magazines, Ltd.)

tearing during muscle action, resulting in cell death and leakage of sarcoplasmic constituents into the body fluids.

The dystrophin gene is even more unusual than the protein, as has been mentioned in Chapter 2. A giant among genes, it extends over more than two million bases of DNA, with less than 1% of the gene coding for the protein (Monaco and Kunkel, 1987). There are approximately 75 exons. We may well wonder whether there is some arcane advantage to this extraordinary organization, or whether it is an example of evolution gone wild, like the oft-cited antlers on the extinct Irish elk. In any case, an abnormal dystrophin gene is the seed of death for one boy in 3500.

Cystic Fibrosis

Approximately one child in 2000–2500 born to Caucasian parents is affected with this serious autosomal recessive disorder. Cystic fibrosis is a disease that is characterized by generalized defects in secretion by exocrine glands. Most troublesome are (1) the blockage of pancreatic enzyme secretion, with adverse effects on digestion; and (2) production of abnormally thick mucus in the lungs. The latter provides a favorable environment for microorganisms. The cells that line the airways gradually die from the effects of chronic infection and obstruction with mucus. Fibrous tissue replaces the epithelial cells and respiratory capacity declines, eventually leading to death.

Cystic fibrosis children used to die within the first year or two of life, but the development of antibiotics and the diligent use of physiotherapy to clear mucus from the lungs have extended the average lifespan to somewhat more than 20 years. However, few CF patients survive beyond age 30.

The frequency of occurrence of the disease and its remorseless progression made it a favorite subject of research for decades. Literally dozens of papers appeared, claiming to have identified the underlying biochemical defect; however, virtually all of them dealt with remote secondary effects at best. There are enough artifacts in the literature on CF to provide material for a PhD thesis on the psychology of scientific folly.

Major progress toward understanding the biochemistry of cystic fibrosis was made by Quinton (1983), who showed that epithelial tissues from CF patients are defective in chloride ion transport. A generalized defect in exocrine gland chloride secretion can explain why the ducts of the pancreas become clogged with mucus, and why the mucus in the airways of the lungs is exceptionally thick. The macromolecules that constitute the bulk of mucus are presumably secreted normally, but in the absence of normal amounts of Cl^- and its counterion, Na^+, relatively little water would enter the ducts of the glands; the mucus would therefore be abnormally viscous.

More recent work has extended Quinton's basic finding, showing that chloride channels from epithelial cells of CF patients function normally in isolated membrane fragments and that the actual defect appears to be somewhere in the regulation of the response to cAMP, which controls chloride secretion. A hypothetical chloride channel is diagrammed in Figure 5-13. Regulation of the channel is complex, involving calcium ions, and probably at least two protein kinases.

Concomitantly with the biochemical studies, several laboratories searched for the CF gene, using the positional analysis strategy. The first task was to map the gene, and because of the commonness of the disease, abundant family material was available for linkage analysis. Many dozens of RFLPs were tested unsuccessfully before L. C. Tsui and his colleagues

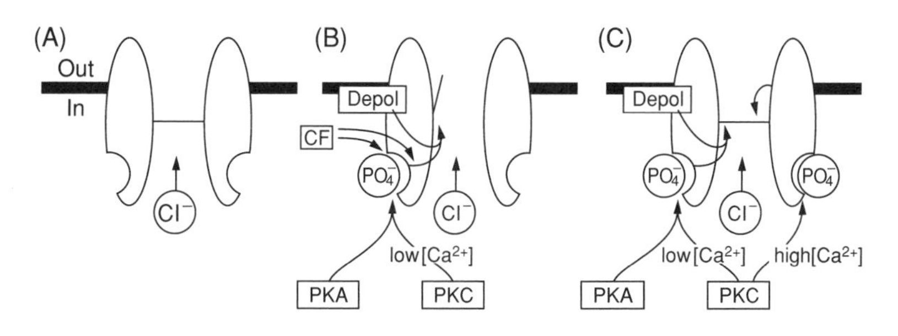

Figure 5–13

A model of chloride channel regulation. Inner and outer surfaces of the cell membrane are labeled in (A). The channel is inactive when the "gate" is closed (horizontal line in the center of panels A and C) and active when open (panel B). "Depol" refers to strong membrane depolarization (+100 to +140 mV). PKA and PKC are protein kinases. CF indicates a possible site of action of the CF gene product. Although the channel is depicted as a single entity, it actually consists of several subunits. (From Li et al., 1989, p. 1355; © 1979 by the AAAS)

found one that appeared to be linked to CF at a distance of 15 cM (Tsui et al., 1985). The probe was quickly mapped to chromosome 7, and although that result was not immediately published (Knowlton et al., 1985), the news spread via the scientific grapevine, leading several large CF laboratories to focus their efforts on chromosome 7, also.

By the fall of 1985, Robert Williamson and his colleagues in London (Wainwright et al., 1985) had shown linkage between CF and D7S8, an RFLP site that was mapped with a panel of somatic cell hybrids to the long arm of chromosome 7. At the same time, Ray White and co-workers in Utah (White et al., 1985) found linkage between CF and the *met* oncogene, which was mapped by *in situ* hybridization to the long arm of chromosome 7, between bands q21 and q31. Both D7S8 and *met* were tightly linked to CF, there being no recombinants at all in the first set of data. These findings led to an international collaboration to map CF more precisely, and in 1986, Beaudet et al. published the results of linkage studies on 200 families, which strongly implied that the CF gene lies between the loci for *met* and D7S8 in an interval no greater than 2 Mbp. The possibility of identifying the CF gene was now tantalizingly clear.

But it was not easy to find the CF gene. There were no known deletions or other rearrangements, which had been so helpful in finding the genes for DMD, chronic granulomatous disease, and in several other successful examples of positional analysis. Ultimately it was brute force that did the job, and the successful gene hunters were a team led by Lap-Chee Tsui (1985) at the Hospital for Sick Children in Toronto and Francis Collins at the University of Michigan (Kerem et al., 1989; Riordan et al., 1989; Rommens et al., 1989).

The basic strategy was first to clone a portion of chromosome 7 that must include the CF locus. They could have targeted the entire region between *met* and D7S8, but first they simplified the problem by finding two more RFLPs that were closer to CF than *met*, thus narrowing the target region to about 1 Mb. Then they began a "chromosome walk" in order to produce an ordered series of overlapping clones spanning that region. A walk is a step-by-step process of using one clone to find overlapping clones from a random clone library, then using those clones to find the next clones in the same direction, and so on. Each step is relatively simple, but the process is slow, because each step must be taken sequentially. Moreover, one cannot take many steps in a mammalian genome without encountering an unclonable sequence. Most unclonable sequences are repetitive DNA, which tends to be unstable in bacteria, but other reasons for non-clonability exist. In any case, if an unclonable region is larger than the capacity of a cosmid, it creates an obstacle that stops a walk.

The Tsui-Collins collaborative group overcame the difficulties of a long-distance chromosome walk by supplementing their efforts with "jumping clones," a technique at which Collins was expert. As diagrammed in Figure 5–14, a jumping clone consists of the two ends of a piece of DNA that may

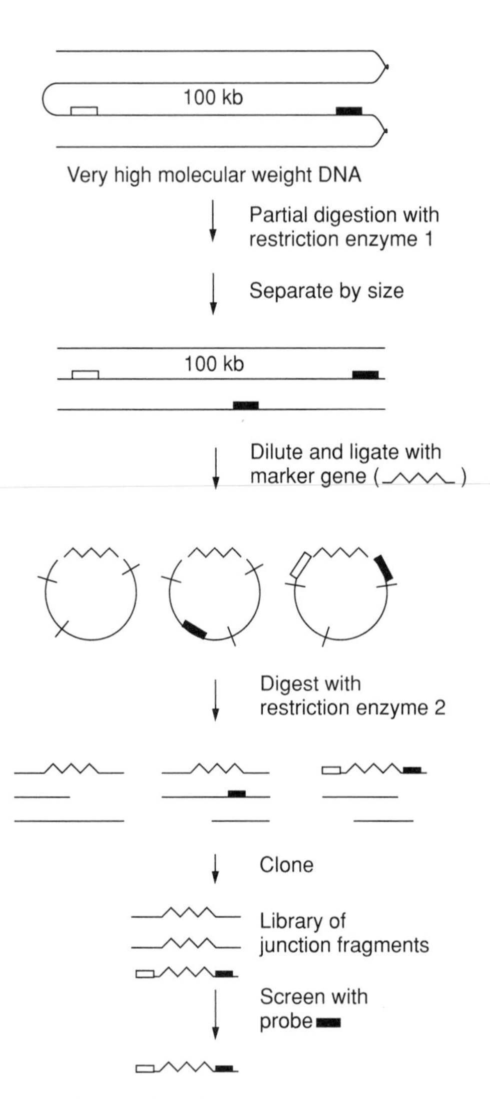

Figure 5–14

Scheme for chromosome jumping. The black box denotes the start site of the jump, for which there must be a probe, and the open box represents the destination some 100 kbp away. Although it is not absolutely necessary, a marker gene (indicated by the wavy line) may be inserted during circularization to make it easier to identify the junction fragments. (From Marx, 1985; © 1985 by the AAAS)

have been 100 kbp or more apart in the genome, but which are brought together in the cloned fragment. One thereby skips over any unclonable region that may have existed between those end pieces. Furthermore, each half of a jumping clone can be used as the starting point for a new walk, thereby bypassing the restriction of taking only one step at a time. When enough steps have been taken from the various starting points, some of the clones will be found to overlap, and the walk segments can be joined into one large, continuous series.

By these methods Tsui, Collins, and collaborators cloned nearly 300 kbp of DNA, beginning at an RFLP site some 500 kbp from *met* and extending toward D7S8 (Figure 5-15). On the way, they passed through IRP, a gene that had been found by the Williamson group two years earlier and which was at first thought to be the CF gene. Tsui and Collins now had a collection of cosmids that might include part or all of the CF gene and they had to answer the question, "How do you know when you have reached your goal?" Their first tactic was to do a zoo blot, looking for sequences that are conserved in several mammals. Three conserved sequences were found.

The next step was to find out whether any of those conserved sequences were expressed in normal tissues of the types that are affected in CF patients. Two of the three had the wrong pattern of expression; that is, they hybridized to messenger RNAs (or cDNAs) from the wrong spectrum

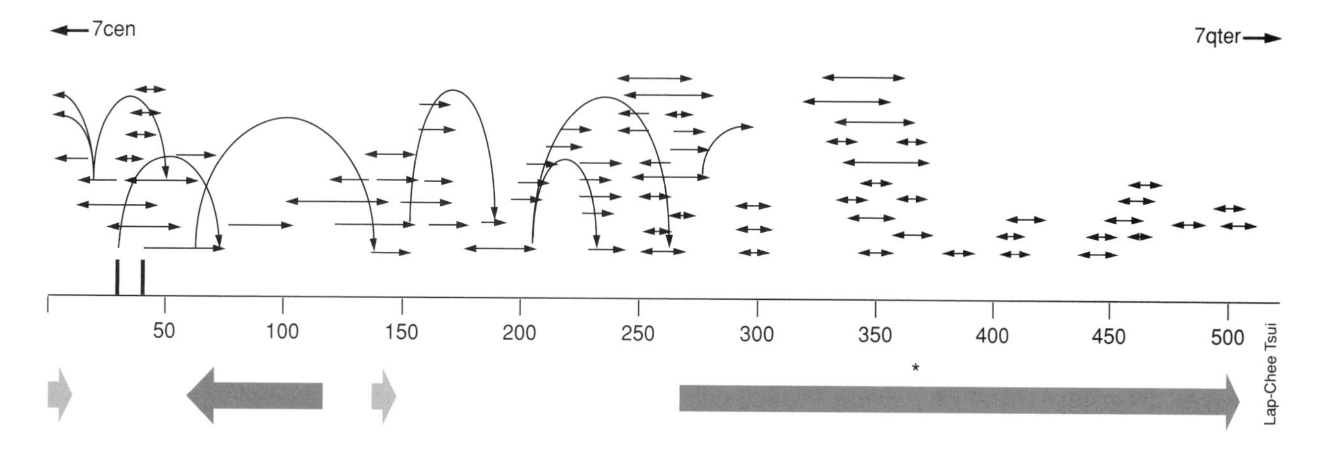

Figure 5–15

The long march to the cystic fibrosis gene. The journey began at a site at the left end of the diagram; this site was known to be close to the CF gene on chromosome 7. The 280 kbp between the start site and the beginning of the CF gene were covered by a combination of chromosome "walking" and "jumping." The straight arrows above the long horizontal line represent the DNA segments cloned by walking and the curved arrows represent the jumps. The long, thick arrow on the lower right indicates the CF gene, which spans about 250 kbp. The asterisk marks the deletion of the codon for phenylalanine 508. The left-pointing thick arrow represents the IRP gene, and the two short, thick, right-pointing arrows mark open reading frames that may or may not be part of protein-coding genes. (From Marx, 1989, p. 25; © 1989 by the AAAS)

of tissues. The third sequence at first seemed not be expressed at all, but finally it found a target in a cDNA library made from messenger RNAs of sweat gland cells, a tissue that is always affected in CF patients. The match between the genomic probe and the cDNA was marginal at first—only 113 bp. Apparently, Tsui and Collins had barely walked far enough along chromosome 7 to reach the beginning of the gene.

One thing leads readily to another in molecular genetics, so the rest of that cDNA from sweat gland cells was used to find genomic sequences downstream, which were used to find more cDNAs and more genomic clones, and so on. Finally, they isolated a gene that covered 250 kbp, which encoded a protein containing 1480 amino acids distributed across 24 exons. The sequence of the protein implied that it might be a membrane protein, possibly part of an ion channel, which was compatible with expectations for the CF gene product.

But was it the CF gene, or another near miss? If it was the genuine CF gene, there should be mutations in it when the corresponding DNA from CF patients was analyzed. This proved to be the case. A deletion of three nucleotides that code for phenylalanine at amino acid position 508 was soon found, and examination of a large series of CF DNAs for that deletion showed that nearly 70% of North American patients had the same mutation. (The other 30% are quite heterogeneous.) The F508 deletion is not an irrelevant polymorphism; it is not found at all in a large sample of normal chromosomes from CF heterozygotes (parents of affected children). The F508 deletion is located near an apparent binding site for ATP, based on what is known about proteins with similar structure, so it may interfere with phosphorylation of the CF protein or some other need for energy provided by ATP.

As more and more mutations in the same gene have been identified in the remaining 30% of CF patients, the evidence that the CF gene has been found has become overwhelming. Identification of the F508 deletion has already had a favorable impact on genetic screening in families where CF is known to occur; it has been suggested that mass screening for heterozygotes should be initiated, but that proposal has met with strong opposition (see Chapter 9). Additional proof that the gene described here is really the CF gene has been provided by transfecting it into CF cells in culture and showing that normal chloride transport is restored.

Evidently, the function of the protein encoded by this gene will now be studied in detail and extensive efforts to find ways of restoring its function or ameliorating the effects of its abnormal function in CF patients will be undertaken. Gene therapy is sure to be considered, also (Chapter 10).

SUMMARY

Genes whose abnormal functions are responsible for genetic diseases were first identified through knowledge of defective proteins. Early examples were the genes for alpha- and beta-globins, mutations in which cause a variety of hemoglobinopathies. The development of recombinant DNA technology and the identification and mapping of numerous polymorphic, anonymous loci (RFLPs, VNTRs) throughout the human genome made it possible to use linkage analysis to map loci responsible for diseases for which the primary biochemical lesion was unknown. Once mapped, an unknown gene can be identified by a variety of techniques subsumed by the term *positional analysis* (or reverse genetics or positional cloning).

Positional analysis typically involves the following steps. (1) The essential first step is to map the gene to a specific chromosomal subregion by linkage analysis. (2) DNA from the region of interest is then obtained as recombinant DNA clones in a suitable microorganism. (3) Putative exons are identified by determining which of the cloned DNA fragments have changed little during mammalian evolution. (4) Candidate gene fragments, identified in step 3, are used to identify mRNAs transcribed in normal tissues of the same types that are abnormal in persons with the disease in question.

It may then be possible to show (5) that at least some affected persons have mRNAs with abnormal size or lack mRNAs corresponding to the candidate gene; or that (6) some affected persons have DNA deletions in the candidate gene. In any case, it is ultimately essential to show (7) that specific mutations in the candidate gene correlate unambiguously with the disease phenotype. Further study of the encoded protein can eventually be expected to provide an understanding of the primary biochemical basis of the disease, which can then be applied to efforts to devise therapy.

Application of these principles is illustrated with four examples. The gene for Huntington disease was mapped to the distal portion of the short arm of chromosome 4 in 1983, but unusual properties of the gene and/or the neighboring region, not yet understood, have prevented identification of the gene. A large DNA deletion in the short arm of the X chromosome was extremely helpful in isolation of the genes for Duchenne muscular dystrophy and for X-linked chronic granulomatous disease. The gene for cystic fibrosis, on the long arm of chromosome 7, was not associated with deletions or missing mRNAs in affected persons; it was identified in 1989 after exhaustive analysis of the region in which it was known to reside.

REFERENCES

Beaudet, A., Bowcock, A., Buchwald, M., et al. 1986. Linkage of cystic fibrosis to two tightly linked DNA markers: joint report from a collaborative study. Am. J. Hum. Genet. 39:681–693.

Bodrug, S. E., Ray, P. N., Gonzalez, I. L., et al. 1987. Molecular analysis of a constitutional X-autosome translocation in a female with muscular dystrophy. Science 237:1620–1624.

Collins, F. S. 1992. Cystic fibrosis: molecular biology and therapeutic implications. Science 256:774–779.

Gusella, J. F., Gilliam, T. C., Tanzi, R. E. et al. 1986. Molecular genetics of Huntington's disease. Cold Spring Harbor Symp. Quant. Biol. 51:359–364.

Gusella, J. F., Wexler, N. S., Conneally, P. M. et al. 1983. A polymorphic DNA marker genetically linked to Huntington's disease. Nature 306:234–238.

Hoffman, E. P., Brown, R. H., Jr., and Kunkel, L. M. 1987. Dystrophin: the protein product of the Duchenne muscular dystrophy locus. Cell 51:919–928.

Jenkins, J. B. and Conneally, P. M. 1989. The paradigm of Huntington Disease. Am. J. Hum. Genet. 45:169–175.

Kerem, B-S., Rommens, J. M., Buchanan, J.A., et al. 1989. Identification of the cystic fibrosis gene: genetic analysis. Science 245:1073–1080.

Knowlton, R. F., Cohen-Haguenauer, O., Cong, N., et al. 1985. A polymorphic DNA marker linked to cystic fibrosis is located on chromosome 7. Nature 318:380–382.

Koenig, M., Hoffman, E. P., Bertelson, C. J., et al. 1987. Complete cloning of the Duchenne muscular dystrophy (DMD) cDNA and preliminary genomic organization of the DMD gene in normal and affected individuals. Cell 50:509–517.

Kunkel, L. M., Monaco, A. P., Middlesworth, W. et al. 1985. Specific cloning of DNA fragments absent from the DNA of a male patient with an X chromosome deletion. Proc. Natl. Acad. Sci. USA 82:4778–4782.

Li, M., McCann, J. R., Anderson, M. P. et al. 1989. Regulation of chloride channels by protein kinase C in normal and cystic fibrosis airway epithelia. Science 244:1353–1356.

Mandel, J. L. 1989. Dystrophin: the gene and its product. Nature 339:584–586.

Marx, J. L. 1985. Hopping along the chromosome. Science 228:1080.

Marx, J. L. 1989. The cystic fibrosis gene is found. Science 245:923–925.

Monaco, A. P. and Kunkel, L. M. 1987. A giant locus for the Duchenne and Becker muscular dystrophy gene. Trends in Genetics 3:33–37.

Monaco, A. P., Neve, R. L., Coletti-Feener, C. A., et al. 1986. Isolation of candidate cDNAs for portions of the Duchenne muscular dystrophy gene. Nature 323:646–650.

Orkin, S. H. 1987. X-linked chronic granulomatous disease: from chromosomal position to the *in vivo* gene product. Trends in Genetics 3:149–151.

Quinton, P. M. 1983. Chloride impermeability in cystic fibrosis. Nature 301:421–422.

Ray, P. N., Belfall, B., Duff, C., et al. 1985. Cloning of the breakpoint of an X;21 translocation associated with Duchenne muscular dystrophy. Nature 318:672–675.

Riordan, J. R., Rommens, J. M., Kerem, B-S., et al. 1989. Identification of the cystic fibrosis gene: cloning and characterization of complementary DNA. Science 245:1066–1073.

Rommens, J. M., Iannuzzi, M. C., Kerem, B-S., et al. 1989. Identification of the cystic fibrosis gene: chromosome walking and jumping. Science 245:1059–1065.

Royer-Pokora, B., Kunkel, L. M., Monaco, A. P., et al. 1986. Cloning the gene for the inherited disorder chronic granulomatous disease on the basis of its chromosomal location. Cold Spring Harbor Symp. Quant. Biol. 51:177–183.

Tsui, L-C., Buchwald, M., Barker, D., et al. 1985. Cystic fibrosis locus defined by a genetically linked polymorphic DNA marker. Science 230:1054–1057.

Vogel, F. and Motulsky, A. G. 1979. Human genetics: problems and approaches. Springer-Verlag Berlin.

Wainwright, B. J., Scambler, P. J., Schmidtke, J., et al. 1985. Localization of cystic fibrosis locus to human chromosome 7cen-q22. Nature 318:384–385.

White, R., et al. 1985. A closely linked genetic marker for cystic fibrosis. Nature 318:382–384.

Worton, R. G., Duff, C., Sylvester, J. E., et al. 1984. Duchenne muscular dystrophy involving translocation of the *dmd* gene next to ribosomal RNA genes. Science 224:1447–1449.

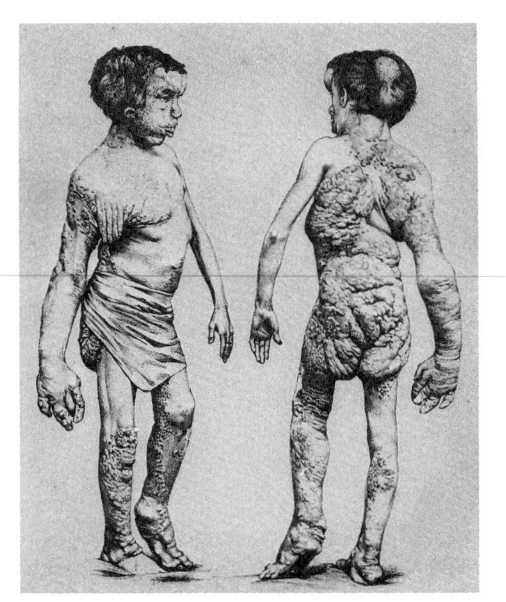

John Merrick, "The Elephant Man," who suffered from an extreme case of neurofibromatosis, probably complicated by some form of bone disorder. (From Treves, 1885)

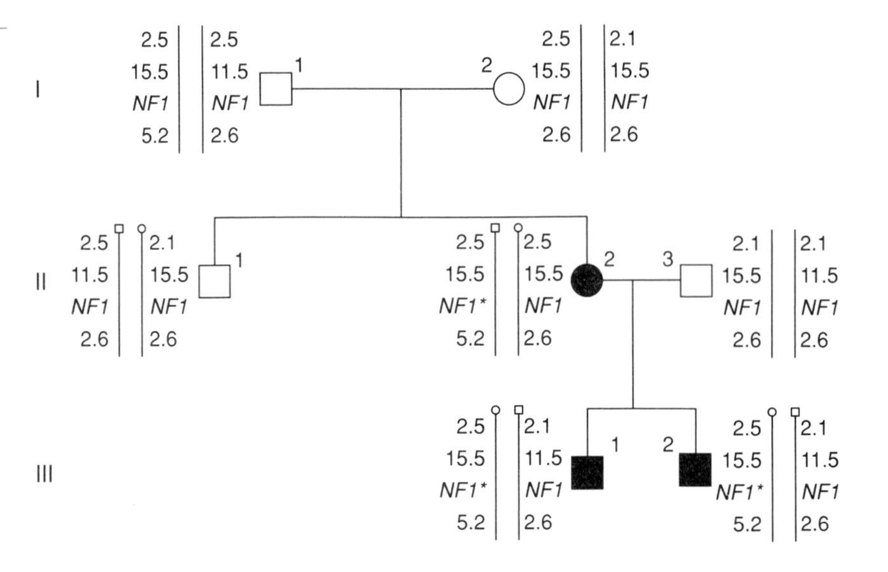

The gene responsible for autosomal dominant Von Recklinghausen neurofibromatosis (NF1) has an exceptionally high mutation rate. This pedigree shows how haplotypes constructed from closely linked polymorphic loci can be used to identify a new mutation and its parental origin. *NF1** indicates the mutant gene; filled symbols represent affected individuals. The presence of one of the haplotypes from I-1 in II-2 and both of her affected children proves that the mutation occurred in the germline of I-1. (Modified from Jadayel et al., 1990)

6 Mutation
QUANTITATIVE ASPECTS

Variation in genes and gene expression provides the materials from which the science of genetics is constructed, and *mutation* is the primary source of variation. Secondary sources of variation are (1) recombination and reassortment of genes during germ cell formation, both of which ultimately depend upon mutation to generate variants; and (2) environmental factors, which may affect the timing and extent of gene expression, sometimes producing *phenocopies* (phenotypes that mimic variants in form or function which can also be caused by specific mutations).

In this chapter we shall consider quantitative aspects of mutation: principally the rate at which mutations occur, the frequency of mutant alleles in the human genome, the overall effect of mutation on disease, and some related topics. In Chapter 7 the causes of mutations and the varieties of mutations that may occur at the DNA and chromosomal levels will be surveyed. Then, in Chapter 8, we shall consider the molecular basis for the classical patterns of dominant and recessive inheritance of mutant alleles and some important variations on those patterns.

Prior to the rise of molecular biology the word *mutation* referred to any observable, spontaneous change in an inherited characteristic, and the word *mutant* referred to an individual carrying the mutation (as it still does). Long before Mendel, animal breeders looked for "sports"; that is, animals with unusual pigmentation or body form. Pigeon fanciers zealously propagated mutants with bizarre feathers or behavior. The early *Drosophila* geneticists collected and conserved mutations that affected eye color, wing number, wing shape, bristle shape or pattern, and segment differentiation.

Nowadays we think of mutations as changes in the base sequence of DNA, ranging from single base substitutions to chromosomal translocations, and we have an impressive variety of techniques for detecting either the DNA changes themselves or their effects on gene products and their function. Nevertheless, much mutational analysis in humans is still done usefully in terms of whether a person is sick or not. We shall begin our discussion of mutations at that level, postponing a survey of the many ways in which a given gene may change into a defective allele until Chapter 7; we shall ask how frequently genes mutate in such a way that they cause recognizable genetic diseases.

Mutation Rates in Germ Cells

General Aspects

Mutation rates vary over a wide range. It is customary to speak of *locus-specific mutation rates*, rather than rates per nucleotide. The locus-specific rate is the sum of the rates for all types of mutations that have been measured for a given genetic locus. Locus-specific rates used to be calculated primarily by measuring the frequency with which functionless (null) alleles or alleles with severely impaired function occurred. That is, any mutation that could alter the expression of an allele in a detectable fashion would be scored. More recently, electrophoretic analysis of polypeptides has made it possible to detect many alleles that are not grossly malfunctional. Still more recently, DNA sequencing and other molecular analyses have made it possible to detect many "silent mutations," which are mutations that have neither a functional nor an electrophoretic effect.

In humans, locus-specific mutation rates vary over at least a 100-fold range, from roughly 10^{-6} to 10^{-4} per generation. Keep in mind that this rate refers to organismal generations, not cell generations. Also note that the rate is calculated per gene copy, not per diploid cell. It is common practice to use the word *allele* as a synonym for "gene copy," even though alleles are traditionally defined as alternative forms of a gene. For example, in a population of 10,000 persons, there are 20,000 copies of any autosomal gene; these may also be referred to as 20,000 alleles. Unless otherwise noted, mutation rates given in this book are expressed in terms of mutations per allele per generation.

Factors That May Influence Mutation Rates at Different Loci

1. Size of the coding region. Genes for large polypeptides should in general be larger targets for mutation than genes for smaller polypeptides.
2. Number and sizes of introns. The more splicing that has to be done to produce a mature messenger RNA, the more possibilities exist for mutations that may affect the process, both qualitatively and quantitatively. In addition, it is intuitively reasonable to expect that very large introns may increase the frequency of abortive transcription or the frequency of incorrect splicing, but this has not been proved.
3. Presence of repetitive sequences. It is possible, but unproved, that unequal crossing-over may be more frequent in genes with larger numbers of reiterated sequences, such as the Alu elements, some of which occur within introns of most human genes.
4. Detailed nucleotide sequence. Mutational hot spots are well known in prokaryotes and bacteriophages; there is no obvious reason why they shouldn't exist in humans. Apparently, some nucleotide sequences are more difficult to replicate accurately or to repair.

Other sequences may be unusually prone to spontaneous chemical changes. The tendency of cytosine to deaminate is the most common example; ordinary cytosine becomes uridine, which should be recognized as abnormal by the repair enzymes, but may occasionally be replicated before it is discovered. If that happens, an AT base pair will be formed where the original DNA had a GC base pair, thus leading to a mutation. Of course, if 5-methylcytosine deaminates, it becomes thymine, and this is guaranteed to be a mutation, because a GC base pair will be replaced by an AT base pair during the next DNA replication. (See Chapter 8 and Figure 8–16.)

Two factors that will affect our ability to measure mutation rates are:

1. Multiple copies of a gene per haploid genome. Although this should not affect the mutation rate per locus, mutations that eliminate the function of an allele will be less readily detected than those for single-locus genes. For example, if a genetic locus is duplicated, four alleles must be inactivated before the organism loses all functional copies of the gene.

2. Method of assay. Assays that detect loss of function will miss many mutations that have little or no effect on function. Assays that detect altered electrophoretic mobility will miss many mutations that affect function, but not mobility. Assays that involve sequencing part or all of a gene will detect many changes that affect neither function nor electrophoretic mobility. It is not valid to compare apparent mutation rates at different loci unless the same types of assays have been used to detect mutations at those loci, or unless appropriate corrections have been made.

Measurement of Mutation Rates In principle, it is possible to measure the mutation rate at any locus either directly or indirectly. Direct methods determine the frequency of *new* mutant alleles in a population in one generation. Direct methods can be subdivided into two categories: (1) assays based on the phenotype of the organism and (2) molecular assays. At the phenotypic level, the direct method requires that individuals with new mutations be distinguishable from individuals who have inherited their mutant alleles from one or both parents. In some cases this is easy, and in others it is quite difficult. When new mutants are identifiable, the mutation rate is obtained simply by calculating the ratio of new mutant alleles/total alleles in the population surveyed. At the molecular level, several techniques for the detection of mutant alleles are available; we shall consider some of them presently.

The indirect method assumes that in a stable population, the frequency of a deleterious allele will increase until the loss of alleles at each generation, because of reduced reproduction of affected persons, is equal to the rate of creation of new alleles via mutation. Thus, the mutation rate can be calculated from the frequency of affected persons. The indirect method for the estimation of mutation rates will be presented after we consider the direct method.

Direct Measurement of Mutation Rates Based on Phenotypes

Autosomal Dominant Conditions If a trait is known to be inherited as an autosomal dominant, the frequency of occurrence of affected children born to unaffected parents is a direct measure of the mutation rate (symbolized by the Greek letter mu, μ). If we express the frequency of affected individuals as I (which stands for "incidence" and is used synonymously with "frequency"), and if the frequency of affected individuals born to unaffected parents is I*, then

$$\mu = (0.5)\mathrm{I}^*$$

The factor of 0.5 is included in the formula because there are two copies of the gene in every individual, whereas μ is expressed as a rate per gene copy per generation, and for a dominant mutation, an affected individual represents only one mutant allele.

A well-studied example of an autosomal dominant syndrome is achondroplastic dwarfism. In a Danish survey (Morch, 1941), there were 10 dwarfs in a total of 94,075 births; eight of the dwarfs were born to non-dwarf parents. This gives

$$\mu = (0.5)8/94{,}075 = 4.2 \times 10^{-5}$$

A similar study in Boston (Nelson and Holmes, 1989) found two apparent new mutant dwarfs in 69,227 births. Here $\mu = 1.4 \times 10^{-5}$. The average for the two studies is 2.8×10^{-5}.

Achondroplasia seems like an ideal case for application of the direct method for calculating mutation rates; one can distinguish dwarfs from non-dwarfs at a glance. However, there are uncertainties that must be taken into consideration in this case, and generally. The major ones are:

Is the legal father the biological father?

Is more than one genetic locus capable of causing the same disease?

Is the mutant allele 100% penetrant? (Penetrance is discussed in Chapter 8; it refers to the frequency with which a mutation is expressed by individuals who have the mutant allele. For an allele that is incompletely penetrant, there will be apparently new mutant individuals who do not actually have new mutant alleles. These will be the affected children of unaffected parents who carry the mutant allele.)

Are the parents really not affected? The level of expression of many mutations varies widely from one individual to another. In special cases (notably Huntington disease and other diseases affecting behavior) the parent's condition may be deliberately hidden from outsiders.

Is the trait ever produced by recessive alleles? If so, affected individuals will usually have unaffected parents.

Do phenocopies exist? That is, can the symptoms arise for nongenetic reasons, such as exposure to toxic drugs or other chemicals during development?

Another large study on dominant mutation rates was carried out over many years at the Jackson Laboratory in Maine, which is one of the world's largest sources of inbred mice. More than 14 million mice were scored for the occurrence of dominant mutations affecting obvious physical characteristics such as coat color; 54 mutants were found, distributed over 12 loci. The average locus-specific mutation rate has been calculated to be 0.4×10^{-6} per allele per generation (Bodmer and Cavalli-Sforza, 1976). This is an interesting result, but we should keep in mind that the average mutation rate in mice is not necessarily the same as the rate in humans.

Autosomal Recessive Conditions Direct determination of autosomal recessive mutation rates is not possible by simply looking at the gross phenotypes of affected persons and their parents. In experimental animals, where the genotypes can be controlled, direct determination is possible in special experimental situations. For example, another study involving nearly a million mice was carried out by the U.S. and U.K. atomic energy programs. Two strains of mice were interbred: one was homozygous recessive at seven loci for easily scored phenotypes, such as coat color and pattern; the other was homozygous wild-type at the same loci. All offspring of these matings should have been heterozygotes, and therefore should have had wild-type phenotypes at all seven loci. But, if a mutation to a recessive allele occurred in a gamete from the wild-type parents, a mouse with recessive phenotype at one locus would be born, as indicated here.

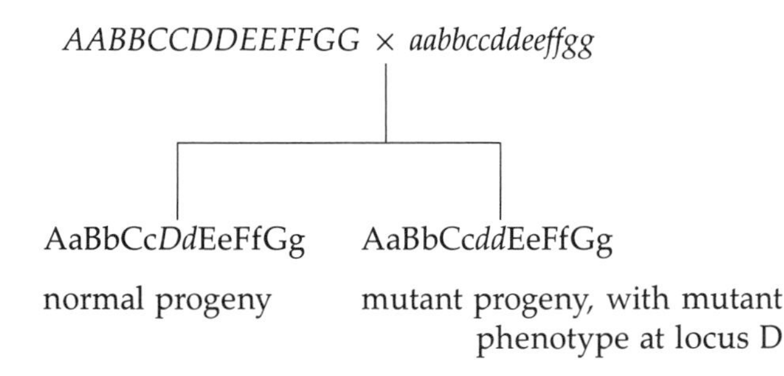

The result was that the locus-specific average mutation rate was 4.75×10^{-6}, when data for both sexes were combined.

Direct Determination of Mutation Rates by Molecular Assays

Modern technology provides several approaches to the direct determination of mutation rates via molecular assays. Here we shall consider two very different approaches. The first is protein electrophoresis, which implies the existence of mutant alleles by virtue of changes in gene products. The second is the Polymerase Chain Reaction, a powerful and versatile new technique for amplification of specific segments of DNA.

Protein Electrophoresis The migration of a protein through a gel in an electric field depends upon the size of the molecule and the net charge on the protein. Every gel, whether it be made of starch, agarose, or acrylamide, has a characteristic range of pore sizes; in any given gel, larger proteins tend to be retarded more than smaller proteins. In buffers containing SDS (sodium dodecyl sulfate), proteins dissociate into their constituent polypeptides, and because the detergent binds more-or-less indiscriminately to all polypeptides, providing a uniform electric charge in proportion to mass, fractionation in an electric field depends almost entirely on size.

On the other hand, in buffers that do not contain ionic detergents, the net charge on a protein can be the major determinant of electrophoretic mobility. The charge is a function of amino acid composition. Most of the net charge represents the balance between basic amino acids (lysine, arginine, and histidine) and acidic amino acids (glutamic and aspartic acids). If a mutation in a gene changes the encoded amino acid sequence in such a way that the net charge of a protein is altered, it is usually possible to separate the products of the wild-type and mutant alleles by electrophoresis. Correspondingly, mutations that change the size of a protein or that alter its configuration may also be detected by shifts in electrophoretic mobility under appropriate conditions.

The largest search for electrophoretically detectable protein variants ("electromorphs") which has been carried out so far was an exhaustive analysis of the children of residents of Hiroshima and Nagasaki, conducted over a period of several decades by a team of American and Japanese scientists, headed by James Neel. The purpose of the survey was to determine whether the amount of radiation received by persons who were proximally exposed (less than 2000 meters from a blast hypocenter) was sufficient to produce a measurable increase in the overall mutation rate. The control group consisted of persons from the same cities who were further than 2000 meters from the hypocenters.

Most of the search for new mutations was done by one-dimensional starch gel electrophoresis on 30 different enzymes and other proteins that were easily detected from blood or blood cells. Assays were done on every available child conceived after the blasts, as well as its parents. When the electrophoretic mobility of a protein from a child was found to be different from the mobility of the corresponding protein from both parents, a large series of additional markers was examined, to be certain that false paternity did not account for the results.

The final report (Neel et al., 1988) summarizes the study and its results. In total, there were 667,404 locus tests on children of proximally exposed persons and 3 mutations were found; 466,881 locus tests on the control group also yielded 3 mutations. (One protein assayed in one child is two locus tests for autosomal genes, because of diploidy). For the original purpose of the study, the most significant conclusion was that there was not a measurable elevation in the mutation rate in persons proximally exposed to the atomic explosions. As Neel et al. (1988) point out, the small number of mutations detected in both exposed and control groups leaves a rather large statistical uncertainty, but the data clearly exclude an order-of-magnitude increase in the mutation rate because of radiation received from the atomic blasts.

Several other large-scale surveys for spontaneous mutations detectable via protein electromorphs have been carried out on non-Japanese populations. Together these add about 600,000 more locus tests and document only one more mutation. Neel et al. (1986) combined all of the available data (excluding the proximally exposed group from Hiroshima and Nagasaki) and arrived at an average *total rate* of 1×10^{-5} mutations per allele per generation.

The term "total rate" includes corrections for two types of mutations that would be missed by the electrophoretic surveys. First, about one-fourth of all single nucleotide substitutions are *synonymous*, insofar as any effect on the encoded polypeptide is concerned; because of degeneracy in the genetic code, they do not change the amino acid specified by the codon in which the mutation occurs. For example, the codon UCU can mutate to UCC, UCA, or UCG without changing serine to any other amino acid; CCU, CCC, CCA, and CCG all code for proline; and so on.

A second correction must be made because amino acid changes that do not alter the charge of the protein are usually not detectable by electrophoresis. For example, any of the four codons for glycine can become valine codons by single base mutations (G to U in the second position of the codon) or glycine can become alanine when C is substituted for G in the second position of any of those four codons. Gly, Val, and Ala are all neutral amino acids.

In fact, only one-third of all possible amino acid substitutions will alter net charge, but because some amino acid substitutions that change the shape of a polypeptide (such as insertion or removal of proline) also change electrophoretic mobility significantly, it is reasonable to multiply the frequency of electrophoretically detectable mutations by two to correct for non-detectable mutations (i.e., electrophoretically "silent" mutations).

Another important number that can be estimated from the preceding data is the *nucleotide substitution rate*; that is, the rate of mutation per nucleotide per generation. Using a conservative estimate that there are 1000 bp in the coding portion of an average gene, Neel et al. (1986) point out that 1×10^{-5} mutations per locus per generation implies approximately 1×10^{-8} mutations per coding nucleotide per generation. If we apply that rate to the entire haploid genome of 3×10^{9} bp, there should be 30 nucleotides that differ from the parental genome in every gamete. However, most of those mutations would not affect gene products. If there are 30,000 genes with an average of 1000 coding nucleotides each, then the total haploid "coding genome" is 3×10^{7} bp, or 1% of total DNA. This implies that there would be less than one nucleotide substitution in coding DNA per gamete per generation.

A related question is whether a total locus-specific rate of 10^{-5} mutations per generation is compatible with an average rate of mutation to deleterious alleles of 10^{-6} per locus per generation. It seems so. About one-fourth of mutations involving coding nucleotides should create synonymous codons, and of the remaining three-fourths, it is plausible that only 10–20% of the resulting amino acid changes would affect the function of the protein in such a way that illness would result. The other side of the same argument suggests that the average per locus mutation rate to deleterious alleles is unlikely to be as low as 1×10^{-7}, which would imply that only 1% of nucleotide substitutions in coding regions reduce protein function to a level that is incompatible with normal health of the individual.

The Polymerase Chain Reaction (PCR) The *Polymerase Chain Reaction* uses DNA polymerases to amplify a nucleic acid segment that is flanked by two short regions of known sequence, for which complementary oligonucleotides can be synthesized, to be used as primers by the polymerase. The method is explained and diagrammed in Box 4–1. Here we shall consider some applications of the PCR to the detection and identification of mutations.

Let us suppose that we want to identify mutations in a gene that is known to be capable of causing a genetic disease when it is abnormally expressed. We could use the PCR to search for those mutations at the genomic level or at the mRNA level. We shall first choose the latter, making the assumption that the mRNA is ordinarily present in white blood cells or some other tissue easily obtained from patients.

RNA would be prepared from a patient's cells and converted to cDNA with reverse transcriptase in the standard manner. PCR primers corresponding to 5′ and 3′ flanking sequences of the translated region of the mRNA would be synthesized. Thus, by choosing primers that will bind to cDNA 5′ to the initiation codon and 3′ to the stop codon, it becomes possible to amplify the entire polypeptide-coding sequence of an average-sized mRNA in one reaction. When the Polymerase Chain Reaction begins, one primer binds to the 3′ position on the single-stranded cDNA and DNA polymerase converts it to double-stranded DNA. After that, both primers will be active and the region between the primers will be amplified as shown in Figure 6–1A.

After the target sequence has been amplified, there are several ways to identify mutations that may be responsible for the patient's illness. Probably the most thorough method is to sequence the amplified DNA. In this way, every difference in coding nucleotides between the patient's gene and the standard normal gene will be identified. Missense mutations, nonsense mutations, frameshifts, and major deletions or insertions will be obvious, as will changes in sequence that do not lead to amino acid replacements. Occasionally, there may be an additional amino acid replacement that is really an innocent polymorphism, but those rare occurrences would not create insurmountable confusion. A good example of this PCR + sequencing strategy for identification of mutations is in Gibbs et al. (1989); they identified nucleotide alterations in 15 independently arising cases of HPRT-deficiency.

The PCR may also be used to search for mutations in genomic DNA. An entire gene cannot be amplified with a single set of PCR primers, because technical factors restrict the maximum distance between primers to about 2000 bp. Therefore, an ideal gene for a mutant search via PCR would be one whose exons had been completely sequenced, and for which about 20 nucleotides flanking the exons at each end of every intron were also known, so that PCR primers could be designed that would bind to the flanking sequences (Figure 6–lB).

In such a situation, every exon could be amplified, independently of

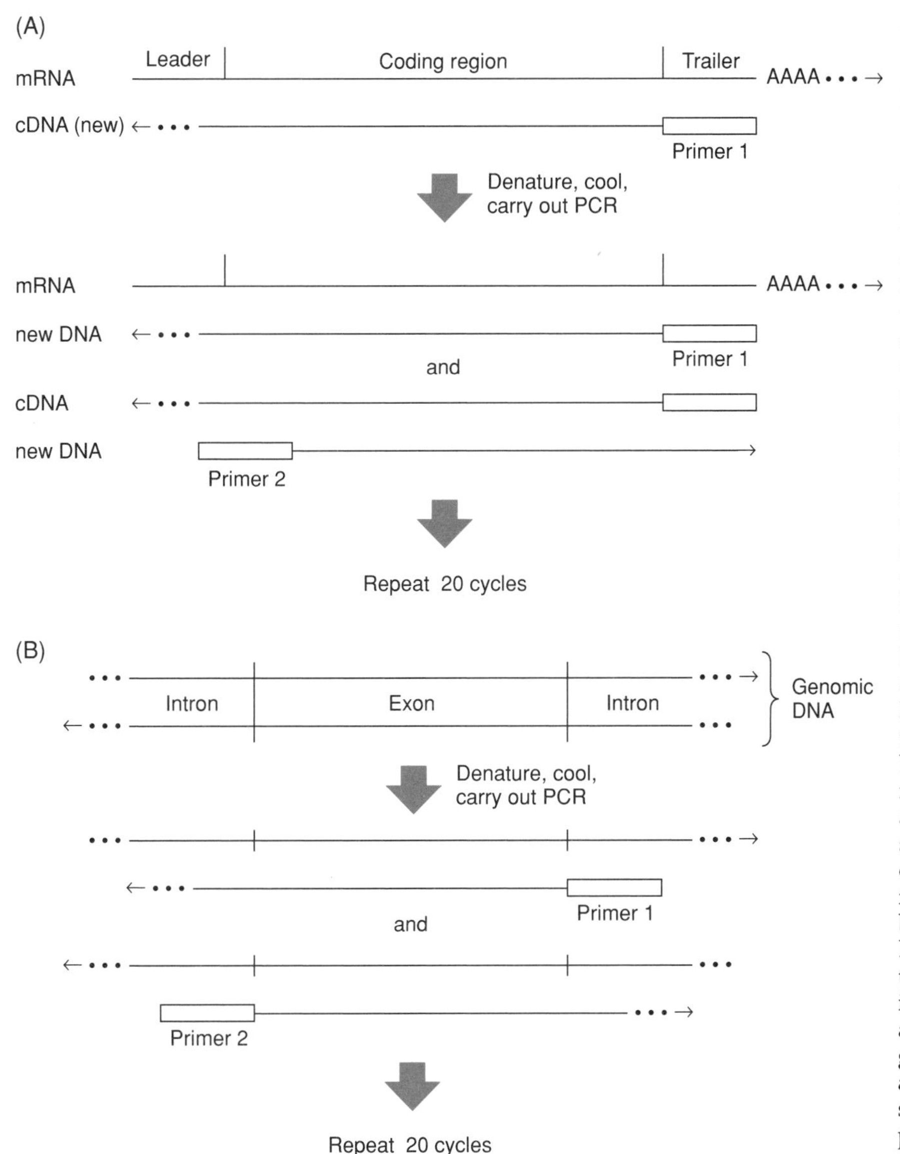

Figure 6–1

Use of the Polymerase Chain Reaction to amplify nucleotide sequences that are to be searched for mutations. (A) Amplification of mRNA. The first stage in the reaction occurs when a primer that is complementary to the sequence 3′ of the stop codon binds to the mRNA and a cDNA is synthesized by reverse transcriptase (the figure shows only the products of that reaction). In the second stage, another cDNA is made from the mRNA and the original cDNA serves as a template upon which a second DNA strand is synthesized, beginning with a primer that is complementary to the sequence that is 5′ of the translation initiation codon. The DNA products are doubled at each of the subsequent cycles. If the mRNA is much larger than 2000 nucleotides, it would be necessary to use two or more pairs of primers, and amplify the mRNA in several sections. (B) Amplification of an exon from genomic DNA. The figure shows one exon, bounded by introns. Genomic DNA is denatured and primers that are complementary to the intron sequences immediately adjacent to the exon bind to the genomic DNA, and DNA copies are synthesized. Each subsequent cycle doubles the product DNA, which consists of the exon plus the intron sequences to which the primers are complementary (about 20 nucleotides on each end).

In both panels, the arrowheads indicate the 3′ end of the nucleic acid strand; dots imply an indeterminate number of nucleotides.

introns. It is possible to amplify several exons in the same reaction ("multiplex PCR"), if the appropriate primer pairs are present. After amplification, nucleotide alterations in the exons could be identified by DNA sequencing, or by a combination of other methods plus sequencing. This strategy would initially miss mutations deep within introns, which might cause RNA processing errors (Chapter 7); it would also miss mutations in transcriptional control regions. However, the absence of mutations from all exons in DNA from an affected person would surely stimulate a more detailed search in the rest of the gene.

Another new technique that is likely to be very helpful in the detection of nucleotide changes, when used in conjunction with the PCR, is single-strand conformation polymorphism (SSCP) analysis (Orita et al., 1989). As mentioned earlier in this section, the migration of a protein through a gel in response to an electric field is a function of mass, net charge on the protein, and conformation. The same rules apply to polynucleotides.

Consider now the special case of a small DNA segment (between a few dozen and a few hundred bp) that has been amplified by the PCR, and then denatured. There will be two single-stranded molecules of identical size and therefore identical charge. (The charge on a polynucleotide at near-neutral pH is almost entirely caused by the phosphate groups, and every nucleotide contributes one phosphate group.)

In a gel of appropriate composition, those two single-strand DNAs will be separable by electrophoresis. Under non-denaturing conditions, single-stranded DNA folds in a complex manner, producing a three-dimensional structure that is stabilized by intrastrand base pairing and other interactions. Complementary sequences will have very different conformations and will move at different rates through a suitable gel. This is not surprising, in view of the extreme difference in sequence represented by the two strands of an originally double-stranded molecule.

But the demonstration by Orita et al. (1989) that a *single base change* in a DNA segment up to at least 200 bp in overall length can be detected by gel electrophoresis is truly astonishing (see Figure 6–2). Clearly, this sensitive new technique, SSCP analysis, has the potential to be an important intermediate step in the detection of mutations. When a large series of samples has been amplified by PCR, those samples that contain one or more nucleotide differences from the reference sequence can be quickly identified by SSCP analysis. Then only the electrophoretically variant samples need to be sequenced, in order to determine whether a mutation is present, and if so, what nucleotide has been changed.

Before we leave the PCR, it is worthwhile to take note of what the reaction *cannot* do. First, the amplification of specific sequences depends upon specific oligonucleotide primers; therefore, the gene to be studied must have already been at least partially characterized in terms of sequence. Second, in regard to mutation studies, the PCR is at present more useful for

Figure 6–2

Single-Strand Conformation Polymorphism (SSCP) analysis of point mutations in exon 1 of the KRAS2 gene. DNA from A549 cells (lane 1), SW480 cells (lane 2), and normal leukocytes (lane 3) was amplified with the PCR, using primers for exon 1 of the KRAS2 gene. A549 lung carcinoma cells contain a G to A transition in codon 12; SW480 colon carcinoma cells contain a G to T transversion in the same codon. Amplified DNA was subjected to electrophoresis after being denatured and the radioactive single-stranded DNA products were located by autoradiography of the dried gel. Lane N contained native DNA. (Redrawn from Orita el al., 1989.)

identifying nucleotide changes in persons who are already known to be carrying one or two mutant alleles at a specific locus, than for mutation rate determinations. Accurate rate measurements would require assays on hundreds of thousands of samples, and the fidelity of the polymerase would have to be substantially greater than one nucleotide error per million nucleotides incorporated. It is not that accurate (Thilly et al., 1988).

Nevertheless, the PCR could be used to help determine autosomal dominant mutation rates in cases where incomplete penetrance, variable expressivity, or other factors make it difficult to know whether a parent of an affected child really has the mutant allele, or whether the child contains a new mutant allele. When the mutation in the child's genome has been characterized in terms of nucleotide sequence, PCR primers for the appropriate part of the gene can be used with DNA from each of the parents. When the amplified segment has been sequenced, there will be no doubt about the presence or absence of the mutation in the parents' genomes.

The Indirect Method for Estimating Mutation Rates

The indirect method of estimating mutation rates depends upon the assumption that the loss of mutant alleles from a population because of nonreproduction or reduced reproduction by affected individuals is balanced by the spontaneous creation of new mutant alleles having the same phenotypic effect (they need not be identical at the nucleotide level). In order for this assumption to be valid, a population must be in a stable state known as "Hardy-Weinberg equilibrium."

Among the first people to consider the implications of Mendel's laws of inheritance for the genetics of populations were the English mathematician G. H. Hardy and the German physician Wilhelm Weinberg, who independently derived a mathematical expression in 1908 which has come to be regarded as the fundamental formula of population genetics. The Hardy-Weinberg Law expresses the relationship between allele frequencies at a given locus and the frequencies of genotypes involving those alleles.

Consider a locus with two alleles, which we will designate as A and A′, in order to avoid any implications about dominance and recessiveness, which are irrelevant to the calculations.

Let p = frequency of allele A
and q = frequency of allele A′
Then $p + q = 1$.

There are three possible genotypes: AA, AA′, and A′A′. The frequency of those genotypes must be p^2, $2pq$, and q^2, respectively. Notice that the frequency of heterozygotes is $2pq$, rather than pq, because there are two ways to make a heterozygote: allele A from the male and allele A′ from the female, or allele A′ from the male and allele A from the female.

The formula $p^2 + 2pq + q^2 = 1$ is the Hardy-Weinberg Law, as applied to a two-allele locus. If there are more than two alleles, the law can be extended to cover them with more classes of genotypes; but we will restrict our examples to two-allele systems. The Hardy-Weinberg Law is an equilibrium relationship; that is, in a population where no other factors act on allele frequencies, the Hardy-Weinberg formula will give the genotype frequencies generation after generation.

What might some of those "other factors" be? First, non-random mating (also called assortative mating) will disturb the Hardy-Weinberg equilibrium. For example, if AA homozygotes preferentially mate with AA homozygotes and A′A′ homozygotes preferentially mate with A′A′ homozygotes, there will eventually be no heterozygotes, even though the frequencies of alleles A and A′ may not have changed.

Second, there must be negligible migration into the population if the Hardy-Weinberg equilibrium is to be maintained, unless allele frequencies among the immigrants are identical to those among the residents. Migration out of the population can also disturb the equilibrium, unless all genotypes leave in proportion to their frequencies.

Third, there must be no selection on any of the genotypes. Selection for or against one of the genotypes can alter the genotype frequencies predicted from the allele frequencies. However, the predicted genotype frequencies can be modified to account for selection, when the effect of selection is known in a particular case and is constant.

Fourth, the population must not be so small that genetic drift (random fluctuations) can significantly alter genotype frequencies from one generation to the next. Of course, over long periods of time, chance will lead to the elimination of virtually any allele from any population, in the absence of selection; however, the Hardy-Weinberg relationship for a given pair of alleles does not have to be valid for eons in order to be useful.

Fifth, in its purest form the Hardy-Weinberg equilibrium requires that allele frequencies should not be changing because of new mutations entering the population. Although we know that new mutations certainly do occur, their frequency is so low that they do not play a major role in altering the frequency of common alleles.

However, in the context of this chapter, we are interested in rare alleles that cause genetic diseases. In a relatively stable population, it is predictable that the production of new deleterious alleles by mutation at a given genetic locus will balance the loss of deleterious alleles because of the reduced fecundity of persons carrying those alleles. If this occurs, the mutation rate can be calculated from the frequency of occurrence of affected persons.

Before we illustrate the indirect method for estimating mutation rates, it is appropriate to point out that modern human populations have become so mobile that they certainly do not meet the criteria for an equilibrium population. Some authors feel that it is a waste of time to apply the indirect method to human mutation rate calculations (e.g., Vogel and Motulsky, 1986). Nevertheless, many estimates of mutation rates by the indirect method are in the same range as direct estimates. For that reason, we shall briefly examine the indirect method.

Autosomal Dominant Conditions In order to apply the indirect method for calculating mutation rates to autosomal dominant conditions, one must know the *reproductive fitness* of the condition. Fitness is defined as the ratio of the average number of offspring produced by affected persons to the average number of offspring produced by the general population. When a condition is incompatible with reproduction, fitness is zero. For autosomal dominant diseases, the relevant formula, with f representing fitness, is:

$$\mu = 0.5(1 - f)\mathrm{I}$$

Notice that this formula uses I, not I*; that is, the indirect method calculates the mutation rate from the frequency of affected persons, whether they have an affected parent or not. Also note that if $f = 0$, the preceding formula becomes the same as the formula for the direct method, because I* becomes I, there being no offspring of affected persons. If $f = 1$, the formula becomes useless. It is also possible for f to be greater than one, indicating enhanced reproduction by affected individuals. In that case, the assumptions underlying the indirect method are violated, and the method is inapplicable.

As an example, consider the data on dwarfs in Denmark given in a previous section, and assume that $f = 0.2$ for achondroplasia. Then

$$\mu = (0.5)(1 - 0.2)(10/94{,}075) = 0.00004$$

This is the same answer that was obtained from the direct method, using the 8 out of 10 dwarfs who were born to non-dwarf parents. Apparently, a fitness of 0.2 is consistent with that calculation.

Autosomal Recessive Conditions Because of the difficulty of identifying heterozygotes for most mutations that cause recessive genetic diseases in humans, the indirect method for estimating mutation rates is usually used. The formula is:

$$\mu = (1 - f)\text{I}.$$

This formula differs from the corresponding formula for autosomal dominant mutations only in the absence of the 0.5 factor. The reason is that for every person affected with a recessive genetic condition, there are two mutant alleles, not one.

Consider an example of a disease that is always fatal before the age of reproduction ($f = 0$), and suppose that the frequency of occurrence of affected persons is 1×10^{-5}. This implies that 2 alleles out of a population of 200,000 alleles are lost every generation, because of non-reproduction. If the population is at equilibrium, mutation will create two new recessive alleles every generation; in other words, the mutation rate must be 2/200,000, which is 1×10^{-5}, the same as the frequency of affected persons. For conditions that allow some reproduction, it is necessary to know the relative fitness in order to use the indirect method for calculating the mutation rate, as stated in the formula.

X-linked Recessive Conditions In order to illustrate the application of the indirect method to the calculation of X-linked recessive mutation rates, we will restrict the discussion to those diseases that have zero or nearly zero fitness, so that we can ignore the possibility of affected females. First, recall that the indirect method depends on the assumptions that the loss of mutant alleles because of non-reproduction of affected persons (only males in this case) is balanced by the spontaneous occurrence of new mutant alleles, and that the size of the population is stable. Then, it is essential to recognize that for every affected person, there must be *three* mutant alleles in the population: one in an affected male and two in carrier females, because females have twice as many X chromosomes as males.

Figure 6–3 presents a numerical example for a population of 2,000,000 people with a recessive lethal X-linked disease that has a mutation rate of 1×10^{-6}. Notice that it is necessary to assume the existence of a specific number of females who are carriers of pre-existing mutations, in order to validate the assumption that equilibrium exists. In particular, there will be twice as many eggs with pre-existing mutations (4), as eggs with new mutations (2). Inasmuch as those 6 eggs will be fertilized indiscriminately by X-bearing and Y-bearing sperm, half of them will become carrier females and half will become affected males.

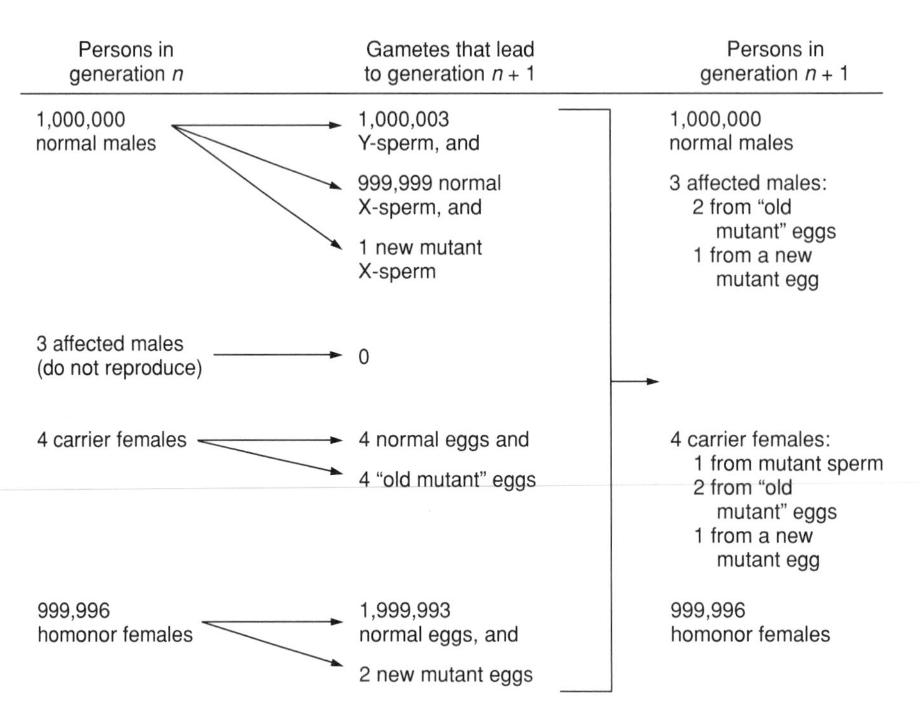

Figure 6–3

Origin of affected males for an X-linked recessive disease in an equilibrium population containing 1,000,000 reproductively active persons of each sex, in which an X-linked reproductive-lethal mutation occurs at a frequency of 1×10^{-6} per locus per generation. In order to make the population in generation $n + 1$ equal the population in generation n, the figure uses a "fudge factor" of three excess Y-bearing sperm and three excess normal eggs. In a real population, there would be fluctuations from one generation to the next. The valid, but remote, possibility that a mutant sperm will fertilize a mutant egg has been ignored. The term "homonor" is an abbreviation for "homozygous normal."

Further study of Figure 6–3 will reveal that two-thirds of the affected males in each generation will arise from eggs that contained pre-existing X-chromosome mutations and one-third will arise from eggs that contained new mutations. Therefore, if the loss of mutant alleles via reduced reproductive fitness of affected males is balanced by new mutations, the mutation rate must be:

$$\mu = (1/3)I^m$$

where I^m is the frequency of affected males, that is, the ratio of affected males/total males. This result has potentially serious consequences for genetic screening programs, because it implies that even if heterozygote females could be detected with 100% efficiency, the rate of occurrence of affected males would still be one-third of its present rate. However, the calculation assumes that mutation rates are the same in male and female germ cells, which is apparently not true for at least some genetic loci (see next section).

Differences in Male and Female Mutation Rates

Many studies in mice have found that the mutation rate at a given locus is generally higher in male animals than in females. The example that was cited in the section on autosomal recessives in mice is pertinent: the average locus-specific mutation rate for females was 1.4×10^{-6}, whereas for males it was 8.1×10^{-6}. Other studies in humans, using less direct methods of analysis for X-linked diseases, have given results consistent with the mouse results.

It is generally assumed that male mammals have higher mutation rates because male gametes are the product of many more cell divisions, on average, than are female gametes. The latter are all formed before the organism is born, whereas sperm are formed continuously throughout adult life. Vogel and Motulsky (1986) estimate that the sperm in a 25-year-old man are the product of 300 cell divisions, compared to the roughly 24 cell divisions that occur between zygote and primary oocyte. Mistakes in DNA replication must be a significant component of the overall mutation rate, and therefore, the more cell generations that have taken place between the zygote and the formation of a mature gamete, the higher the probability that a mutation has occurred.

A molecular method for measuring the mutation rate at X-linked loci, which is capable of distinguishing between mutations that arose in male versus female gametes, now exists. That method is the Polymerase Chain Reaction, which was described Box 4–1 and mentioned earlier in this chapter. Consider the mother of a boy affected with a recessive lethal X-linked disease. As we saw in the preceding section, there is a significant probability that the boy carries a new mutation, which would necessarily have come from his mother. If the boy's mutation can be characterized at the molecular level, application of the PCR to the corresponding portion of his mother's genome, followed by sequencing or any other suitable method of detecting the mutation, should reveal whether one of her chromosomes contains the mutation.

If the mother has the same mutation found in the boy, DNA from her parents can also be examined, using the same technology. In those cases where it turns out that she received a new mutation from one of her parents, the origin of the mutation can be determined by *haplotype analysis.* This strategy depends upon the assumption that there are enough polymorphic loci closely linked to the disease locus to create a unique haplotype for every X chromosome in the family. Figure 6–4 diagrams a simple, hypothetical example. When the mother's DNA or the grandmother's DNA is analyzed, one does not get haplotypes directly; these are deduced from the information on the son and grandfather, and there can be some uncertainty because of the possibility of recombination. Ignoring that, we see that in this example, the son has received his maternal grandfather's X chromosome, and therefore the mutation occurred in the grandfather's germ line.

If the son's haplotype had also been present in the grandmother, no conclusion could have been reached; but if the son's haplotype had been present in the grandmother only, it would have been clear that the mutation occurred in her germ line. Given that several polymorphic restriction enzyme cleavage sites are available near the disease locus, most families should allow determination of the grandparental source of the mutation.

If a large series of similar families were examined, in which it could be shown that the mother of an affected boy received a new mutation from one of her parents, the sex-specific mutation rate at that locus could be determined. A variety of X-linked disease-causing loci are currently being studied with that goal in mind; examples include Duchenne muscular dystrophy and the hemophilias.

We can now return to a question raised in connection with indirect estimates of mutation rates. If molecular analysis shows that fewer than one-third of affected boys receive new mutant X chromosomes from their mothers, can a sex-specific difference in mutation rate explain the depar-

Son	Mother		Grandfather	Grandmother	
1	1	2	1	2	2
2	2	2	2	2	1
m	*m*	+	+	+	+
1	1	1	1	1	1
2	2	1	2	1	1

Figure 6–4

Use of haplotypes to identify the source of a new mutation in an X-linked gene. Each column represents a hypothetical haplotype for four RFLP loci, each with two alleles (indicated by 1 or 2); and the disease locus, where + indicates the normal allele and *m* the mutant allele. It is assumed that the presence or absence of the mutant allele can be detected by some direct molecular assay, such as hybridization to an allele-specific oligonucleotide or PCR amplification of a portion of the gene, followed by sequencing. In either case, knowing that the mutation is present in the mother but absent in both of her parents does not tell us which of her parents was the source of the mutant gamete. Haplotype analysis, using closely linked polymorphic loci, solves that problem. In this example, it is clear that the affected boy has his grandfather's X chromosome; therefore, the mutation that he and his mother possess must have originated in his grandfather's germ cells.

ture from the simple equilibrium prediction? Yes, it can. Let's imagine another population of two million persons, but this time we shall assume that the female mutation rate is 1×10^{-6} per allele per generation and male mutation rate is 5×10^{-6} per allele per generation.

Figure 6–5 shows what would happen in this population, if all other factors allowed equilibrium between loss of mutant alleles in affected males and their precise replacement by new mutations. The conclusion from Figure 6–5 is that only 1/7 of affected males would be the result of new mutations. We can generalize this relationship. If the rate of mutation at a given X chromosome locus is higher in males than in females, then in a population at equilibrium, less than 1/3 of the affected males will be new mutants.

The Impact of Mutation on Disease

In this section we shall consider two questions. First, how many potentially disease-causing recessive alleles are carried in the average human genome; and second, what is the impact of mutation on mortality and morbidity?

Genetic Load

The average number of recessive lethal alleles carried in the genome of the average individual is called the *genetic load*. An allele is considered to be a *recessive lethal* if, in individuals homozygous for that allele, death occurs before the age of reproduction. The total genetic load includes *lethal equivalents*; that is, alleles that cause death less than 100% of the time, when homozygous, are counted as fractional lethals, and the fractional lethals are summed to yield lethal equivalents. For example, if allele A at locus 1 causes lethality in homozygotes 67% of the time, allele B at locus 2 causes lethality in 42% of homozygotes, and allele C at locus 3 causes lethality in 91% of homozygotes, the three alleles together represent two lethal equivalents.

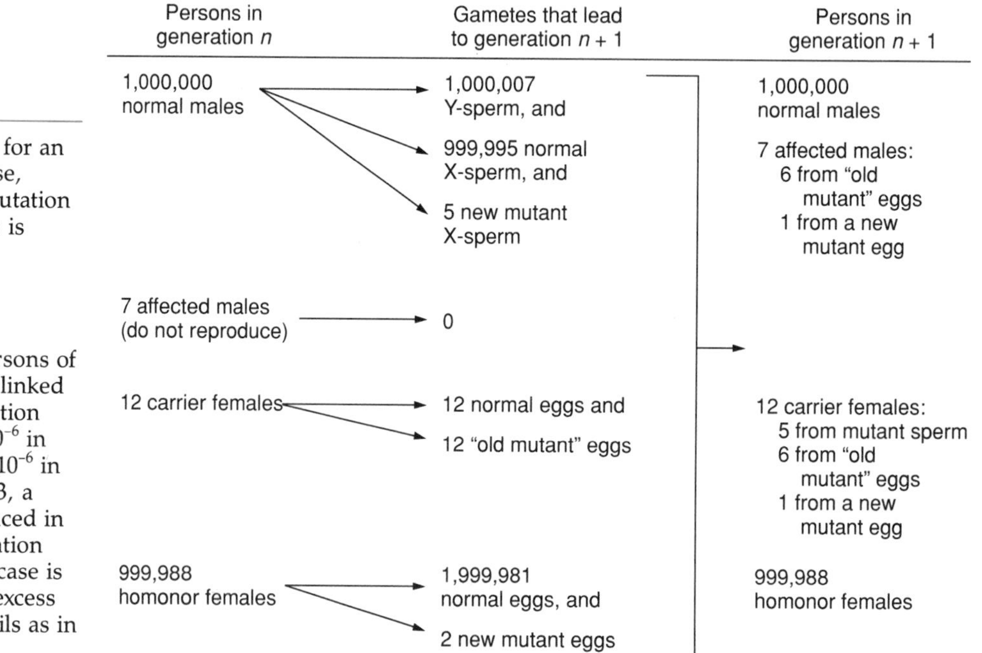

Figure 6–5

Origin of affected males for an X-linked recessive disease, when a higher rate of mutation in males than in females is taken into account. The example assumes an equilibrium population containing 1,000,000 reproductively active persons of each sex, in which an X-linked reproductive lethal mutation occurs at a rate of 5×10^{-6} in males and a rate of 1×10^{-6} in females. As in Figure 6–3, a "fudge factor" is introduced in order to keep the population stable; the factor in this case is 7 excess Y-sperm and 7 excess normal eggs. Other details as in Figure 6–3.

Genetic load is estimated by comparing mortality among the progeny of consanguineous matings to mortality among the progeny of matings between unrelated individuals. The reason for using consanguineous matings is that there is a significant probability of an allele being "homozygous by descent" (a concept that we met in Chapter 4 in connection with linkage analysis). The higher the degree of consanguinity, the higher the probability of homozygosity by descent.

In laboratory animals, genetic load estimates are based on brother-sister matings, where the probability that a recessive allele present in one of the parents of the siblings who are mated will be homozygous in one of their progeny is 1/16. As shown in Figure 6–6, the probability that any allele present in generation I will be present in both the brother and sister in generation II is 1/2 × 1/2 = 1/4. Then the probability that an offspring of that mating will receive the recessive allele from both parents is 1/4. The probability of homozygosity by descent *for a specific allele* in an individual in generation III is therefore 1/4 × 1/4 = 1/16. The overall probability of homozygosity by descent for any of the four gene copies at that locus in the grandparents is 4 × 1/16 = 1/4. The latter quantity is called the *coefficient of inbreeding*.

Brother-sister matings are exceedingly rare among humans, being proscribed in virtually all societies. However, first cousin matings, although they are much less frequent in most of the world than they used to be, still occur often enough to provide statistically significant data. Consider the pedigree of a first cousin marriage in Figure 6–7. First cousins have one set of common grandparents—the middle couple in generation I of Figure 6–7. Therefore, it is possible for both cousins to carry any given recessive allele

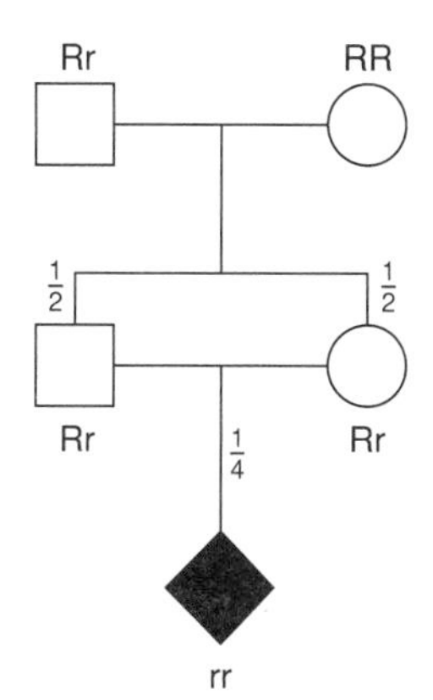

Figure 6–6

A pedigree showing a brother-sister mating. The probability that a recessive allele (r) that is present in one of the parents of the siblings who mate will be transmitted to both siblings and subsequently transmitted in homozygous form to an offspring of the sibling mating is 1/2 × 1/2 × 1/4 = 1/16.

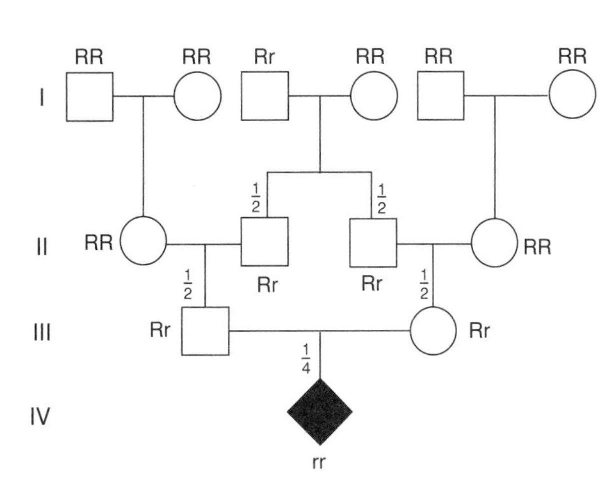

Figure 6–7

A pedigree of a first cousin mating. The two central individuals in generation I are common grandparents of the cousins in generation III. Any allele present in one of the common grandparents has the probabilities of being transmitted to individuals in subsequent generations shown in the diagram. The overall probability of a child of the first cousin mating receiving two copies of allele r from one of the common grandparents is 1/2 × 1/2 × 1/2 × 1/2 × 1/4 = 1/64.

that happened to be present in one of those shared grandparents (indicated by r in the figure). The probability that one of the cousins has allele r is $1/2 \times 1/2 = 1/4$; the probability that both cousins have allele r is $1/4 \times 1/4 = 1/16$. If both cousins are heterozygotes for the same allele, the probability that any given child will receive both copies of that allele is 1/4; therefore, the overall probability that a child born to first cousin parents will be homozygous for a specific allele present in one of the common grandparents of the cousins is $1/16 \times 1/4 = 1/64$, which can also be expressed as 1.56%.

Because the children of first cousin marriages will also be exposed to all the risks of death to which the children of unrelated parents are exposed, both genetic and non-genetic, the effect of consanguinity must be measured as *excess mortality*. The total frequency of early death among the progeny of first cousin marriages (i.e., death prior to birth, in neonates, and in juveniles) is determined, then the total frequency of early death among the progeny of unrelated parents is subtracted, and the remainder is the excess mortality. In carrying out studies of this sort, it is essential that the control group be strictly comparable to the consanguineous group. Any major differences in socioeconomic status or lifestyle could affect the risk of death from non-genetic causes and thereby lead to an erroneous result.

Each recessive lethal allele or lethal equivalent in the combined genomes of the common grandparents of first cousins who marry will contribute 1.56% excess mortality to their offspring, on a statistical basis. However, genetic load is expressed in terms of recessive lethals *per individual*, so we must introduce a factor of two. That is, for every recessive lethal allele per individual, there will be $2 \times 1.56\% = 3.12\%$ excess mortality in the progeny of first cousin matings. We shall round this off to 3%; we shall also ignore the small probability that both of the grandparents are heterozygotes for recessive lethal alleles at the same locus. Notice that genetic load, unlike almost everything else in genetics, is not expressed as units per haploid genome.

What is the extent of the human genetic load? A variety of studies on first cousin matings have been carried out, and although different investigators get somewhat different results, 12–15% excess mortality is a representative finding. On the basis of the calculations in the preceding paragraph, this implies an average genetic load of 4 or 5 recessive lethal alleles and/or lethal equivalents.

Surveys of excess mortality do not measure early embryonic deaths; therefore, the true genetic load is greater than the measured genetic load. The effects of deleterious recessive alleles that become homozygous by descent and lead to reduction of fecundity and/or to premature adult death are also omitted from the typical calculation of genetic load.

Before leaving the subject of genetic load, it is worth asking whether the preceding estimate of 4 to 5 recessive lethals per person is consistent with the estimates of mutation rates that we considered earlier in this chapter. Suppose the average mutation rate is 10^{-6} per locus per generation. In equilibrium populations, the frequency (q) of a recessive lethal allele is the square root of the mutation rate (μ), because the incidence of affected individuals (I) $= q^2 = \mu$. Therefore, if the average frequency of recessive lethal alleles is 10^{-3}, what would the average total number of such alleles be per diploid genome (i.e., per person)?

In order to answer that question, we must make an assumption about the total number of genes that can mutate to recessive lethal alleles. It is surely less than the total number of genes, because many genes are dispensable. For example, we can survive quite well without the gene for brown eye color or the gene for retention of hair in elderly males or the gene for tasting PTC. There is no way to make a truly informed estimate about the number of essential genes, so let's choose 1×10^{-4}, just for the sake of argument. If we multiply 10^{-3} (the average frequency of recessive lethal alleles assumed above) by 10^{4} (the assumed number of essential genes), we find that there would be 10 recessive lethals per haploid genome; that is, the genetic load would be 20 recessive lethals per diploid genome.

Might the average genetic load in humans be as high as 20, instead of 5, the estimate based on observable excess mortality in the progeny of first cousin matings? One way to evaluate this possibility is to estimate what the excess mortality would be if there are really 20 recessive lethals per person. To make that estimate, we cannot multiply 3% by 20, because that would ignore the statistical certainty that many offspring of first cousin marriages would be homozygous for more than one lethal allele; and humans, unlike the nine-lived cats of folklore, can only die once.

The proper question is, what percent of the progeny of first cousin matings would be homozygous for *none* of the 40 recessive lethal alleles in the combined genomes of the cousins' common grandparents? Inasmuch as the probability of a child not being homozygous for one recessive allele is 63/64, the excess mortality expected for a genetic load of 20 per person is given by:

$$1 - (63/64)^{40} = 1 - 0.53 = 0.47$$

Therefore, we predict a total excess mortality of 47% among the progeny of first cousin matings, given a *total* genetic load of 20 recessive lethals per person. This is about 3X the observed genetic load, which is based on fetal, neonatal, and juvenile deaths. If the estimate is accurate, most of the non-observed excess mortality would have to occur during embryogenesis.

We can conclude that the observed genetic load (4 to 5 recessive lethals per person) is roughly consistent with an average mutation rate of 1×10^{-6}, assuming approximately 10,000 essential genes. Alternatively, a better fit between observation and prediction could be obtained if the average mutation rate were lower than 1×10^{-6}, or there were fewer than 10,000 essential genes. Note that we are dealing here with mutations to autosomal recessive lethal alleles only. Dominant mutations of all kinds, recessive mutations that do not affect the function of a gene seriously enough to cause death of homozygous individuals at any time, and totally benign mutations are not included in these calculations.

The Impact of Mutation on Mortality and Morbidity

If we knew the average mutation rate per locus and the number of essential genetic loci, we could calculate the contribution that mutation makes to death and serious disease in the human population. Although such calculations must be based on rather uncertain numbers, rough estimates can be made. For the calculations that follow, we shall continue to use a mutation rate of 1×10^{-6} per allele per generation and 10,000 genes whose loss leads to death prior to adulthood.

Let's first consider dominant lethal mutations, where all cases necessarily represent new mutations. Assuming that a zygote contains 20,000 alleles that can give rise to dominant lethal alleles (10,000 from each parent), the probability that a zygote contains no dominant lethal is

$$(0.999{,}999)^{20{,}000} = 0.98$$

This tells us that 2% of all conceptuses might be lost because of new dominant lethal mutations. Inasmuch as roughly 40% of all human conceptuses fail to be born, for all genetic and nongenetic reasons, dominant lethal mutations are unlikely to be a major cause of premature death.

If we extend the preceding calculation to recessive lethals, we must subdivide the question, because affected individuals must contain two copies of the recessive allele at a given locus. First, the probability that a zygote contains two *new* recessive lethals is negligible (10^{-12}). Next, given a mutation rate of 1×10^{-6}, and therefore an allele frequency of 1×10^{-3}, the probability that a zygote will receive one pre-existing recessive allele is approximately 2×10^{-3}. The probability that the zygote has one pre-existing and one new mutant allele is $2 \times 10^{-3} \times 10^{-6} = 2 \times 10^{-9}$, which is also a negligible number.

If a zygote has two recessive alleles at a particular locus, it is highly likely that it received pre-existing alleles from both parents. As we have seen, a mutation rate of 1×10^{-6} implies that the frequency of homozygotes will also be 1×10^{-6}. The frequency of non-homozygotes will therefore be 0.999,999 and if we raise that number to the 10,000th power, we get 0.99. In other words, only 1% of all conceptuses would be homozygotes for recessive lethal alleles, if the given assumptions apply. Most of these would probably be lethal before birth; accordingly, the frequency of children born with potentially fatal recessive genetic diseases should be substantially less than 10 per 1000, which is consistent with Table 1–1.

Somatic Mutation in Humans

Germ cells are not unique in being susceptible to genetic change; mutations of many types occur in somatic cells, also. The rate of somatic mutation is of interest for several reasons. If it varies from individual to individual, it may be indicative of a person's inherent risk of developing cancer. It may also be possible to estimate a person's exposure to mutagenic chemicals by measuring somatic cell mutation rates. Finally, knowledge obtained from studying somatic mutations may be useful in understanding the causes of germ cell mutations, even though there is no reason to expect overall rates to be the same in the two classes of cells.

Fibroblasts and lymphocytes have been used in several attempts to measure the somatic mutation rate in humans, because they can be cloned in culture. The enzyme HPRT has been the most frequently studied system, because the HPRT locus is on the X chromosome and because mutant cells lacking HPRT can easily be selected. The X chromosome location means that cells from males are hemizygous for the HPRT gene; therefore, only one mutation is necessary to produce a cell with a mutant phenotype.

A typical experiment would be designed as follows. Prepare a culture of normal male cells in the presence of thioguanine (TG), a toxic purine analog. Cells that are $HPRT^+$ will incorporate TG into their DNA and die. Cells that are $HPRT^-$ will not be able to use the TG; each will prosper and form a colony in the culture vessel (Figure 6–8). If one million cells are put into the vessel initially and 10 TG-resistant colonies form, then the frequency of $HPRT^-$ cells is 1×10^{-5}.

One recent study on lymphocytes from 14 normal males (Morley et al., 1983) found that the average frequency of TG-resistant colonies was 3.0×10^{-6}. However, there was a lot of variation; the 95% confidence limits were 1.6×10^{-5} to 5.4×10^{-7}. Even so, the result corresponds satisfactorily to the mutant frequency measured for the same locus in fibroblasts.

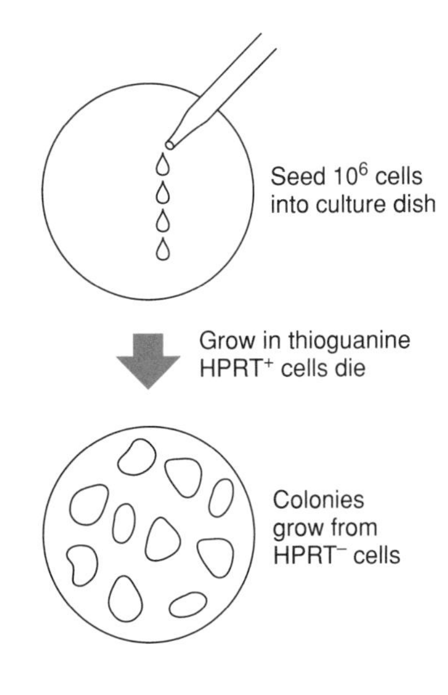

Figure 6–8

Determination of the frequency of spontaneous mutants at the HPRT locus in somatic cells. Cells such as fibroblasts or lymphocytes are placed into culture under conditions where wild-type cells (HPRT$^+$) are unable to survive. HPRT$^-$ cells cannot incorporate the toxic compound thioguanine into their DNA; accordingly, they survive and grow into individual colonies that can be easily counted.

Another good experimental system involves the glycophorin A (GPA) locus. GPA is the major cell surface sialoglycoprotein of red blood cells, present at about 500,000 copies/cell. Despite its prominence, its function is dispensable; there are rare, but apparently normal people who have no detectable GPA. The M and N blood groups are the products of allelic variants at the GPA locus; the proteins differ in two amino acids (positions 1 and 5 from the amino terminus). The alleles are almost equally abundant, and heterozygotes represent nearly 50% of the population.

There are now several monoclonal antibodies (MAbs are described in Box 12–1) that recognize either the M or N variant of GPA specifically. These antibodies make it possible to detect variant RBCs in MN heterozygotes; that is, cells that have become hemizygous or homozygous for expression of one allele. The assay can be done in several ways. For example, an M-specific MAb can be labeled with one type of fluorescent compound and an N-specific MAb with a different fluorescent compound. These are mixed with a sample of blood from a heterozygote and, after unbound antibody has been removed, the RBCs are passed through a fluorescence-activated cell sorter. The machines are sufficiently sensitive to distinguish cells that have bound both types of antibodies from those that have bound only one type, using the difference in emission wavelengths of the two types of fluorescent dyes coupled to the antibodies. Moreover, the machine can distinguish a cell with hemizygous expression of GPA from a cell that apparently expresses a homozygous level of GPA.

After analyzing blood from 15 normal donors, one group of investigators (Langlois et al., 1986) concluded that the average frequency of hemizygous RBC variants (either M- or N-) was 9.4×10^{-6}. Hemizygous expression of one allele implies that a null allele has been created at the other locus. In the context of an antibody assay, a null allele probably results from a mutation that effectively eliminates synthesis of the corresponding polypeptide. This would include many nonsense and frameshift mutations, as well as deletions and insertions. Most missense mutations would probably not be detected, because they would not prevent synthesis of a polypeptide that the antibody could recognize.

A somewhat surprising observation was that apparently homozygous variants occurred at an average frequency of 17.3×10^{-6}, or roughly twice the hemizygous frequency. Of course, RBCs have no nucleus and cannot be multiplied in tissue culture, so it is impossible to obtain direct proof of homozygosity. Nevertheless, dosage compensation at the level of gene expression is very rare in humans. A fact relevant to this particular situation is that there are some individuals who have inherited one null allele at the GPA locus, and these persons all express the remaining allele at the hemizygous level.

Homozygosity-generating mechanisms have been documented in numerous cases, especially in tumors (Chapter 11). These mechanisms include nondisjunction followed by duplication of the surviving chromosome, mitotic recombination, and gene conversion. Therefore, the occurrence of homozygosity at the GPA locus is not surprising in principle, but its relatively high frequency of occurrence is noteworthy, and it raises the question of whether homozygosity from mutation is similarly frequent at other loci.

The glycophorin A study included measurement of variants in RBCs of cancer patients who had received chemotherapy for an average of 20 weeks. Most were studied immediately after cessation of therapy. They had an average frequency of hemizygous variants of 30.7×10^{-6}, a 3–4 fold increase over the control value. Three patients who were sampled 12–26 weeks after cessation of therapy had hemizygous variant levels of 26, 40, and 60×10^{-6} respectively, indicating that the elevated rate of variant RBC production does not decline rapidly after the termination of chemotherapy.

Another study that used fluorescent antibodies to detect rare hemoglobin variants is of interest. In this case, the investigators had antibodies that were monospecific for Hemoglobins S or C (Stamatoyannopoulos, 1979). Each antibody would bind only to hemoglobin of that one type, not to normal hemoglobin or to any other variants tested. Each of those two variant classes of hemoglobin is caused by one specific nucleotide change. The results of this study showed an average mutant frequency of 1.1×10^{-7}, with a range of 4×10^{-8} to 3×10^{-7}.

It is difficult to compare the preceding data with the nucleotide substitution rate for germ cells (1×10^{-8} per allele per generation) estimated earlier in this chapter, for two reasons. First, the number of cell generations that the somatic cells in various experiments have undergone may not have been similar to the number of cell generations that germ cells undergo. Second, the studies cited in this section all measured the frequency of mutant cells, not the rate of production of new mutants. A valid comparison would require that the rate of mutation for both somatic and germ cells be expressed as a rate per cell generation.

Uncertainties notwithstanding, and with approximately 2.5×10^{13} red blood cells in the human body at any given moment, it is clear that each of us probably carries at least a million cells that express any known hemoglobin variant, including those that are associated with sickle-cell anemia, methemoglobinemia, and various thalassemias. The prospect is more amusing than threatening.

SUMMARY

Mutations are changes in the base sequence of DNA. Some mutations affect the structure and function of proteins and some of those effects lead to disease. This chapter focuses on the rate of mutation to disease-causing alleles, a rate that varies generally from 10^{-6} to 10^{-4} per gene copy per generation.

Mutation rates may be calculated by direct or indirect methods. In humans, the rate of mutation to autosomal dominant conditions can be obtained by measuring the frequency with which affected children are born in families where neither parent is affected.

Direct determination of the rate of mutation to autosomal recessive alleles in humans has depended primarily upon electrophoretic assays for variant proteins. Those studies yield an estimate of 10^{-5} mutations per gene copy per generation, on average.

The indirect method for estimating mutation rates depends upon the assumption that there is equilibrium between the rate of loss of mutant alleles from reduced reproduction of affected individuals and the rate at which new mutant alleles arise. There is general agreement between estimates based upon the indirect and the direct methods.

Mutation rates to recessive alleles on the X chromosome can be estimated from the frequency of affected males in the population. However, mutation rates at all loci (autosomal and sex chromosomal) tend to be several-fold higher in human males than in females, and that difference affects the calculation of a rate based on the incidence of males affected with a particular disease.

Genetic load is the average number of recessive lethal alleles carried in an individual diploid genome. Genetic load is usually estimated by determining the "excess mortality" among progeny of first cousin matings. The minimum number of recessive lethal alleles per person appears to be 4 or 5, and the total number may be several-fold higher.

Mutation rates can also be studied in somatic cells, such as fibroblasts and lymphocytes, which can be grown as cell clones in culture.

REFERENCES

Bodmer, W. F. and Cavalli-Sforza, L. L. 1976. Genetics, evolution, and man. W. H. Freeman, San Francisco.

Gibbs, R. A., Nguyen, P-N., McBride, L. J., et al. 1989. Identification of mutations leading to the Lesch-Nyhan syndrome by automated direct DNA sequencing of *in vitro* amplified cDNA. Proc. Natl. Acad. Sci. USA 86:1919–1923.

Jadayel, D., Fain, P., Upadhyaya, M., et al. 1990. Paternal origin of new mutations in Von Recklinghausen neurofibromatosis. Nature 343:558–559.

Langlois, R. G., Bigbee, W. L., and Jensen, R. H. 1986. Measurements of the frequency of human erythrocytes with gene expression loss phenotypes at the glycophorin A locus. Hum. Genet. 74:353–362.

Morch, E. T. 1941. Chondrodystrophic dwarfs in Denmark. Munksgaard, Copenhagen.

Morley, A. A., Trainor, K. G., Seshadri, R., and Ryall, R. G. 1983. Measurement of *in vivo* mutations in human lymphocytes. Nature 302:155–156.

Neel, J. V., Satoh, C., Goriki, K., et al. 1986. The rate with which spontaneous mutation alters the electrophoretic mobility of polypeptides. Proc. Natl. Acad. Sci. USA 83:389–393.

Neel, J. V., Satoh, C., Goriki, K., et al. 1988. Search for mutations altering protein charge and/or function in children of atomic bomb survivors: final report. Am. J. Hum. Genet. 42:663–676.

Nelson, K. and Holmes, L. B. Malformations due to presumed spontaneous mutations in newborn infants. New England J. Med. 320:19–23.

Novitski, E. 1982. Human genetics. 2nd ed. Macmillan, New York.

Orita, M., Suzuki, Y., Sekiya, T., and Hayashi, K. 1989. Rapid and sensitive detection of point mutations and DNA polymorphisms using the polymerase chain reaction. Genomics 5:874–879.

Stamatoyannopoulos, G. 1979. Possiblities for demonstrating point mutations in somatic cells, as illustrated by studies of mutant hemoglobins. In "Genetic Damage in Man Caused by Environmental Agents", K. Berg, ed., pp. 49–62. Academic Press, New York.

Thilly, W. G., Liu, V. F. Brown, B. J., et al. 1989. Direct measurement of mutational spectra in humans. Genome 31:590–593.

Treves, F. 1885. A case of congenital deformity. Trans. Pathological Soc. London 36:494–498.

Vogel, F. and Motulsky, A. G. 1986. Human genetics: problems and approaches. 2nd ed. Springer-Verlag, Berlin.

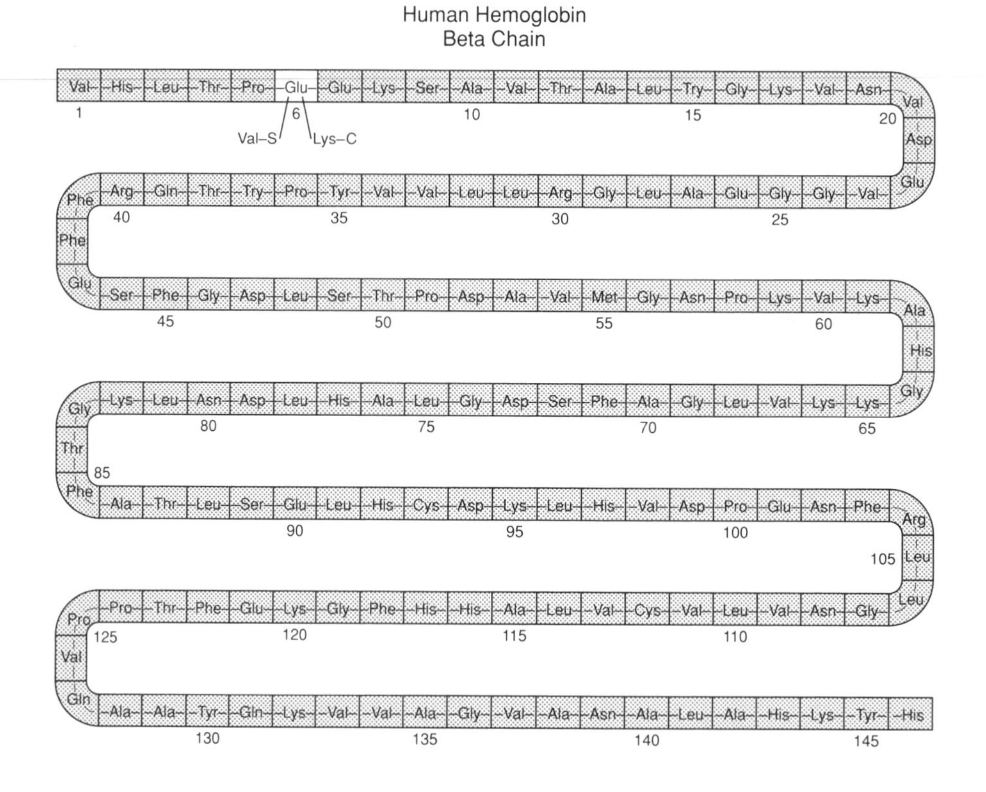

The amino acid sequence of human beta-globin. Mutations in codon 6 are responsible for sickle cell anemia (Glu—Val) and for Hb C disease (Glu—Lys).

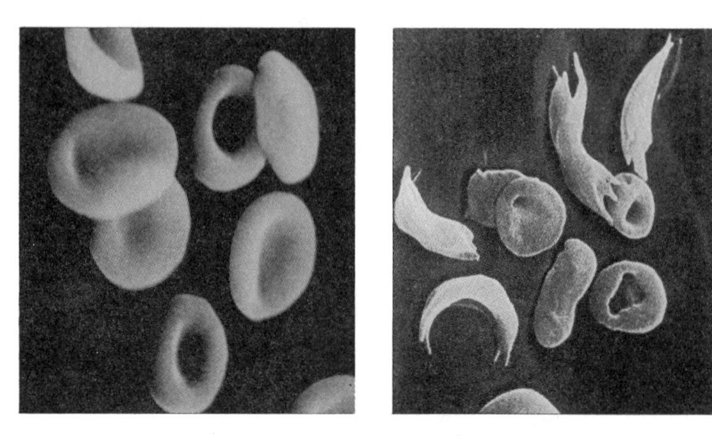

Normal red blood cells (left panel) and sickled red blood cells (right panel). (Left: © Omikron, Science Source/Photo Researchers; right: © Nigel Calder, Science Source/Photo Researchers)

7 Mutation

QUALITATIVE ASPECTS

One of the major generalizations that has emerged from molecular analysis of mutant genes is that most genetic diseases have heterogeneous origins. Two types of heterogeneity exist. The first is well known; it is *interlocus heterogeneity*. In other words, the same disease (defined in terms of clinical symptoms) may result from mutations in either of two or more loci. Examples include albinism, which is the end result of interrupting any one of several steps in the synthesis of melanin; and phenylketonuria, where a small fraction of affected individuals have normal phenylalanine hydroxylase, but lack the ability to make a cofactor necessary for that enzyme's activity. In general, whenever a disease results from lack of the end product of a metabolic pathway, mutations in any of the genes that govern the various steps in that pathway may produce the same clinical symptoms (Figure 7–1)

Identification of interlocus heterogeneity may occur in several ways. If a patient fails to respond to a treatment known to be effective in some cases with similar clinical symptoms, more detailed biochemical analysis of that patient's metabolism may reveal a distinct lesion. If a genetic defect is assayable in cultured cells, complementation analysis by cell fusion can be undertaken. For example, the autosomal recessive disease, xeroderma pigmentosum (XP), involves defects in repair of UV-induced damage in DNA—a complex process that is only partially understood at the biochemical level. Patients are abnormally sensitive to sunlight, developing skin cancer after relatively brief exposure. By fusing fibroblasts from various patients with XP, scientists have found at least seven complementation groups, implying seven genes. The assay consists essentially of asking whether irradiated, non-growing cells will incorporate radioactive DNA precursors; cells from any one patient will not, but if hybrid cells from two patients come from different complementation groups, they will carry out DNA repair (Figure 7–2). Linkage studies also can reveal the existence of two or more loci capable of causing the same disease, as we saw in Chapter 4. In some cases, this may provide clues to previously unrecognized steps in a biochemical pathway.

Intralocus heterogeneity is the second class of genetic disease heterogeneity. The importance of this class of variation is now becoming more widely appreciated, as the molecular bases of more and more diseases are documented. For many years it was conventional to assume that a disease that shows a single-gene pattern of inheritance was caused by a specific type of mutation. The case of sickle-cell anemia was a misleading precedent. Sickle-cell disease is caused by a single nucleotide substitution in the sixth codon of the beta-globin gene, changing a glutamic acid to a valine; and the consequences of that change lead to an easily recognized clinical picture, distinct from all the other anemias. But the truth is that sickle-cell anemia is an exceptional situation. It is much more common for the same clinical symptoms to arise from a variety of mutations in one genetic locus, any one of which prevents or alters gene expression.

We are dealing here with a striking example of the notorious Murphy's law—anything that can go wrong, will go wrong (at least in someone, somewhere on earth). The old idea that a deleterious mutation abolishes the normal function of a protein by substituting one amino acid for another is sometimes true, but is too simplistic. A large number of other types of

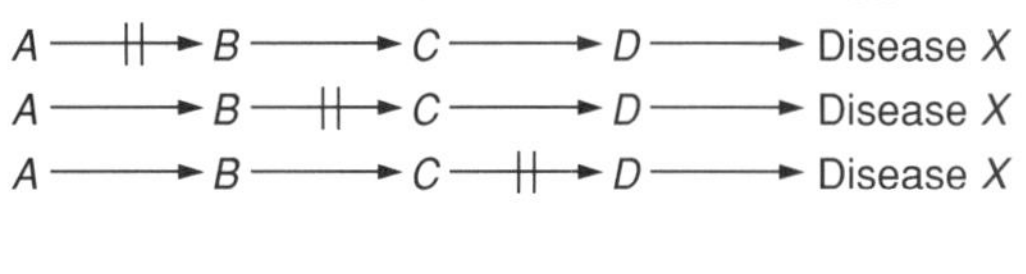

A → *B* → *C* → *D* → Health
A –||→ *B* → *C* → *D* → Disease *X*
A → *B* –||→ *C* → *D* → Disease *X*
A → *B* → *C* –||→ *D* → Disease *X*

Figure 7–1

A hypothetical multistep biochemical sequence leading to endproduct *D*. When any step in the synthesis of *D* is blocked, as indicated by the vertical lines, disease *X* results.

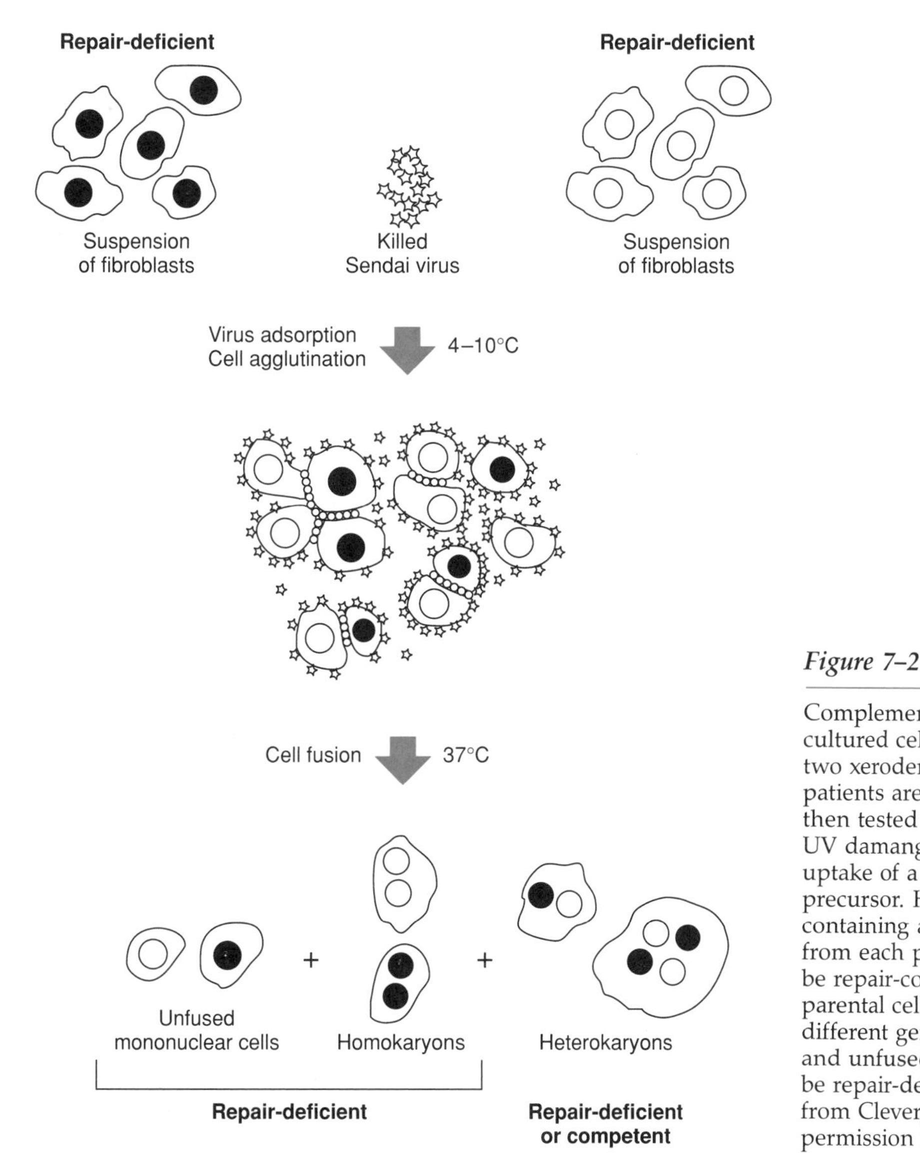

Figure 7–2

Complementation analysis with cultured cells. Fibroblasts from two xeroderma pigmentosum patients are fused in culture, then tested for ability to repair UV damange to DNA, via uptake of a radioactive DNA precursor. Hybrid cells containing at least one nucleus from each parental cell line will be repair-competent, if the parental cells had mutations in different genes. Homokaryons and unfused parental cells will be repair-defective. (Modified from Clever, 1983; by permission of McGraw-Hill)

changes are possible, which may affect control of transcription, processing of pre-mRNA, control of translation, and stability of mRNA or protein. Any or all of these changes may alter the expression of a gene in ways that will make a person sick.

Specific examples of almost all theoretically possible types of mutations have already been documented. Among the largest variety of mutations known for any given human genes are those described for the alpha- and beta-globin genes. The structure of those genes is diagrammed in Figure 7-3, so that the reader may refer to them in the context of specific mutations. A comprehensive review of the globin genes and their molecular biology is given by Collins and Weissman (1985).

Many of the following examples pertain to patients with some form of thalassemia, usually manifested as debility caused by insufficient red blood cells, which in turn is caused by a subnormal amount of alpha- or beta-globin. The word *thalassemia* means "sea blood," a reference to the tendency for those diseases to occur with high frequency among the inhabitants of port cities around the Mediterranean Sea. In fact, the sea had nothing to do with the situation; it was the mosquitoes that lived in the

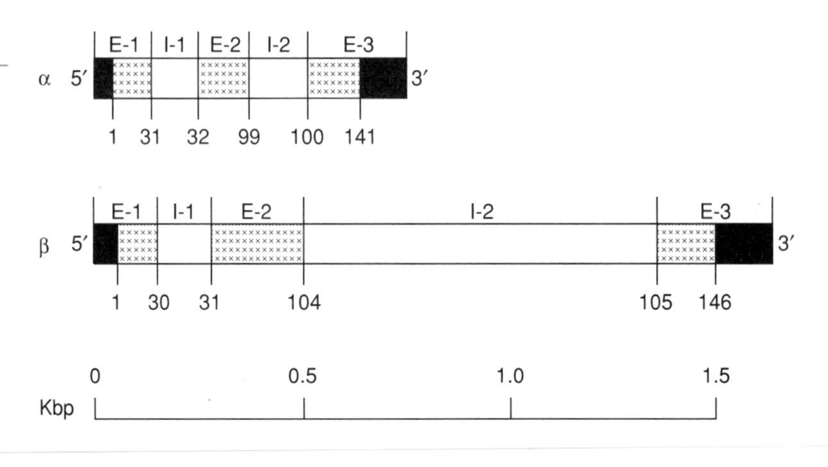

Figure 7–3

The structure of human alpha-globin and beta-globin genes. Transcription begins at the 5′ end and continues to the 3′ end. E-1, E-2, and E-3 are exons (sequences present in mature mRNA); I-1 and I-2 are introns, sequences removed from pre-mRNA during intranuclear processing. The nontranslated leader sequence in E-1 and the nontranslated trailer sequence in E-3 are shown as filled spaces. The remainder of each exon codes for the polypeptide sequence. The numbers below each bar diagram indicate the encoded amino acids.

marshes near the river mouths, where port cities were often established, that were the critical factor. Although the relationship is not as clearly established as it is for sickle-cell anemia, most thalassemias appear to confer some resistance to malaria.

The beta thalassemias are divided into beta-zero (no normal beta-globin) and beta-plus (some normal beta-globin). As we shall see, there are very clear molecular explanations for these categories, which were originally established to define purely clinical findings. Analogous defects occur in the genes for alpha-globin, but in addition there are some common whole-gene deletions that will be discussed in the section on deletions and insertions.

Mutations Involving One or a Few Nucleotides

Missense Mutations

Substitutions of one amino acid for another may or may not affect the function of a protein. Some substitutions will be *null alleles;* that is, the gene product will be totally functionless (thus confirming the conventional simplistic view about mutations). Other substitutions will lead to a gene product with partial activity, while occasionally a missense mutation may increase the catalytic activity of an enzyme. Still other substitutions may have no detectable consequences under ordinary circumstances, but may create a protein that is inadequate in situations involving unusual stress, such as extreme fatigue, starvation, exposure to unusually high or low temperatures, or exposure to toxic substances. Missense mutations may affect any aspect of a protein's function, including catalytic activity, binding of cofactors, interactions with other macromolecules, and stability. Finally, some single amino acid changes may be truly neutral.

There are more than 100 known single amino acid substitutions in beta-globin, and dozens more in alpha-globin. Some of these have obvious consequences for the health of the carrier. For example, replacement of histidine 58 in alpha-globin causes methemoglobinemia, a drastic reduction in the ability of hemoglobin to bind oxygen. The effect is dominant; homozygotes are unknown and would probably be lethal. By contrast, substitution of the adjacent amino acid (glycine 57) by aspartic acid has no apparent effect (Harris, 1980).

Numerous amino acid substitutions in alpha- and beta-globin produce unstable proteins, which in turn reduce the half-life of red cells, yielding a syndrome called "chronic hemolytic anemia." Note that the term *thalasse-*

mia is not applied to these situations. Many other mutations in the globin genes have been discovered because they alter the electrophoretic mobility of hemoglobin without affecting the health of the carrier in any obvious way.

A special case of missense mutations are "anti-nonsense" mutations, where the normal termination codon of an mRNA is converted to an amino acid codon. The possible alterations of the stop codon UAA by single-base substitutions are diagrammed in Figure 7–4. Hemoglobin Constant Spring is the best known example. A single base change has caused codon 142 to be read as glutamine, instead of "stop," and 31 amino acids are thereby added to the carboxyl end of the protein, because the first stop codon in the 3′ "trailer" sequence is the 32nd triplet past the normal termination codon. This leads to a mild form of thalassemia, at least in persons with two copies of the Constant Spring allele, which apparently arises from reduced stability of the mRNA and/or the abnormally long alpha polypeptide.

Nonsense Mutations

When an amino acid codon has been replaced by a termination codon, a nonsense mutation has occurred. These usually produce null alleles, because a truncated protein will be synthesized, which is unlikely to be functional. The closer to the amino terminus of a protein the nonsense mutation occurs, the more likely it is to abolish all function. Occasionally, a nonsense mutation near the carboxy terminus may allow partial or even full activity of the protein.

Nonsense mutations may arise directly by nucleotide substitution or indirectly as a result of a frameshift. In the latter case, deletion or addition of 1 or 2 nucleotides (or a multiple thereof) changes the reading frame, so that the amino acid sequence becomes totally different from the normal sequence downstream of the mutation. In most cases, a stop signal will be encountered within a few dozen codons.

Two examples of direct nonsense mutations involving beta-globin occur at codon 17 in Chinese populations and at codon 39 in Mediterranean populations. These lead to beta-zero thalassemia in homozygotes. Several small frameshift mutations in beta-globin are known, and these also produce beta-zero thalassemia (Orkin, 1983).

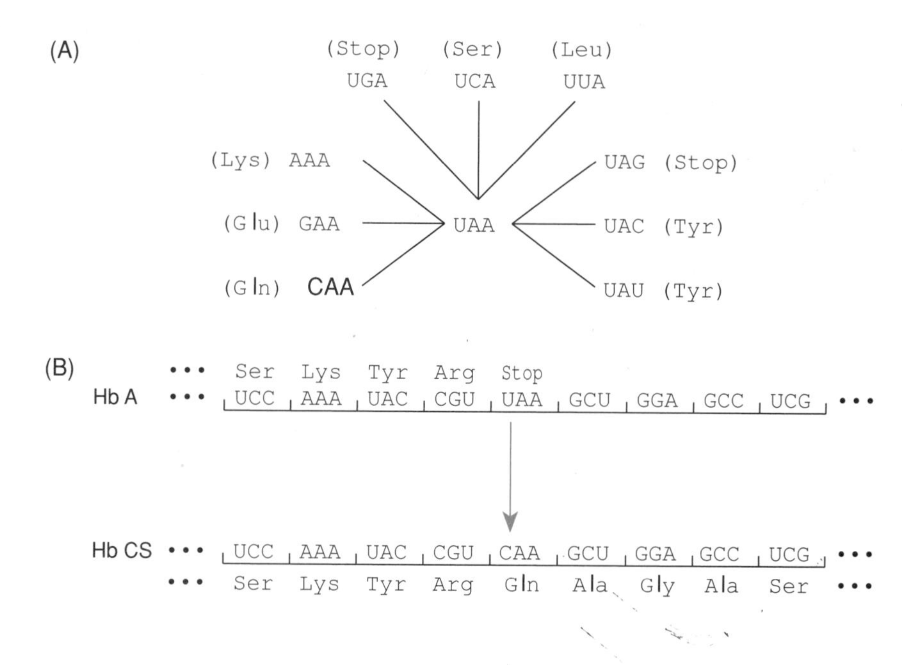

Figure 7–4

(A) The nine possible single-base changes that the UAA termination codon can undergo. Hemoglobin Constant Spring (see text and panel B of this figure) is the result of the UAA–CAA transition. Most of the other possible alpha-globin anti-termination mutations have been found in at least one person. Similar sets of nine single-base changes could be constructed for the other stop codons, UAG and UGA. (B) The Constant Spring anti-termination mutation of an alpha-globin gene (Hb CS). The nucleotide sequence near the normal stop codon is shown in the upper line (Hb A) along with the last four amino acids. The Constant Spring mutation from U to C changes the stop codon into a Gln codon. This allows translation of the mRNA to extend into the 3′ trailer region (lower line), until the next stop codon is reached (not shown).

Transcriptional Mutations

Changes in 5′-flanking sequences of a gene may increase or decrease the rate of transcription of that gene. Such changes may occur in promoters, thus directly affecting the binding of RNA polymerase; or they may occur in regulatory sequences, thus affecting the binding of proteins that influence the accessibility of RNA polymerase to the promoter. Other changes may occur at the 3′ end of a gene, leading to abnormal termination of transcription.

An example of a transcriptional control mutation in the beta-globin gene occurs at position −28. In this individual, the TATA box sequence has been changed from ATAAAA to ATACAA, resulting in a "promoter down" phenotype. The patient has severe beta-plus thalassemia. He is actually a compound heterozygote; his other beta-globin gene has a different mutation.

In most mammalian genes, the sequence AATAAA occurs near the 3′ end of the gene, where it acts as a signal to terminate and add poly(A) nearby. There is a patient from Saudi Arabia who has the sequence AATAAG in two of his alpha-globin alleles. The level of mRNA produced from these genes in the patient is 15% of normal. When the mutant gene was incorporated into a vector and transfected into tissue culture cells, most of the mRNA produced from it was longer than normal alpha-globin mRNA, indicating that the mutant sequence was usually not read as a termination signal (Higgs et al., 1983). It isn't clear why the long mRNAs do not accumulate in the patient's erythroblasts.

Mutations that Abolish Splice Sites in Pre-mRNA

Virtually all introns have the dinucleotide GT at the 5′ end and the dinucleotide AG at the 3′ end. Substitution of any other base at any of those four positions totally aborts normal processing of that intron (Figure 7–5). In most cases, the resulting abnormal mRNA will be incapable of yielding a functional protein. At best, the mutant protein will contain a block of amino acids that it isn't supposed to have. More likely, the intron that has not been spliced out will contain one or more stop codons. The presence of the intron may also shift the reading frame, resulting in the creation of new stop codons in the coding sequence. Moreover, the mutant mRNA may be unstable. Mutations of this class will almost always be null alleles.

In the beta-globin gene, there is a mutation from G to A in position 1 of intron 1, and in another individual, there has been a change from G to A in position 1 of intron 2. Some splicing of pre-mRNA may occur at alternative sites, but no normal mRNA is produced. These mutations cause beta-zero thalassemia in homozygotes (Orkin, 1983).

Consensus Sequence Substitutions

Although the GT and AG dinucleotides are essential for correct splicing, more information is required. Several nucleotides on either side of the GT and AG have a measurable effect on splicing. These nucleotides are not rigidly specified, but certain combinations occur much more frequently than others; they are called *consensus sequences* and are illustrated in Figure 7–5.

Changes in any of the consensus nucleotides, other than the required GT and AG pairs, can have a quantitative effect on RNA processing. Two examples are a change from G to C in position 5 of the 5′ consensus sequence of intron 1 of beta-globin (numbering from the exon/intron bound-

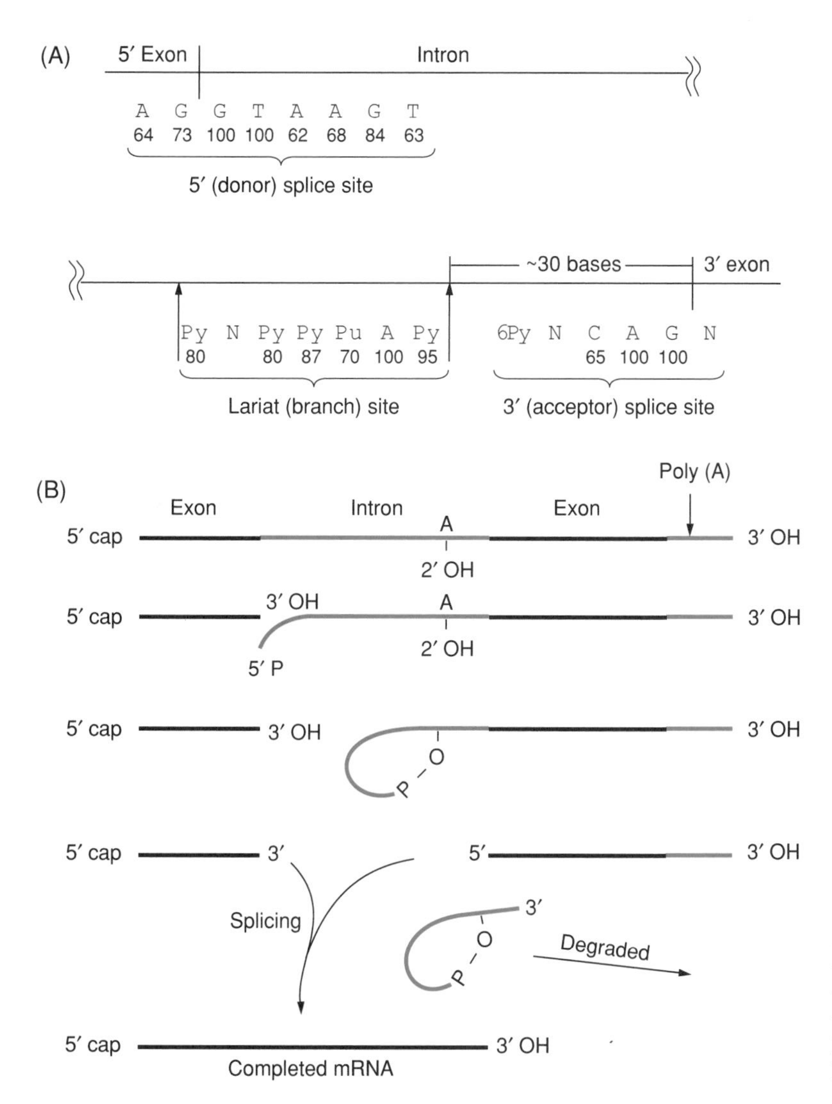

Figure 7–5

(A) Diagram of an intron and flanking exons, showing the consensus sequences at the splice sites and the lariat site (see panel B). Numbers below each base give the frequency of occurrence (data from Lewin, 1990). Vertical lines indicate the exon-intron boundaries. (B) The lariat is an intermediate step in intron excision. After cleavage of the pre-mRNA at the 5′ splice site, the 5′ end of the intron is joined to an A residue near the 3′ end of the intron, forming a loop with a stem consisting of about 30 nucleotides between the base of the loop and the 3′ end. Subsequently, the 3′ splice site is cleaved, the exons are ligated together, and the intron, now in the form of a lariat, is released, usually to meet its fate at the hands of nucleases. (From Prescott, 1988, p. 279)

ary) and a change from T to C in the sixth nucleotide of the same intron (but in a different individual). Both cause beta-plus thalassemia in homozygotes, so there appears to be some normal processing (Orkin, 1983).

More recently, a third consensus sequence has been described. It involves a region about 20 to 30 nucleotides upstream from the 3′ end of an intron, where a lariat structure is formed as an intermediate in intron excision (see legend to Figure 7–5). For globin genes, the consensus is: Py-N-Py-PY-Pu-A-Py (Figure 7–5). In principle, changes in that sequence could affect the efficiency of RNA processing.

Creation of New Splice Sites

A single base change in an exon or intron may create an alternative splice site. Obviously, there will be numerous short sequences that are similar to the splicing consensus sequences scattered throughout any typical gene. Sometimes only one base change is enough to cause the splicing enzymes to make mistakes and cut the pre-mRNA where it should not be cut. We don't yet know all the information that these enzymes respond to; base sequence is presumably not enough. There may be a need for specific tertiary structures of the RNA in the vicinity of the consensus sequence.

If a new splice site arises within an intron, mRNAs with extra nucleotides will be produced. If the number of extra nucleotides is a multiple of three, new amino acids will be inserted into the protein within the usual sequence, and these will have unpredictable consequences. Alternatively, the extra nucleotides may contain a stop codon, resulting in truncation of the polypeptide. If the extra nucleotides are not a multiple of three, a frameshift will occur, which will usually create a null allele. If a new splice site arises within an exon, nucleotides will be deleted from the mRNA. The possible consequences are the same as for the addition of nucleotides. In any case, splicing at the new site is not likely to replace splicing at the normal site completely, although cases are known where the new site is strongly preferred. Nevertheless, the normal site still exists, and some fraction of the pre-mRNAs will be processed correctly. Thus, some normal protein will be made, although it may only be a small percent of the usual amount.

An example of a new splice site has been found in intron 1 of the beta-globin gene (Forget et al. 1983). The normal 3′ splice site for that intron has the sequence TTAGGCT, and there is a nearby upstream sequence that is TTGGTCT within the intron (Figure 7-6). This differs from the normal splice site by two nucleotides. In some patients with rather severe beta-plus thalassemia, this downstream site has become TTAGTCT, differing from the normal site by only one nucleotide. Apparently the mutant site is more attractive to the processing enzymes than the normal site, because very little normal beta-globin mRNA is produced by patients homozygous for this mutation. To prove this hypothesis, the patient's gene was cloned, inserted into a shuttle vector, amplified, and transfected into monkey kidney cells. Analysis of the RNAs produced by these cells showed that 90% of the mRNAs from the beta-thalassemia gene were 19 nucleotides longer than normal beta-globin mRNAs, which were only 10% of the total. Note that addition of 19 nucleotides creates a frameshift, which implies that translation will terminate at a nearby stop codon.

Creation of new splice sites within exons is exemplified by a fairly common beta-globin variant that produces Hemoglobin E (Orkin et al., 1982). In exon 1, amino acid 26 changes from Glu to Lys because of a codon change from GAG to AAG. Of course, this protein has a missense mutation, but the protein is made at very low levels because the same nucleotide change is also a processing mutation. The sequences are:

normal exon	GGTGAGG
mutant exon	GGTAAGG
normal 5′ splice site	GGTTGGT

Most of the transcripts from the mutant gene are processed within exon 1, leading to a truncated protein, which is functionless. Those transcripts

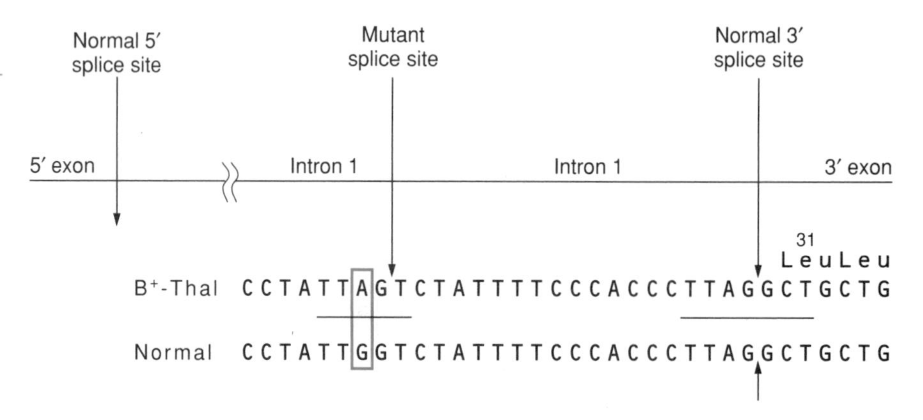

Figure 7–6

The nucleotide sequence of the human beta-globin gene near the 3′ end of intron 1 (IVS-1), showing the normal splice site and an alternative splice site created by a nearby single base change (in box) in a beta-thalassemia patient. The normal and mutant consensus sequences are underlined. (Modified from Forget et al., 1983)

that are processed normally produce a protein with a missense mutation, which seems to be functional. It is not obvious why the processing enzyme finds the mutant exon more appealing than the normal exon. This reminds us that there is more involved in splice site recognition than we know.

Mutations Affecting Translation

In principle, mutations that reduce or enhance the efficiency with which ribosomes bind to an mRNA, or mutations that alter any other aspect of polypeptide chain initiation, could have a large enough effect on gene expression to produce ill health. Initiation of translation varies greatly from one mRNA to another, but there is evidence for a consensus sequence immediately upstream from the initiator AUG, which affects initiation rate for most higher eukaryotes. The sequence is CC(A or G)CCAUG (the initiator codon is underlined). A person homozygous for a deletion of the two nucleotides at postions −2 and −3 prior to the AUG codon is known; this person has alpha-thalassemia, caused by very limited translation of the alpha-globin mRNAs that contain the deletion (Morlé et al., 1985).

At least two other possibilities for mutations affecting protein synthesis can be envisioned. First, there may be synonymous codon mutations (i.e., codons that code for the same amino acid) that result in reduced protein synthesis from a given mRNA, because the cognate tRNA for the mutant codon is not as abundant as the tRNA for the normal codon. Second, base changes in the 3′ noncoding sequence of an mRNA may have an effect on mRNA stability, or possibly on the efficiency of translation. We don't really know what these "trailers" do; if there are regulatory molecules that interact with the trailers, mutations in the trailers may affect binding sites for the regulators. At present, we can only list these as speculative possibilities.

Deletions and Insertions

This is a heterogeneous category, ranging from the removal or insertion of one nucleotide, through total deletion or total duplication of a gene, to the excision of several million nucleotides from one chromosome and sometimes its transfer to another chromosome. Small deletions and/or insertions may produce frameshifts; larger changes may produce almost any of the effects described in the preceding categories. As more and more genes are cloned and defined at the molecular level, it is becoming apparent that deletions are a non-trivial cause of most genetic disease, accounting for 5–10% of the mutants in many cases. In some genes, deletions appear to be the major cause of mutation. It is already apparent that deletions account for at least 50% of Duchenne muscular dystrophy patients (Chapter 5), and the frequency may prove to be even higher when more detailed analyses are performed. Another example is steroid sulfatase deficiency, which leads to one form of the disease called ichthyosis (scaly skin); in one study, 14 out of 15 apparently unrelated patients had deletions (Bonifas et al., 1987).

Probably the most famous deletions are those that cause the large majority of alpha-thalassemia cases. The alpha-globin genes are unusual in that there are two loci, tandemly situated on chromosome 16, and those loci produce identical polypeptides. Thus, a normal individual has four alpha-globin alleles. Two types of deletions are known; one of them deletes a single alpha-globin allele, the other deletes both alleles on one chromosome. It is therefore possible to produce individuals who have lost one, two, three, or all four alpha-globin alleles. That leads to the following genotypes and their corresponding syndromes.

genotype	*phenotype*
+ +/+ +	Normal
+ +/+ −	Essentially normal; slight imbalance of alpha- and beta-globin. "Alpha-thal. 2"
+ +/− −	Moderate anemia. "Alpha-thal. 1"
+ −/+ −	Moderate anemia. "Alpha-thal. 1"
+ −/− −	Severe anemia. "Hb H disease"
− −/− −	Fatal at or before birth. "Hydrops fetalis"

Hemoglobin H disease and hydrops fetalis are common in southeast Asia, where both types of alpha-globin deletions occur. The single-allele type of deletion is also common in parts of Africa, but the two-allele deletion is not. Thus, blacks often suffer from mild-to-moderate alpha-thalassemia, but Hb H disease and hydrops fetalis are virtually unknown among them. In all of these situations, the deletions are most common in areas heavily afflicted with malaria, and it is believed that heterozygotes must have relatively high resistance to infection, although the biochemical basis for the resistance is not clear.

Mechanisms of Deletion/Insertion Production

The most plausible and well-documented mechanism for the creation of deletions and insertions is by *unequal crossing-over* (Tartof, 1988). For example, after duplication of the alpha-globin genes occurred, it became possible for mispairing to arise at meiosis, with the 5′ member of the duo on one chromosome paired with the 3′ member of the duo on the homologous chromosome. A crossover could then lead to one chromatid with only one alpha gene and another chromatid with three alpha genes, as diagrammed in Figure 7–7. Indeed, individuals with three alpha-globin alleles on one chromosome have been found, and the reciprocal product—a deletion of one alpha-globin gene—has already been mentioned.

A classic case of unequal crossing-over among the genes of the beta-globin complex is Hb Lepore, which results from a mispairing of the beta- and delta-globin loci at meiosis, as shown in Figure 7–8. Hb Lepore is the product of a gene with the 5′ portion of the delta-globin locus and the 3′ portion of the beta-globin locus. This compound gene has lost the sequences that control the expression of beta-globin; instead, it is controlled by sequences flanking the delta-globin locus. Accordingly, individuals who are homozygous for Hb Lepore produce very little beta-type hemoglobin as adults; they have thalassemia. The reciprocal of Hb Lepore is Hb anti-Lepore, which has little or no effect on the phenotype of the individual.

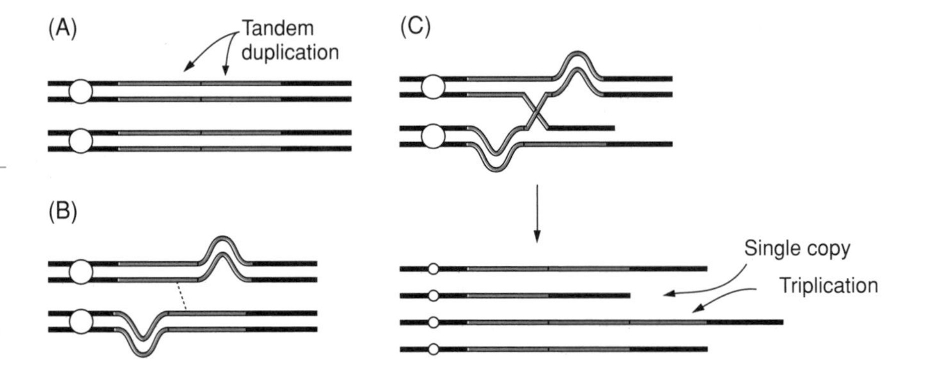

Figure 7–7

Unequal crossing-over caused by misalignment of tandemly repeated sequences leads to chromosomes with different numbers of the repeated sequences. (Snyder et al., 1985, p. 209)

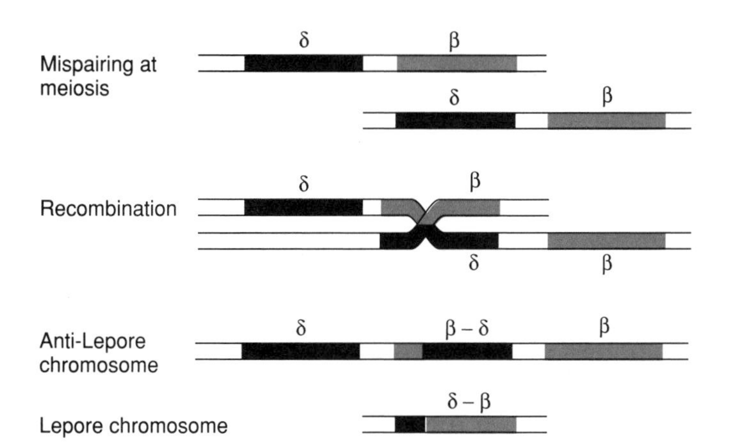

Figure 7–8

Unequal crossing-over within the beta-globin gene cluster leads to Lepore and anti-Lepore globin genes.

Another potential source of unequal crossing-over is the existence of repetitive DNA sequences within introns of a complex gene, which raises the possibility of internal mispairing during meiosis. In principle, this should be more likely for larger genes than for smaller ones, and it may help explain the frequency of deletions in the giant gene for dystrophin. However, unequal crossing-over within a gene should lead to one allele that contains an insertion for every allele that has suffered a deletion, just as unequal crossing-over between genes should lead to one chromosome containing an extra gene for every chromosome that has lost a gene, as we saw in the globin examples cited previously. Although a few cases of insertions within genes have been described, they are conspicuously rare, compared to the frequency of deletions. Since there is no reason to expect insertions to be any less deleterious than deletions, we are left with a puzzle. If unequal crossing-over is the source of most deletion mutations, what happens to the reciprocal products? Is there some other mechanism that might account for the production of deletions?

Wolff et al. (1989) deepened this mystery by asking whether unequal crossing-over accounted for the production of new alleles at VNTR loci (which were described in Chapter 4). Recognizing that there should be recombination of flanking markers whenever a crossover has occurred, they studied the highly variable locus known as λMS1, where individuals heterozygous for closely linked markers on both sides of the VNTR locus were available. They found that unequal crossing-over at meiosis was not the major source of new alleles at the λMS1 locus.

We are left with the conclusion that there must be deletion-generating mechanisms that do not require unequal meiotic crossing-over. One possibility is replication-slippage, a term that implies that DNA polymerase sometimes detaches from the template strand and resumes replication some distance away, with the ends of the gap being joined later.

Another possibility is excision of material between repeated sequences. As Figure 7–9 shows, it is possible for intrastrand pairing of inverted repeated sequences to occur, forming a loop from intervening DNA, which might then be excised by enzymes. This is a known phenomenon for some transposable elements (see next section), and it might also occur with DNA in general, if the requisite enzymes were present.

Transposable Elements

Segments of DNA that are potentially capable of moving from one region of the genome to another are called *transposable elements*. They have been extensively studied in bacteria, in maize, and in *Drosophila*. Although it has

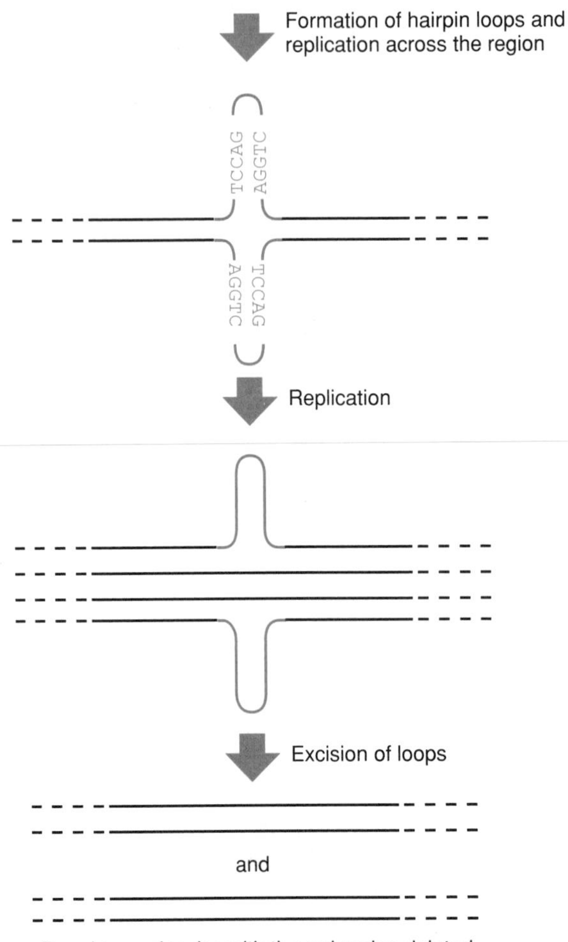

Figure 7–9

A deletion resulting from replication across the base of a loop formed by inverted repeats, followed by excision of the loops in the parental DNA strands. (Modified from Snyder et al., 1985, p. 368)

been difficult to prove the existence of active transposable elements in humans, there is compelling evidence for their presence during human evolution. Two of the repetitive DNA families that were mentioned in Chapter 2, LINEs and SINEs, have members with the diagnostic features of transposable elements; and the common occurrence of processed pseudogenes testifies to the occasional activity of enzymes capable of converting RNA into DNA, followed by integration of the DNA into the cellular genome.

Transposable elements (also called *transposons*) may be based entirely on DNA or they may include RNA as a transient stage in their life history. DNA-limited transposons are well known in bacteria, in maize, and in fruit flies. For some elements, transposition is conservative; that is, a transposon is excised from its current location in the genome and is inserted elsewhere. In other cases, transposition is replicative. That is, an existing transposon is copied and the copy is inserted at a new location. Replicative transposition increases the number of transposable elements in the genome; conservative transposition does not.

Retrotransposons utilize RNA as an intermediate stage in their replication. The most familiar retrotransposons are the retroviruses. Virions of retroviruses contain RNA, which typically codes for three major proteins: an envelope protein, a protein that binds tightly to the RNA in the virion, and a reverse transcriptase (see Chapter 11). The reverse transcriptase makes a DNA copy of the virion RNA, which is then converted into a double-stranded DNA molecule and inserted more-or-less randomly into the host genome. This DNA form of the viral genome, also called a *provirus*, is replicated whenever the host cell replicates. Usually, it is also transcribed, the result being that the infected cell continually produces new viruses, which bud from the cell membrane.

There is another class of retrotransposons, similar in organization to retroviruses, which apparently never produces viruses. Some examples are the Ty elements of yeast and the copia elements of *Drosophila*, both of which are known to transpose, even though the detailed mechanisms of transposition have not yet been elucidated.

In humans, the L1 family of long interspersed repetitive sequences has similarities with retroviruses. There are thousands of L1 sequences in the human genome, but most of them are truncated at the 5′ end, suggesting that they may be "fossil" transposable elements. The larger members of this family are about 6500 bp long and contain two open reading frames, one of which potentially codes for a protein with homology to retroviral reverse transcriptases.

For many years there was no proof that such an enzyme was actually expressed by any L1 element, but in late 1991, Mathias et al. showed that the putative reverse transcriptase gene from one L1 element, when transfected into yeast cells, had reverse transcriptase activity. The same group also described two patients with hemophilia A (an X-linked disease described in Chapter 13), in which the factor VIII gene was mutated by insertion of a truncated L1 element. These were new mutations, not present in any parent of either child. Analysis of the parents of one of the affected boys identified the probable source of the transposed sequence, a full-length L1 element on chromosome 22 (Dombroski et al., 1991).

If some L1 elements do code for reverse transcriptase and integrase enzymes, it is also possible that those enzymes occasionally make a mistake and convert another mRNA into a processed pseudogene. The same enzymes may mediate transposition of Alu elements, which do not contain enough information to encode their own enzymes.

The Alu family of repetitive DNA consists of about 300,000 copies of a sequence approximately 300 bp long; these are widely dispersed throughout the human genome. There is considerable variation in details of the sequence from one copy to another, but overall, there is enough similarity to define a family of related sequences (Chapter 2). Many of them are not transcribed at all, others are transcribed by polymerase III, and still others occur within the introns of structural genes, where they are transcribed by polymerase II. Typically, an Alu sequence is flanked by short direct repeats at both ends, a characteristic of known transposable elements.

Strong evidence for transposition of Alu sequences as a cause of genetic disease was provided recently by a report of a new mutation associated with neurofibromatosis type 1 (a disease described in Chapter 11). The effect was shown to be caused by insertion of an Alu element into an intron, which caused deletion of a downstream exon and shifted the reading frame, resulting in a nonfunctional polypeptide (Wallace et al., 1991).

The conclusion that transposable elements have played an important role, not only in human evolution, but also as a source of genomic variation at the present time, is inescapable. Insertion of a transposable element into a gene could cause abolition or modification of that gene's expression. Excision of a transposable element might activate or inactivate a gene. Another interesting possibility is that insertion or excision of a transposable element from the control sequences of a gene might alter the timing or location of its expression in such a way that a drastic effect on phenotype would result.

Chromosomal Mutations

Genomic accidents that alter the number of chromosomes or that produce cytologically detectable variations in one or more chromosomes are called *chromosomal mutations*. Of course, this is only a nomenclatorial convenience. There is no sharp dividing line between mutations that produce microscopically visible alterations in chromosomes and those that can only be detected by molecular methods.

Polyploidy

The most extreme chromosomal variations are those that affect entire sets of chromosomes, producing embryos that are polyploid, instead of the usual diploid condition. In humans, the most common polyploidy is triploidy (three times the haploid number of chromosomes found in normal gametes). In various studies, it has been reported to account for roughly 8–12% of spontaneous abortions during the first trimester of pregnancy. Occasionally, triploid embryos will survive longer, and a few live births have been recorded, none of which survived more than a few days. Tetraploid embryos are quite rare, and higher ploidies are virtually non-existent. Apparently, polyploidy in general is not compatible with life in humans.

The origin of polyploid embryos can often be attributed to dispermy (two sperm fertilize one egg) or it may result from any event that prevents the separation of entire sets of chromosomes at either meiosis I or meiosis II (e.g., failure of synapsed homologous chromosomes to separate at anaphase of meiosis I, failure of chromatids to separate at anaphase of meiosis II, or failure of the mitotic spindle to form and function normally). Failure of chromosomes to separate (i.e., disjoin) at anaphase and move to opposite poles of the spindle is called *nondisjunction*. It may happen to single chromosomes or it may involve the entire set of chromosomes.

In either case, if a diploid sperm were to fertilize a normal haploid egg, a triploid embryo would be created. The reverse situation, with the egg being diploid and the sperm being haploid, could also occur. Nondisjunction of the entire chromosome complement at the first mitotic division of a diploid zygote would produce one tetraploid cell and one cell with no chromosomes; the latter would undoubtedly die, leaving a tetraploid embryo.

Nondisjunction after the first cleavage division would create a mosaic; that is, some tissues would be tetraploid and some would be diploid. At least one such infant has been described. Most triploid embryos that survive until birth or nearly to birth have also been shown to be mosaics. However, it is difficult to attribute mosaic triploidy to nondisjunction after zygote formation. It would require a bizarre event: one member of each pair of chromosomes would have to disjoin normally, while the other member of that pair underwent nondisjunction!

Mosaic triploid embryos are more likely to be chimeras; that is, they may represent the fusion of two embryos, one of which was diploid and the other triploid. Another possibility is that a polar body was fertilized, it underwent several cell divisions, and its descendant cells were incorporated into the embryo. If either the main embryo or the polar body were triploid, a mosaic embryo would result.

Aneuploidies

The term *euploid* refers to the normal set of chromosomes that is characteristic for a given species. Haploid, diploid, and polyploid genomes are all euploid. If a genome contains anything other than a multiple of the basic euploid number (the "n" number, which is 23 in humans), that genome is *aneuploid*. The best known aneuploidies in humans are *trisomies*; that is, one chromosome is present in three copies per cell, instead of the usual two copies. *Monosomies* (cases where a particular chromosome is present in only one copy per cell) occur, but except when the X or Y chromosome is involved, they almost always fail to develop beyond early embryonic stages.

The most common cause of single chromosome aneuploidies is nondisjunction. Figure 7-10 illustrates the consequences of nondisjunction for sperm formation. If there is nondisjunction at meiosis I, two sperm with two copies of a given chromosome and two with no copy of that chromosome will result. If there is nondisjunction at meiosis II, two normal sperm, one disomic sperm, and one sperm that is nullisomic for the given chromosome will be produced.

The most common aneuploidies in humans involve the sex chromosomes (also known as *gonosomes*). One of the most frequent of these is Kleinfelter syndrome (47,XXY), which occurs about once in 1000 births. These people are phenotypically male, although they are usually sterile, with small testes, rounded hips, and some breast development. They tend to be tall. Effects on intelligence are highly variable; some males with Kleinfelter syndrome are apparently normal and their genotype may never be recognized, unless they go to an infertility clinic. Other males with Kleinfelter syndrome are somewhat retarded and may have emotional problems. It is noteworthy that the incidence of Kleinfelter syndrome in mental institutions is about ten times higher than in the general public.

The 47,XYY karyotype is approximately as frequent as the Kleinfelter karyotype. These men are usually well over six feet tall and are fertile. The incidence of 47,XYY men in some penal institutions has been reported to be as high as one inmate in 30, an observation that has engendered a large amount of naive speculation on the effects of an extra Y chromosome on aggressive behavior. At present, most geneticists favor the hypothesis that males with the XYY karyotype tend to be more impulsive than XY males, and this sometimes leads them to commit violent acts against other persons. Claims that XYY males also have somewhat below average intelli-

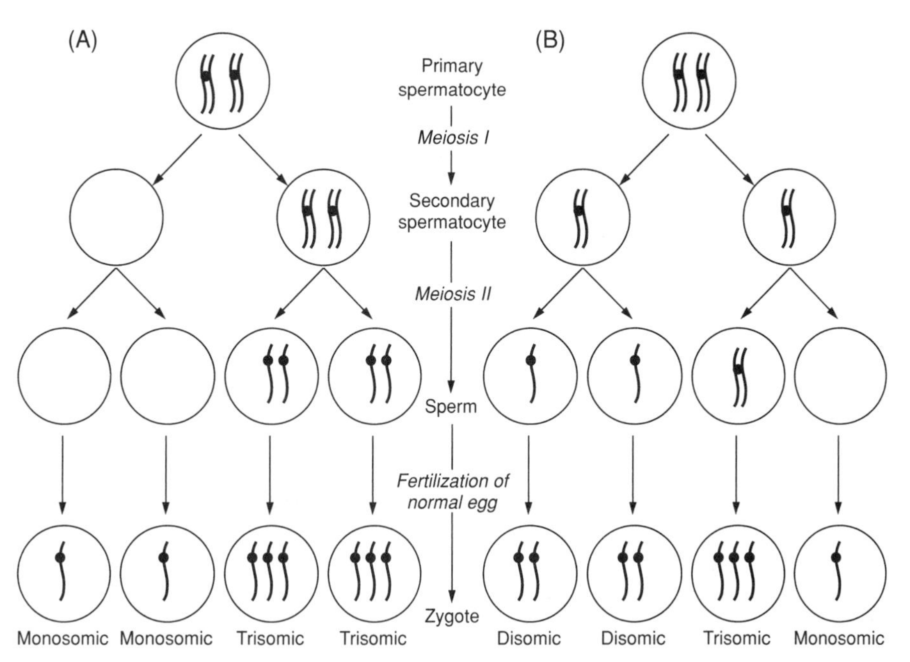

Figure 7-10

Meiotic nondisjunction can occur at the first division (panel A) or the second division (panel B). The diagram illustrates nondisjunction involving only one pair of chromosomes. Nondisjunction of the entire set of chromosomes, which can occur if the spindle fails to form or function normally, could lead to a triploid embryo.

gence are controversial. What *is* clear is that the large majority of XYY males lead normal lives, and their unusual karyotype is never discovered (Evans et al., 1990).

Turner syndrome (45,X) is the only monosomy that is compatible with survival past birth. It occurs once in several thousand female births. The Turner female is quite short, usually under five feet tall when adult, and is infertile. Secondary sexual characteristics are poorly developed. The neck is broad and webbed. Many other symptoms occur with variable frequency. Most have normal intelligence. There is evidence that nearly 98% of all embryos with the 45,X karyotype are aborted during the first three months of pregnancy; it is the most common karyotype in spontaneous abortions. This presents a paradox: if the 45,X genome is so lethal most of the time, why are the few postnatal survivors relatively normal?

Numerous other, but rarer, gonosome aneuploidies have been reported. These include 47,XXX and 48,XXXX females, as well as 48,XXXY and 48,XXYY males, to mention only a few. All whole-body aneuploidies, whether they involve the sex chromosomes or autosomes, can be accounted for by nondisjunction of single chromosomes during gamete formation. Mosaic aneuploidies also occur; these presumably arise from nondisjunction of a chromosome in one of the early cleavage blastomeres.

Sex chromosome aneuploidies are far more numerous among live births than are autosomal aneuploidies. The relatively benign effects of sex chromosome aneuploidies can plausibly be attributed to two facts: (1) there are very few genes on the Y chromosome and (2) the normal process of X chromosome inactivation leaves only one active X chromosome (there are a few active genes on the inactive X, most of which are on the short arm in the pseudoautosomal region; see Chapter 13). Thus, in an individual with Kleinfelter syndrome, most of the extra X chromosome will be genetically silent; and a female with Turner syndrome lacks only the expression of those same few active genes on the "inactive" X chromosome that are expressed in excess in the male with Kleinfelter syndrome.

The autosomal aneuploidies include Down syndrome (trisomy 21), which is probably the genetic defect best known among laymen. Estimates of the overall frequency of trisomy 21 range from one in 700 to about one in 1000 live births. It occurs in all human populations and the equivalent defect has also been reported in chimpanzees. A correlation between the age of the mother and the frequency of Down syndrome was recognized many years before the chromosomal basis of the syndrome was identified by Lejeune et al. in 1959. After women reach age 35, the incidence of Down syndrome increases rapidly; women who are 45 years old have 50–60 times higher risk of bearing a Down syndrome child than women who are 20 years old.

The availability of haplotypes based on RFLPs has made it possible to distinguish the maternal chromosome 21 from the paternal chromosome 21, and thereby to determine which parent was the source of the gamete with two copies of chromosome 21. From this type of analysis we have learned that paternal nondisjunction accounts for about one-fourth of Down syndrome cases. There does not appear to be a strong correlation between paternal age and the probability of fathering a child with Down syndrome.

Down syndrome patients are well known for their mental retardation, with I.Q.s usually in the range of 40–50. Previously, they were confined to mental institutions, but most are now cared for by their families or in special homes. With patient and loving care, the Down syndrome individual can achieve far more than was previously thought to be possible.

Among the physical aspects of the syndrome are short stature, broad face, protruding tongue, slanted eyelids, and an unusual pattern of hand creases. Heart and endocrine gland defects are common. Leukemia is as much as 20 times more frequent than in the general population. Males are always sterile; some females are fertile and a few have become mothers. Many Down syndrome babies die within the first year of life, but the mean life expectancy is now greater than 30 years.

Trisomy 21, like the 45,X genotype, is more common among spontaneous abortions than among live births. We have no biochemical explanation for that fact, nor do we understand why the presence of three copies of chromosome 21 leads to the phenotype of Down syndrome. More than 20 genes have been mapped to chromosome 21 and several of them have been implicated as possible causative agents for one or more Down syndrome characteristics, but as yet there is no generally accepted cause and effect relationship. The chromosome is under intensive study in several laboratories, and because it is the smallest human chromosome, it may be the first one to be sequenced. Then, perhaps we may understand Down syndrome at the molecular level, but perhaps not. The path from gene to clinic can be long and devious.

There are only two other autosomal trisomies that occur with significant frequency in newborns. These are trisomy 13 and trisomy 18. Estimates of their frequency range from one in 5000 to one in 15,000 live births. There appears to be some correlation between maternal age and frequency of these trisomies, but the effect is much less pronounced than for trisomy 21. In both trisomy 13 and trisomy 18, the infants have multiple, severe malformations and usually survive only a few months.

Other aneuploidies can be found in abortuses with variable frequencies. It is not clear whether some chromosomes are more prone to undergo nondisjunction than others, or whether some trisomies usually cause death during early embryogenesis and are therefore unobserved. Monosomies in autosomes are never born; most of them appear to be lethal quite early in development. The reason why monosomies are in general more deleterious than trisomies is not understood. It is customary to say that the effects of having only 50% of the normal level of a whole set of gene products must be more serious than the effects of having 150% (as would be expected for trisomies), but this falls far short of a satisfactory explanation.

Rearrangements

There are two broad categories of chromosomal rearrangements: *deletions* and *translocations*. Deletions can be *terminal,* which means that some genetic material has been lost from one end of a chromosome, or *interstitial,* which implies loss of an internal section of a chromosome. In either case, the missing material may be truly gone or it may have been moved to another chromosome (see the discussion of translocations).

The most common terminal deletion involves part of the end of the short arm of chromosome 5. Patients who are 5p− have the "cat-cry" syndrome, so called because newborns make a sound similar to a kitten. The syndrome is characterized by severe mental retardation and a variety of minor physical abnormalities. Affected individuals often survive to adulthood. The incidence of the condition is about one in 50,000 births.

Wolf-Hirschhorn syndrome is caused by a deletion of the tip of chromosome 4p. It is a "defect of midline fusion" with abnormalities of the nose, lips, palate, penis, and some internal organs. Death usually occurs in infancy.

Interstitial deletions should be rarer than terminal deletions, inasmuch as the chromosome must be broken in two places in order for an internal segment to be lost. However, true terminal deletions require replacement of telomere sequences (see Chapter 2) in order for the remaining chromosome to be stable. It is not obvious how this might occur. Therefore, many apparent terminal deletions may really be interstitial deletions, one end of which involved a break close to a telomere. Two of the best known interstitial deletions are in chromosome 13q and in chromosome 11p. The former is associated with retinoblastoma and the latter with Wilms tumor (both are described in Chapter 11).

Translocations occur when there have been breaks in two non-homologous chromosomes followed by joining of the broken ends in new combinations. If no genetic material is lost, the translocation is reciprocal, and a person who possesses both of the rearranged chromosomes has a balanced translocation, which usually produces a normal phenotype. Alternatively, if a translocation interrupts a gene, an abnormal phenotype may result. The abnormality may be caused by the absence of a function or a change in the expression of a gene. Some translocations are major factors in the genesis of tumors (Chapter 11).

A special class of translocations involves the acrocentric chromosomes 13, 14, 15, 21, and 22. The short arms of these chromosomes contain moderately repetitive DNA that codes for ribosomal RNAs plus some heterochromatin. There is a tendency for these chromosomes to break at the centromere and for a reciprocal exchange to occur, joining the two long arms in one product and the two short arms in the other product (Figure 7-11). These are called *Robertsonian translocations,* in reference to a scientist who studied them extensively. The conjoined short arms are usually lost in a subsequent mitosis, but the conjoined long arms are stable. The most frequent combinations are between chromosomes 14 and 21, designated (14q21q), and between the two chromosomes 21, designated (21q21q). Individuals with a Robertsonian translocation are normal, the loss of the short arms having no apparent effect (additional copies of ribosomal RNA genes are present on the other acrocentric chromosomes).

The deleterious effects of balanced translocations become evident when such an individual produces gametes. The presence of two normal chromosomes and two translocated chromosomes makes pairing of homologous chromosomes at meiosis complicated, as shown in Figure 7-12. The cross-shaped structure that results can be separated in three possible ways at anaphase of the first meiotic division, and only one of those three patterns leads to gametes with a complete complement of genes. The other patterns

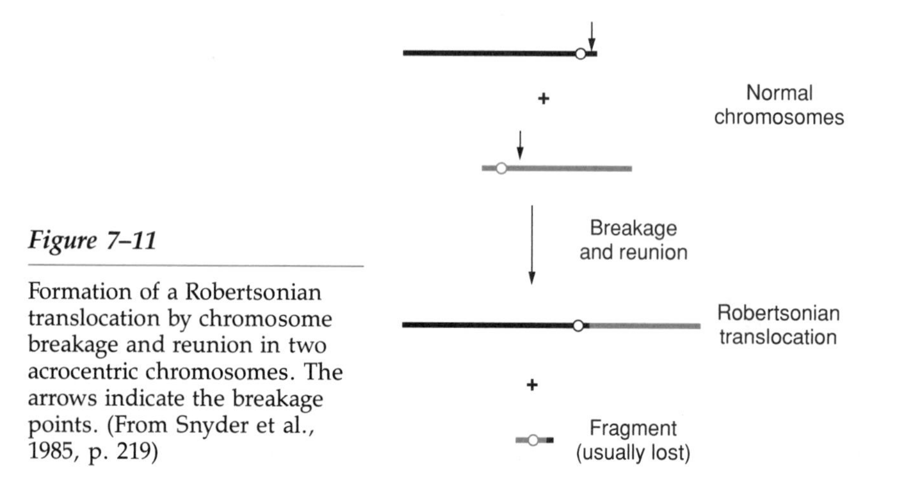

Figure 7–11

Formation of a Robertsonian translocation by chromosome breakage and reunion in two acrocentric chromosomes. The arrows indicate the breakage points. (From Snyder et al., 1985, p. 219)

produce incomplete genomes which will lead to non-viable embryos. In many families with a history of repeated spontaneous abortions, a balanced translocation in one of the parents has been found to be the underlying cause.

A few percent of Down syndrome cases are caused by the presence of a Robertsonian translocation in one parent. As Figure 7–13 shows, one of the six possible gamete classes produced by such an individual will have two copies of chromosome 21q. If such a gamete joins with a normal gamete to form a zygote, trisomy 21 will be produced. Note also in Figure 7–13 that the other three chromosomally unbalanced classes of gametes will be two nullisomics, and one disomic (for chromosome 14 in the example). Embryos formed from such gametes will not be viable.

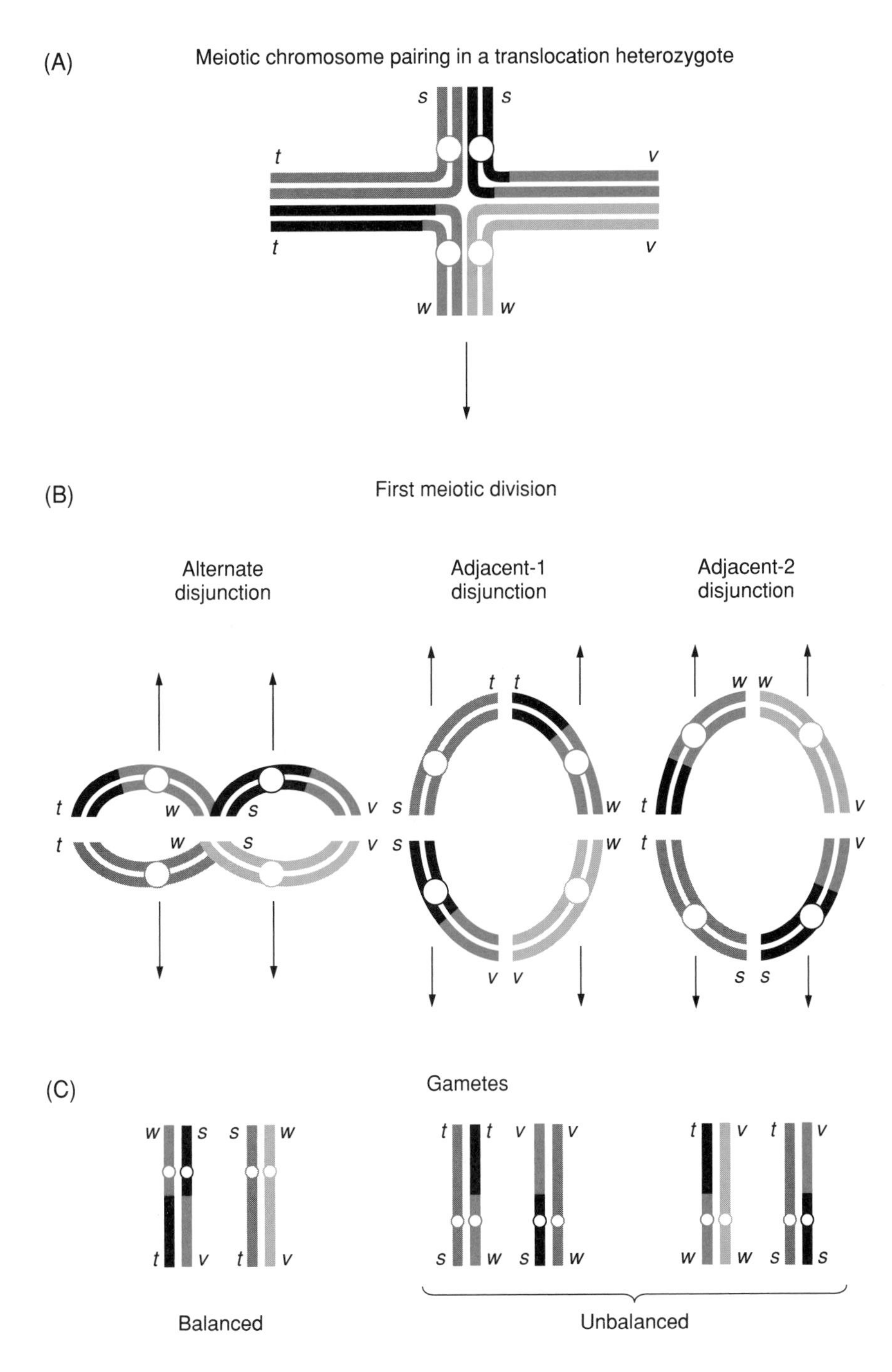

Figure 7–12

Meiosis in a translocation heterozygote. (A) The structure resulting from chromosome pairing. (B) Three possible orientations in metaphase I of meiosis which determine the chromosomes that will go together at anaphase. (C) Chromosomes resulting from the three modes of segregation. (Modified from Snyder et al., 1985, p. 217)

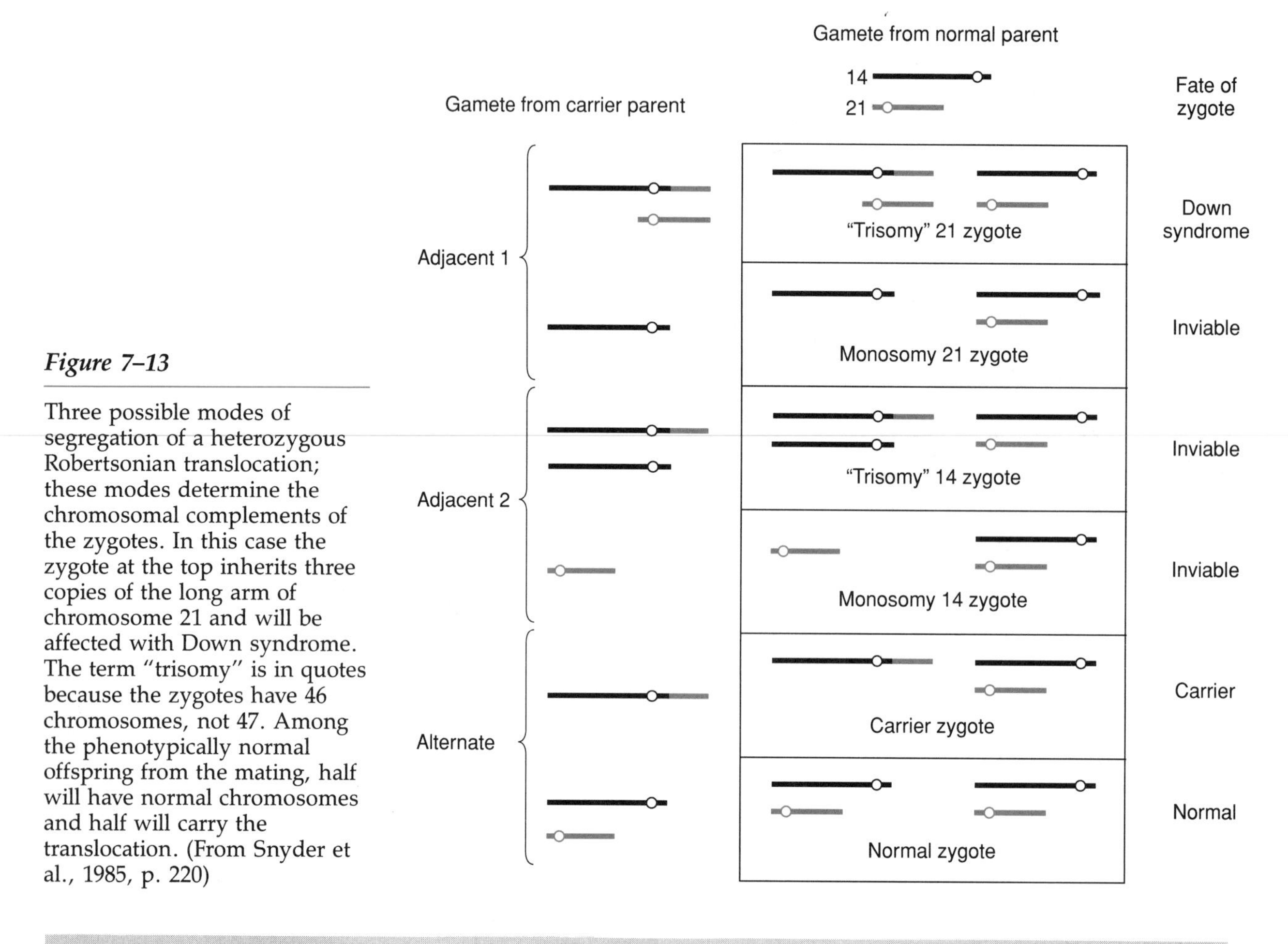

Figure 7–13

Three possible modes of segregation of a heterozygous Robertsonian translocation; these modes determine the chromosomal complements of the zygotes. In this case the zygote at the top inherits three copies of the long arm of chromosome 21 and will be affected with Down syndrome. The term "trisomy" is in quotes because the zygotes have 46 chromosomes, not 47. Among the phenotypically normal offspring from the mating, half will have normal chromosomes and half will carry the translocation. (From Snyder et al., 1985, p. 220)

SUMMARY

Many types of changes in the sequence of DNA nucleotides can affect gene function. Almost every human genetic disease, as defined at the clinical level, is caused by a variety of mutations. Although some heterogeneity is the result of mutations at more than one locus, most is caused by intralocus heterogeneity. Molecular analysis of variation in many genes has provided examples of virtually every conceivable change.

Single nucleotide changes can substitute one amino acid for another (missense) or cause premature termination of polypeptide synthesis (nonsense); they may affect the amount of transcription of a gene, or they may alter splice sites for RNA processing, often with drastic effects on gene expression.

Deletions and insertions are another important source of genetic variation. Both may occur as a result of unequal crossing-over, but other mechanisms, not yet clearly understood, are apparently also involved.

Transposable elements, which have been an important source of genetic variation during human evolution, continue to generate mutations at the present time.

Chromosomal mutations represent gross changes that can be detected by microscopic examination of metaphase or late prophase chromosomes. They include polyploidy (extra sets of chromosomes), aneuploidy (any number of chromosomes other than a multiple of the haploid number), and rearrangements.

Polyploidy is quite rare and is not compatible with extended postnatal life in humans. Aneuploidies of the sex chromosomes are relatively common; some, like XO, XXY, and XYY have limited effects on phenotype. The most common autosomal aneuploidy is trisomy 21 (Down syndrome). Most autosomal aneuploidies have severe effects and few affected fetuses are born alive.

Common rearrangements include deletions and translocations. Balanced translocations arise when there have been breaks in two non-homologous chromosomes, followed by joining of the broken ends in new combinations. Balanced translocations often have no deleterious effects on persons who carry them, but they are responsible for abnormal gamete formation, which may lead to abnormal off-spring or multiple spontaneous abortions.

REFERENCES

Bonifas, J. M., Morley, B. J., Oakey, R. E., Kan, Y. W., and Epstein, E. H. 1987. Cloning of a cDNA for steroid sulfatase: frequent occurrence of gene deletions in patients with recessive X chromosome-linked ichthyosis. Proc. Natl. Acad. Sci. USA 84:9248–9251.

Clever, J. E. 1983. Xeroderma pigmentosum. In "The Metabolic Basis of Inherited Disease," 5th ed. J. B. Stanbury et al.,eds. pp. 1227–1248. McGraw Hill, New York.

Collins, F. S. and Weissman, S. M. 1985. The molecular genetics of human hemoglobins. Prog. Nucleic Acid Res. and Mol. Biol. 31:315–462.

Dombroski, B. A., Mathias, S. L., Nanthakumar, E., Scott, A. F., and Kazazian, H. H., Jr. 1991. Isolation of an active human transposable element. Science 254:1805–1808.

Edlin, G. 1990. Human Genetics: A Modern Synthesis. Jones and Bartlett, Boston.

Evans, J. L., Hamerton, J. L., and Robinson, A. (eds.) 1990. Children and young adults with sex chromosome aneuploidy. Birth Defects: Original Article Series, Vol. 26, No. 4.

Forget, B. G., Benz, E. J, Jr., and Weissman, S. M. 1983. Normal human globin gene structure and mutations causing the beta-thalassemia syndromes. In "Recombinant DNA Applications to Human Disease," C. T. Caskey and R. L. White, eds., pp. 3-17. Cold Spring Harbor Laboratory.

Harris, H. 1980. The principles of human biochemical genetics, 3rd ed. Elsevier/North Holland Biomedical Press, Amsterdam.

Higgs, D. R., Goudbourn, S. E. Y., Lamb, J., et al. 1983. Alpha thalassemia caused by a polyadenylation signal. Nature 306:398–400.

Lewin, B. 1990. Genes IV. Oxford University Press, Oxford, England.

Mathias, S. L., Scott, A. F., Kazazian, H. H., Jr., Boeke, J. D., and Gabriel, A. 1991. Reverse transcriptase encoded by a human transposable element. Science 254:1808–1810.

Morlé, F., Lopez, B., Henni, T., and Godet, J. 1985. Alpha thalassemia associated with the deletion of two nucleotides at position -2 and -3 preceding the AUG codon. EMBO J. 4:1245–1250.

Orkin, S. 1983. A review of beta-thalassemias: the spectrum of gene mutations. In "Recombinant DNA Applications to Human Disease," C. T. Caskey and R. L. White, eds., pp. 19-28. Cold Spring Harbor Laboratory.

Orkin, S. H., Kazazian, H. H., Jr., Antonarakis, S. E., et al. 1982. Abnormal RNA processing due to the exon mutation of betaE-globin gene. Nature 300:768–769.

Prescott, D. M. 1988. Cells. Jones and Bartlett, Boston.

Snyder, L. A., Freifelder, D., and Hartl, D. L. 1985. General genetics. Jones and Bartlett, Boston.

Tartof, K. D. 1988. Unequal crossing-over then and now. Genetics 120:1-6.

Wallace, M. R., Andersen, L. B., Saulino, A. M., et al. 1991. A *de novo* Alu insertion results in neurofibromatosis type 1. Nature 353:864–866.

Wolff, R. K., Plaetke, R., Jeffreys, A. J., and White, R. 1989. Unequal crossing-over between homologous chromosomes is not the major mechanism involved in the generation of new alleles at VNTR loci. Genomics 5:382–384.

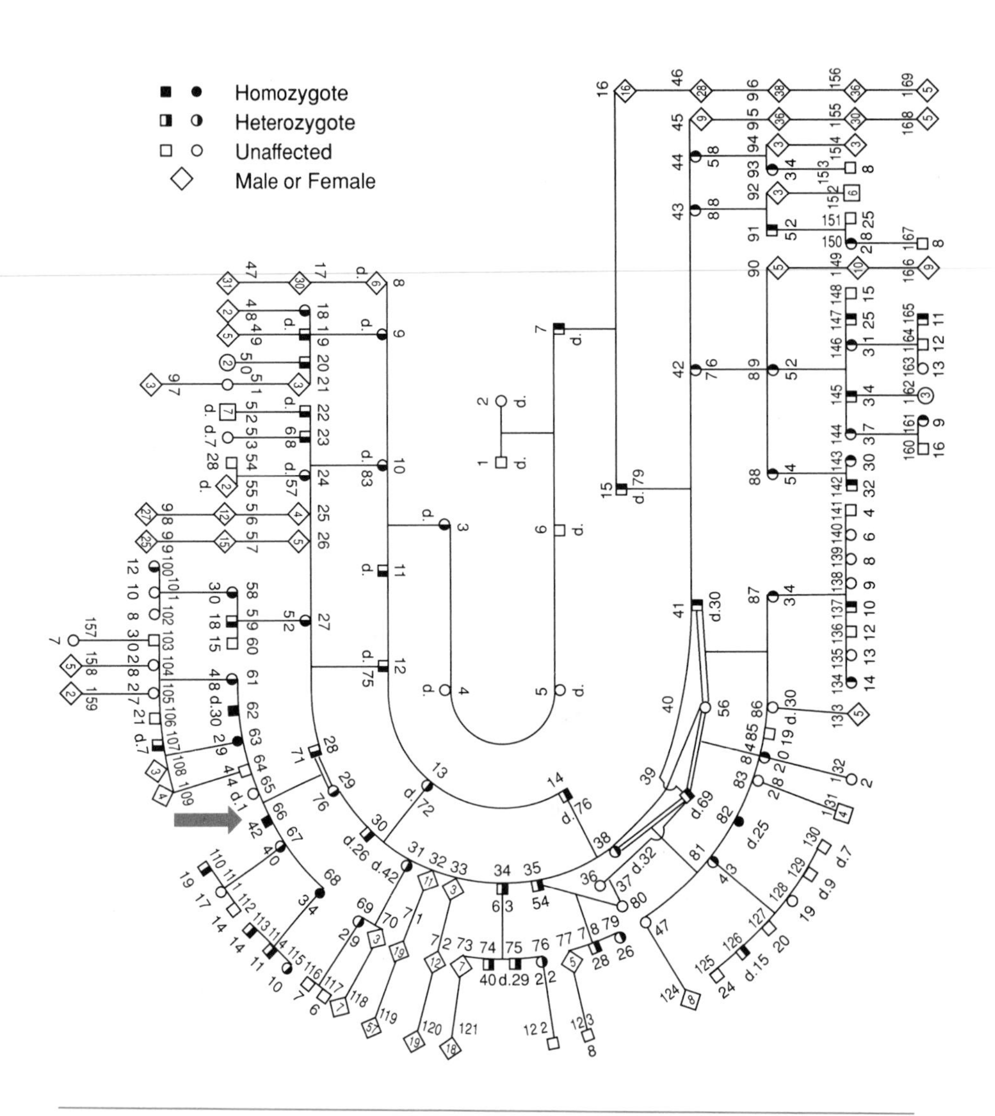

Pedigree of an autosomal dominant gene for Charcot-Marie-Tooth disease (peroneal muscular atrophy) in a large kindred of French Acadian descent. Number above each symbol identifies the symbol; number within symbol indicates number of individuals represented by that symbol; number below symbol indicates age at study or at death. Note four presumed homozygotes (filled symbols) in family indicated by the arrow, resulting from a first-cousin mating. (From Killian and Kloepfer, 1978)

8 Dominance and Recessiveness, Penetrance and Imprinting

SOME MOLECULAR EXPLANATIONS

Ever since the classic experiments of Mendel on peas, the concepts of dominance and recessiveness have been part of the fundamental framework of genetic theory. In the Mendelian sense, a recessive allele has no influence on the organism's phenotype if only one copy of the allele is present in the genome; two copies of a recessive allele are needed in order for the characteristic phenotype to be expressed. Nowadays virtually every high school student who has had a biology course knows that the gene for white flowers is recessive to the gene for purple flowers in peas, that blue eye color in humans is recessive to brown eye color, and that the inability to taste the bitter compound PTC is recessive to tasting ability. In contrast, a dominant allele will determine a phenotype when only one copy is present in an organism's genome; and in its simplest sense, the concept of genetic dominance implies that homozygosity for a dominant allele produces the same phenotype as heterozygosity.

Classification of an allele as dominant or recessive depends on the pattern of inheritance of the phenotype associated with that allele, as revealed by *pedigree analysis*. In this chapter, we will limit ourselves to a consideration of pedigrees of autosomal genes. Pedigrees are a diagrammatic presentation of the occurrence of an inherited condition in two or more generations of a family or group of interrelated families.

Autosomal dominant inheritance has the following characteristics:

1. Affected individuals have at least one affected parent.
2. Matings between a normal person and a heterozygous affected person have a 50% chance of producing an affected offspring and a 50% chance of producing a normal offspring with each pregnancy.
3. Males and females are affected in roughly equal numbers.
4. Both males and females transmit the phenotype.

A classic pedigree of brachydactyly (short digits), showing dominant inheritance, is in Figure 8–1.

Autosomal recessive inheritance has the following characteristics:

1. Affected individuals usually have two normal parents.
2. Matings between heterozygotes (both phenotypically normal) have a 75% chance of producing a normal offspring and a 25% chance of producing an affected offspring with each pregnancy. Thus, on a population basis, the ratio of normal to affected offspring will be 3:1.
3. Matings between affected persons produce only affected children.
4. Males and females are affected in roughly equal numbers.
5. Both males and females transmit mutant alleles.

Figure 8–2 shows a hypothetical four-generation pedigree in which an autosomal recessive disease appears in the last generation. Note the total absence of affected persons elsewhere in the pedigree. In this example, the parents of the affected child are unrelated.

When large, multi-generation families are available, it may not be difficult to classify a disease-causing allele as dominant or recessive to wild type. Unfortunately, the classification is often unclear because the available families are small. One common difficulty is unrecognized consanguinity, which may produce a pedigree that is indicative of dominance, when the mutation is acutally recessive. Figure 8–3 shows two stages of pedigree analysis of a family with several cases of alkaptonuria. In panel A, the evidence for dominance seems obvious, despite the fact that alkaptonuria is

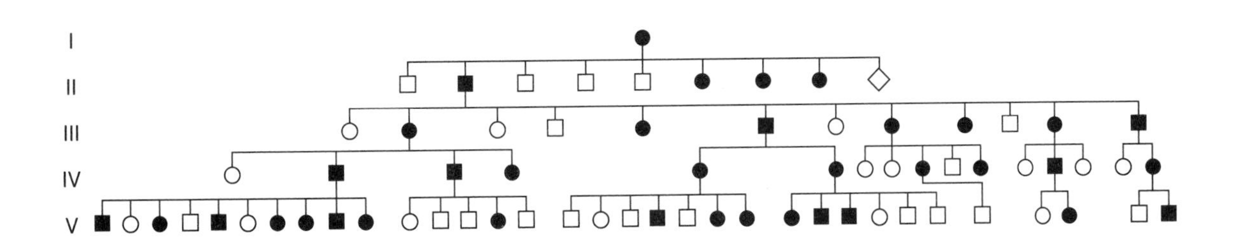

Figure 8–1

A classic example of autosomal dominant inheritance; a brachydactyly pedigree first published in 1905. Filled symbols indicate affected individuals. (From Vogel and Motulsky, 1979, p. 84; by permission of Springer-Verlag)

generally recognized as a recessive disease. However, further analysis of the same family reveals consanguinity in several matings, thereby justifying the interpretation of recessive inheritance (see Chapter 6 and Figures 6–6 and 6–7 for a discussion of homozygosity by descent).

Recessiveness and dominance, like all of the fundamental early concepts of genetics, were originally theoretical; the material basis for the various patterns of inheritance was unknown. Some of the early steps in the actualization of genetics were the recognition that genes are carried in chromosomes and Garrod's concept of "inborn errors of metabolism," which connected genetic disease in humans with biochemical abnormalities. But it was not until the rise of molecular biology, beginning in the 1950s, that it became possible to envisage a detailed understanding of genetic phenomena in terms of molecular structures and processes. Although much remains to be learned, we can now interpret patterns of inheritance, normal and abnormal genes, and gene function in terms of the nucleotide sequence of DNA, the proteins encoded by DNA, and the proteins that control transcription of DNA.

Molecular Explanations of Dominance and Recessiveness

Recessive Alleles Are Biochemically Benign

Why are some genetic diseases dominant, while others are recessive? An explanation for many recessive conditions is straightforward. Consider the reaction

$$S \xrightarrow{E} P$$

which indicates that enzyme E catalyzes the conversion of substrate S to product P. A heterozygote for a null allele in the E gene will generally have

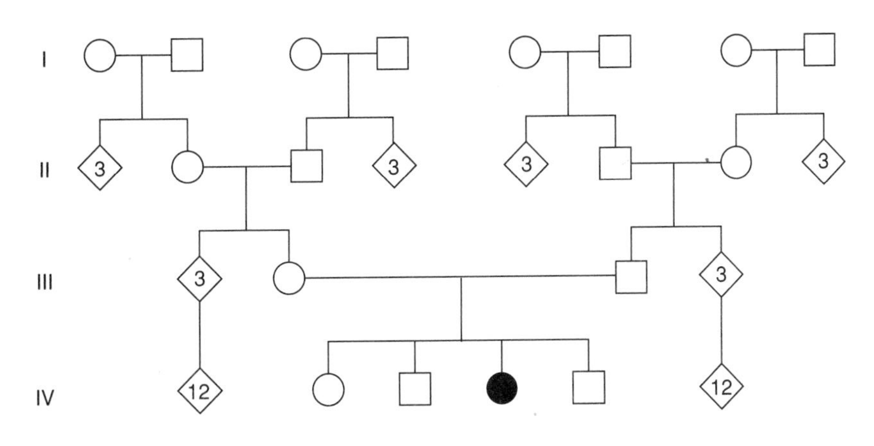

Figure 8–2

A hypothetical pedigree showing how an autosomal recessive disease (e.g. cystic fibrosis, PKU, or Tay-Sachs) can appear among the offspring of unrelated parents, although there has been no occurrence of the condition among the numerous relatives of the affected child in previous and current generations. Numbers in the diamonds indicate individuals of either sex. The entire pedigree consists of normal individuals, except for the affected girl (filled circle) in generation IV.

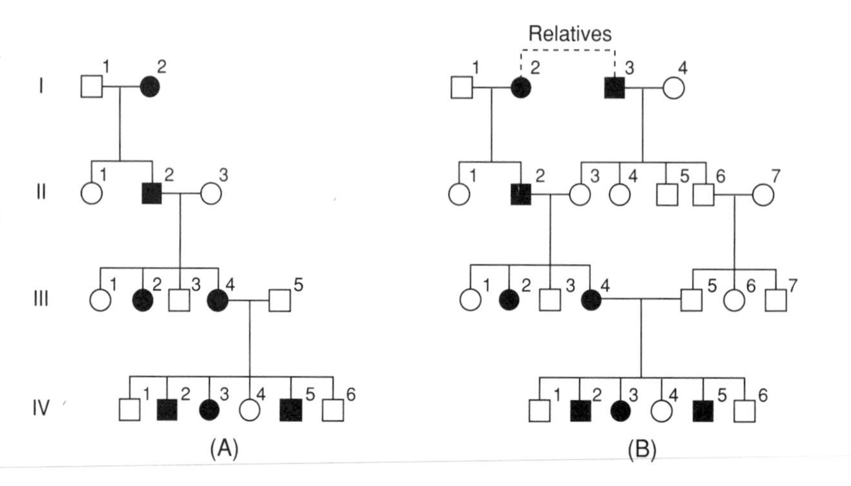

Figure 8–3

Psuedodominant inheritance of a recessive disease, such as alkaptonuria. Panel A suggests dominance of the disease-causing allele, based on incomplete information. Panel B shows more information on the same family. Consanguinity between I-2 and I-3 (e.g., cousins or siblings) accounts for the affected individuals in generations III and IV as homozygous recessives. In order to account for the affected individual in generation II, one would have to assume that his father was a heterozygote (e.g., I-1 and I-2 might have been cousins).

half as much of enzyme E as a homozygous normal person. The reason is that there is no general mechanism for dosage compensation; there is no feedback mechanism that tells a normal allele that its counterpart is non-functional, so the normal allele continues to produce the usual amount of mRNA.

In the most common situation, enzyme E will not be functioning at maximal velocity at normal concentrations of its substrate; therefore, in the heterozygote for a null allele, the concentration of S will build up until it is high enough to drive enzyme E at twice the usual rate, thus yielding the same amount of P as in a homozygous normal person, as indicated here.

$$2S \xrightarrow{0.5E} 1P$$

If there are no complicating factors, null alleles at this locus will be completely recessive to normal alleles.

It is generally believed that most recessive genetic diseases fall into the enzyme-deficiency category. Well-known examples in which heterozygotes are clinically normal are Tay-Sachs Disease (absence of hexosaminidase A), phenylketonuria (absence of phenylalanine hydroxylase), and galactosemia (absence of galactose-1-phosphate uridyl transferase).

Dominant Diseases Have a Variety of Molecular Explanations

Poisonous Products from Mutant Alleles One of the conditions for recessiveness is that the product of a mutant allele, if it is present at all, must not interfere with expression of the normal allele. However, if the product of the mutant allele prevents the function of the normal gene product, directly or indirectly, the mutation can have a *dominant negative* effect, because it acts as a *poisonous product*. A simple, hypothetical case could involve any homotetrameric enzyme. If the presence of one or more mutant polypeptides is sufficient to render the whole tetramer non-functional, then in theory, only 1/16 of the tetramers will contain four normal subunits and be functional (Figure 8–4). In many cases, this level of activity would be insufficient for normal cellular function and would result in a disease.

Hemoglobin A, the predominant form of hemoglobin in adults, provides an interesting example of a dominant disease with some features of the poisonous product category. Hemoglobin A is a tetramer, consisting of

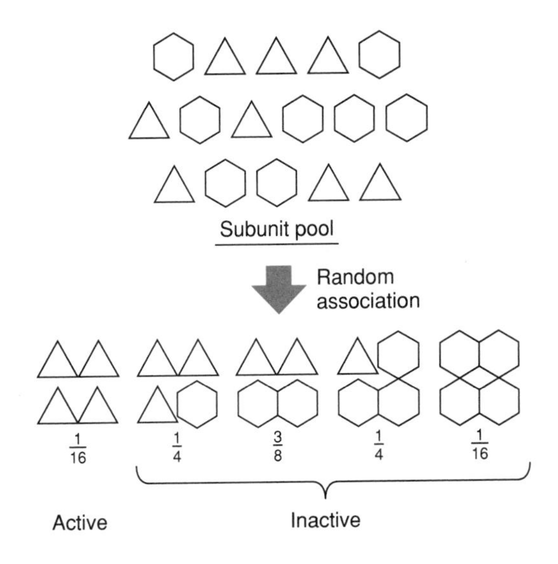

Figure 8–4

Diagram showing the possible combinations of normal polypeptides (triangle) and mutant polypeptides (hexagon), if there is random association to form a tetrameric enzyme. Frequencies of each type of tetramer are shown below the corresponding drawing. If the presence of one or more mutant polypeptides prevents the entire tetramer from functioning, only 1/16 of the tetramers will be active.

two alpha-globin and two beta-globin subunits. Each subunit binds one molecule of heme and one atom of iron. In order to bind oxygen, the iron must be in the ferrous or reduced state (Fe^{++}); however, the iron spontaneously becomes oxidized to the ferric state (Fe^{+++}), at which time the whole complex of proteins, heme, and iron is called *methemoglobin*. There is an enzyme, called *methemoglobin reductase,* that converts ferric iron to ferrous iron. Genetic defects in this enzyme are expressed as recessives.

However, there are missense mutations known in the alpha- and beta-globin genes that lead to changes in the shape of alpha- or beta-globin, which have the effect of making it difficult for methemoglobin reductase to act on ferric iron bound to those subunits. These mutations are inherited as genetic dominants, because every red cell has a substantial portion of its hemoglobin with one or two iron atoms in the ferric state; and because of the cooperativity of oxygen binding in hemoglobin, the overall effect is a large reduction in oxygen-carrying capacity. In theory, one-fourth of the hemoglobin tetramers should contain no mutant subunits, but this is apparently insufficient for normality.

Heterozygotes for these mutations are true "blue-bloods," because ferric hemoglobin absorbs red light more strongly than ferrous hemoglobin, thereby enriching the reflected light for the blue end of the spectrum. Unfortunately, the condition is no asset; the patients' tissues are perpetually starved for oxygen. Infants with this condition are sometimes misdiagnosed as having congenital heart disease, which also may produce a bluish color to the skin.

Additional examples of poisonous gene products can be found among other non-enzymic proteins. If the presence of an abnormal polypeptide prevents the assembly of normal intracellular structures, such as microtubules or other cytoskeletal elements, the corresponding mutation may produce a dominant genetic disease.

Haploid Insufficiency Another important category of dominant genetic diseases arises from mutations in enzymes where half the normal level of activity is not enough to produce a normal phenotype; that is, there is *haploid insufficiency*. Many metabolic pathways have a rate-limiting step, which is catalyzed by an enzyme that functions at or near its maximum velocity in persons with two normal alleles. If a mutation abolishes the production of active enzyme from one allele, the remaining enzyme cannot produce more of its product, even though the concentration of its substrate rises substantially.

In humans, a well known example is acute intermittent porphyria, where a crucial enzyme in the heme pathway (uroporphyrinogen synthase, also known as porphobilinogen deaminase) is saturated by normal levels of its substrate (Figure 8–5). In persons with one non-functional allele of the uroporphyrinogen synthase gene, approximately half as much heme is made as in homozygous normals; this leads via feedback to overproduction of heme precursors, which in some manner produces episodes of colic, partial paralyis, and mental confusion.

In addition, mutations in genes that encode structural proteins and other non-enzymic proteins may fall into the haploid insufficiency category. Perhaps the best known example is the receptor for low-density lipoprotein (LDL), which is crucial for the control of cholesterol levels in human blood (Figure 8–6). Mutations in the LDL-receptor gene that reduce the number of functional receptor molecules on the cell surface by 50% lead to the dominant disease, familial hypercholesterolemia. Persons with this condition develop atherosclerosis in early adulthood and are at greatly increased risk for heart attacks.

Increased Enzyme Activity Mutations that increase activity of an enzyme beyond limits that are compatible with health can also cause dominant diseases. In the simplest situation, an increase in enzyme activity will bring the concentration of its substrate down to a new steady-state level, and the product of the reaction will be generated at the normal rate. Such a mutation will often have no phenotypic effect, even if present in the homozygous state. However, if the substrate is used by more than one metabolic pathway, the reduced level of the metabolite in question may interfere with normal functioning of another pathway, and behave as a dominant defect.

There is a family in which 11 out of 23 members had a 45–70-fold increase in red cell adenosine deaminase activity (Valentine et al., 1977). The affected persons had mild anemia and enlarged spleens. Studies on the purified enzyme from affected persons showed that it had the same biochemical properties as enzyme from normal persons; apparently it was simply made in greatly increased amounts. The reason for the anemia appeared to be a decreased lifetime for red cells, which may have been caused by a deficiency of ATP. Figure 8–7 provides a speculative, but plausible, explanation for the observed effect. The idea is that the purine salvage pathway is important in maintaining the ATP level in red blood cells, and that

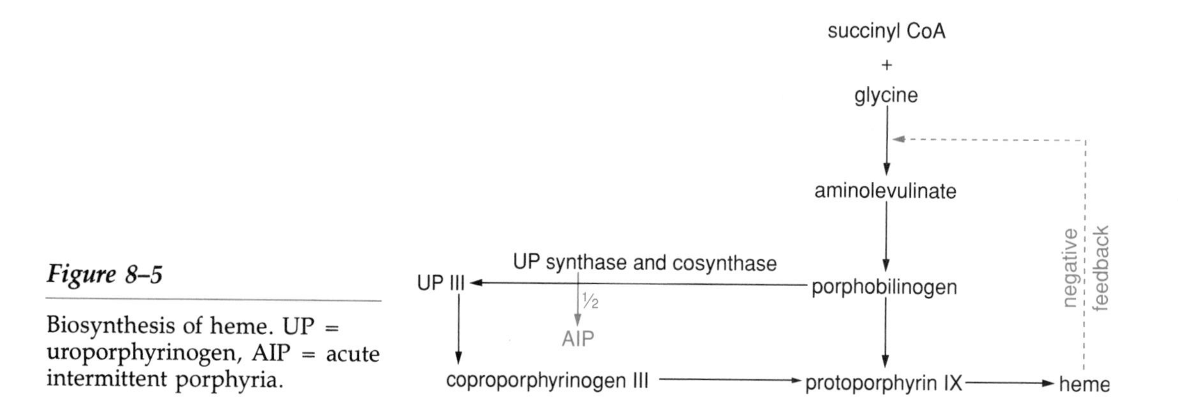

Figure 8–5

Biosynthesis of heme. UP = uroporphyrinogen, AIP = acute intermittent porphyria.

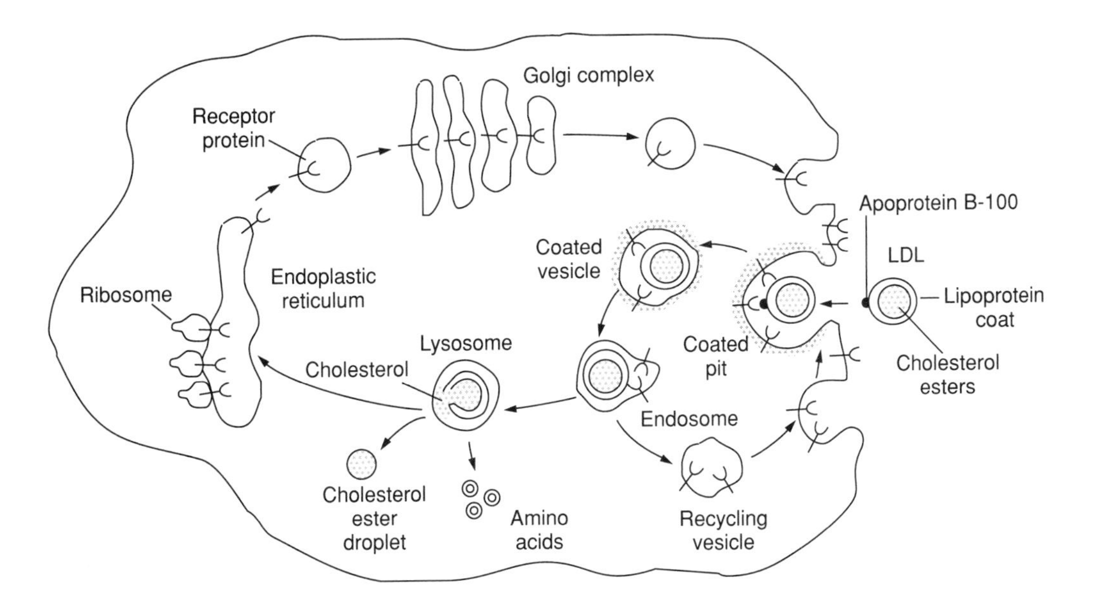

Figure 8–6

Itinerary of the LDL receptor in mammalian cells, from its site of production in the endoplasmic reticulum, to the Golgi complex, to the cell surface, back into the cell via coated pits and endosomes, then back to the surface. Cholesterol brought in by the LDL receptors and released from lysosomes has several effects: esterification of excess cholesterol for storage is stimulated, production of new cholesterol within the cell is inhibited, and production of new LDL receptors is inhibited. (From Mange and Mange, 1990, p. 319; by permission of Sinauer Associates)

increased adenosine deaminase activity removes most of the adenosine before the salvage pathway can capture it.

Mutations that increase enzyme activity may produce increased levels of the product of the enzymic reaction. If the increased product level upsets a feedback circuit, the mutation may have deleterious consequences and be expressed as a dominant condition. An example is provided by the X-linked enzyme PRPP synthetase, which controls the rate of purine synthesis (Figure 8–8). A family is known with members who have about 3X the usual PRPP synthetase activity, apparently caused by increased specific activity of the enzyme (Becker et al., 1973). The mutation is dominant and affected persons have gout. It seems that the increased level of PRPP leads to more production of purines than can be used for anabolic processes, so the excess is degraded to uric acid faster than the uric acid can be converted to urea. Uric acid is quite insoluble; the inflammation and pain of gout are caused by precipitation of needle-like crystals of uric acid, principally in joints and the peripheral capillaries.

Gain-of-Function Mutations A potential source of dominant genetic diseases is represented by mutations that allow normal expression of a gene in the usual tissue at the usual time, but also cause *inappropriate expression*. Generally, these are mutations that affect transcriptional control elements in a gene, rather than the structure of the encoded polypeptide. These mutations result either in *ectopic expression* (synthesis of the gene product in the wrong place; that is, in cells that do not ordinarily express that gene at all and therefore function abnormally as a result of expression of that gene) or

$$\text{Adenosine} + \text{AKP} \xrightarrow[\text{kinase}]{\text{adenosine}} \text{AMP} + \text{ADP}$$

$$\text{(A)} \quad \text{AMP} + \text{ATP} \xrightarrow[\text{kinase}]{\text{adenylate}} 2\,\text{ADP}$$

$$3\,\text{ADP} + 3\,\text{Pi} \xrightarrow{\text{glycolysis}} 3\,\text{ATP}$$

$$\text{(B)} \quad \text{Adenosine} + H_2O \xrightarrow[\text{deaminase}]{\text{adenosine}} \text{inosine} + NH_3$$

Figure 8–7

Competing reactions that utilize adenosine. Excess adenosine deaminase may reduce the supply of ATP in red blood cells, which have no mitochondria.

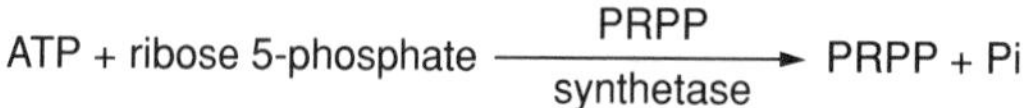

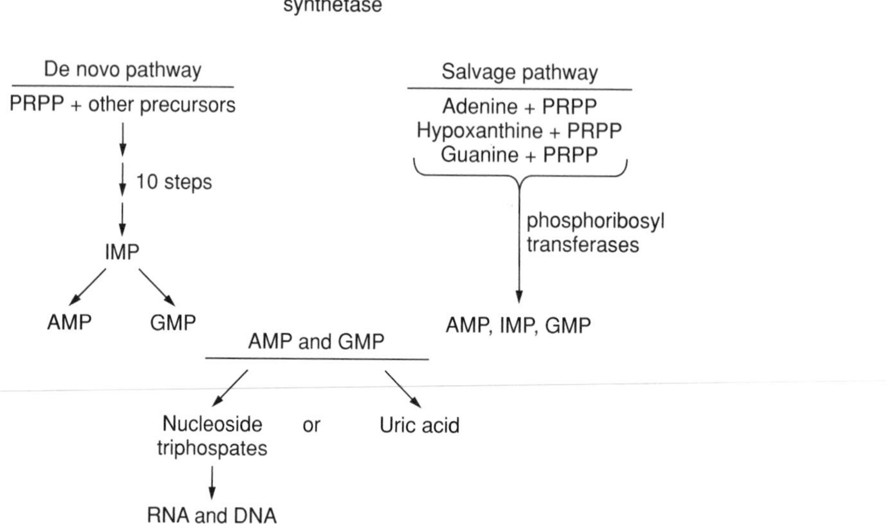

Figure 8–8

Sequence of steps leading to overproduction of uric acid in persons with unusually high PRPP synthetase activity. Excess PRPP drives the production of AMP and GMP to levels higher than needed for nucleic acid synthesis, despite the presence of feedback interactions. (AMP, GMP, and IMP are inhibitors of several steps in their own biosynthesis.) The resulting high level of uric acid in serum leads to uric acid crystals in joints and capillaries, causing gout.

in *ecchronic expression* (synthesis of the gene product in cells that normally express the gene, but at the wrong time). There may appear to be some overlap between the inappropriate expression category and the excessive activity category; however, the distinguishing feature of the concept of inappropriate expression is that the mutant gene functions normally in the usual place at the usual time, whereas its deleterious effect comes from the fact that it also is expressed in the wrong place or at the wrong time in the right place (or both).

The classic examples of gain-of-function mutations are the homeotic mutations in the fruit fly, *Drosophila*, which produce a more-or-less normal structure in an abnormal location. For example, some mutations in the *bithorax* gene complex result in a fly with four wings, instead of the usual two. Other mutations in the *Antennapedia* gene complex cause a leg to be formed on the head, where an antenna would ordinarily develop. An interesting example of the latter class of homeotic mutation is in Frischer et al. (1986). They describe a dominant allele that is an inversion that brings most of the coding region of the *Antennapedia* gene under the control of the 5′ region of another gene that is normally expressed in the head during development. Apparently, the *Antennapedia* gene issues a command to form a leg, which overrides the normal instructions to form an antenna. Normal legs are formed in the thoracic segments of the fly, presumably because the normal allele of the *Antennapedia* gene is active there and the mutant allele is not.

As yet, inappropriate expression is only a theoretical source of dominant genetic diseases in humans, but in view of our rapidly increasing understanding of transcriptional control mechanisms, it is quite likely that concrete examples of inappropriate expression will be available in the near future. Perhaps the most likely class of inappropriate expression mutations will be changes in DNA methylation of control elements. These may affect the binding of transcriptional activators and/or repressors, thus causing ectopic or ecchronic expression. Alternatively, methylation mutations may affect the process of genomic imprinting (see the last section of this chapter), which may then lead to aberrant transcription.

Position Effects The early *Drosophila* geneticists observed a number of genetic loci whose expression varied according to their chromosomal location. In particular, when a chromosomal rearrangement such as an inversion placed a gene close to heterochromatin, the activity of that gene was often suppressed, partially or totally. This phenomenon is known as a *position effect*. Its molecular basis is not yet understood, but it is presumed to be the result of the topology of chromatin in regions where there are large tandem arrays of highly reiterated DNA sequences. Probably genes at the margin of a highly coiled heterochromatic segment of a chromosome are not as accessible to RNA polymerase and/or transcriptional activators as they would be in a euchromatic region.

Most position-effect mutations show recessive inheritance, as would be expected for a structural change that affects only one chromosome, but a few examples of dominant position effects have been described in *Drosophila*. Charles Laird (1990) has recently presented a thought-provoking hypothesis, in which he suggests that Huntington disease (HD) may be the result of a dominant position effect mutation similar to *brown-dominant* in *Drosophila*.

Brown-dominant is unusual in that it inactivates not only the allele on the chromosome carrying the mutation, but the allele on the homologous chromosome as well. How might *trans-inactivation* occur? It apparently requires some form of somatic pairing between homologs in the affected region (Figure 8–9). Somehow, pairing must lead to proteins from the heterochromatic region of the mutant chromosome spreading to the normal chromosome and organizing heterochromatin there, also.

Laird's (1990) hypothesis suggests that the mutation that causes HD is an alteration of chromosome 4 near a structural gene, so that the gene is

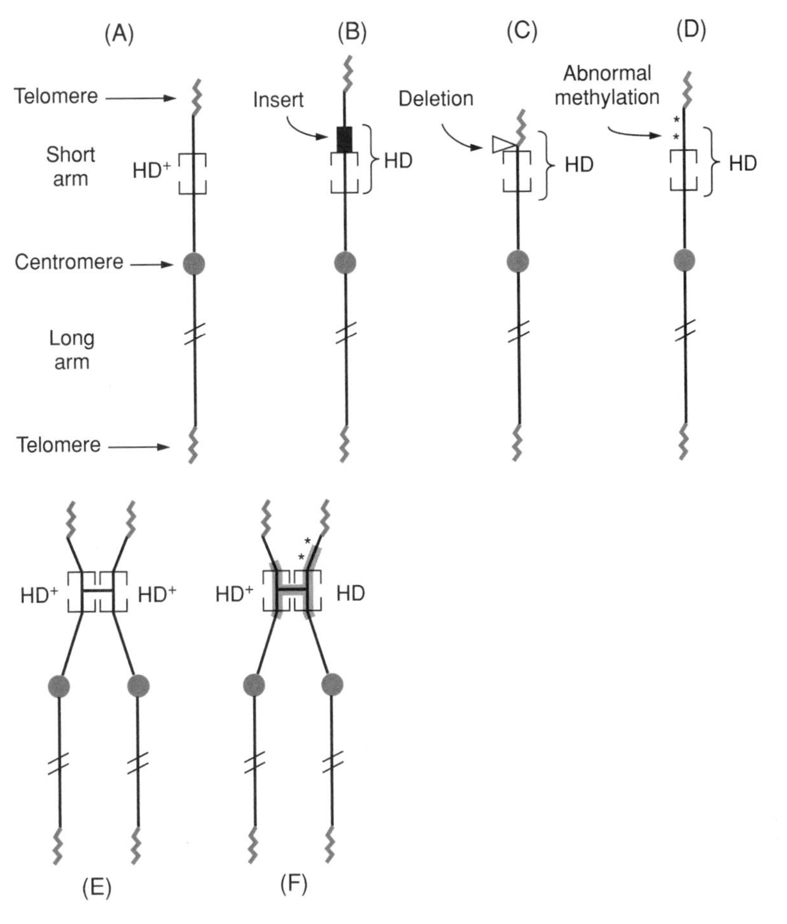

Figure 8–9

Dominance arising from a position effect, as applied to Huntington disease. (A) The normal allele, located near the telomere of the short arm of chromosome 4. (B) Insertion of heterochromatin near the HD locus. (C) Deletion of DNA between HD and the telomere. (D) Methylation of DNA between HD and the telomere. Any of these events, as well as a microinversion, might inactivate the HD locus. (E) and (F) illustrate the hypothesis of trans-inactivation (see text), which may explain how position-effect variegation can display dominant inheritance. (From Laird, 1990)

brought into juxtaposition with heterochromatin. Possible mechanisms include a microinversion, insertion of heterochromatin next to the HD gene, deletion of euchromatin between the HD gene and the telomeric heterochromatin, or creation of facultative heterochromatin next to the HD gene by some event, such as extensive methylation (Figure 8–9).

An interesting feature of Laird's hypothesis is the prediction that the mutation that causes the disease may not be in the HD gene itself, but in neighboring DNA. And another aspect of position effects that may be relevant to HD is mosaic inactivation of the affected locus (position-effect variegation); that is, inactivation may vary from individual to individual or from cell to cell, and the time of inactivation during development may vary. Similar effects could help to explain the wide range in time of onset of HD (Figure 5–1) and the variability of symptoms.

Dominance and Recessiveness Vary According to the Level of Observation

In the early days of genetics, when our knowledge of biochemistry was extremely limited, dominance and recessiveness could only be observed in terms of the gross phenotype of organisms. Mendel's peas were either tall or short, their seeds were either wrinkled or round, their flowers were either purple or white, etc. It was not long before someone noticed that a cross between white-flowered and red-flowered snapdragons produced an F1 generation with pink flowers, and so it became apparent that heterozygotes could be intermediate in phenotype between the homozygotes. This phenomenon is variously called *incomplete dominance, semi-dominance,* or *partial dominance*. Its counterpart in human genetic disease falls into the category of haploid insufficiency. More specifically, many dominant genetic diseases are less severe in the heterozygote than in the homozygote. This fact will be considered in more detail in the next section.

Heterozygotes for recessive genetic diseases may have the wild-type phenotype when observed at the clinical level, but when examined at the biochemical level, they are frequently intermediate. A good example is Tay-Sachs disease, which is caused by the absence of hexosaminidase A activity. Heterozygotes are indeed clinically normal, but when the level of hexosaminidase A in their blood is measured, it is found to lie between the level found in homozygous normals and the zero or near-zero level in infants affected with the disease. This fact is very useful for genetic screening (Chapter 9), because a simple, cheap assay can be used to identify couples at risk for giving birth to an affected child.

Codominant alleles are expressed independently of one another. Good examples may be found among the blood group substances, such as the MN locus. When appropriate antisera are used, the presence of the M and N alleles in any combination can be detected. If we go one level deeper to the DNA itself, we see that any assay that directly detects the nucleotide differences between alleles defines those alleles as codominant at that level of observation. Thus, all RFLPs create codominant alleles. (The consequences of the RFLPs may be recessive or dominant when observed at the organismal level, but when we assay for the restriction fragments themselves, we see them as codominant alleles.) Also, any electrophoretic assay that separates the products of different alleles reveals those alleles as codominant entities.

The sickle-cell mutation in beta-globin illustrates how complex the definitions of dominance and recessiveness can become. At the clinical level and under ordinary conditions of life, the sickle-cell allele is recessive to the wild-type allele of beta-globin; heterozygotes have normal health. However, under conditions of moderate oxygen deprivation, such as one encounters at high altitudes, heterozygotes may develop abdominal pain and bloody urine. If the heterozygote is observed at such a time, one would be obliged to conclude that the wild-type allele is only partially dominant.

There have been a few reports suggesting that when a sickle-cell heterozygote is subjected to extreme physiological exertion, as during an athletic contest, the exceptional demand for oxygen may lead to a sickle-cell crisis, occasionally with a fatal outcome. In such a case, the sickle-cell allele could be considered as a full dominant, but whether the presence of the heterozygous state was really the primary cause of the fatal or near-fatal response to stress, or whether sickling occurred *ex post facto* in the few reported cases, is a moot point.

When the blood of a sickle-cell heterozygote is examined by electrophoresis under appropriate conditions, two forms of hemoglobin can be identified; one contains wild-type beta-globin, while the other contains sickle-cell beta-globin. Alternatively, there is a restriction endonuclease that will cut the normal beta-globin gene, but will not cut the sickle-cell beta-globin allele at the site of the mutation. Both the protein assay and the nuclease assay reveal the sickle-cell allele as codominant with the normal allele.

Genetic Diseases Rarely Show Complete Dominance at the Phenotypic Level

Homozygotes for dominant genetic diseases are hard to find. That is not surprising, given the facts that affected heterozygotes are not numerous and that their abnormality often reduces the likelihood of their becoming parents. However, matings between heterozygotes for dominant genetic diseases occasionally occur, usually within an extended family. Unrelated heterozygotes may be brought together by the "misery loves company" phenomenon, known more formally as *assortative mating*. In either case, the potential for homozygous progeny exists. Pauli (1983) compiled data on a dozen dominant genetic diseases, where one or more matings between heterozygotes were known; some examples were achondroplasia, brachydactyly, and aniridia. In every case, one or more of the children were more severely affected than either of the parents; some of them died in infancy. Several other dominant genetic diseases not covered by Pauli, such as the Marfan Syndrome (Chemke et al., 1984), also showed exceptionally severe manifestations in putative homozygotes. Although there was no formal proof, either by linkage analysis or molecular analysis, that the severely affected offspring of these heterozygote-heterozygote matings were actually homozygous for the mutant allele, the assumption is unlikely to have been universally wrong.

Thus, for some years it appeared that complete dominance rarely or never applied to human genetic diseases. Then, in 1987, Wexler et al. reported on four homozygotes for Huntington Disease in the large Venezuelan kindred. These were found in a family of 14 children from a mating between relatives, both of whom were subsequently affected with the disease, and their genotypes were established by molecular analysis. The fascinating result of their study of this family was that the homozygotes were not more severely affected than their heterozygous sibs or other affected persons in the family. There was no apparent difference in the age of onset of symptoms, either. Here at last was the first well-documented case of a genetic disease that shows classical Mendelian dominance; that is, the phenotype of the homozygote is not different from that of the heterozygote. Two years later, Myers et al. (1989) described more probable homozygotes for HD and confirmed Wexler et al.'s conclusions.

What can we learn from these observations? Is the rarity of complete dominance in human genetic diseases understandable in molecular terms? Let us consider the five classes of dominant mutations described earlier in this chapter.

1. Poisonous product. If the product of the mutant allele completely abolishes the function of its normal counterpart or causes the death of the cell in which it is expressed, complete dominance will ensue; that is, the presence of a second mutant allele could not cause any further reduction of function.

2. Haploid insufficiency. The definition of this class of dominant mutations implies that heterozygotes have only one allele that is functioning normally. Therefore, it is to be expected that homozygotes for mutant alleles will have more serious effects. This implies incomplete dominance.

3. Excessive production and/or activity of product. If one mutant allele produces enough of its product to cause clinical symptoms, then it seems likely that two mutant alleles would lead to more severe symptoms. In order for the mutation to be completely dominant, it would have to cause total inhibition of some metabolic step, or possibly cell death, when only one allele was present.

4. Inappropriate expression. Mutations in this class could have effects at all levels of severity, ranging from barely detectable diminution of health to lethality. Therefore, all degrees of dominance are possible.

5. Trans-inactivation via a position effect. Inasmuch as one mutant allele will silence both copies of a gene subject to the position effect, trans-inactivation mutations should be fully dominant.

The preceding considerations imply that there are several possible molecular explanations for complete dominance. In any case, in order for complete dominance to be observed, it must be present in a function whose absence is deleterious enough to cause illness or clinically apparent abnormality, but not a function whose absence is lethal (unless lethality occurs after reproductive age). This surely restricts the number of candidate functions.

Huntington Disease is a good candidate for an inappropriate expression mutation that allows normal function for a large fraction of the life span, but then begins to be expressed aberrantly, leading to disease. It could be an allele that is turned off in mid-life, leading to delayed-onset haploid insufficiency, or it could be a gain-of-function mutation that causes the gene to be turned on ectopically or ecchronically. If the expression of the mutant gene in certain neurons leads to cell death, then the presence of two copies in an HD homozygote would not be expected to produce more severe symptoms than are shown by the heterozygote; a dead cell is a dead cell. The timing of inappropriate expression of the HD allele may well depend upon other genes (e.g., transcriptional regulators), which vary from individual to individual. Laird's (1990) suggestion that HD is caused by a dominant position effect, with trans-inactivation of the normal allele, is another way of explaining the fact that HD homozygotes have the same phenotype as heterozygotes.

Molecular Explanations of Incomplete Penetrance and Variable Expressivity

Incomplete Penetrance

Dominant and recessive alleles in humans are identified by their pattern of inheritance, which is revealed by pedigree analysis, as explained in the introduction to this chapter. When the standard Mendelian rules are obeyed and when sufficiently large pedigrees are available, it is not difficult to distinguish recessive traits from dominant ones. Unfortunately, pedigree analysis in humans is often complicated by the phenomenon of *incomplete penetrance* of dominant alleles; that is, some individuals who must be assumed to have a disease-causing allele (if the hypothesis of dominance is correct) do not show any clinically detectable expression of it. This leads to the presence of "skipped generations" in pedigrees. A classic example is the pedigree of polydactyly shown in Figure 8–10, where individuals II-6, II-10, and III-13 must be carriers of the polydactyly allele, but they have the normal number of digits.

In the past, it has been customary to account for incomplete penetrance by reference to variations in the environment or the "genetic background." No doubt this explanation is true, but it does not go far enough. Leaving aside the matter of environmental influences, particularly the prenatal environment, it is timely to ask whether there are molecular genetic mechanisms that could lead to nonexpression of dominant mutations in occasional individuals.

Three sources of incomplete penetrance can be postulated on theoretical grounds. First, mutations that are expressed as changes in transcription of a given gene may be overridden by transcription factors that are not sensitive to the alteration in the control region that has been created by the mutation. The more that we learn about transcription factors, the more complicated the story becomes. Many of them occur as families of closely related proteins, with overlapping specificities. It is likely, although unproven, that some transcription factors will be found to be polymorphic; that is, that different individuals will have different forms of the same transcription factor because of allelic differences in the corresponding gene, and that these polymorphisms make subtle differences in the binding specificities of the factors.

In such circumstances, it seems virtually certain that some transcriptional control mutations will cause aberrant expression in some individuals

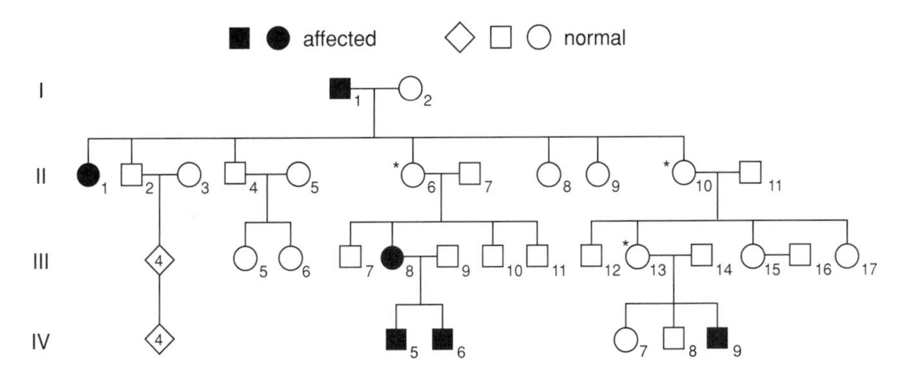

Figure 8–10

Incomplete penetrance in a pedigree of polydactyly. The three persons indicated by asterisks must have carried the dominant allele for polydactyly, but did not express it. (From Neel and Schull, 1954, p. 50; by permission of University of Chicago Press)

and normal expression will be allowed in other individuals. Thus, if the spouse of a person who is affected with a dominant disease has a gene for a variant transcription factor, then those progeny who inherit the gene for the disease and the gene for the variant transcription factor may not develop the disease. This could lead to incomplete penetrance of dominant mutations, which cause disease because of haploid insufficiency, excessive production of product, or inappropriate expression.

Another cause of incomplete penetrance may be protein-protein interactions that suppress the "poisonous product" phenomenon. The simplest situation would be a form of *interallelic complementation*, which is the complete or partial restoration of enzyme activity through the noncovalent interaction of polypeptides in a multisubunit protein. A good example of interallelic complementation has been documented for the alcohol dehydrogenase gene (ADH) in *Drosophila* (Hollocher and Place, 1987). Using 12 null activity mutants and making all possible pairwise crosses, they found a large number of crosses that produced some activity of this dimeric enzyme. Apparently, conformational abnormalities in one mutant often complement conformational abnormalities in another mutant, so that some enzyme activity is restored in the heterodimers, even though both homodimers are inactive.

To understand how interallelic complementation might produce incomplete penetrance of a dominant allele, consider a human gene that is polymorphic; that is, there are several alleles that produce more-or-less normal levels of the dimeric enzyme encoded by that gene. Assume that a dominant mutation (A^D) occurs, which makes a polypeptide that is inactive as a homodimer and which prevents activity of the normal polypeptide when it forms a heterodimer with the product of allele A^1. Let there be another wild-type allele (A^2), whose product is not inactivated by combination with the A^D polypeptide. The possible dimers that could form from these hypothetical polypeptides are diagrammed in Figure 8–11(A). When affected individuals (A^1A^D) pass their A^D allele to offspring who receive A^2 from the other parent, the A^D allele will not be expressed at a clinically obvious level, and incomplete penetrance will result (Figure 8–11(B)).

The third source of incomplete penetrance is balanced translocations. A very interesting pedigree for the inherited childhood tumor, retinoblastoma (RB), was described by Strong et al. (1981). Molecular analysis of this family showed that there was a balanced interstitial translocation in some phenotypically normal members. A portion of the long arm of chromosome 13 containing the RB gene had been translocated to the short arm of chromosome 3. In persons containing both modified chromosomes, there was no genetic deficiency and no tumor formed. However, there is a 25% probability that gametes formed by balanced translocation carriers will contain the deficient chromosome 13 and the normal chromosome 3 (Figure 8–12(A)). These individuals will develop retinoblastoma. Figure 8–12(B) is a hypothetical pedigree that illustrates the principles involved.

Readers should note that this form of incomplete penetrance has special characteristics. Although non-affected individuals have affected children, there cannot be any affected person in the direct line of descent of the non-affected carrier. That is, once the two components of the balanced translocation are separated by meiotic segregation, they cannot come together again (inbreeding excepted), and a person who has RB because he or she has inherited the deletion variant of chromosome 13 cannot have children who carry the deletion chromosome without expressing the disease.

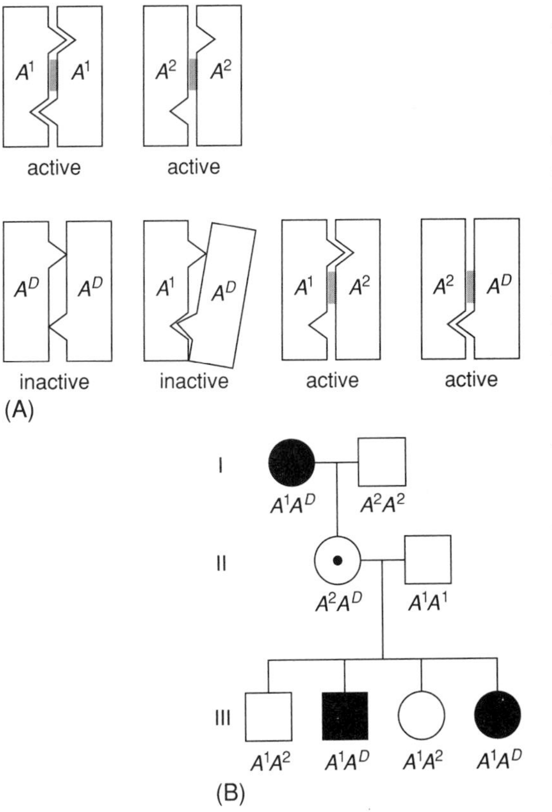

Figure 8–11

(A) Interallelic complementation via polypeptide interactions (the pit and bump hypothesis). Three types of subunits are proposed, which must interact in closely apposed pairs to form a catalytic site (indicated by the shading between subunits). Type A^1 subunits have pits and bumps; they can form active homodimers. Type A^2 subunits have pits, but no bumps; they can also form active homodimers. Heterodimers of A^1 and A^2 are active, because the bump on the former fits into the pit in the latter. Type A^D subunits, however, have lost the pit and kept the bump; homodimers cannot approach each other closely enough to form a catalytic site, and the same is true for the A^1A^D heterodimer. Heterodimers of A^2 with A^D are active, because the bump on the latter fits into the pit in the former, allowing close interaction of the surfaces that form the catalytic site. (B) Hypothetical pedigree showing incomplete penetrance caused by interallelic complementation. Filled symbols indicate affected individuals. The female in generation II does not express the A^D allele, because heterodimers formed between the A^2 and A^D subunits of the enzyme are functional, in contrast to the inactive dimers formed between A^1 and A^D subunits.

Variable Expressivity

Many genetic diseases show *variable expressivity*. Most patients display enough of the characteristic symptoms to permit diagnosis, but the symptoms are not all the same in every patient. There may also be persons who carry mutations at a given disease-causing locus, but who are not severely enough affected to be accurately diagnosed or even to come to the attention of physicians. The most extreme form of variable expressivity is *incomplete penetrance,* where there is no phenotypic abnormality at all, despite the presence of a mutant allele that causes severe disease in some individuals.

A good example of variable expressivity is the autosomal dominant disorder of connective tissue known as Marfan syndrome. The underlying genetic defect in most cases is in a gene that encodes fibrillin (see McKusick, 1991). The effects of mutations are just beginning to be analyzed at the molecular level, but at the organismal level they are highly *pleiotropic;* that is, many structures and functions are abnormal. Table 8–1 lists many of the known manifestations of the Marfan syndrome (Pyeritz, 1989). Patients with Marfan syndrome are usually tall and thin, with exceptionally long limbs and digits. They may have a variety of other skeletal abnormalities, they may have ocular defects, and they usually have some cardiovascular disorder. No patient shows all of the symptoms listed in the table.

At the molecular level, two classes of phenomena are known that can account for variable expressivity. The first of these is multiple disease-causing mutations at a single locus. The many ways in which a gene can be altered were surveyed in Chapter 7. One example that has emerged from molecular analysis of the dystrophin gene concerns the difference between classical Duchenne muscular dystrophy (Chapter 5) and a much milder, later-onset form known as Becker muscular dystrophy. It is now clear that

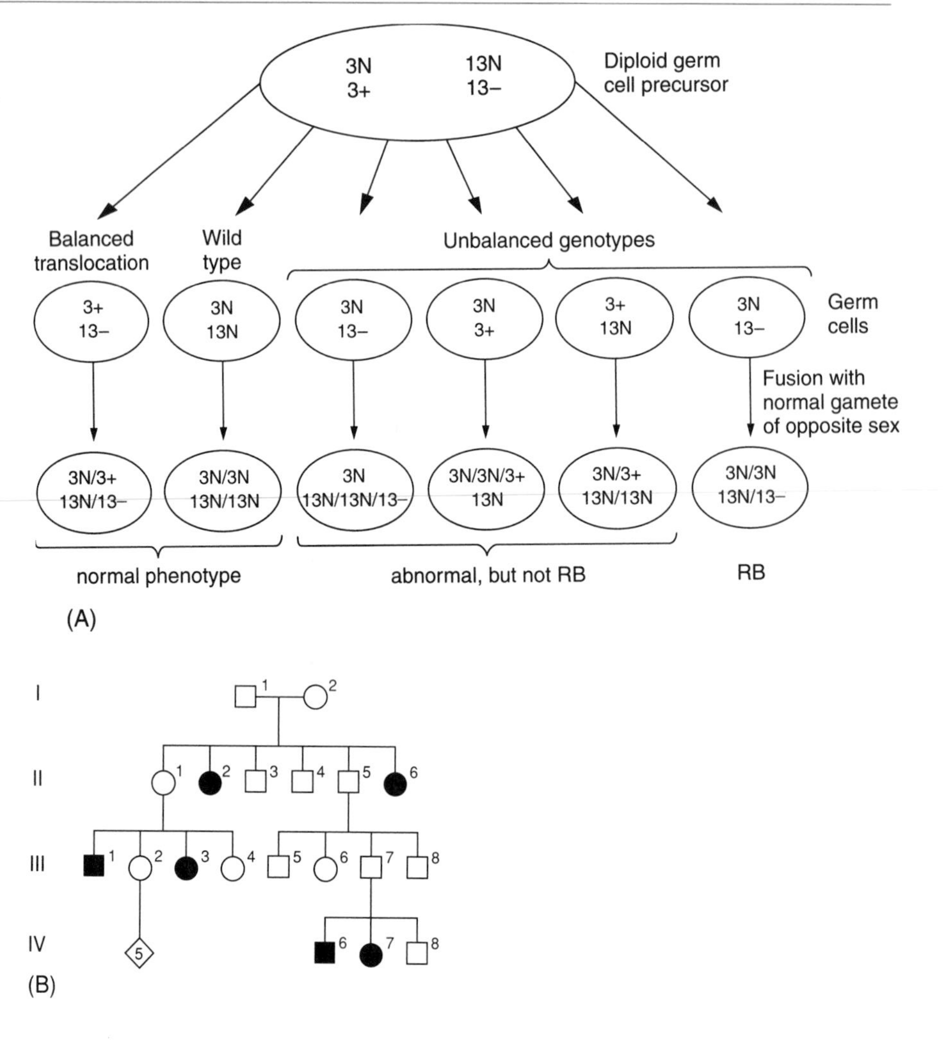

Figure 8–12

(A) The four types of germ cells that can be produced by an individual with a balanced translocation. This example involves transfer of a portion of chromosome 13 to chromosome 3, but it can be generalized to any balanced translocation. 13N and 3N are the normal chromosomes; 13– has a deletion and 3+ has an insertion. (B) Hypothetical pedigree showing apparent incomplete penetrance of the dominant childhood tumor, retinoblastoma (RB). None of the six affected individuals (filled symbols) has an affected parent. Spouses of II-1, II-5, III-2, and III-7 are assumed to be normal. Note II-1, who has two affected siblings and two affected children. Also note III-7, who has two aunts with RB, a niece and a nephew with RB, and two children with RB. Although new mutations cannot be excluded as the source of any of these affected individuals, the probability that all of them are caused by new mutations is negligible. A balanced translocation in one member of generation I, and in individuals II-1, II-5, and III-7 can account for the pedigree.

both Duchenne and Becker dystrophies are caused by mutations at the same locus, and that most of the Becker cases arise from mutations that do not disrupt the reading frame of the protein-coding portion of the gene. The effect of small deletions that remove one or a few amino acids, while allowing most of the usual amino acid sequence of the large dystrophin protein to be synthesized, is evidently less deleterious than most of the mutations that change the reading frame downstream from the mutation site. The latter typically produce the more serious Duchenne form of muscular dystrophy.

Table 8–1 Variable Expressivity in the Marfan Syndrome

MUSCULOSKELETAL SYSTEM	OCULAR SYSTEM	CARDIOVASCULAR SYSTEM	INTEGUMENT
Arachnodactyly	Ectopia lentis	Aortic regurgitation	Inguinal hernias
Dolichostenomelia	Myopia	Aortic root aneurysm	
Tall stature	Retinal detachment	Mitral regurgitation	
Vertebral column deformities		Congestive heart failure	
Abnormal joint mobility			

Source: modified from Pyeritz, 1989.

Cystic fibrosis also shows variable expressivity in several ways. The severity of pulmonary obstruction varies extensively. Patients can also be subdivided into pancreatic-sufficient and pancreatic-insufficient cases. The current explosion of activity that has followed the identification of the CF gene (Chapter 5) has already made it clear that some of the many alleles that exist at this locus cause pulmonary malfunction without seriously affecting the output of pancreatic enzymes, while other alleles affect both lungs and pancreas.

However, there must be many ways in which variable expressivity can really be attributed to the "genetic background." At one level, this is a trivial statement. Many genetic defects have repercussions thoughout the organism; some lead ultimately to death. It is futile to seek a molecular analysis of all the remote effects of one malfunctioning gene. But it is important to realize that there can be *modifier loci* that may directly affect the consequences that a mutant allele at another locus has on the health of an individual.

We do not have any fully analyzed examples of modifier loci that affect human genetic diseases at the moment. The standard examples refer to experimental organisms. One of them is coat color in mice and some other mammals, which is determined by several genes. There is another locus where an allele called *dilute* reduces the intensity of the coat color, whether it be black, brown, or yellow.

This brings us to *epistasis*, which refers to the fact that alleles at one locus may determine whether alleles at another locus are expressed or not, or the extent to which the genotype at the second locus is expressed. For example, a person who is a true albino may have one or two wild-type alleles at the locus that determines brown eye color, but that person will not have brown eyes.

In general, if the product of one gene is necessary for the function mediated by a second gene, the first gene will be epistatic to the second gene. The relationship is easiest to understand in terms of enzymes, where one enzyme catalyzes the synthesis of the substrate for another enzyme; but epistasis need not be restricted to enzyme-coding genes. In the context of this chapter, we should recognize that epistasis can, in principle, be responsible for some cases of variable expressivity or incomplete penetrance. That is, allelic variations in a gene that controls one step in a pathway can mask the presence of deleterious alleles at another locus that controls a later step in the same pathway.

Genomic Imprinting

One of the major exceptions to the patterns of inheritance predicted by simple Mendelian dominance and recessiveness is variation in the expression of a gene or group of genes which depends on whether that gene came from the male or female parent. That type of variation has been documented relatively recently, and it is being recognized in a increasing number of cases. It is generally called *genomic imprinting*, in reference to the fact that the process of gametogenesis in one sex apparently marks some genetic material as being different from its counterpart supplied by the opposite sex.

The most straightforward evidence for genomic imprinting comes from work on mouse embryos where both haploid chromosome complements have been derived from the same sex parent. This can achieved by nuclear transplantation (Surani et al., 1984). The male and female pronuclei remain separate for some time in the newly fertilized egg and they can be distinguished from one another in the microscope. One pronucleus can be removed, and another pronucleus of the same parental origin as the one remaining in the egg can be obtained from another zygote and inserted into the first egg (Figure 8–13).

Embryos with two maternally derived nuclei (gynogenetic zygotes) have poorly developed extra-embryonic membranes, although the embryos themselves are initially more-or-less normal. Embryos with two paternally derived nuclei (androgenetic zygotes) have relatively normal membranes initially, but abnormal embryonic structures. Both conditions are lethal. We are forced to conclude that *both* a maternal and a paternal pronucleus are necessary for normal development. An explanation for that conclusion is that some genes may be differentially inactivated (imprinted) during gametogenesis in the separate sexes.

There are several other sources of evidence for differential functioning of paternal and maternal genomes during embryonic development (Hall, 1990). One source is human triploids, which arise naturally in two ways. Diandric triploids have two chromosome complements from the father and one from the mother; they generally develop a large, but abnormal placenta. Digynic triploids, which have two chromosome complements from the mother and one from the father, rarely survive long enough to be observed, possibly because the placenta is severely underveloped.

Uniparental disomies are euploid genomes where both copies of one chromosome or a major portion of a chromosome originated from one parent, the other parent having contributed nothing to that portion of the embryo's genome. In humans, uniparental disomies are believed to arise when a gamete that contains two copies of one chromosome (presumably as a result of nondisjunction during meiosis) joins with a gamete of the opposite sex that is nullisomic for the same chromosome (also the result of nondisjunction).

Uniparental disomies have helped to define the regions of the mouse genome that are susceptible to imprinting. These usually involve only one arm of a chromosome or a large segment thereof, derived from a reciprocal translocation, followed by controlled breeding (Figure 8–14). The genome is complete and balanced, but both copies of the disomic segment come from one parent. Seven regions of the mouse genome have been shown to have differing effects on survival, growth, or behavior, depending on the parental origin of the disomic segment.

In humans, several genetic diseases are associated with heterozygous deletions of specific chromosomal regions. Prader-Willi syndrome often appears to be caused by a deletion in 15q1, but in some patients with no visible deletion, *maternal* disomy for chromosome 15 has been documented. Conversely, Angelman syndrome can also be caused by a deletion of 15q1, which is microscopically indistinguishable from the deletion that causes Prader-Willi syndrome. Non-deletion Angelman syndrome can be the

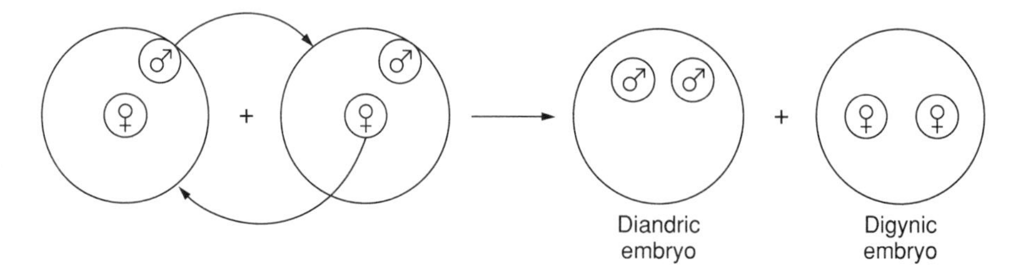

Figure 8–13

Production of diandric and digynic mouse embryos by transplanation of pronuclei.

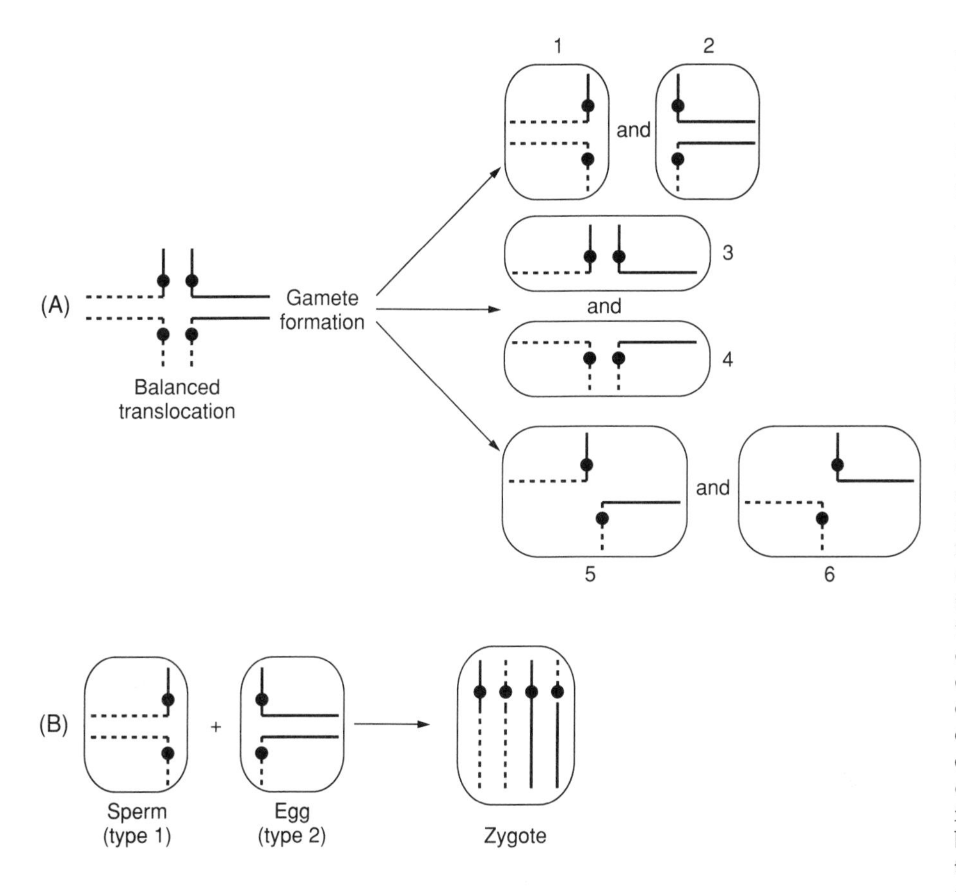

Figure 8–14

Creation of uniparental disomies in mice. (A) The figure shows the six types of gametes that can be produced by animals with balanced translocations. When two animals with the same translocation are bred, all possible combinations of gametes can occur. Progeny that arise from the union of a type 1 gamete with a type 2 gamete, and progeny that arise from the union of a type 3 gamete with a type 4 gamete, will have uniparental disomies. (B) Diagram of uniparental disomy resulting from the fertilization of a type 1 male gamete with a type 2 female gamete. Note that the zygote has two complete copies of the dashed-line chromosome and two complete copies of the solid-line chromosome, but most of both copies of the long arm of the dashed-line chromosome came from the father, while most of both copies of the long arm of the solid-line chromosome came from the mother.

result of *paternal* disomy for that portion of chromosome 15. The two syndromes are clinically distinct. It is likely, therefore, that there is differential imprinting of a cluster of genes in the 15q1 region and that both a paternally derived chromosome and a maternally derived chromosome are required for normal development (Knoll et al., 1989). There is growing evidence that the effects of deletions in several other human chromosome regions, including 4p−, 5p−, 8q−, 17p−, 18p−, 18q−, and 22q−, vary according to the parental origin of the deleted chromosome.

Clearly, genomic imprinting can account for many cases of incomplete penetrance, and conversely, incomplete penetrance may help to identify genes subject to imprinting. Consider a case of maternal imprinting, where some event during female gametogenesis modifies a gene so that it will not be expressed in the next generation (Figure 8–15). If such a woman is a known heterozygote for a dominant deleterious allele (because of her family history and because she is affected with the corresponding disorder), her offspring will be non-expressing (assuming that the father of her children is homozygous normal). However, if male gametogenesis does not "imprint" that gene, half of her sons may have affected offspring, of either sex. Her daughters will have only non-expressing offspring, even though half of them may carry the dominant allele. The reciprocal situation would occur at a locus where there is paternal imprinting.

The mechanism of imprinting is not yet understood, although *methylation of DNA* appears to be involved in at least some cases. We must first review the basic facts about DNA methylation and its probable effects on gene expression. In vertebrates, methylation of DNA is almost exclusively found on cytosine residues, where a methyl group can be attached to the carbon atom at position 5 (Figure 8–16). In mammals, a few percent of total DNA cytosines are methylated. The methyl groups are mostly found on

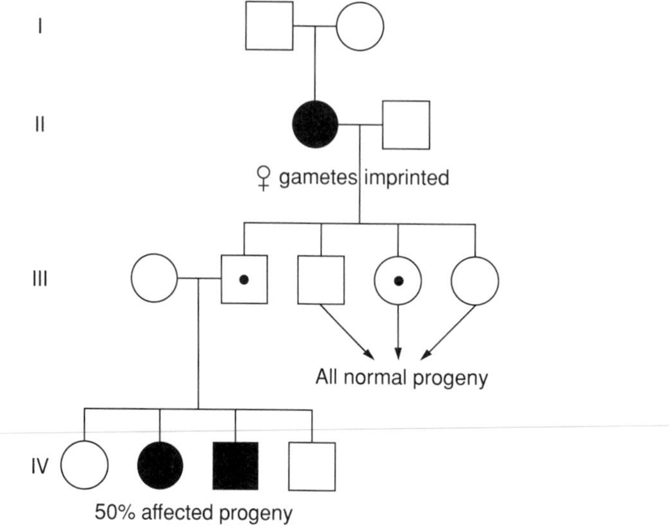

Figure 8–15

Incomplete penetrance resulting from imprinting. It is assumed that the gene in question is inactivated (imprinted) during female gametogenesis. The female in generation II is affected with a disorder caused by an autosomal dominant allele. 50% of her offspring carry the deleterious allele, but do not express it, because of imprinting (carriers are indicated by unfilled symbols with a dot in the middle). One of her sons transmits the allele to 50% of his progeny, where it is expressed.

cytosines that are followed by guanines; these are called CpG dinucleotides, to indicate that the G nucleotide is on the 3′ side of the C nucleotide. As much as 75–80% of the CpG dinucleotides in a mammalian genome may be methylated on the cytosine moiety.

Although CpG nucleotides are widely dispersed throughout the genome, it is a striking fact that regulatory sequences of actively expressed genes are often unmethylated or hypomethylated. In tissues in which a gene is not expressed, it is usually found that the putative regulatory sequences (a few hundred or thousand nucleotides that are immediately 5′ to the transcription start site) are methylated.

A causal role of DNA methylation in gene expression has been strongly supported by transfection experiments, in which methylated copies of several genes, such as actin and insulin, have been introduced into various cells. If the recipient cell is of a type that would normally express the transfected gene, it is demethylated and activated. If the recipient cell is of a type where the transfected gene would not ordinarily be expressed, it remains methylated and is not expressed.

In female mammals, one X chromosome is fully active and the other X chromosome is almost completely inactive in every somatic cell (see Chapter 13). Several genes have been found to be methylated on the inactive X and demethylated on the active X chromosome. Another phenomenon of interest is the gradual loss of methyl groups from DNA in fibroblasts as they age in culture and from aging mouse tissues. Spontaneous conversion of inactive genes to active genes on the inactive X chromosome has also been studied. Although the rate is low, it is higher in old animals than in

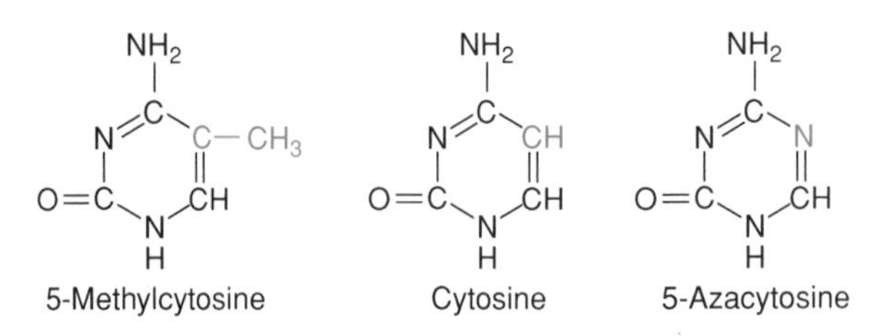

Figure 8–16

Formulas of cytosine, 5-methylcytosine, and 5-azacytosine. 5-methylcytosine is formed *in vivo* by a methyltransferase that acts on newly replicated DNA. 5-azacytosine is a synthetic product, which can be incorporated into DNA when supplied to mammalian cells in the form of the nucleoside, 5-azacytidine.

young animals. It has been speculated that the loss of methylation in older animals is responsible for this effect.

Methylation of cytosines in DNA must be a relatively stable condition; the methyl content of specific genes tends to remain the same throughout development. Holliday and Pugh (1975) and Riggs (1975) have independently proposed that methylation patterns are retained as cells proliferate by the action of a "maintenance methylase," an enzyme that adds methyl groups to newly replicated DNA to reproduce the pattern that existed before replication. They point out that CpG pairs are symmetrical; since C always pairs with G and G always pairs with C in the Watson-Crick model of double-stranded DNA, a CpG pair on one strand implies a CpG pair on the other (Figure 8–17). (Remember that the two strands have opposite polarity.)

The maintenance methylase hypothesis assumes the existence of an enzyme that can recognize the location of a methylated C on an old DNA strand and add a methyl group to the corresponding C on the new strand (Figure 8–17). We are obliged to assume that specific proteins must exist when the methylation state of a gene in a particular cell type needs to be changed. Thus, activation of a previously inactive gene could be achieved if a protein bound to newly replicated DNA at the appropriate spot before the maintenance methylase could do its work. After one more round of cell division, both strands of the double helix would be non-methylated, and the gene could be expressed. Some form of reciprocal action could explain the shift from the active to the inactive state.

Methyl groups are gradually lost from DNA when cells are treated with 5-azacytidine (Figure 8–16), which can be incorporated into DNA in place of cytidine. Azacytidine cannot be methylated; it also has a direct inhibitory effect on the maintenance methylase. Cells that have been exposed to azacytidine have been shown to express several genes that were previously not expressed. Another relevant observation concerns the differentiation of fibroblasts in culture after exposure to azacytidine; some of these relatively

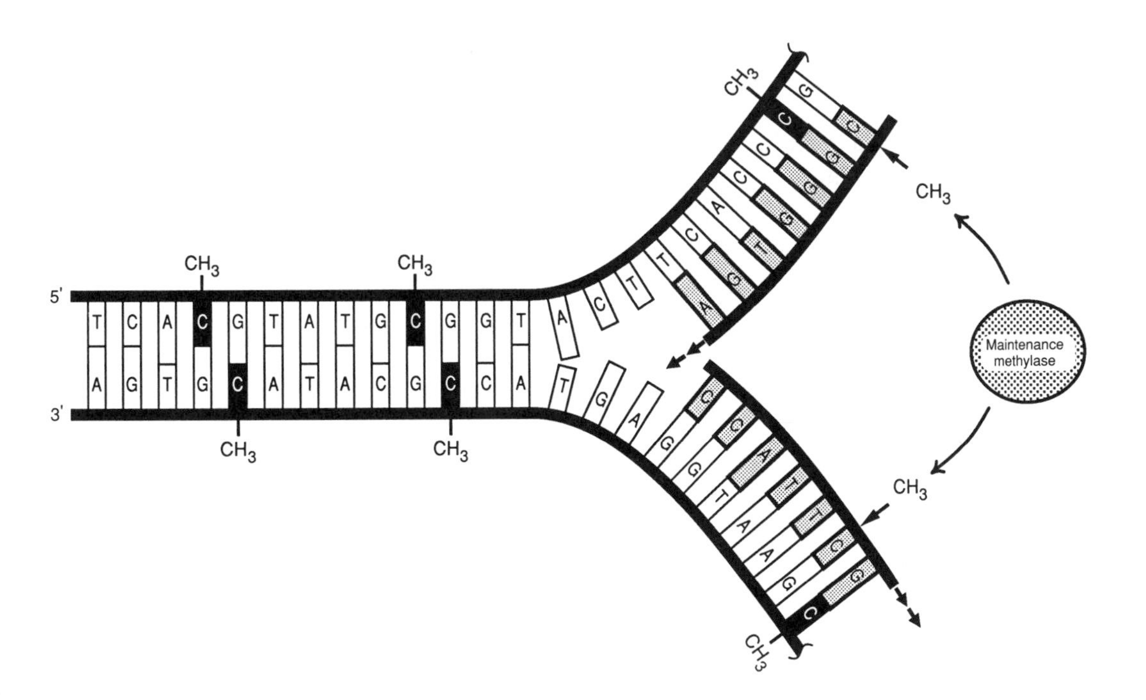

Figure 8–17

Transmission of the DNA-methylation pattern in one cell to its daughter cells is accomplished by an enzyme known as a maintenance methylase (circle). The result of such transmission can be an identical pattern of protein binding and thus of gene activity and inactivity. When the DNA replicates, the methylase puts new methyl groups on the newly synthesized strands of DNA opposite the methyl groups on the old strands, by acting on CpG doublets. (From Freifelder and Malacinski, 1993, p. 367)

undifferentiated connective tissue cells became myoblasts, while others became adipocytes (fat-storing cells).

These and other data suggest that methylation of specific regions of at least some genes is an important transcriptional control mechanism. It is important to note that methylation may be only one step in transcriptional activation or deactivation, and as yet, there has been no direct proof that the presence of methyl groups interferes with the binding of proteins necessary for transcription of specific genes. Nevertheless, the indirect evidence that supports the hypothesis is impressive.

A variety of experiments involving transgenic mice have implicated methylation in genomic imprinting. Transgenic mice are created by injecting cloned DNA into the male pronucleus of a zygote shortly after fertilization. The male pronucleus is closer to the periphery of the egg than the female pronucleus, which makes it more accessible for injection. In some fraction of injected eggs, one or more copies of the injected DNA will be integrated at random into the host genome, where they will be replicated along with the host DNA during development and can be passed along to the descendants of the original transgenic animal.

In about one-fourth of the cases, it has been found that transgenes are highly methylated when inherited from a female parent and are less methylated when inherited from a male parent. In most cases, expression of a transgene that shows the parental origin effect also is correlated with the level of methylation of that gene. However, the methylation level may be low in all tissues, even though the gene is expressed in only one or a few tissues. This emphasizes again that there may be several levels of regulation involved in gene expression, of which methylation is only one.

SUMMARY

Mendel's concepts of dominant and recessive alleles, when applied to humans, refer to patterns of inheritance of genetically determined characteristics. When a condition is determined by an autosomal dominant allele, affected individuals usually have at least one affected parent; whereas, for autosomal recessive conditions, affected individuals usually have two normal parents. Matings between a homozygous normal person and a person heterozygous for an autosomal dominant allele produce an average of 50% affected and 50% normal children; matings between two persons heterozygous for autosomal recessive alleles produce an average of three normal children for each affected child. (Additional characteristics are listed in the text.)

Recessive alleles produce products that do not interfere with the activity of the normal allele's product. Recessive alleles often reduce or eliminate enzyme activity by the product of the mutant allele in situations where the activity of the enzyme produced by one normal allele is sufficient for normal function of the organism.

Diseases caused by dominant alleles have a variety of molecular explanations. They include: the product of the mutant allele interferes with the activity of the product of the normal allele (i.e., it is a dominant negative or poisonous product); the output of product by one normal allele is insufficient for normal health (haploid insufficiency); the mutation increases enzyme activity so much that other biochemical pathways are disturbed; and the mutation causes expression of a normal gene product in the wrong cells or tissues or in the usual place at the wrong time (gain-of-function mutations).

Dominance and recessiveness vary according to the level of observation. For example, an allele that is recessive at the level of clinical observation can be codominant when analyzed at the level of the polypeptide or the gene.

Genetic diseases rarely show complete dominance at the level of the organism. Most dominant alleles cause more severe effects when homozygous than when heterozygous.

Many dominant alleles show incomplete penetrance; that is, the absence of symptoms in persons who are known to be carrying the mutant allele, because of pedigree data. Molecular explanations for incomplete penetrance are just beginning to be developed.

Variable expressivity refers to inconstant effects

of mutations in the same gene in different individuals. Some variable expressivity is caused by the fact that different mutant alleles have different effects on the expression of the same gene, but molecular explanations of most examples of variable expressivity are not yet available.

Genomic imprinting is a process whereby some genes are differentially inactivated during gametogenesis in the two sexes, so that their activity in the embryo depends upon the parent from which they were derived. The earliest manifestation of genomic imprinting is the requirement for one pronucleus from each parent; two female pronuclei or two male pronuclei will not allow normal development.

Many examples of genomic imprinting involving limited regions of specific chromosomes are known. It is believed that one phase of genomic imprinting is caused by methylation of DNA on cytosine residues, but the whole process is probably more complex.

REFERENCES

Becker, M. A., Meyer, L. J., Wood, A. W., and Seegmiller, J. E. 1973. Purine overproduction in man associated with increased PRPP synthetase activity. Science 179:1123–1126.

Brown, M. S. and Goldstein, J. L. 1986. A receptor-mediated pathway for cholesterol homeostasis. Science 232:34–47.

Chemke, J., Nisani, R., Feigl, A., et al., 1984. Homozygosity for autosomal dominant Marfan syndrome. J. Med. Genet. 21:173–177.

Friefelder, D. and Malouinski, G. M. 1993. Essentials of Molecular Biology, 2nd ed. Jones and Bartlett Publishers, Boston.

Frischer, L. E., Hagen, F. S., and Garber, R. L. 1986. An inversion that disrupts the *Antennapedia* gene causes abnormal structure and localization of RNAs. Cell 47:1017–1023.

Hall, J. G. 1990. Genomic imprinting: review and relevance to human diseases. Am. J. Hum. Genet. 46:857–873.

Holliday, R. and Pugh, J. E. 1975. DNA modification mechanisms and gene activity during development. Science 187:226–232.

Hollocher, H. and Place, A. R. 1987. Partial correction of structural defects in alcohol dehydrogenase through interallelic complementation in *Drosophila melanogaster*. Genetics 116:265–274.

Killian, J. M. and Kloepfer, H. W. 1979. Homozygous expression of a dominant gene for Charcot-Marie-Tooth neuropathy. Annals of Neurology 5:515–522.

Knoll, J., Nicholls, R., Magenis, R., Graham, J., Lalande, M., and Latt, S. 1989. Angelman and Prader-Willi syndromes share a common chromosome deletion but differ in parental origin of the deletion. Am. J. Med. Genet. 32:285–290.

Laird, C. D. 1990. Proposed genetic basis of Huntington's disease. Trends in Genetics 6:242–247.

Levitan, M. 1988. Textbook of human genetics, 3rd ed. Oxford University Press, New York.

Mange, A. P. and Mange, E. J. 1990. Genetics: human aspects, 2nd ed. Sinauer Associates, Sunderland, MA.

McKusick, V. A. 1991. The defect in Marfan syndrome. Nature 352:279–281.

Myers, R. H., Leavitt, J., et al. 1989. Homozygote for Huntington disease. Am. J. Hum. Genet. 45:615–618.

Neel, J. V. and Schull, W. J. 1954. Human Heredity. University of Chicago Press, Chicago, IL.

Pauli, R. M. 1983. Dominance and homozygosity in man. Am. J. Med. Genet. 16:455–458.

Pyeritz, R. E. 1989. Pleiotropy revisited: molecular explanations of a classic concept. Am. J. Med. Genet. 34:124–134.

Riggs, A. D. 1975. X inactivation, differentiation, and DNA methylation. Cytogenet. Cell Genet. 14:9–25.

Strong, L. C., Riccardi, V. M., Ferrell, R. E., and Sparkes, R. S. 1981. Familial retinoblastoma and chromosome 13 deletion transmitted via an insertional translocation. Science 213:1501–1503.

Surani, M. A. H., Barton, S. C., and Norris, M. L. 1984. Development of reconstituted mouse eggs suggests imprinting of the genome during gametogenesis. Nature 308:548–550.

Valentine, W. N., Paglia, D. E., Tartaglia, A. P., and Gilsanz, F. 1977. Hereditary hemolytic anemia with increased red cell adenosine deaminase activity (45-70 fold) and decreased ATP. Science 195:783–785.

Vogel, F. and Motulsky, A. G. 1979. Human genetics: problems and approaches. Springer-Verlag, Heidelberg.

Wexler, N. S., Young, A. B., Tanzi, R. E., et al. 1987. Homozygotes for Huntington's disease. Nature 326:194–197.

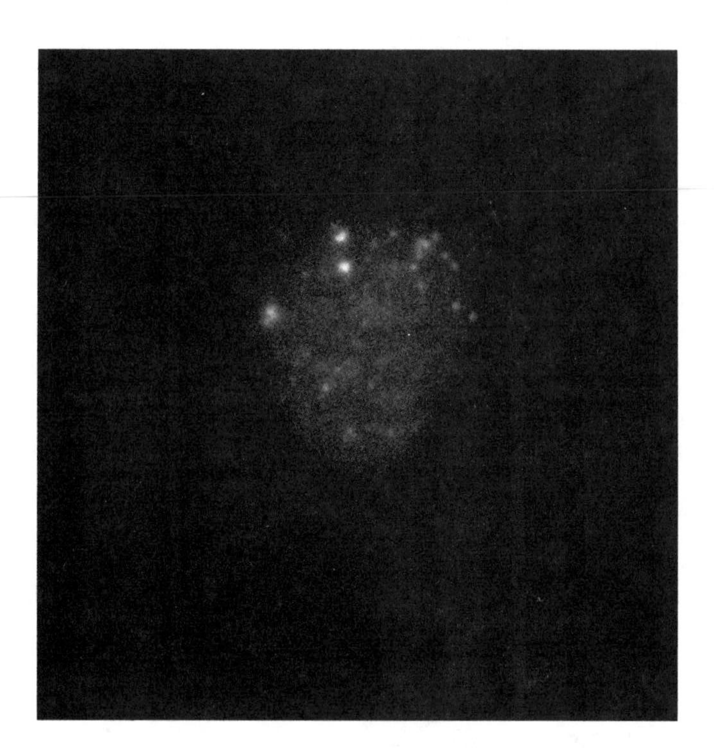

Demonstration of trisomy 21 in an interphase nucleus by FISH. The three bright spots in the upper left quadrant are the chromosomes 21. The probe was a biotinylated chromosome 21 library. (Courtesy of Joe Gray and Wen-Lin Kuo)

9 Genetic Screening and Prenatal Diagnosis

When all or most of the members of a population are tested for the presence of a genetic disease or the presence of particular alleles at a known locus, the process is known as *genetic screening*. The population that is screened may be everyone of reproductive age who lives in a country, everyone with a given ethnic background, everyone exposed to a specific environmental risk (such as employees in an industry where mutagenic chemicals or radioactive materials are used), or some more limited group. Screening may be voluntary or mandatory, depending upon the disease and the society in which screening is done.

Prenatal diagnosis is usually done only on fetuses known to be at risk for a specific genetic defect, based on some information about the genotypes of the parents. However, genetic screening and prenatal diagnosis are not mutually exclusive, as exemplifed by widespread screening of fetuses carried by women over 35 years of age, who are at increased risk for trisomy 21.

This chapter begins with a survey of three classes of screening programs: prenatal, neonatal, and adult screening. The general purposes of such programs and specific examples will be presented, together with an explanation of the major methods that are used. We shall then consider prenatal diagnosis and the techniques that are currently in use for that purpose.

Although this chapter will adhere to the emphasis on molecular biology that pervades the entire volume, it is neither possible nor desirable to ignore the profound implications of genetic screening and prenatal diagnosis for individual ethical decisions and for society in both ethical and legal contexts. The reader is urged to think deeply about these matters and to read more about the non-technical consequences of our increasing knowledge of human genetics.

Genetic Screening

Prenatal Screening

Two methods for obtaining samples of fetal cells without significant risk to the fetus or the mother are in widespread use. They are *amniocentesis* and *chorionic villus sampling* (CVS). Amniocentesis traditionally involves removal of about 30ml of amniotic fluid by abdominal puncture between the sixteenth and twentieth weeks of pregnancy (Fuchs, 1980). Although a few assays can be done on the fluid, most information is obtained from the fetal cells that accompany the fluid (Figure 9–1). These cells are either shed from the amnion (an extra-embryonic membrane that originates from the fetus, rather than the mother) or from various parts of the fetus itself.

Most amniotic fluid cells are nonviable, but some will grow in culture. Some assays, including most of the newer ones that employ DNA probes for specific genes, can be done on the cells in the original sample, but other assays require that the viable cells be multiplied in culture. The latter is true of karyotyping, which must be done when assaying for Down syndrome and other gross chromosomal abnormalities. It often takes two to three weeks' growth in culture before enough cells are available to permit reliable chromosomal analysis. This leaves little time during which a therapeutic abortion can be decided on and performed, if the fetus should prove to be affected with a serious genetic disorder.

In 1975, a method for sampling cells of the chorionic villi was introduced in China; it reached the United States in 1983 (Doran, 1990). The chorion is another extra-embryonic membrane of fetal origin, which an-

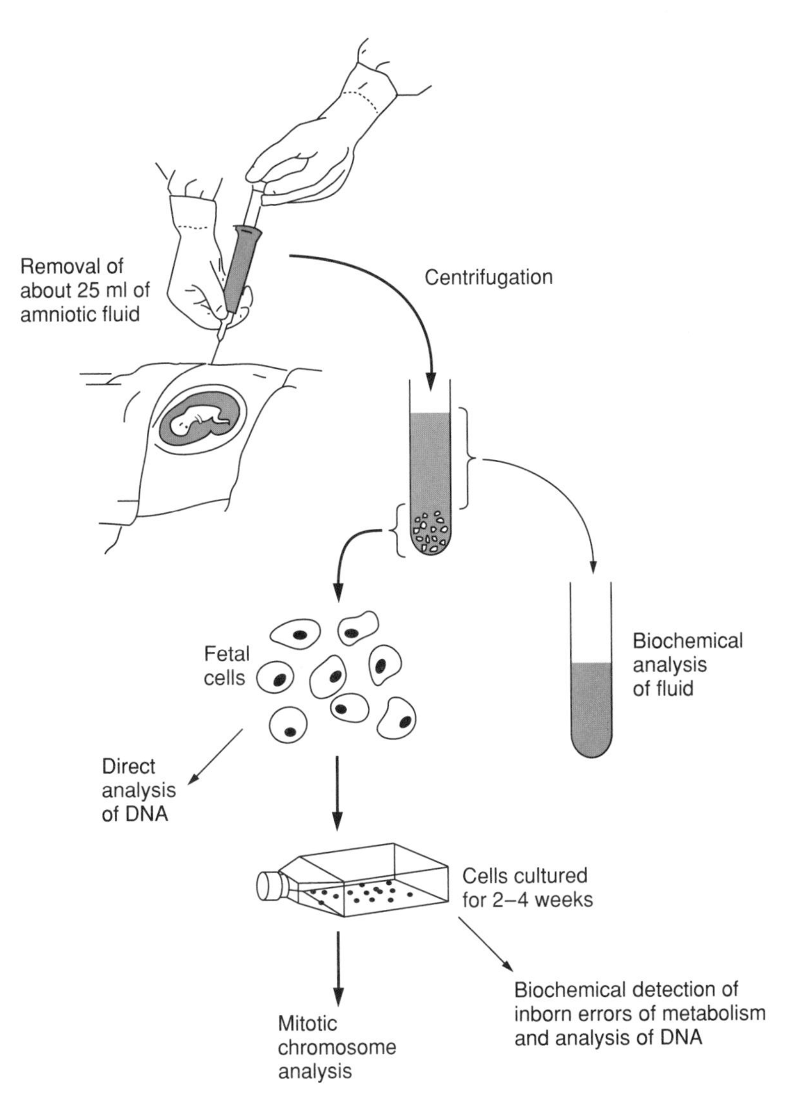

Figure 9–1

Steps in the prenatal diagnosis of disease after amniocentesis. The bold arrows show the most common pathway of analysis—karyotyping in order to detect Down syndrome or other chromosomal abnormality. (From Mange and Mange, 1990, p. 503; by permission of Sinauer Associates)

chors the fetus to the uterine wall and participates in nutrient exchange, prior to development of the definitive placenta. It is possible to obtain small pieces of chorionic villi between the ninth and twelfth weeks of pregnancy, either by inserting a catheter through the cervix (Figure 9–2) or by puncturing the abdominal and uterine walls, in much the same manner as for amniocentesis.

The main advantage of chorionic villus sampling (CVS) is clear: diagnosis is obtained during the first trimester, when therapeutic abortion can be performed with less emotional trauma and less physical danger to the mother. It permits the decision regarding possible termination to be a private one; that is, no one outside of the immediate family need be aware that a pregnancy occurred. There is also an advantage to the geneticist, because the villus cells are virtually all in an active phase of growth, and in many cases they do not need to be cultured before assays are performed.

CVS has vocal proponents and detractors. One disadvantage appears to be a slightly increased frequency of induced miscarriage compared to amniocentesis (somewhat greater than 1% for CVS versus less than 0.5% for amniocentesis). Some practitioners feel that the miscarriage rate associated with CVS is quite acceptable, when compared to the advantages of early diagnosis (Rhoads et al., 1989). Another disadvantage is that the cho-

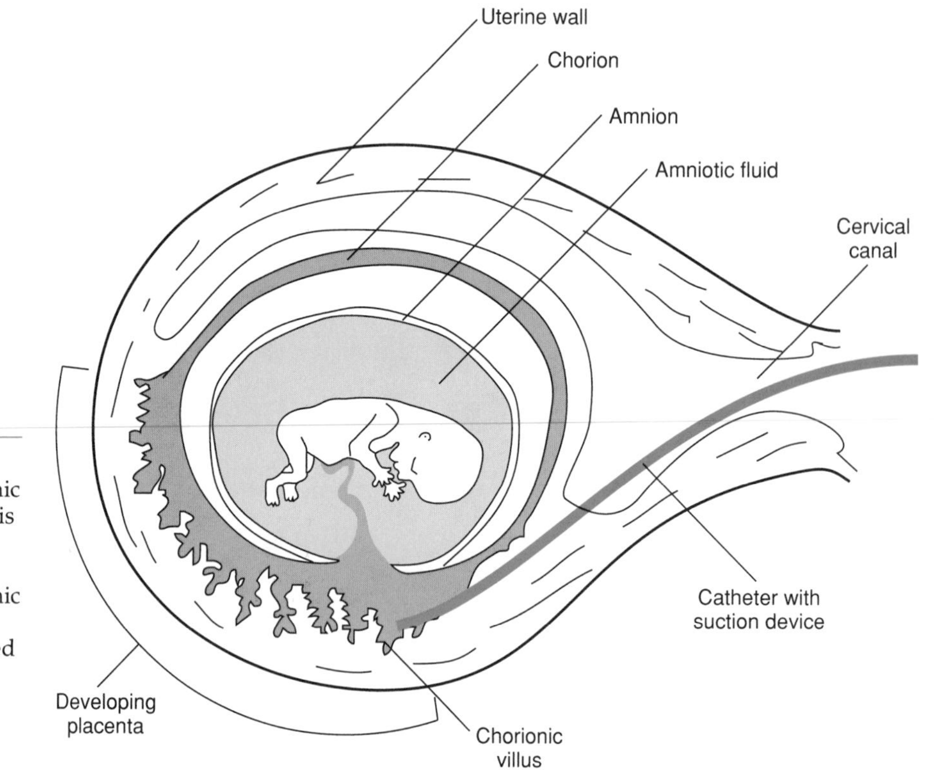

Figure 9–2

The transcervical method for obtaining fetal cells by chorionic villus sampling (CVS), which is done at the ninth to twelfth weeks of pregnancy. It is also possible to sample the chorionic villi by transabdominal puncture with a needle, guided by ultrasonography. (From Mange and Mange, 1990, p. 504; by permission of Sinauer Associates)

rionic villus samples contain both fetal and maternal tissue. The latter must be carefully dissected away from the villus before diagnostic tests are performed. A third problem with CVS is that trophoblast tissues (from which the villi originate) are often mosaic for chromosomal abnormalities, which may not be present in the fetus itself.

Recently, it has been shown that amniocentesis can be performed as early as the thirteenth or fourteenth week of pregnancy (Johnson and Godmilow, 1988), without increased risk to the fetus, if the sampling needle is expertly guided into the amniotic cavity by ultrasonography. At that stage of pregnancy, only about 10 ml of amniotic fluid can be removed, but there are more viable cells than in later samples, and it is possible to get enough cell growth within about 8 days to do a reliable karyotype analysis (Henry et al., 1985).

Current Prenatal Screening Programs Although prenatal diagnosis of fetuses at risk for a specific genetic disorder is an increasingly important application of genetic knowledge, there are few reasons for carrying out screening, in the strict sense. Obviously, we are not interested in identifying heterozygotes who carry deleterious autosomal recessive alleles at that stage of life. Nor is there much point in mass screening of fetuses to determine whether they are affected with a particular genetic disease, in the absence of any information on parental genotypes. Fetal diagnosis is too expensive to be done on a mass basis.

The best example of a genetic screening program for fetuses is the effort to identify Down syndrome babies in women over 35. In this case, and for microscopically detectable chromosomal aberrations in general, it is usually not helpful to know the parents' genotypes; the fetus's condition is the result of a new mutation. The frequency with which trisomy 21 occurs in fetuses in women over 35 is about 1/1000, on average. This justifies a screening program.

In some countries and in some states within the United States, screening for alpha-fetoprotein (AFP) is routinely offered to pregnant women during the sixteenth to eighteenth weeks of pregnancy. AFP is a prominent serum protein of fetuses that leaks into the amniotic fluid in large quantities when a neural tube defect (NTD) is present. NTDs result from a failure of the neural folds to join and form a tube, which is the precursor of the brain and spinal cord in early development. If there is an anterior opening, it can produce *anencephaly* (extreme disorganization or virtual absence of the brain), which generally is not compatible with postnatal life for more than a few weeks. If the opening is more caudal, it leads to *spina bifida*. Children with this defect may live many years, but they usually are paralyzed below the level of the opening, and are frequently afflicted with incontinence, fluid accumulation in the brain, and various other disabilities.

When a neural tube defect exists, some AFP finds its way into the maternal blood, and so the initial screen for NTDs is done on maternal serum. If an elevated concentration is found, then amniocentesis is performed and AFP is measured in the fluid. The fetus is also examined in detail by ultrasonography. In most cases, elevated AFP in maternal serum does not predict an abnormal fetus; but if there is an NTD, then there will be elevated fetal AFP. NTDs occur in about one in 500 to 1000 births in the United States. Women who have had one child with an NTD are at greatly increased risk for another, but it is not yet clear whether there is a genetic basis for this phenomenon.

Future Possibilities In principle, fetal screening would be appropriate for any genetic condition where a significant fraction of affected individuals represents new mutations. Other than chromosomal aneuploidies and rearrangements, the conditions that fit into that category are the X-linked diseases, where as many as 1/3 of affected boys may be the result of new mutations (see Chapter 6), and autosomal dominant diseases that are reproductive lethals or near-lethals. In actuality, most autosomal dominant diseases are too rare to justify fetal screening, but some of the X-linked diseases may be common enough.

For example, one can reasonably argue that screening all fetuses for Duchenne muscular dystrophy (DMD), where new mutations may cause one male birth in 10,000 to be affected, would be socially desirable. The problem is, detection of DMD alleles at the dystrophin locus by DNA-based assays is unlikely to be effective in many cases, because there are too many alleles. This problem is discussed more fully in the context of cystic fibrosis, later in this chapter. Despite the power of multiplex PCR (amplification and analysis of many DNA segments in a single assay), which may detect a majority of mutant alleles for DMD and CF, many mutant alleles will be missed, at least with current technology and costs.

Science fiction devotees may claim that some day it will be possible to induce amnion cells to activate the dystrophin locus, so that the presence of an abnormal protein or the complete absence of normal dystrophin can be detected by relatively simple protein-based assays. Perhaps so, but readers should not hold their breath in anticipation of that miraculous achievement.

Neonatal Screening

The primary purpose of genetic screening during the neonatal period (roughly one month after birth) is to identify infants affected with specific genetic diseases, in order to initiate therapy and prevent the consequences of untreated disease. At present, neonatal screening is usually done either by measuring altered metabolite concentrations in a blood sample or by enzyme assays, although the use of DNA assays for confirmation of an initial positive result is becoming routine for some genetic disorders.

The PKU Screening Program The largest genetic screening program in most developed countries has the detection of infants with *phenylketonuria* (PKU) as its primary goal. This autosomal recessive genetic disease occurs in about one infant in 10,000 to 15,000 in the United States. It is caused by the absence of *phenylalanine hydroxylase* (PAH) activity, which normally converts phenylalanine to tyrosine. Untreated infants begin to suffer neurological damage within a few weeks, and although they may survive for decades, they are severely retarded. The underlying biochemical cause of retardation is not yet known, but inasmuch as tyrosine is a precursor of several neurotransmitters (Figure 9–3), it is not implausible that a deficit of tyrosine should have an effect on development of the central nervous system. This may be only one of several reasons why retardation occurs, however.

Homozygotes for PAH mutations have normal serum concentrations of phenylalanine at birth, because excess phenylalanine that arises in the fetus from the catabolism of proteins passes across the placenta into the maternal circulation. The mother, an obligate heterozygote, can convert phenylalanine into tyrosine in her liver. After birth, however, phenylalanine in the infant's blood rises rapidly, reaching about 15X to 30X the normal level within one week.

Elevated concentrations of phenylalanine are detected by the *Guthrie test* (Guthrie, 1973). A drop of blood is removed from each infant's heel

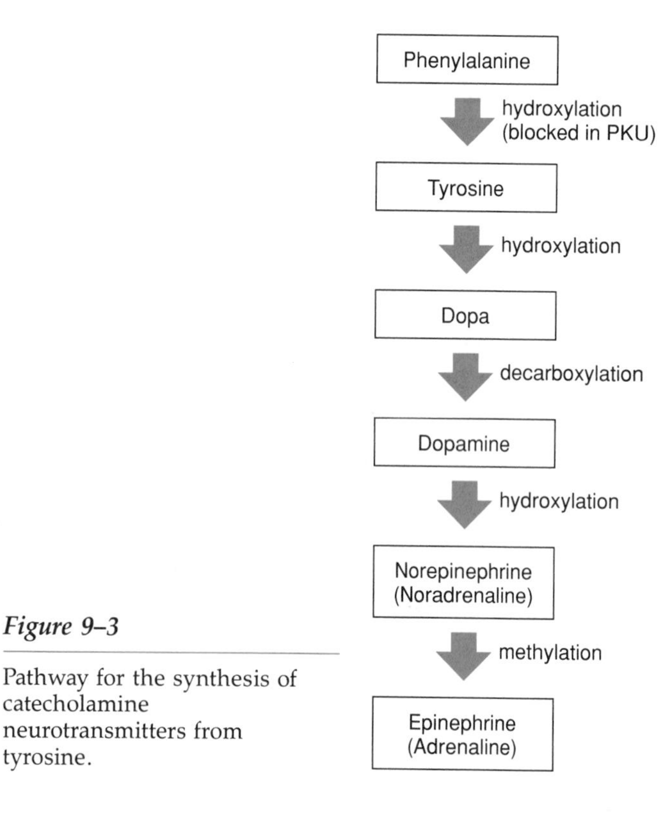

Figure 9–3

Pathway for the synthesis of catecholamine neurotransmitters from tyrosine.

shortly before the child is taken home from hospital. The blood is dried on a small filter paper disk and mailed to a central laboratory where hundreds or thousands of samples can be processed at once. The simple but ingenious Guthrie test employs a bacterium that requires phenylalanine for growth. Spores of the bacterium are mixed with nutrient agar lacking phenylalanine. The agar is spread on trays that can accept a large number of disks containing infant blood, plus a series of control disks with known amounts of phenylalanine. The agar also contains a nonmetabolizable competitive inhibitor whose concentration is set at such a level that the amount of phenylalanine in normal blood will not support noticeable bacterial growth.

The sample trays are then placed in an incubator overnight. The bacterial spores germinate, and wherever there is a disk with an unusually large amount of phenylalanine, bacterial growth ensues. The next day, those disks can be recognized at a glance because they are surrounded by a halo of bacteria, which gives the agar a whitish appearance (Figure 9–4).

However, a positive Guthrie test is not diagnostic of PKU by itself; follow-up tests are necessary. One difficulty is that most newborns are now released from the hospital within two days, and usually even sooner, at which time the phenylalanine concentration in blood of an infant with PKU is only moderately elevated. This requires that the Guthrie test be calibrated so that infants with moderately elevated phenylalanine will give a positive result, but because normal concentrations of phenylalanine vary substantially, most of the positive disks in the Guthrie test will be from normal infants. Additional measurements on infants with initially positive Guthrie tests, done on blood samples taken about one week later, separate the normals from the true PKU cases.

All states in the United States also screen for congenital hypothyroidism, using a dried blood spot. Hypothyroidism occurs in about one in 4000 births. Most states also screen for one or more hemoglobinopathies.

Several genetic diseases that are much rarer than PKU, and for which independent screening programs could not be justified economically, are often carried out in conjunction with the PKU program (Guthrie, 1973;

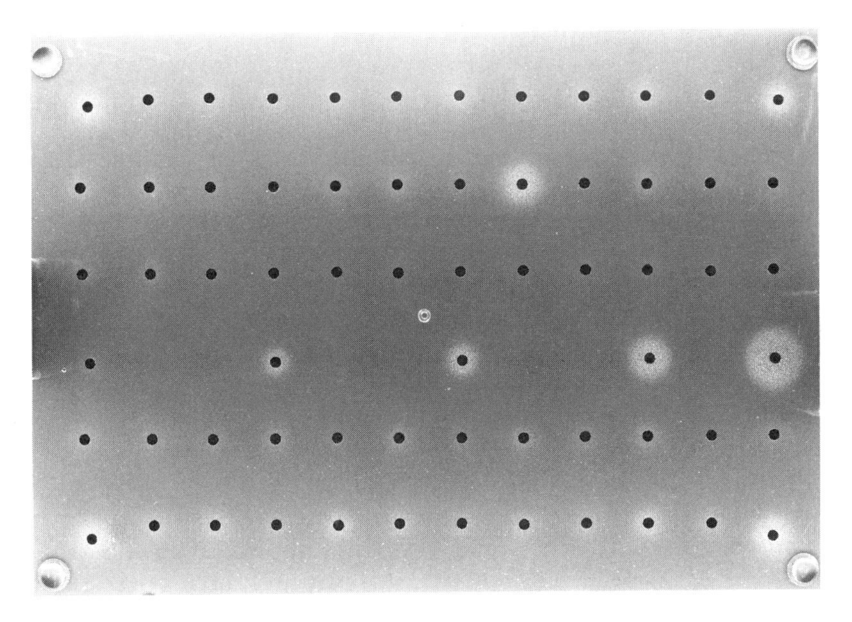

Figure 9–4

A Guthrie test plate for PKU screening. Filter paper disks containing dried blood from a large number of newborns were placed on agar containing bacterial spores and nutrients (top three rows and bottom two rows). The bacteria will only grow if a certain level of phenylalanine is present. The fourth row of disks contains a series of controls, with increasing concentrations of phenylalanine; the four corner spots are also controls. Notice the test disk that is surrounded by a distinct halo of bacteria, fifth from the right end of the second row. This sample came from a child subsequently diagnosed as having PKU. (Photo courtesy of Colorado Department of Health Laboratory and William Seltzer)

American Academy of Pediatrics, 1989). Examples include (i) galactosemia, which occurs at a frequency of one in 60,000 to 80,000 births in the United States; (ii) maple syrup urine disease (also known as branched chain amino acid disease), with a frequency of about one in 250,000 births; and (iii) homocystinuria, with a frequency of one in 50,000 to 150,000 births. Assays on dried blood spots, collected at the same time as the PKU test sample, are used for all of these conditions. Some measure metabolite levels in the blood; others, such as the galactosemia assay, measure the activity of an enzyme.

An important feature of all the genetic diseases that are detected by the preceding screening programs is that the patients will benefit from identification of their condition soon after birth. Reduction of phenylalanine in the diet makes the difference between severe mental retardation and near-normal intellectual development for PKU babies. Similarly, removal of lactose and other sources of galactose from the diet of infants with galactosemia changes their prognosis from probable death to normality; removal of branched amino acids (leucine, isoleucine, and valine) from the diet of infants with maple syrup urine disease avoids death if begun early enough, and ameliorates most of the morbidity associated with the condition. Hypothyroidism is effectively controlled by oral intake of synthetic thyroid hormone.

Future Possibilities At present, there are no neonatal screening programs based on DNA assays. It is conceivable that a situation may arise in which there is a genetic disease (such as Huntington disease) that does not manifest itself until some months or years after birth, and that we learn how to cure or greatly ameliorate the disease, *if* the mutant gene is detected shortly after birth. In such a case, a DNA-based neonatal screening program would be desirable, but whether it would be feasible would depend upon the number of mutant alleles in the population. If we could not detect nearly 100% of affected fetuses at an acceptable cost, the program would not be feasible.

Adult Screening

Heterozygote detection is the main purpose of genetic screening programs for adults. Identification of heterozygotes is often desirable, so that couples at risk for the birth of a child affected with a serious autosomal recessive disorder may be made aware of that possibility.

Application of the word "carrier" to individuals who are heterozygotes for deleterious recessive alleles is an undesirable, albeit common, practice. No matter how much the public may be educated about a particular disease, it is impossible to separate "carrier" from pejorative implications for some people, possibly because it is the same word used to describe persons who harbor an infectious organism, and are therefore a threat to the health of others. In addition, many individuals who know themselves to be "carriers" of an undesirable gene are unable to escape feelings of guilt and inferiority. Those feelings may be repressed and become an insidious threat to an individual's self-image.

Professional geneticists and physicians should avoid the use of the word "carrier" as much as possible. It is probably too much to expect laymen to accept "heterozygote," but the abbreviation "het" deserves to be tried. It is simple and it is free of negative associations.

Examples of Adult Screening Programs At present, adult screening programs are voluntary and are limited to high-risk groups. Probably the best example involves *Tay-Sachs disease,* which occurs with a frequency of about one in 3000 among persons of Ashkenazi (Eastern and Central European) Jewish descent. Tay-Sachs disease results from absence of *hexosaminidase A,* an enzyme that is part of the catabotic pathway for membrane lipids known as gangliosides (Figure 9-5).

In babies who are homozygous recessives for Tay-Sachs alleles, the buildup of gangliosides in lysosomes gradually reduces cell metabolism and eventually leads to cell death. Neurons are most seriously affected. Tay-Sachs infants begin to show abnormal motor responses within the first half year of life; although they may crawl and sit, they rarely become able to walk. After 18 months, generalized paralysis, progressive deafness, blindness, and convulsions develop. Death usually occurs by three years of age, from bronchopneumonia. There is no treatment.

Fortunately, a relatively straightforward biochemical assay for hexosaminidase is available, so that the amount of activity in blood serum can be readily measured. Even more fortunately, hexosaminidase A is an enzyme where heterozygotes have a level of activity that is clearly different from the level present in homozygotes (Figure 9-6). [It is worth mentioning in passing that the level of activity for many enzymes in homozygotes is distributed over such a wide range that there is substantial overlap with the amount found in heterozygotes, which makes identification of the latter difficult. There are both physiological and genetic reasons for such variability in homozygous "normal" individuals.]

Voluntary screening programs for Tay-Sachs heterozygotes are usually organized by Jewish religious groups. The primary targets are "expectant fathers," because if the husband of a pregnant woman has the homozygous wild-type level of hexosaminidase A, there is no need to subject the woman to the stress of having an assay. When the father is a heterozygote, it becomes imperative for the mother to have the test. Couples who are both heterozygotes have a 25% chance of the fetus being affected with Tay-Sachs disease, for every pregnancy. The enzyme activity in the fetus can be measured in amniotic fluid, but it is more reliably measured in cultured amniotic fluid cells. Affected fetuses are almost always intentionally aborted, because of the hopelessness of the disease.

Tay-Sachs screening has been quite successful, with the frequency of affected infants among Ashkenazi Jewish groups in North America being reduced by more than 65%, when 1980 births were compared with 1970

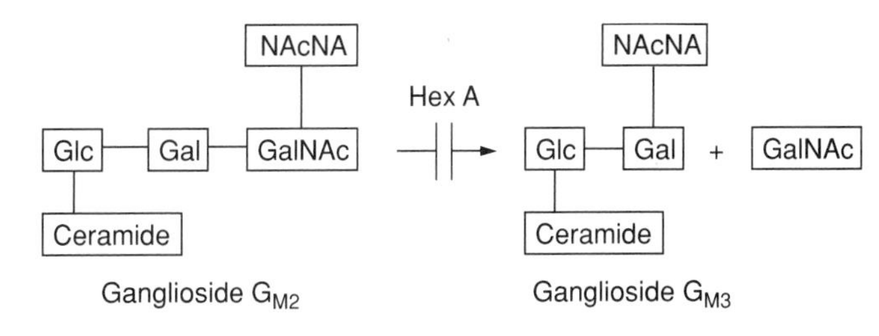

Figure 9–5

The step in the degradation of ganglioside G_{m2} that is carried out by hexosaminidase A and blocked in persons with Tay-Sachs disease. Ceramide is a relatively complex molecule containing an 18-carbon alcohol (sphingosine), to which a long-chain fatty acid is attached. GalNAc is *N*-acetylgalactosamine, Gal is galactose, Glc is glucose, and NAcNA is *N*-acetylneuraminic acid (sialic acid).

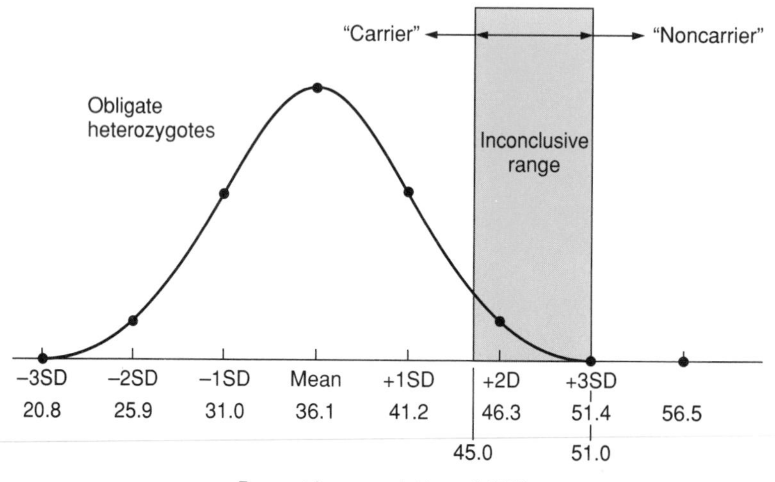

Figure 9–6

Hexosaminidase A activity in serum of 43 obligate heterozygotes (parents of Tay-Sachs children). Hexosaminidase A activity is calculated as the heat labile fraction of total hexosaminidase activity (52°C versus 37°C), using a synthetic substrate. Non-carriers (homozygous normals) are defined as those persons whose hexosaminidase A activity is greater than three standard deviations above the mean of activity found in obligate heterozygotes (standard deviation is a statistical measure of variation around a mean). The inconclusive range (shaded area) contains those hexosaminidase activities that cannot be interpreted as representing either heterozygotes or normal homozygotes. (From Kaback et al., 1977, p. 269; by permission of John Wiley)

births (O'Brien, 1983). The tests have generally been accompanied by intensive educational efforts, aimed at reducing stress for those individuals who learn that they are heterozygotes.

In the 1960s, one of the peripheral consequences of the civil rights movement in the United States was that a number of states passed laws requiring black people who were planning to marry to be screened for Hemoglobin S, the product of the *sickle-cell* allele. In some cases, even school children were required to be screened. These programs were counterproductive. Inevitably, they were interpreted as racial discrimination, even though some of the laws mandating screening had been proposed by black legislators with purely altruistic intentions. Moreover, the screening programs were not accompanied by adequate education efforts. In particular, heterozygotes for the sickle-cell allele were labeled as "carriers," or they were described as having "sickle-cell trait," and were stigmatized by widespread misunderstanding of the meaning of the terms.

Another undesirable consequence of the sickle-cell screening programs was that couples who were told that they were at risk for the birth of an affected child had no way to use the information constructively. At that time there was no assay for the sickle-cell allele of beta-globin in amnion cells, which do not express the gene, and although it was possible to obtain a blood sample from a fetus, it was an expensive and risky procedure. Thus, heterozygote couples were either forced to decide on having no children or to live for the rest of their reproductive lives with the tension generated by the knowledge that any newborn child had a 25% probability of being affected with a dreaded disease.

Eventually, all of the laws that mandated screening of black people for the sickle-cell allele were repealed, and identification of adult sickle-cell heterozygotes is now done only in private clinics, at the request of informed individuals who know about the high frequency of the mutation among certain racial groups. When two heterozygotes wish to become parents, a simple restriction enzyme assay can be applied to DNA from fetal cells to determine whether the fetus is homozygous for the sickle-cell allele (Figure 9–7).

More recently, mandatory screening of all newborns for sickle-cell anemia has been established in some states, because early identification of affected infants can significantly reduce mortality in infancy. Few would argue with that goal, but the screening program inescapably identifies heterozygotes, also. What use should be made of that information? Should the parents of the child be told; should the child be told when he or she reaches a certain age; and should the information be retained in files, to be made available to individuals only upon their own request? These are complex issues that have not yet been resolved.

Cystic Fibrosis and DNA-based Screening When the cystic fibrosis gene was identified in 1989 (see Chapter 5), it was reported that approximately 70% of CF alleles had the same mutation—a deletion of a codon for phenylalanine at position 508 in the protein (abbreviated ΔF_{508}). This led some geneticists to suggest that mass screening for CF alleles at the DNA level might be feasible.

Screening for mutant alleles at the DNA level poses a problem that is not present when screening for a protein. The problem arises from the fact that most genetic diseases are the result of any one of a variety of mutations in a given genetic locus (see Chapter 7), but DNA assays are allele-specific. Therefore, one must have a specific DNA probe for every allele known to cause the disease; otherwise, false negative results will be obtained. By contrast, enzyme assays are not allele-specific. Different alleles may produce different amounts of enzyme activity, but if the activity is below some arbitrarily defined limit, the presence of a mutant allele can be inferred, regardless of the change that has occurred in the gene.

It is easy to be misled by the example of sickle-cell anemia, where a recognizable clinical syndrome is caused by one and only one nucleotide change in the gene for beta-globin. Screening for sickle-cell heterozygotes and homozygotes could be done at the DNA level, because the Glu to Val mutation (GAG to GTG) alters a restriction enzyme cleavage site (Figure 9-7). However, it is highly unusual for a clinically distinct disease to be caused by only one type of mutant allele.

In the case of cystic fibrosis, the high frequency of the ΔF_{508} allele raised hopes that a small number of alleles would be found to underlie virtually all cases of CF. But those hopes were not fulfilled. As laboratories around the world began to analyze the CF gene in different groups of pa-

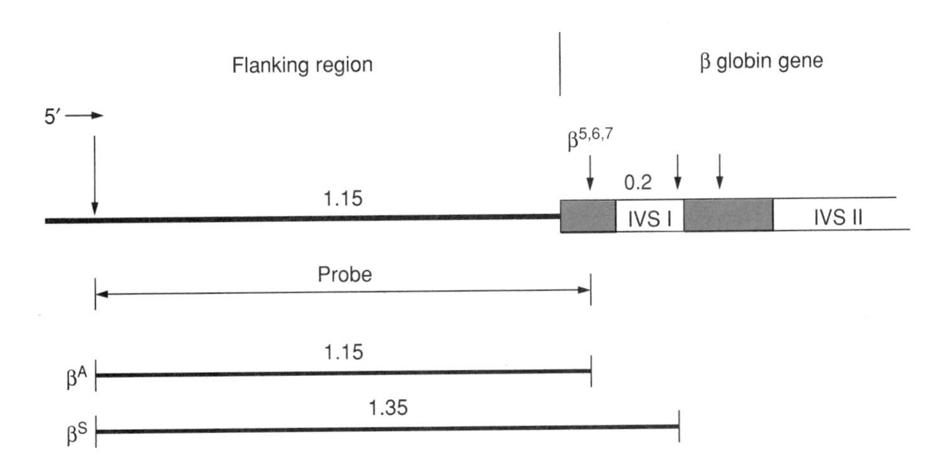

Figure 9–7

Detection of the sickle-cell allele of beta-globin with the restriction enzyme Mst II. Arrows indicate the Mst II sites in the gene and its 5′ flanking region. One site involves the codons for amino acids 5, 6, and 7; it is present in the normal allele and absent in the sickle-cell allele. In the normal genome, a 1.15 kbp fragment is generated, which can be detected on Southern blots with the probe indicated in the center of the diagram. When a sickle-cell allele is present, the probe will reveal a 1.35 kbp fragment. IVS I and IVS II are the introns within the beta-globin gene. (From Chang and Kan, 1982)

tients, many rare alleles were identified. At this time, more than 100 different alleles have been found, and more can be expected.

This presents us with a dilemma. Mass screening for heterozygotes cannot be done if dozens of alleles have to be assayed individually, because the cost of the assays would be prohibitive (at the present time). Mass screening could be done for a few common alleles, if false negatives from rare alleles were acceptable. But, how does one define "acceptable"? Is it 1%, is it 5%, or what?

It is useful to imagine the situation of a young couple who have given birth to a child affected with cystic fibrosis, even though both husband and wife had been screened for common CF alleles before the pregnancy occurred. If they are well-informed about CF, they may understand that they participated in a genetic lottery and lost. They may realize that no one is to be blamed for their misfortune. But how many such couples will be present in a large population, even in a technically sophisticated culture like the United States? In many cases, the parents of the CF child will feel betrayed and angry, and some of them may seek vengeance in the courts.

The proposal to undertake mass screening for CF alleles has revealed another complication of DNA-based assays, with the inevitable concomitant of false negatives. Some couples will be found where one member is a heterozygote for a common CF allele and the other member is not. It has been claimed that such couples are at "increased risk," relative to the theoretical risk that existed before the screening tests were done (Beaudet, 1990).

The concept of *increased risk* can be understood by considering the following numbers. Prior to screening and in the absence of any family history of the disease, a Caucasian couple in the United States has a risk of about one in 2000 of having a CF child. This implies that CF alleles, in total, have a frequency of 1/22. (The probability of both husband and wife being heterozygotes is then $1/22 \times 1/22 = 1/484$; and the probability of a child receiving both CF alleles is $1/484 \times 1/4 = 1/1936$.)

Now suppose that the couple has been tested for the ΔF_{508} allele and the husband's test was positive, but the wife's test was negative. Because the ΔF_{508} allele is 70% of all CF alleles, the probability that the wife also has one CF allele (other than ΔF_{508}) is $0.3/22 = 1/73.3$. For this couple, the probability of having a CF child becomes $1/4 \times 1/73.3 = 1/293.2$, a substantial increase over the probability of 1/2000 that applied before the assay for the ΔF_{508} allele was done.

This example illustrates a general problem associated with DNA-based screening programs. Is an increased perception of risk for some individuals acceptable to the public, to health professionals, and to lawyers? If not, then it would be necessary for the DNA assays to be able to identify enough mutant alleles so that no couple was at apparently increased risk of having an affected child as a result of the screening. Table 9–1 shows that the relevant number for cystic fibrosis is 96% detection of CF alleles, a level of accuracy that would be difficult to achieve at this time at an acceptable cost for this disease, because of the large number of rare alleles. Nevertheless, pilot screening programs sponsored by the National Institutes of Health are underway, and a number of clinics offer CF allele testing on a fee-for-service basis.

Before closing this section, it is appropriate to point out that mass screening for cystic fibrosis is also beset by non-quantitative problems (Wilfond and Fost, 1990). What is to be gained by identifying heterozy-

Table 9–1 Effect of Rate of Mutation Detection on Carrier Testing

% MUTATION DETECTION	RISK OF CF IF ONE PARENT IS CARRIER AND OTHER HAS NEGATIVE TEST	% OF AT-RISK COUPLES DETECTED
75	1/396	56
85	1/661	72
95	1/1,964	90
99	1/9,814	98

Source: from Beaudet, 1990, p. 604.

gotes? Even if we make the dubious assumption that most persons who are found to be heterozygotes will learn enough about the meaning of their condition to avoid feelings of guilt and inferiority, what use will they make of that knowledge?

Some couples in which both husband and wife are known heterozygotes will be grateful for the knowledge, and will plan to abort any fetuses who are CF homozygotes. Other couples, even if they have no moral objection to abortion in principle, will find it very difficult to decide whether to let a CF fetus come to term. There is no comparison between the fate of a Tay-Sachs child and a CF child. The former is doomed to death in infancy, following a period of behavioral degeneration that is a heart-rending experience for the parents. The CF child can usually expect to survive at least 20 years, and there is no intellectual or behavioral impairment. Moreover, now that the gene responsible for the disease has been identified, there is a clear possibility that a cure for the disease or greatly improved treatment will be available within a few years (Chapter 10).

We cannot escape the fact that mass screening proposals for genes that cause diseases like cystic fibrosis have eugenic overtones, even when the screening will be voluntary. The mere existence of the program would imply that society has decided that it is wrong to be born with the disease. Even more disturbing is the possibility that screening might be mandatory.

Inevitably, there are powerful commercial pressures to establish mass screening programs, because of the huge profit potential. It is unfortunate that those pressures may influence decisions that should be made only after extensive and objective consideration of the individual and social benefits and liabilities that may ensue when any new screening program is sanctioned by the government or by the medical profession.

Prenatal Diagnosis

The purpose of prenatal diagnosis is to detect fetuses that will be born with or will subsequently develop a serious genetic disease. The use to which knowledge of the fetus's genotype will be put is a personal decision, made by each couple on the basis of their emotional and economic needs, as well as their ethical and religious convictions. Usually, a couple will not ask for prenatal diagnosis unless they are seriously considering abortion of an affected fetus. However, some couples who know they are at risk for birth of an affected child simply want to be forewarned about the baby's genotype, so that they can be emotionally prepared to deal with a stressful situation after the birth.

At present there is no genetic disease that can be better treated before birth than after birth, so prenatal diagnosis is not done for the purpose of treating the fetus. It is conceivable that prenatal gene therapy will be possible at some future date. For example, if it were possible to introduce the gene for hexosaminidase A into a Tay-Sachs fetus at an early stage of development, while the central nervous system was still forming, a viable child might be born. However, such scenarios belong to the realm of fantasy at this time.

Prenatal diagnosis currently uses all the methods that are available for genetic screening. These include karyotype analysis for chromosomal abnormalities; enzyme assays (e.g., Tay-Sachs disease); and a rapidly increasing collection of allele-specific DNA assays that employ restriction enzymes, oligonucleotide probes, and DNA sequencing. Many of the DNA-based assays make use of the PCR (see Chapter 6) at some stage, to amplify small regions of specific genes, so that mutations may be more easily detected.

RFLPs and Prenatal Diagnosis

One technique that is often useful in prenatal diagnosis is prediction of a fetus's genotype by analysis of RFLPs that are closely linked to the disease locus. The need for this type of assay arises because there is a transitory stage in the analysis of many genetic diseases when the gene has been mapped to a small region of a specific chromosome, but not yet identified. Cystic fibrosis was in that stage from 1986 to 1989, and the Huntington Disease gene is still unknown, although its location near the telomere of chromosome 4p is well established.

If one or more highly polymorphic restriction enzyme cleavage sites are known to be very close to a mapped gene, so that recombination between an RFLP locus and the disease is a rare event, many families at risk for the birth of an affected child can be offered fetal diagnosis. However, it is necessary that the genotypes of the parents and at least one child be *informative*.

To explain what geneticists mean by an "informative" family, we shall consider autosomal dominant diseases, which are the simplest case. Figure 9-8 diagrams a nuclear family, where the father has an autosomal dominant disorder. The father is also heterozygous at a tightly linked marker (RFLP) locus. The affected parent *must* be a double heterozygote (that is, heterozygous at the disease locus and the marker locus), and for maximum information content, the other parent must be unaffected (homozygous wild-type at the disease locus) and homozygous at the marker locus.

Notice the genotype of the child in Figure 9-8(A). It is evident that the child received both the disease allele and allele M1 at the RFLP locus from its father. This establishes the *linkage phase* (see Chapter 4) in the father; that is, the child's genotype tells us the affected parent's *haplotypes*. We now know that the father has allele D and allele M1 on one chromosome, with allele d and allele M2 being on the other copy of that same autosome.

Now that we know the father's haplotypes, we can predict the genotypes of subsequent fetuses at the disease locus after we determine which alleles they have at the marker locus. In the present case, any fetus that is M2/M2 is likely to be normal, and any fetus that is M1/M2 is likely to be affected. This family is *informative*.

Notice that for autosomal dominant diseases, an unaffected child is just as useful as an affected child, because we are also able to deduce the father's haplotypes from the genotype of the unaffected child (Figure 9-8(B)). If there were no previous children, the family in Figure 9-8 would not

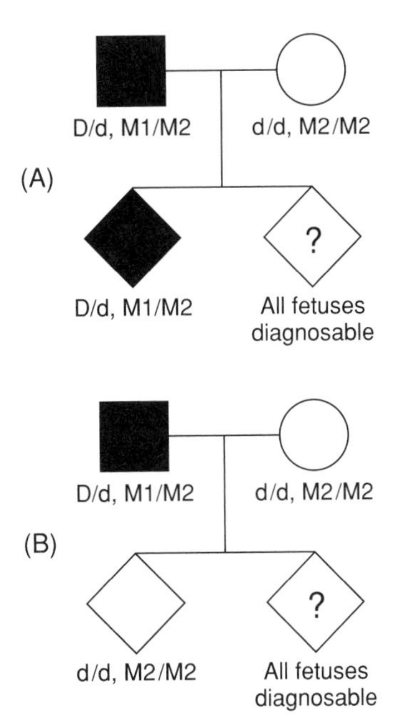

Figure 9–8

Diagrams of a family in which the father is heterozygous for an autosomal dominant allele that produces a genetic disease (locus D), and heterozygous at a nearby polymorphic marker locus (locus M). The mother is homozygous normal at the disease locus and homozygous for one of the alleles at the marker locus. In (A), the first child is affected with the same disease as the father; in (B), the first child is normal. In both cases, the genotype of the first child at the marker locus, together with the child's phenotype at the disease locus, implies the linkage phase of the alleles in the father. This information permits a diagnosis of all subsequent fetuses, in terms of being affected or normal, after their genotypes at the marker locus have been determined by DNA analysis.

be informative unless we had enough information about the parents or other relatives of the affected parent to permit us to deduce his or her linkage phase.

Recombination and Prenatal Diagnosis via RFLPs

Prenatal diagnosis based on closely linked polymorphic marker loci is subject to some uncertainty, because recombination may occur between the marker locus and the disease locus. To illustrate the problem, let's suppose that a marker locus is 2 cM distant from a disease locus. We then expect that in 2% of meioses, there will be a crossover between the two loci, which can cause a genetic counselor to make an incorrect prediction.

In fact, the uncertainty is nearly twice 2%, for a one-child family like the example in Figure 9–8. It is easy to recognize that the second child (the fetus to be diagnosed) will be a recombinant in 2% of such cases; but it is also true that the first child will be a recombinant 2% of the time, leading us to deduce the father's linkage phase incorrectly. Accordingly, a correct diagnosis of the second child's genotype at the disease locus will only occur if both children are non-recombinants or both are recombinants. For a large series of families identical to this example, the error frequency will be 0.02 + 0.02 − 0.0004 = 0.0396, or nearly 4%. In general, the error frequency (E) for families of this type will be

$$E = (2RF) - (RF)^2$$

The preceding formula does not apply to families in which there are two or more children prior to the fetus who is to be diagnosed. The more children there are, the smaller is the probability of making an incorrect diagnosis because of unrecognized recombination.

The uncertainty level can also be minimized by using two marker loci, one on each side of the disease locus. When the affected parent is heterozygous for both *flanking markers*, the genotype of a fetus at those loci will reveal whether recombination has occurred between them (except for the negligible probability of double crossovers), because a recombinant child will have a new combination of alleles at the flanking marker loci. However, when a recombinant fetus is identified, it is not possible to know whether

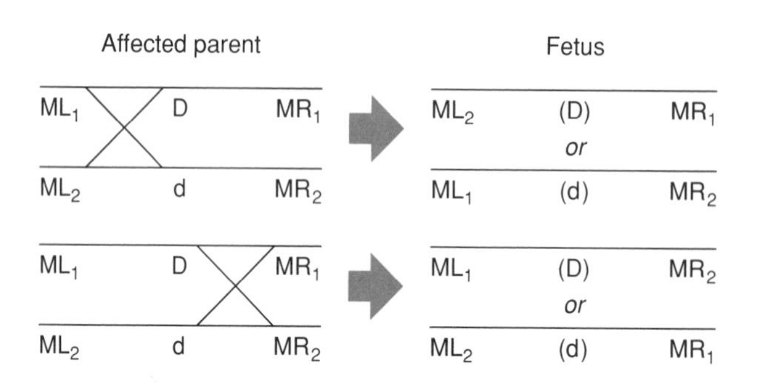

Figure 9–9

Diagram showing the consequences of recombination between marker loci (ML and MR) flanking a disease locus (D). Assume that the haplotypes shown on the left are present in an affected parent and were implied by the genotype of the firstborn child. If the second fetus has one of the marker locus combinations shown on the right (ML2, MR1 or ML1, MR2), it is clear that either the first child or the second child is a recombinant, but we cannot decide which is which. If two or more previous children are consistent with the haplotypes shown on the left, then it is extremely likely that the fetus (with marker genotype ML2, MR1 or ML1, MR2) is a recombinant, but we cannot decide whether it carries the D or the d allele. Those alleles are placed in parentheses in the right-hand diagrams to emphasize the uncertainty.

the crossover occurred between the left marker and the disease locus or between the right marker and the disease locus (Figure 9–9). The genetic counselor must tell the client that a diagnosis of the fetus is not possible with the available information.

All of the preceding considerations emphasize the need to find as many polymorphic loci near an unidentified disease locus as possible, in order to minimize the uncertainty of fetal diagnosis. Most people are willing to make important decisions on the basis of 98 or 99% probability in their favor; but when the odds are substantially lower, many people find decision making to be highly stressful. Evidently, prenatal diagnosis via linked RFLPs is likely to be a transitory stage in genetic practice. As genes responsible for diseases are isolated, it becomes possible to identify the precise mutation in any given family, which then allows the geneticist to predict the phenotype of an unborn child with great confidence.

SUMMARY

Genetic screening is a process of testing all available members of a defined population for a specific genetic condition. Prenatal diagnosis is mostly done on a more limited basis, on fetuses known to be at high risk for a specific genetic defect.

Two methods of obtaining fetal material are amniocentesis (removal of amniotic fluid and cells therein, usually at the sixteenth to twentieth weeks of pregnancy) and chorionic villus sampling (removal of a piece of the chorion, usually between the ninth and twelfth weeks of pregnancy).

Pregnant women over the age of 35 are encouraged to have their fetuses tested for trisomy 21, because the incidence of that aneuploidy increases rapidly with maternal age. Neural tube defects, which are only partially genetic in origin, can be detected by assaying maternal serum for alpha fetoprotein, followed by testing of fetal tissue in those cases where the maternal serum level is high.

Neonates are screened for phenylketonuria (PKU) in developed countries, using a dried sample of blood obtained when the infant is a few days old. Affected infants will thrive if placed on a low phenylalanine diet before neurological damage occurs. Additional blood samples are often used to test for hypothyroidism and several other inherited metabolic deficiencies that can be corrected with controlled diets. Screening of newborns for one or more hemoglobinopathies is also required in some areas.

The main purpose of adult genetic screening programs is detection of heterozygotes, so that couples at risk for birth of a child affected with an autosomal recessive disease can be informed. Voluntary screening programs organized by the Jewish community have been very successful in reducing the incidence of infants with Tay-Sachs disease.

Identification of the cystic fibrosis gene and its variants has led to proposals that mass screening of Caucasian populations for heterozygotes at that locus should be established, because one person in 20 to 25 is a heterozygote. However, the very large number of different mutant alleles makes it economically impractical to test for all of them at the present time. If the screening program only detects a few of the more common alleles, some individuals will test negative when they are actually carriers of rare recessive alleles. There is a controversy over whether that situation is sufficiently undesirable to postpone screening for common alleles.

Prenatal diagnosis is usually done on fetuses in families where one affected child has already been born. Techniques for prenatal diagnosis of a wide variety of genetic diseases are now available.

When the gene responsible for a particular disease has not yet been identified, it is often possible to predict the genotype of the fetus at the disease locus from the genotype at one or more closely linked marker loci, such as RFLPs. The possibility of recombination between the disease locus and the marker locus always makes such predictions less than 100% reliable.

REFERENCES

American Academy of Pediatrics, Committee on Genetics. 1989. Newborn screening fact sheets. Pediatrics 83:449–464.

Beaudet, A. L. 1990. Invited editorial: carrier screening for cystic fibrosis. Am. J. Hum. Genet. 47:603–605.

Chang, J. C. and Kan, Y. W. 1982. A sensitive new prenatal test for sickle cell anemia. New Engl. J. Med. 307:30–32.

Doran, T. A. 1990. Chorionic villus sampling as the primary diagnostic tool in prenatal diagnosis. Should it replace genetic amniocentesis? J. Repro. Med. 35:935–940.

Fuchs, F. 1980. Genetic amniocentesis. Sci. Am. 242 (June):47–53.

Guthrie, R. 1973. Mass screening for genetic disease. In "Medical Genetics," V. A. McKusick and R. Claiborne, eds., pp. 229–236. HP Publishing Co., New York.

Henry, G., Peakman, D., Winkler, W., and O'Conner, K. 1985. Amniocentesis before 15 weeks instead of CVS for earlier prenatal cytogenetic diagnosis. Am. J. Hum. Genet. 37(Suppl):A219.

Johnson, A. and Godmilow, L. 1988. Genetic amniocentesis at 14 weeks or less. Clin. Obstet. Gynecol. 31:345–352.

Kaback, M. M., Shapiro, L. J., Hirsch, P., and Roy, C. 1977. Tay-Sachs disease heterozygote detection: a quality control study. In "Tay-Sachs Disease: Screening and Prevention," M. M. Kaback, ed., pp. 267–279. Alan R. Liss, Inc., New York.

Mange, A. P. and Mange, E. J. 1990. Genetics: human aspects, 2nd ed. Sinauer Assoc., Sunderland, MA.

Rhoads, G. G., Jackson, L. G., Schlesselman, S. E., et al. 1989. The safety and efficacy of chorionic villus sampling for early prenatal diagnosis of cytogenetic abnormalities. New Engl. J. Med. 320:609–617.

Stryer, L. 1988. Biochemistry, 3rd ed. W. H. Freeman, New York.

Wilfond, B. S. and Fost, N. 1990. The cystic fibrosis gene: medical and social implications for heterozygote detection. JAMA 263:2777–2783.

The small, white, "knockout" mouse is a transgenic animal that possesses two inactive copies of the cystic fibrosis gene. Its large, dark-haired sibling is a heterozygote, with one normal CF gene. Both animals were produced by homologous recombination (See Figure 10-2). Knockout mice are being widely used to develop treatments for human genetic diseases. (Photo courtesy of Beverly Koller)

10 Treatment of Genetic Disease

Efforts to understand the biochemical basis of a genetic disease are always accompanied by the hope that the knowledge gained will lead to a cure for the disease, or at least to significant amelioration of symptoms. Nowadays there is extensive publicity about gene therapy, so it seems appropriate to survey that topic first. Publicity notwithstanding, substantially more morbidity is avoided at present by non-gene-level treatments than by gene therapy. Accordingly, a brief summary of traditional treatment modalities for genetic diseases will be given in the second section of this chapter.

Gene Therapy

Treatment of disease by replacement or correction of defective genes, or addition of functionally normal gene copies to a genome, is termed *gene therapy*. Virtually all of this chapter will deal with *somatic cell gene therapy*; that is, protocols where the target tissues do not include the germ cells or their progenitors. The effects of somatic cell gene therapy are limited to one individual. The chapter will also include a brief discussion of *germ-line gene therapy*, where the effects of genomic alterations may be passed down to subsequent generations.

The possibility of treating genetic disease by correcting a defect at the level of the gene itself, rather than attempting to compensate for abnormal or missing gene functions and their metabolic consequences, has been evident for at least two decades (e.g., Friedmann and Roblin, 1972). Transforming viruses, which add their own genetic information to that of a host cell in an inheritable fashion, provided a paradigm that molecular biologists hoped to emulate.

The development of recombinant DNA technology has provided the weaponry necessary to achieve that goal. In the early 1980s, it became evident that genetically engineered viruses could be used to transfer genes into cultured bone marrow cells of mice, and the prospect of refining the techniques to the point where human gene therapy trials might begin was under active discussion. One of the pioneer thinkers in this field was W. French Anderson, who outlined the technical criteria that would have to be met before any attempt to carry out gene therapy in humans could be undertaken (Anderson, 1984).

Criteria for Effective Gene Therapy

1. The first requirement is that the gene to be transferred to the patient must be available. The development of recombinant DNA technology has taken care of that need. Hundreds of cloned human genes are already available, and the Human Genome Project (Chapter 15) can be expected to make every human gene obtainable in cloned form eventually.

2. The second requirement is for an effective method of introducing the gene to be transferred into the recipient cells. The most efficient and widely used method at this time is to incorporate the gene to be transferred into a genetically engineered retrovirus (see Chapter 11). Other methods, such as electroporation, are also employed. The essential facts about technology will be summarized in the next section of this chapter.

3. The third requirement for effective gene therapy is that the target tissue or cells must be accessible to the gene transfer procedure. Although the use of retroviral vectors has made gene transfer into cells in culture an efficient process, much less is known about direct transfer of genes into cells in the organism. For this reason, current prospects for gene therapy are focused on diseases that affect hematopoietic cells, whose progenitors are found in bone marrow. It is feasible to remove large samples of bone marrow from a patient, treat them with a retrovirus carrying the gene that is to be transferred, and return them to the body.

Some other readily accessible cells are fibroblasts and endothelial cells. Although no major genetic diseases are primarily expressed by fibroblasts or endothelial cells, the cells can be given ectopic functions by gene transfer, then returned to the donor's body, where they may ameliorate symptoms by performing functions that they do not normally carry out, such as production of phenylalanine hydroxylase (lacking in PKU) or clotting factors (lacking in the hemophilias). It may even become possible to use *heterologous* cells (not from the patient), encapsulated in some matrix that protects them from contact with immune system cells, but allows free entry and exit of metabolites.

Hepatocytes are also being tested for their potential usefulness in gene therapy, using nonhuman experimental animals. A rather large section of liver can be removed (without permanent damage to the donor) and dissociated into a suspension of cells. The hepatocytes can be transfected with a retrovirus carrying a gene that is to be transferred, and the genetically altered cells can be returned to the donor by one of several routes, usually the spleen or hepatic portal vein. In either case, hepatocytes take up residence in the liver. Some of the diseases that are currently being studied as candidates for hepatocyte-mediated gene therapy are familial hypercholesterolemia, ornithine transcarbamylase deficiency, hemophilia B, and PKU.

The day of the therapeutic virus that will cure genetic disease after one or a few injections of a high titer suspension into the patient is somewhat farther away, but the possibility is not being ignored. The most obvious target disease for gene therapy *in vivo* is cystic fibrosis, where one may hope to deliver a gene-transfer virus to the lung epithelium in the form of an aerosol that can be inhaled. However, many technical problems must be overcome before that vision can be realized.

4. The fourth requirement is that the gene therapy procedure must not harm the patients. Two potential sources of harm are obvious and are of great concern; both arise from the fact that transfecting retroviruses integrate at random throughout the genome.

The first potential problem is that there may be *inappropriate expression* of the transferred gene in cells that have a different developmental fate from the target cells. For example, if a retrovirus carrying the gene for beta-globin is used to transfect bone marrow cells from a patient with beta thalassemia, will the globin gene be expressed in the patient's white blood cells as well as in the red blood cells, and will that have an adverse effect on the patient? This problem may gradually disappear as we learn more about DNA sequences that are responsible for tissue-specific expression of individual genes. Recent progress with the beta-globin gene has been encouraging. A short DNA sequence derived from a region about 10,000 bp upstream of the gene has been shown to confer regulated, high-level expression to the human beta-globin gene in transgenic mice (Ryan et al., 1989; Novak et al., 1990).

Another predictable effect of random integration by retroviral vectors is *insertional mutagenesis*. If a retrovirus integrates into an active gene, it may abort the function of that gene in that cell. When this happens, it is not likely to have much effect on a patient, because very few cells will lose the same function. However, if the disrupted function results in loss of growth control, a tumor may result. It is not yet clear whether insertional mutagenesis will be a significant problem in human gene therapy.

5. The fifth requirement for effective gene therapy in humans is that the procedure must make a significant improvement in the patient's health. If the cells that have been transfected represent only a small fraction of the target cells, or if the transgene is not expressed at a high level, the patient may receive little benefit. No doubt this problem will be easier to deal with as clinical experience with specific systems accumulates, but in the current, exploratory stages of gene therapy, it is as difficult to predict that the patient will be helped as it is to guarantee that the patient will not be harmed.

Technical Aspects of Gene Therapy

The technical problems referred to in the preceding section apply specifically to therapy of recessive genetic disorders. By definition, a recessive allele produces a product that either lacks function completely or has too little function to produce a normal phenotype when present in homozygous state; the product of a recessive allele also must not interfere with the function of a normal allele, when present in heterozygous condition. Therapy of recessive disorders, therefore, does not require that the abnormal alleles be removed or altered. It suffices to introduce a normal allele into the genome at any point, provided that the transgene will be expressed actively. In contrast, dominant genetic disorders cannot be cured by gene therapy unless the abnormal allele is silenced or changed into a normal allele. Technical aspects of gene therapy have been covered in several recent reviews (e.g., Friedmann, 1989; Cournoyer et al., 1990).

Therapy of Recessive Genetic Diseases Not many years ago, introduction of DNA into mammalian cells seemed to be a hopeless enterprise. Mammalian cells simply would not take up naked DNA, integrate it into their genomes, and express it, as many bacteria would do. This obstacle was overcome when it was discovered that DNA enmeshed in tiny crystals of calcium phosphate was avidly engulfed by mammalian cells in culture (Graham and Van der Eb, 1973). However, the method has little promise for gene therapy because of its low efficiency; only about one cell in 10^5 becomes transformed.

It is also possible to transfer cloned genes by microinjection. Most of this work has been done with mouse embryos, where it is possible to inject

Quiz 5
CLOSE BOOK (5 min)

8

BIO 530
Human Genetics
July 16, 1997

Name: Marguerite McCarthy

1. Either define Reproductive Fitness

Or State the nucleotide base change that leads to Sickle Cell Anemia.

GAG → GTG

(9)

Quiz 5
OPEN BOOK (10 min)

BIO 530
Human Genetics
July 16, 1997

Name: Marguerite McCarthy

2. If there is one live birth with Cystic Fibrosis per 2,500 individuals, what proportion of individuals in the population will be lacking the CF genes?

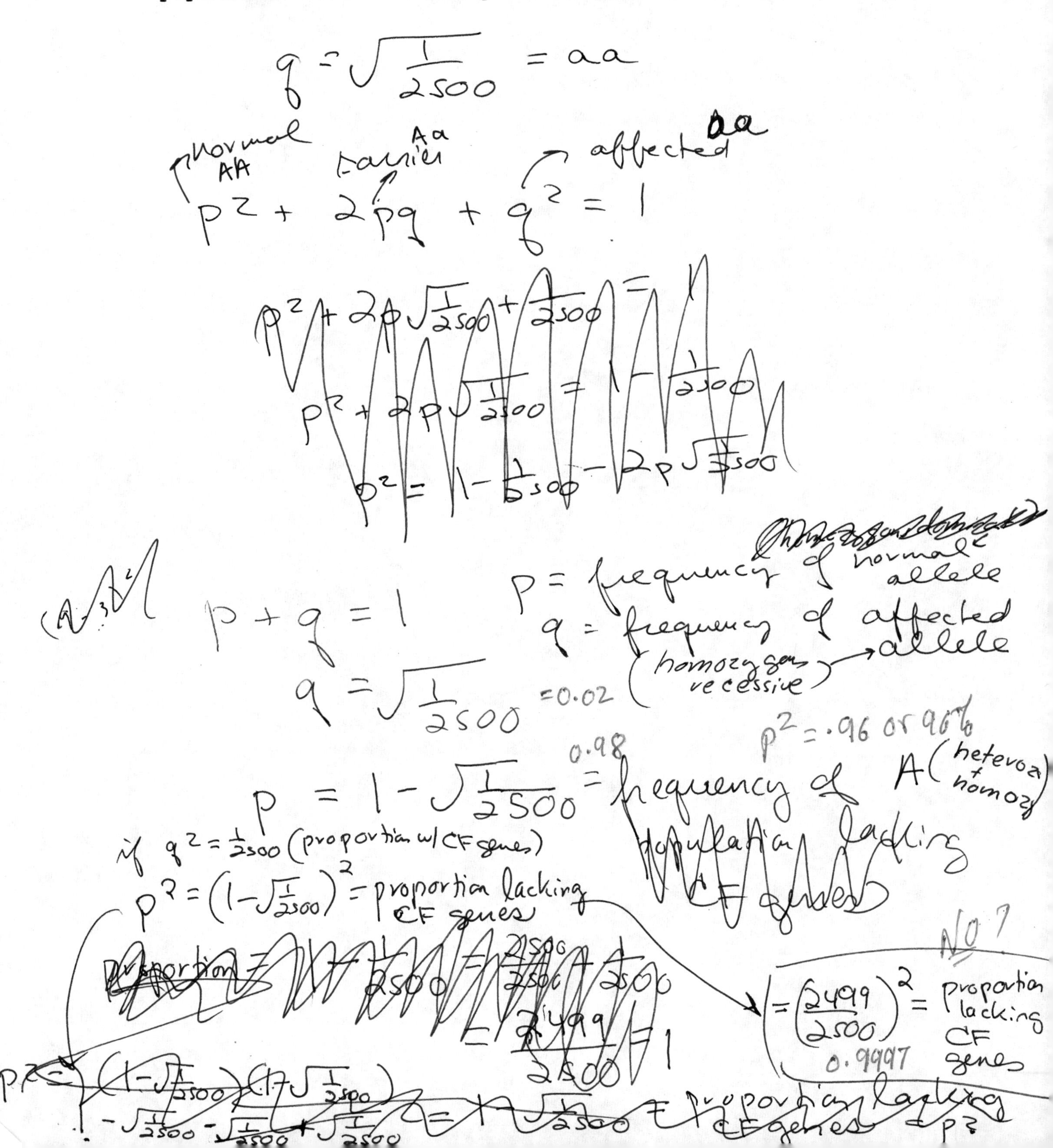

DNA into the male pronucleus, which enlarges after fertilization, prior to fusion with the egg nucleus (Gordon and Ruddle, 1981). One of the more newsworthy applications of this technique produced giant mice, after the embryos had been injected with the human growth hormone gene (Palmiter et al., 1983). The gene had been attached to the metallothionein gene promoter, so that a high level of expression could be achieved in response to heavy metal administration.

However, microinjection is unlikely to be used for somatic cell gene therapy. Although an experienced technician can inject several hundred fibroblasts in a day, effective therapy would require transformation of millions of cells. No doubt the original transformants could be multiplied in culture, but more effective methods of gene transfer are available.

Another useful technique is *electroporation* (Neumann et al., 1982). Short pulses of high-voltage electricity open pores in cell membranes transiently, through which DNA from the surrounding medium enters efficiently. Several million cells in suspension can be treated at one time.

Some novel approaches for gene transfer *in vivo* have been tried recently. DNA complexed with a polylysine-asialoglycoprotein conjugate was infused intravenously into rats, in the hope that liver cells would internalize the complex via their receptors for asialoglycoproteins (Wu et al., 1989). Evidence that a reporter gene was expressed by liver cells was presented. If partial hepatectomy was performed shortly after gene transfer to stimulate cell division in the liver, it was found that some of the transferred DNA was incorporated into the host genome. Another new technique with potential applications for *in vivo* transfer is the "biolistic" device, which shoots DNA-coated microprojectiles into target cells or tissues (Moffat, 1990). It is too soon to evaluate the efficacy of these techniques, but the need for methods to transfer genes into target tissues without removing them from the host is obvious.

For gene transfer into cells in culture, which may subsequently be returned to the host, highly efficient retroviruses have been developed by elegant application of genetic engineering techniques. Viruses in general are adept at injecting their genomes into suitable host cells. Retroviruses, after gaining entry into a cell, copy their RNA genomes into DNA with reverse transcriptase (an enzyme carried by the virus and encoded by the viral genome). The DNA is able to integrate efficiently and randomly into the host genome, where it may remain and be expressed indefinitely. Retroviral genomes are described in more detail and illustrated in Chapter 11.

Retroviruses can also transfer a non-viral gene, as was made apparent by the recognition of the *src* gene of Rous sarcoma virus (Chapter 11). The possibility of using retroviruses for gene therapy was evident to a number of investigators some years ago, but several problems had to be solved. First, the viral oncogene (if there was one) had to be removed. Second, the virus had to retain the ability to infect cells and integrate its altered genome into the host. Third, there should be no replication of the virus in the host, and for this purpose, replication-defective gene-transfer retroviruses were created. These viruses are assembled in "helper" cell lines, which have been constructed to provide the viral proteins that are no longer encoded by the gene-transfer virus, but which are necessary for viral genome synthesis and virion assembly (Figure 10–1).

Currently, retroviruses are the agent of choice for introducing genes into mammalian cells, but some problems remain. First, the carrying capacity of a retrovirus is not much more than 8 kbp, which is too little to accommodate the largest mRNAs, let alone genes complete with introns. Second,

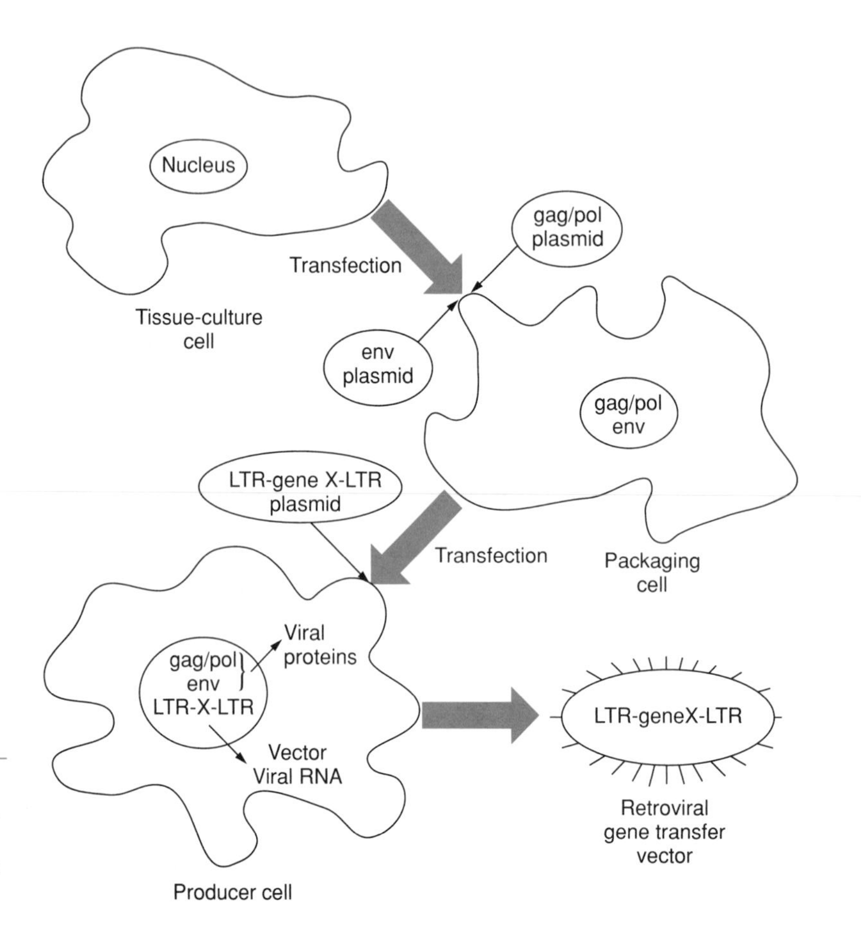

Figure 10–1

Sequence of events for the generation of cell lines that can produce a gene-transfer retroviral vector. (Modifed from McLachlin et al., 1990, p. 101)

the possibility of insertional mutagenesis cannot be denied. Third, production of infectious virus by recombination of the gene transfer virus with endogenous retroviral sequences in the host genome is a possibility. Fourth, retroviral genome integration into the host cell genome does not take place unless the cells are replicating. This is more of a limitation to the use of retroviruses for gene transfer *in vivo*, where most differentiated cells do not divide, than in cell culture, where mitogenic conditions can easily be supplied.

Therapy of Dominant Genetic Diseases In theory, it might be possible to cure a dominant genetic disease by *homologous recombination* of a transgene with the endogenous genes of the same type. When homologous recombination between exogenous and endogenous DNA occurs, the endogenous sequence is snipped out of the genome and the exogenous sequence is inserted. Homologous recombination occurs commonly in microorganisms, including yeast, but it is rather rare in mammalian cells, where non-homologous recombination takes place about 1000 times more often than homologous recombination.

Recent efforts to exploit homologous recombination in mammalian cells have employed a strategy known as *gene targeting*. First, it is necessary to point out that retroviruses are not suitable for homologous recombination experiments, because of their inherent capacity to integrate anywhere in the host genome, rather than at specific sites. DNA for homologous recombination studies can be injected, or transferred by electroporation.

The basic idea behind gene targeting is that the DNA to be transferred carries one or more markers which will allow the investigator to distinguish cells that have integrated the exogenous DNA via homologous recombination from those that have integrated it via non-homologous recombination (Capecchi, 1989). Figure 10–2 illustrates a system known as positive-negative selection. All cells that have integrated the exogenous DNA are selected by virtue of their resistance to G418 (a neomycin analog). Then, the cells that have integrated the exogenous DNA by non-homologous recombination are killed, because of their sensitivity to the antibiotic gancyclovir. Cells that have undergone homologous recombination are not affected by gancyclovir, because the process of homologous recombination excludes the gene that confers gancyclovir sensitivity.

Most of the work on gene targeting has been done with mice. The general approach is to transfer marked DNA into a population of embryonic stem cells, obtained from dissociated embryos; to select those cells that have integrated the desired gene by homologous recombination; and then to create chimeric embryos by fusing the transfected cells with whole embryos from the same inbred strain of mouse. This approach is proving to be highly valuable for the production of mice with specific mutations and for the creation of mouse models of human diseases.

So far, gene targeting systems have not been developed to the point where they would be useful for human gene therapy, but there is no doubt that the approach will receive intense effort with that goal in mind. At present, the obstacles are formidable. In the first place, the goal of gene therapy for dominant diseases is to *remove* a mutant allele, whereas most of the gene targeting systems that have been used in mice are intended to *introduce* mutant alleles.

Another question is whether there would be any point in using homol-

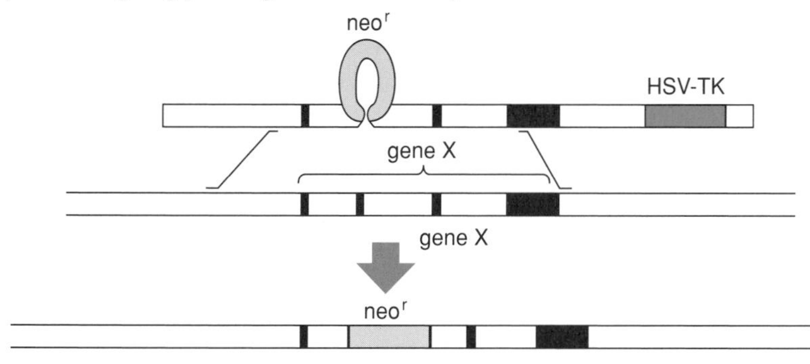

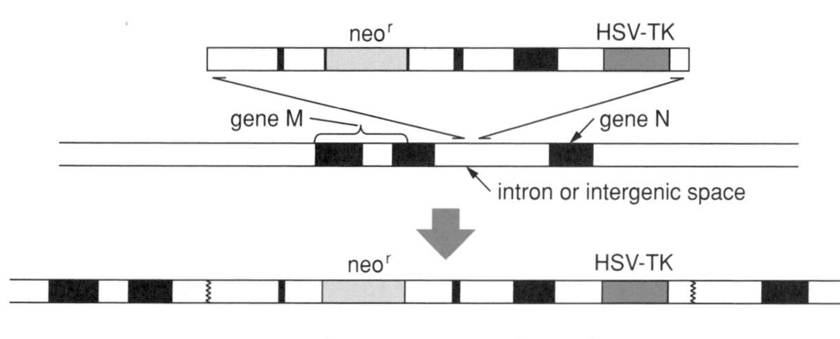

Figure 10–2

The positive-negative selection procedure used to enrich for embryonic stem cells containing a targeted disruption of an identifiable gene (gene X). Gene X is presumed to be on the X chromosome, and therefore is present in only one copy in males. Panel A shows a vector that contains an insertion of the neo^r gene in an exon of gene X (which renders gene X inactive) and a linked HSV-tk gene. The vector pairs with a chromosomal copy of gene X, and homologous recombination results in the insertion of the defective gene X into the chromosome, but the exclusion of the HSV-tk gene. Such cells will lack gene X function, but will be resistant to both G418 and gancyclovir. Panel B illustrates non-homologous recombination, which occurs through the ends of the exogenous vector DNA, resulting in the incorporation of the entire vector. Such cells will retain gene X function (via the original chromosomal copy); they will be resistant to G418, but sensitive to gancyclovir, because they have the HSV-tk gene. (From Capecchi, 1989, p.75)

ogous recombination to correct a dominant defect in a minority of the cells affected by a given disease. How could an entire tissue or organ be subjected to gene targeting outside of the patient? It is possible that therapeutic applications of gene targeting may not be feasible unless some way to increase the frequency of homologous recombination by several orders of magnitude is devised, but there will surely be much ingenuity applied to this challenge.

Current Gene Therapy Projects

In 1990, actual gene therapy trials in humans began. After more than a decade of technical development, accompanied by extensive discussions of the ethical aspects of gene therapy, the National Institutes of Health finally authorized a clinical trial in response to a protocol submitted by Drs. W. French Anderson and R. Michael Blaese (Culliton, 1990). The patient was a four-year-old girl with adenosine deaminase (ADA) deficiency, an extremely rare autosomal recessive disorder. Patients without adenosine deaminase accumulate abnormally high blood levels of adenosine, which is continually produced from the intracellular catabolism of nucleic acids. Both T and B lymphocytes are killed by excess adenosine. The patients, without either humoral or cell-mediated immunity (see Chapter 12) have Severe Combined Immunodeficiency Disease; they succumb to infections early in life, unless they are kept in a germ-free environment.

Attempts to supply the missing enzyme by injecting it into the bloodstream were a failure, because it was degraded within minutes. Subsequently, a preparation of adenosine deaminase coated with a polymer (PEG-ADA) was found to last for several days (Markert et al., 1987), the result being that patients who received weekly injections were no longer confined to their germ-free bubble chambers. The girl who was the subject of the first gene therapy trial in September, 1990, was also receiving PEG-ADA, and she continued to receive it after the gene therapy began.

The major steps in the protocol were: isolation of a large sample of T lymphocytes from the patient's blood, multiplication of the T-cells in culture, transfection with a mouse leukemia virus that carried the human ADA gene, and return of the T-cells to her circulatory system. The patient was given 8 infusions of gene-corrected T lymphocytes over a period of approximately eleven months. She was taken off gene therapy for six months and then placed on a maintenance program of infusions of gene-corrected lymphocytes at three- to five-month intervals. A second child with adenosine deaminase deficiency began receiving infusions of her own gene-corrected T-cells in January, 1991. Both patients have responded favorably; they attend public schools and have no more than the average number of infections (Anderson, 1992).

A second type of gene therapy is underway at the National Institutes of Health, under the direction of Dr. Steven Rosenberg. The goal of this project is to cure cancer, specifically malignant melanoma. For a number of years, Dr. Rosenberg and his colleagues have been working with Tumor Infiltrating Lymphocytes (TILs), T-cells (see Chapter 12) that attack tumors. When part of a tumor has been removed surgically, TILs can be isolated from it. In a feasibility study (Rosenberg et al., 1990), the Rosenberg group transfected TILs with a retrovirus carrying a marker gene (neomycin resistance), returned the TILs to a patient, and proved that the marker cells returned to the tumor and survived in the patients for at least several weeks (Figure 10-3).

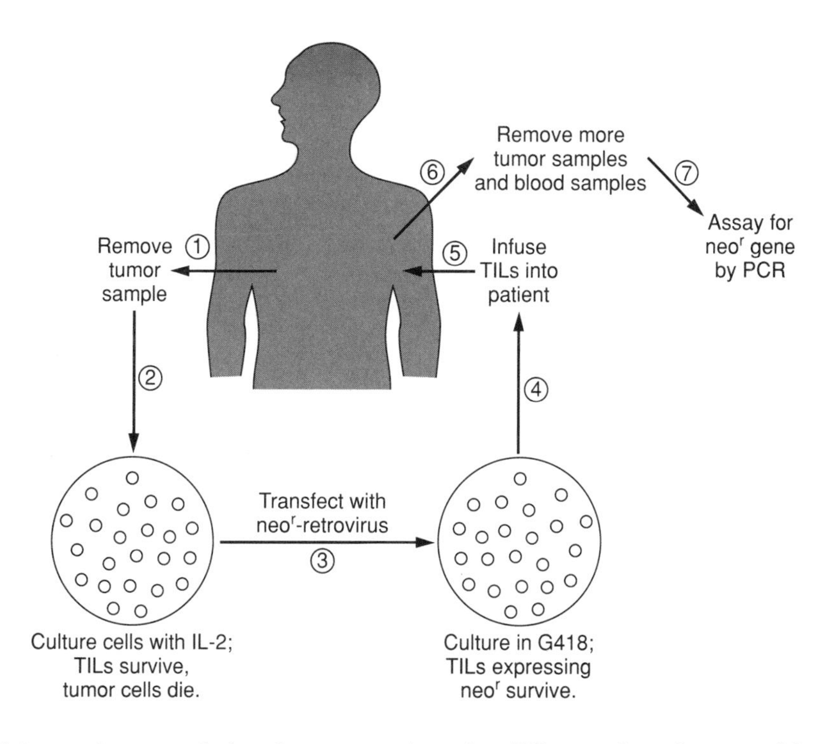

Figure 10–3

A potential gene therapy protocol for treatment of cancer. A sample of the tumor is removed surgically, dissociated, and placed into culture with interleukin-2 (IL-2). In time, the tumor cells die, while the tumor-infiltrating lymphocytes (TILs) multiply. The TILs are then transfected with a retroviral vector carrying the gene for neomycin resistance (neo^r), and the cells that have incorporated the vector into their genomes can be selected via resistance to the neomycin analog, G418. These cells, along with a large number of non-transfected TILs from the same patient, are infused back into the patient. At various times thereafter, the presence of the transfected TILs in the tumor and in the blood can be demonstrated by the use of the polymerase chain reaction (PCR) to assay for the neo^r gene. Future studies will use TILs transfected with genes that may prove to be selectively toxic to the tumor. (Based on data in Rosenberg et al., 1990)

This result opened the door to engineering TILs so that they would become more effective in destroying tumor cells. The strategy that will be employed is to transfect TILs with the gene for tumor necrosis factor (TNF), a protein that is a very effective cytotoxic agent against tumor cells. It cannot be injected directly into a patient, because it would have too severe toxic side effects on normal cells, but if only TILs produce the protein, its toxicity can, in theory, be directed specifically against the tumor. A clinical trial of this protocol is in progress as this is being written.

Many more gene therapy trials are certain to follow these pioneering studies. In the United States, plans have already been formalized for feasibility studies or actual treatments involving familial hypercholesterolemia, ovarian cancer, and several forms of leukemia. An attempt to control hemophilia B with gene-corrected fibroblasts is underway in China.

Among the attractive candidates for gene therapy is cystic fibrosis (see Chapter 5). Most of the morbidity associated with this disease is a result of abnormal lung secretions, and the airway cells are accessible. It is hoped that it may be possible to administer a gene transfer virus directly to cystic fibrosis patients in the form of an aerosol that can be inhaled. Considerable encouragement was provided by the recent demonstration that abnormal chloride transport in CF airway epithelial cells in culture could be corrected by transfection with a virus carrying the cystic fibrosis gene (Quinton, 1990). In addition, Rosenfeld et al. (1991) have already shown that an adenovirus vector, carrying the gene for human alpha-1-antitrypsin, transferred the human gene into epithelial cells of rat lungs, where it was expressed.

All proposals for gene therapy receive exceptionally thorough review before being authorized. At NIH they are first scrutinized by the Human Gene Therapy Subcommittee, which passes its recommendations on to its parent, the Recombinant DNA Advisory Committee (RAC). Both groups include ethicists, legal scholars, and other non-scientists, as well as scientists with medical and research backgrounds, among their members. If both groups approve a proposal, it is sent to the Director of NIH, who must also give approval before a gene therapy trial can commence. In addition, Institutional Review Boards at hospitals where the work will be carried out must grant approval. The current gene therapy trial on the girl with ADA deficiency underwent at least 15 different reviews over a period of several years before it received final authorization.

Needless to say, the information media have devoted much attention to the recently initiated clinical gene therapy trials. Most major newspapers have reported the events, and more general descriptions of gene therapy and its possibilities for treatment of disease have appeared in popular magazines; examples include *Time* (Jaroff, 1990) and *Discover* (Montgomery, 1990). There appears to be general public acceptance of somatic cell gene therapy as an exciting new addition to the repertoire of the medical profession. Most of the ethical concerns that were expressed earlier have been satisfied, although there are still a few vocal critics who are opposed to any form of gene therapy in principle.

Germ-Line Gene Therapy

In contrast to the situation with somatic cell gene therapy, the prospect of deliberately making *inheritable* alterations in the human genome has raised a hurricane of fear and criticism. Theologians in impressive numbers have joined forces to speak against it, while presidential commissions and congressional committees have concluded, after lengthy deliberations, that germ-line gene therapy should be labeled unethical and probably should be banned outright. As yet, there are no laws proscribing gene therapy in the United States, but NIH has a policy of refusing to consider grant applications dealing with the subject.

In part, the virtually unanimous disapproval of germ-line gene therapy is based on the fear of new eugenics programs, especially those arising from political motivation, as was the case in Nazi Germany. But there also appears to be a widespread intuitive feeling that altering the genome in a manner that will be passed on indefinitely to an individual's descendants is trespassing on sacred ground. In some difficult-to-define, but deeply felt way such changes seem to be a threat to our humanity.

It is true that the human gene pool will change as the centuries pass, whether scientists intervene or not. It is also true that many of the triumphs of medical genetics, which enable individuals with previously lethal or incapacitating genetic defects to survive and reproduce, are gradually changing the gene pool. But the helter-skelter incorporation of cloned recombinant DNAs into a person's germ line is a more immediate threat; and the threat is magnified because we do not know where it ends.

The phrase "germ-line gene therapy" implies that the "patient" will be an embryo with some serious genetic defect, which can be corrected by incorporating cloned genes capable of providing the missing function into the DNA of all or most of the cells in that embryo. If the treatment is effective, a normal child will be born. The transgene is likely to be included in the germ cells as they form in the fetus, even though the purpose of the gene therapy may have been to correct a defective liver, pancreas, brain, etc. Therefore, any genetic alterations resulting from gene therapy on one embryo will be potentially transmissible to future generations.

The technical power of the Polymerase Chain Reaction (see Box 4–1) has brought the prospect of germ-line gene therapy much closer to reality. In principle, it is possible to remove one cell from a blastocyst that has been produced by *in vitro* fertilization, and use the PCR to determine whether the embryo will be affected with a disease to which it is known to be at risk.

Then a retrovirus carrying a normal allele of the defective gene could be used to transfect the rest of the blastocyst and it could be returned to the mother to be carried to term. This scenario is oversimplified and unrealistic at the moment, but it is not total fantasy.

In what circumstances would an embryo be a candidate for gene therapy? As an example, let us consider Tay-Sachs disease. As long as couples at risk are willing to abort affected fetuses, there will be no need for embryonic gene therapy. However, if a couple at risk were adamantly opposed to abortion, then an embryo affected with Tay-Sachs disease would be a potential candidate for gene therapy, inasmuch as somatic cell gene therapy of this central nervous system defect after birth would be out of the question.

As another example, consider a couple, both of whom are homozygotes for PKU. They will be essentially normal physically and mentally if they were placed on a low-phenylalanine diet from infancy onwards. All of their offspring will also have PKU. Suppose the couple decide that they do not want their children to spend their lives on a severely restricted diet, with all the physical and emotional stresses that accompany it. Do they have a legal right to have normal children, if the technology to correct the genetic defect in embryos is available, and somatic cell gene therapy of newborns has not been successful?

These and similar dilemmas will be with us for a long time, no matter how clever our gene transfer technology becomes. Indeed, the only certainty is that technology *will* become more powerful. If gene therapy of human embryos who are affected with serious genetic defects becomes socially and medically acceptable, how long will it be before eugenics advocates become vocal, calling for various "improvements" of the species, such as replacement of "poorly designed" genes like dystrophin with modern, genetically engineered versions in normal human embryos?

Traditional Treatment Modalities

In its broadest sense, some form of treatment is available for virtually all genetic diseases. At worst, treatment may be only palliative, such as sedation for patients in advanced stages of Huntington Disease. An intermediate level of treatment is exemplified by antibiotic therapy and physiotherapy to remove lung secretions for cystic fibrosis patients. However, clinical protocols are beyond the scope of this book. Here we will only mention treatments that are directed at the primary biochemical defects resulting from mutations in specific genes.

Nutritional Treatments

In several well-known genetic diseases, essentially normal health can be achieved by altering the patient's diet. Probably the best-known example is control of PKU by restricting intake of phenylalanine. Another example is prevention of galactosemia by removal of sources of galactose from the diet. Reduction of cholesterol intake significantly improves the prognosis for persons with familial hypercholesterolemia, both in the heterozygous and homozygous conditions. Intake of synthetic thyroid hormone effectively eliminates the symptoms of congenital hypothyroidism.

As we learn more about the biochemical basis of genetic diseases, we can expect that many of those diseases will be at least partially treatable by alterations in diet or the patient's environment, or by more specific and sophisticated pharmacological regimens.

Protein Replacement Therapy

In general, injection of purified proteins into the body, usually via the circulatory system, has not been effective. Most injected proteins are destroyed by proteases in a very brief time, and repeated injections can generate an immunological response that may be life threatening. The case of ADA replacement therapy was mentioned earlier; the enzyme was degraded within minutes when injected alone, and when it was protected by binding to a non-biological polymer (PEG), its half-life was extended to no more than a few days.

Transfusion of purified clotting factors into hemophiliacs represents an exception to the preceding generalization. Here we are dealing with proteins that are normal constituents of blood, so it is logical that they should not be attacked by serum proteases. However, in some hemophiliacs the mutant allele does not produce a protein at all, or it produces only a small fragment of the usual protein, because of a nonsense mutation near the beginning of the translated sequence. In such individuals, the immune system may regard exogenous clotting factor as a foreign protein, with the patient eventually becoming unable to tolerate it.

Another general problem that interferes with attempts at protein replacement therapy is the fact that most proteins function inside of cells, not in the blood or lymph. The technical problems of designing a protein that can survive long enough to transit the circulatory system, a protein that will be taken up by specific target cells and not destoyed in lysosomes, and a protein that still retains the desired function, are substantial.

Nevertheless, there are a variety of enzymes that normally function in lysosomes and whose absence leads to severe genetic disease. Gaucher disease is the result of one of those enzyme deficiencies (absence of glucocerebrosidase). Currently, Gaucher disease is treated by repeated intravenous injections of glucocerebrosidase, prepared from human placenta. The enzyme is avidly engulfed by macrophages, where it enters lysosomes and functions very well. Significant improvement in many symptoms (e.g., spleen size, red cell count, and general vigor) have been reported. It is expected that the high cost of this treatment will decline when enzyme synthesized by recombinant microorganisms becomes available. Even so, as we saw in the first section of this chapter, the best long-term solution appears to be to supply the missing gene and let the needed protein be synthesized *in situ*.

SUMMARY

Gene therapy refers to the addition of normal gene copies to a genome lacking functional genes of a specific type, as well as replacement or correction of defective genes. Somatic cell gene therapy is done on any body cells other than the germ cells or their progenitors and is therefore not transmissible to future generations.

Germ-line gene therapy refers to genetic alterations that will be included in the germ cells. Gene changes made in early embryos, even though the primary target is somatic cells that will differentiate later, are likely to affect the germ cells as well. Germ-line gene therapy causes genomic changes that are inheritable.

Effective somatic cell gene therapy has the following requirements: the gene to be transferred must be available as a recombinant clone, a method for efficient introduction of the gene into recipient cells must be available, the target tissue must be accessible, there must be no deleterious side effects of the procedure, and the procedure must make a significant improvement in the patient's health.

Current efforts are focused on therapy of recessive genetic disease, where it is not necessary to remove the mutant genes in order to restore normal function. Most gene transfer vectors are altered retroviruses, modified so that they cannot produce infectious virus, but are still able to transfer their genetic information into the genome of recipient cells, where it will be expressed in a stable manner.

Gene therapy of dominant diseases is not yet possible, but much effort is being expended on controlling homologous recombination, which results in the replacement of an endogenous gene by a transgene. Model systems in mice employ the strategy of gene targeting, which allows investigators to distinguish homologous recombination events from the more numerous non-homologous recombinations.

The first human gene therapy trial began in 1990 with a girl who has adenosine deaminase deficiency. Early reports indicate at least partial success in restoring her immune system function. Another project involving genetically engineered lymphocytes targeted to kill specific cancer cells is beginning. Many other gene therapy projects are being planned.

There is no compelling reason to consider germ-line gene therapy trials at the present time.

Traditional treatment modalities are effective for many genetic diseases, including phenylketonuria, hypercholesterolemia, and hypothyroidism. Protein replacement therapy has had limited success. In most cases, injected proteins are degraded rapidly; but in a few diseases, such as hemophilia A and Gaucher disease, the exogenous protein is relatively stable and functional.

REFERENCES

Anderson, W. F. 1984. Prospects for human gene therapy. Science 226:401–409.

Anderson, W. F. 1992. Human gene therapy. Science 256:808–813.

Capecchi, M. R. 1989. The new mouse genetics: altering the genome by gene targeting. Trends in Genetics 5:70–76.

Cournoyer, D., Scarpa, M., and Caskey, C. T. 1990. Gene therapy. Current Opinion Biotech. 1:196–208.

Culliton, B. J. 1990. Gene therapy: into the home stretch. Science 249:974–976.

Friedmann, T. 1989. Progress toward human gene therapy. Science 244:1275–1281.

Friedmann, T. and Roblin, R. 1972. Gene therapy for human genetic disease? Science 175:949–954.

Gordon, J. W. and Ruddle, F. H. 1981. Integration and stable germ line transmission of genes injected into mouse pronuclei. Science 214:1244–1246.

Graham, F. L. and Van der Eb, A. J. 1973. A new technique for the assay of infectivity of human adenovirus 5 DNA. Virology 52:456–467.

Jaroff, L. 1990. Giant step for gene therapy. Time, Sept. 24, 1990, pp. 74–76.

Markert, M. L., Herschfield, M. S., Schiff, R. I., and Buckley, R. H. 1987. Adenosine deaminase and purine nucleoside phosphorylase deficiencies: evaluation of therapeutic interventions in 8 patients. J. Clin. Immunol. 7:389–399.

McLachlin, J. R., Cornetta, K., Eglitis, M. A., and Anderson, W. F. 1990. Retroviral-mediated gene transfer. Prog. Nuc. Acid Res. Mol. Biol. 38:91–135.

Moffat, A. S. 1990. Animal cells transformed *in vivo*. Science 248:1493.

Montgomery, G. 1990. The ultimate medicine. Discover, March, 1990, pp. 60–68.

Neumann, E., Schaefer-Ridder, M., Wang, Y., and Hofschneider, P. H. 1982. Gene transfer into mouse lyoma cells by electroporation in high electric fields. EMBO J. 1:841–845.

Novak, U., Harris, E. A. S., Forrester, W., Groudine, M., and Gelinas, R. 1990. High level beta-globin expression after retroviral transfer of locus activation region-containing human beta-globin gene derivatives into murine erythroleukemia cells. Proc. Natl. Acad. Sci. USA 87:3386–3390.

Palmiter, R. D., Norstedt, G., Gelinas, R. E., Hammer, R. E., and Brinster, R. L. 1983. Metallothionein-human GH fusion genes stimulate growth of mice. Science 222:809–814.

Quinton, P. M. 1990. Righting the wrong protein. Nature 347:226.

Rosenberg, S. A., Aehersold, P., Cornetta, K., et al. 1990. Gene transfer into humans: immunotherapy of patients with advanced melanoma using tumor infiltrating lymphocytes modified by retroviral gene transduction. New Engl. J. Med. 323:570–578.

Rosenfeld, M. A., Siegfried, W., Yoshimura, K., et al. 1991. Adenovirus-mediated transfer of a recombinant alpha-1-antitrypsin gene to the lung epithelium *in vivo*. Science 252:431–434.

Ryan, T. M., Behringer, R. R., Martin, N. C., et al. 1989. A single erythroid-specific DNAse I super-hypersensitive site activates high levels of human beta-globin gene expression in transgenic mice. Genes Dev. 3:314–323.

Wu, C. H., Wilson, J. M., and Wu, G. Y. 1989. Targeting genes: delivery and persistent expression of a foreign gene driven by mammalian regulatory elements *in vivo*. J. Biol. Chem. 264:16985–16987.

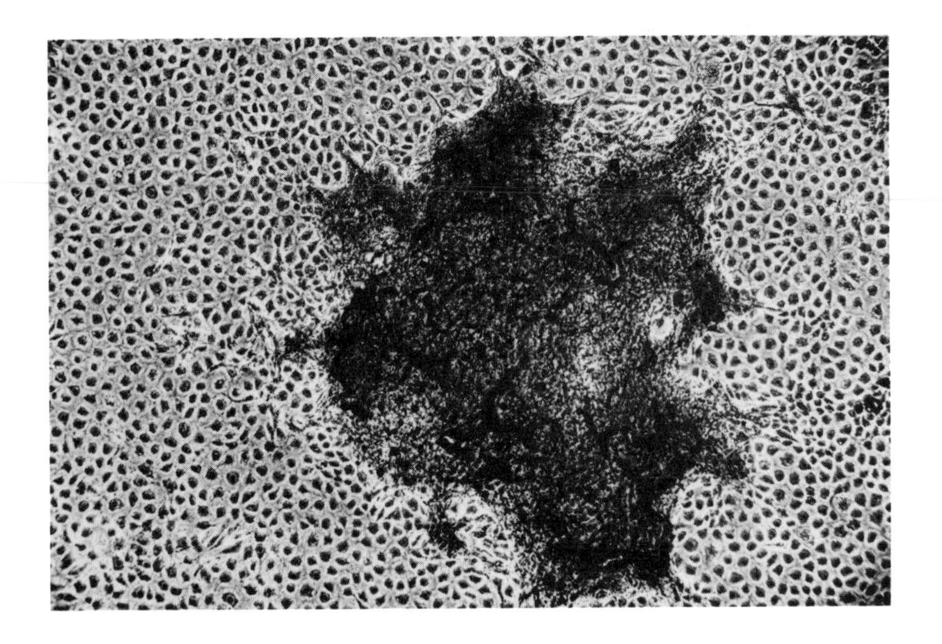

Light micrograph of a monolayer of normal animal cells in which a transformed cell has arisen. The normal cells remain arrested in GO, but the transformed cell has proliferated, giving rise to an irregular mass of piled-up cells. (From Prescott, 1988; courtesy of Robert E. Scott)

11 The Genetic Basis of Cancer

The central theme of this chapter is that cancer is fundamentally a genetic disease. As far as we know, all tumors result from alterations in DNA, which ultimately have the effect of producing unregulated growth. Most tumors appear to be the result of somatic cell mutations, but a very important group of tumors are caused, at least in part, by inherited mutations; that is, by mutations transmitted through the germ cells from one human generation to the next.

The possibility that epigenetic changes also contribute to tumorigenesis cannot yet be discounted. Potential examples include disturbances of DNA methylation (Hoffman, 1990) or environmentally produced shifts in patterns of gene expression which define a specific differentiated state. Nevertheless, one of the major conclusions arising from research on the molecular biology of cancer during the past two decades is that cancer is not *primarily* an epigenetic phenomenon. Cancer cells contain mutations that enable them to grow and multiply at inappropriate times and places; those mutations will be inherited through an indefinite number of cell generations.

Most tumors are clonal in origin; an entire tumor usually consists of the descendants of one cell that began to multiply without restraint. This should not be taken to imply that a single mutation leads to all the phenotypic characteristics of a malignant cancer. There is ample reason to believe that at least two mutations are required for initiation of unregulated growth in some specific tumors. Indeed, studies have shown that tumor formation in humans increases with age at a rate that implies that 4 to 6 mutations are required to produce a tumor that is capable of growing indefinitely and spreading to new locations within its host (Peto, 1977; Dix, 1989).

Tumor development is often described in terms of stages known as *initiation, promotion,* and *progression*. Initiation refers to the first mutagenic event that gives a cell the potential for unrestrained growth and multiplication. Depending upon the type of cell in which it occurs, the initiating mutation may be expressed immediately (as might be the case in a continuously proliferating stem cell population) or after a latent period of many months or years. In the latter case (which might apply to a quiescent cell in a tissue where there is very little cell division), expression of the initial oncogenic mutation may require "promotion," that is, receipt of a mitogenic signal; the latter might come from natural causes, such as the death of nearby cells, or from artificial causes, such as exposure to an organic chemical that has tumor-promoting properties.

A cell that enters into a state of uncontrolled proliferation will produce a benign tumor, but it does not necessarily produce a cancer. Progression to the fully malignant state presumably requires additional mutations. Acquisition of *invasiveness* allows a tumor to extend into the surrounding normal tissue and ultimately to interfere with normal function in that area. Tumor growth, invasive or not, may also depend upon angiogenesis, the development of a blood supply within the tumor.

The final stage in tumor progression is *metastasis,* the property of cells from the original tumor to migrate to new sites and establish secondary tumors there. The new sites may be in tissues quite distinct from the one in which the primary tumor arose.

Tumor progression is not yet understood at the molecular level, but it is surely a multistep process that may involve several somatic mutations before a full-blown metastatic cancer arises. The necessary mutations may be different in tumors that arise in different cell types.

A paradoxical aspect of tumor progression is the characteristic chromosomal instability of cancer cells (Yunis, 1983; Rowley, 1990). Karyotypic abnormalities are routine in metastatic tumors, and most of the changes seem to be random, although there are important exceptions (as explained later). The causes of this chromosomal instability are unknown. In most cases, the genes whose malfunctions are thought to be responsible for tumorigenesis have no obvious relationship to the maintenance of chromosomal integrity.

These cautionary statements serve to remind us that the molecular biology of cancer is still a developing subject, albeit a very active field. Some examples of recent reviews are Knudson, 1985; Levine, 1990; Sager, 1989; Steel, 1990; and Weinberg, 1990. At least three book-length surveys of cancer genetics were published in 1990 (Carney and Sikora, 1990; Cooper, 1990; and Weinberg, 1990).

One of the major generalizations that has emerged in the past decade is that there are two major classes of cancer-causing genes. First, there are *oncogenes*, or *growth-promoting genes*, whose normal activity is necessary for cells to grow and divide. Mutations in oncogenes tend to be phenotypically dominant; that is, a mutant oncogene will have an effect on the growth of a cell, despite the presence of a normal allele.

The second class of cancer-causing genes is a group of *tumor suppressor genes*, sometimes referred to as *anti-oncogenes*. These are genes whose normal function is to prevent cells from multiplying, as is usually necessary for most cells in a fully differentiated organ. In most cases, mutations in tumor suppressor genes behave as genetic recessives, the presence of one normal allele being sufficient for growth control. However, as we saw in Chapter 8, the presence of an abnormal gene product sometimes interferes with the function of a normal gene product, and the possibility that some somatic mutations in tumor suppressor genes behave as phenotypic dominants cannot be excluded.

The Oncogenes: Growth Promoters

The first molecular studies on genes capable of causing cancer were done on viral oncogenes. Although it had been known since Peyton Rous' work in 1911 that a virus could cause cancer in chickens, it was more than half a century later that *v-src*, the gene responsible for the transforming ability of Rous sarcoma virus (RSV), was identified.

In the interim, other viruses were shown to produce tumors in birds or mammals, and for some years there was rampant speculation that most human tumors were caused by viruses. That turned out to be a hasty generalization, but it was not totally wrong. We now know that hepatitis B virus induces carcinoma of the liver, that Epstein-Barr virus is involved in Burkitt's lymphoma, and that HTLV-I and HTLV-II cause some forms of leukemia.

Tumor viruses include both DNA viruses and RNA viruses. Although some of the DNA tumor viruses have been analyzed extensively, their oncogenes have complex functions. The oncogenes of RNA tumor viruses, on the other hand, have obvious relationships to normal cell genes, and the study of RNA viral oncogenes has been very helpful in analyzing normal controls of cell growth. For these and other reasons, we will limit our coverage of DNA tumor viruses to an outline of the various types and will focus our attention on the RNA tumor viruses.

DNA Tumor Viruses

Viruses have been shown to cause tumors in a wide variety of vertebrate organisms, including frogs, reptiles, birds, and mammals. There are six known families of tumor-causing DNA viruses. In order of increasing genome size, they are as follows.

1. Hepatitis B viruses (genome size about 3 kbp) cause hepatic carcinomas in several mammals, including humans. Tumors appear in chronically infected individuals after a latent period of two to four years. The mechanism of tumor induction is unknown.

2. SV40 and polyomavirus (genome size about 5 kbp) have been extensively studied. In their natural hosts (monkeys for SV40 and mice for polyoma), these viruses produce lytic infections only, but they are capable of transforming cultured cells from heterologous hosts (they are not known to cause human tumors). Transformation depends upon the gene for large T antigen in the case of SV40, whereas transformation by polyomavirus requires genes for large T and middle T antigen. Both viruses produce another gene product known as small T antigen, which is not required for transformation. The biochemical activities of the transforming proteins have not yet been fully defined, but they clearly are multifunctional proteins.

3. Papillomaviruses (genome size about 8 kbp) occur in a large variety of types in mammals, including humans. Some cause warts; others are routinely found integrated into the genome of cervical carcinoma cells in humans. Two genes required for transformation of cultured cells have been identified.

4. Adenoviruses (genome size about 35 kbp) also are a diverse group. Like SV40 and polyomavirus, they cause only lytic infections in their natural hosts, but can transform cells or produce tumors in heterologous hosts (not humans, as far as is known). Transformation depends upon expression of the E1A and E1B genes.

5. Herpesviruses (genome size 100–200 kbp) sometimes cause tumors in their natural hosts. In humans, Epstein-Barr virus is of interest, because it appears to be involved in the induction of Burkitt's lymphoma and nasopharyngeal carcinoma.

6. Poxviruses (genome size about 200 kbp) cause benign neoplasms in some mammals (but probably not humans). Unlike all the other tumor-causing viruses, they replicate in the cytoplasm of host cells.

Retroviruses and Their Oncogenes

The retroviruses are a class of RNA viruses that often cause tumors in suitable hosts. Infection of a susceptible cell is followed by synthesis of a double-stranded DNA from the single-stranded genomic RNA (reverse transcription). The DNA integrates at random locations with the host cell genome; it is then referred to as a *provirus*. Proviral genes may be transcribed continuously, leading to concomitant production of new viral particles, without lysis of host cells.

The structure of a typical retrovirus genome is diagrammed in Figure 11–1. There are three genes, but each gene encodes a polyprotein, which is cleaved by a protease (also encoded by one of the viral genes) into several smaller proteins. The *gag* gene encodes the proteins that associate most directly with the RNA of the viral genome to form the nucleoprotein core or capsid; the *pol* gene encodes the reverse transcriptase and integrase activities that are responsible for converting the virion RNA into double-stranded DNA and integrating it at random locations within the host genome; and the *env* gene produces proteins that help form the envelope of the virion, along with components captured from the host cell membrane. The protease is either translated as part of the *gag* protein, or is translated independently between the *gag* and *pol* segments of the RNA.

Retroviral genomes begin and end with long terminal repeats (LTRs, Figure 11–1), which contain sequences that function in the integration of proviral DNA into host cell genomes, as well as transcriptional regulatory sequences. Other control sequences (not shown in Figure 11–1) are a site required for packaging RNA into virions and sites for initiation of DNA synthesis.

Based on their oncogenic potential, retroviruses can be classified in two groups. *Acutely transforming* viruses induce tumors in animals within a few weeks and transform cells in culture efficiently. *Weakly oncogenic* viruses induce tumors in animals only after long latent periods (months or years) and rarely or never transform cells in culture. Rous sarcoma virus (RSV) is the

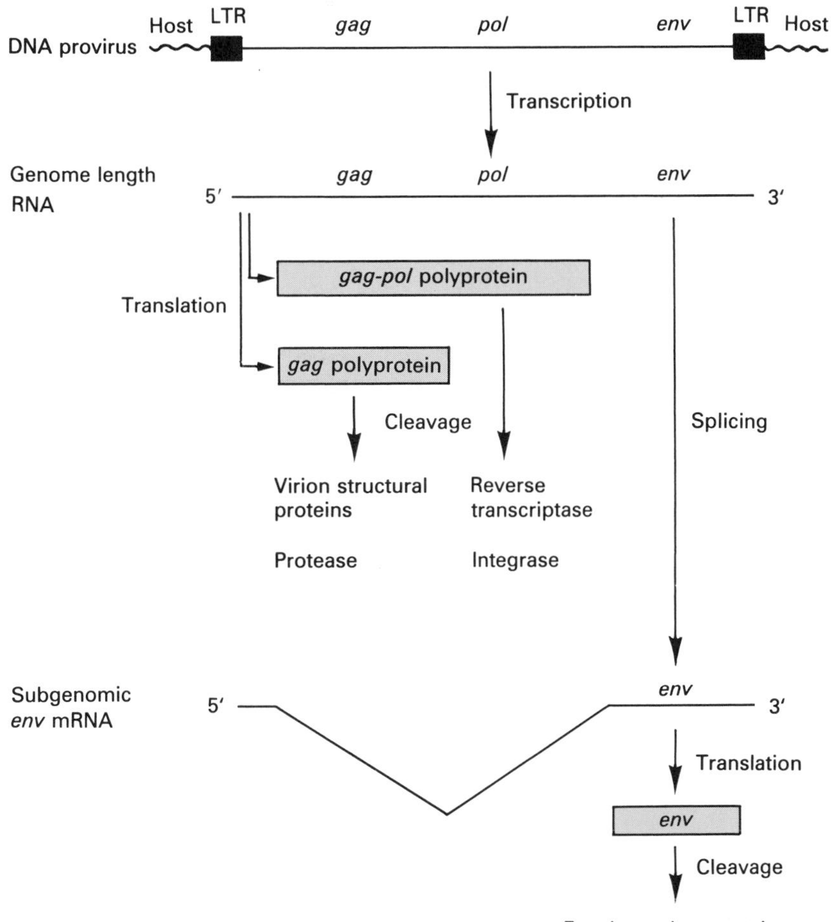

Figure 11–1

Retrovirus genome organization and expression. ALV is shown as an example of a typical nontransforming retrovirus. Proteins are depicted as filled boxes. The DNA provirus, integrated into the host chromosome, is transcribed to yield full length viral RNA. This primary transcript is the genomic RNA for progeny virus particles, as well as serving as the mRNA for the *gag* and *pol* proteins. In addition, the primary transcript is spliced to yield mRNA for *env*. (From Cooper, 1990, p. 41)

prototype acutely transforming retrovirus and avian leukosis virus (ALV) is the prototype weakly oncogenic virus.

Discovery of Retroviral Oncogenes Oncogenes were discovered as a result of efforts to determine the genetic basis of the difference in oncogenic activity of RSV and ALV. An early observation was that the genome size of RSV is about 10 kbp, whereas that of ALV is only about 8.5 kbp. This suggested that RSV contains genetic information that is not present in ALV.

Deletion mutants of RSV, lacking about 1500 bp, were described by Vogt in 1971. These mutants were capable of replicating in appropriate host cells, but they could not transform cells. This result showed that the transforming potential of RSV was not included among the properties of the genes needed for viral replication. Subsequently, the isolation of temperature-sensitive mutants of RSV, which could transform cells at 35 °C but not at 41 °C (although they could replicate at either temperature), led to the conclusion that transformation required the continuous expression of the gene containing the temperature-sensitive mutation.

The preceding results implied that RSV contains an extra gene—a gene not needed for viral replication, but required for transformation of host cells. Because the tumors induced by RSV are sarcomas, the name *src* (pronounced "sark") was given to that gene. Further analysis showed that *src* lies between the end of the *env* gene and the 3′ LTR (Figure 11–2).

Remarkably, RSV is the only known retrovirus that carries a complete set of normal retroviral genes and a separate oncogene; RSV is *replication competent*. All other oncogenic retroviruses are *replication defective*; that is, they require the presence of a helper virus to provide one or more missing functions. In the most common situation, one of the retroviral genes has been partially deleted, and a gene that is necessary for transforming ability has been joined to it. The product of such a gene is a *fusion protein* (Figure 11–3).

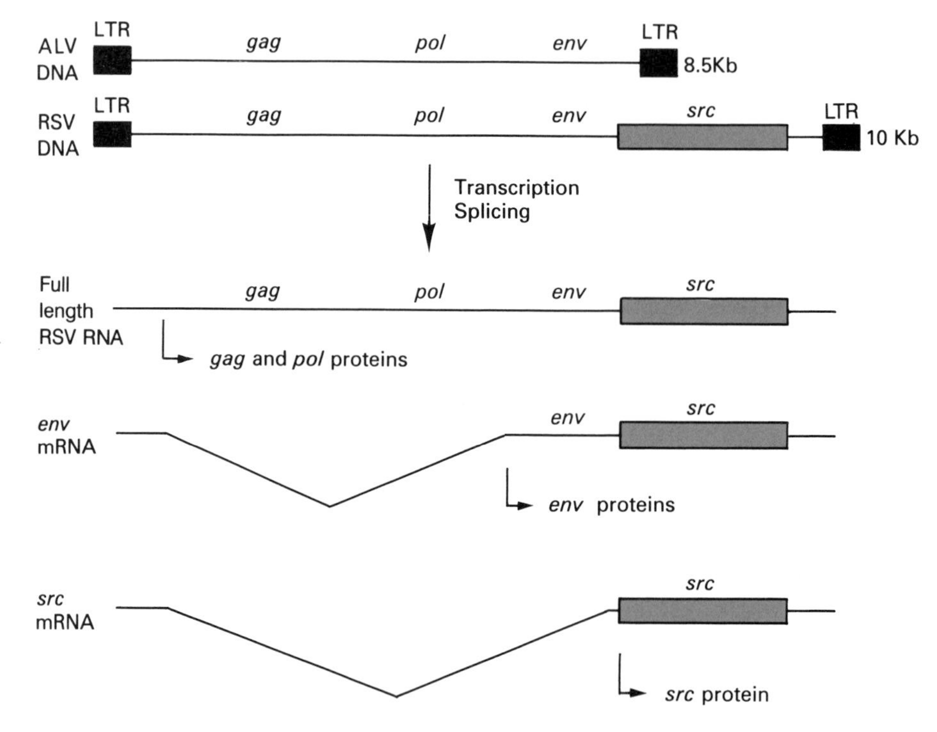

Figure 11–2

Organization and expression of the RSV genome. RSV contains an additional gene, *src*, that is not present in ALV. The RSV provirus is transcribed to yield a full length RNA, which serves as progeny genomic RNA and mRNA for *gag* and *pol*. The primary transcript is also spliced to yield two subgenomic mRNAs, one for *env* and the other for *src*. (From Cooper, 1990, p. 43)

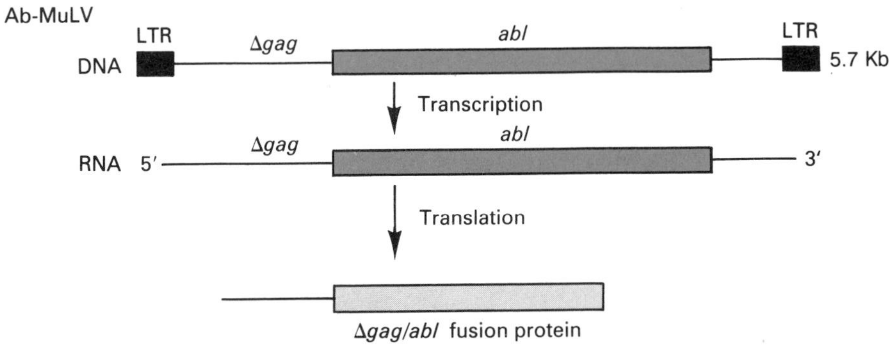

Figure 11–3

Organization and expression of the replication-defective genome of Abelson murine leukemia virus (Ab-MuLV). The *abl* oncogene is expressed as a fusion protein formed by recombination with viral *gag* sequences. (From Cooper, 1990, p. 46)

Origin of Retroviral Oncogenes

The fact that retroviral oncogenes are not required for viral replication raised the question of their origin and the manner in which they became incorporated into retroviruses. Evidently, the host cell genome was the most likely source of non-viral genetic information.

The possibility that a viral oncogene was related to a gene in normal cells was first tested by Varmus and Bishop (Stehelin et al., 1976) with the *src* gene of RSV. Taking advantage of the existence of deletion mutants, they used nucleic acid hybridization to subtract everything except *src* sequences from RSV cDNA, and thus obtained an *src*-specific probe. When that probe was tested with DNA from normal chickens, it hybridized quite well. This showed that normal DNA contains sequences closely related to a retroviral oncogene.

Subsequently, all retroviral oncogenes that have been tested have been shown to have homologs in normal host cell DNA. They have been given the name *proto-oncogenes*. A plausible scenario that explains the origin of retroviral oncogenes is diagrammed in Figure 11–4. The figure shows a retrovirus that happens to have integrated into the host genome adjacent to a proto-oncogene. Either because of a DNA deletion or because of aberrant transcription and RNA splicing, a fusion transcript is produced. If the fusion transcript is packaged into a virion together with a normal retroviral transcript (retrovirus virions typically contain two genomic RNAs), recombination may occur after the virus infects an appropriate host.

If the captured host gene fragment had oncogenic potential, at least when under the control of a viral promoter, then an oncogenic virus would have been created. This virus could multiply if it were accompanied by a replication-competent virus when it infected the next host. Oncogenic activity might also be enhanced as a result of the acquisition of mutations in the oncogene during viral replication.

An alternative hypothesis for creation of a viral oncogene proposes that a cellular mRNA for a proto-oncogene is sometimes packaged into a virion, along with a copy of the normal retroviral RNA. When this virus infects another host cell, recombination between the viral genome and the mRNA could produce an oncogenic retrovirus.

Cellular Oncogenes

The existence of genes in normal cells that are similar to retroviral oncogenes raised the possibility that tumors could arise by alterations in the

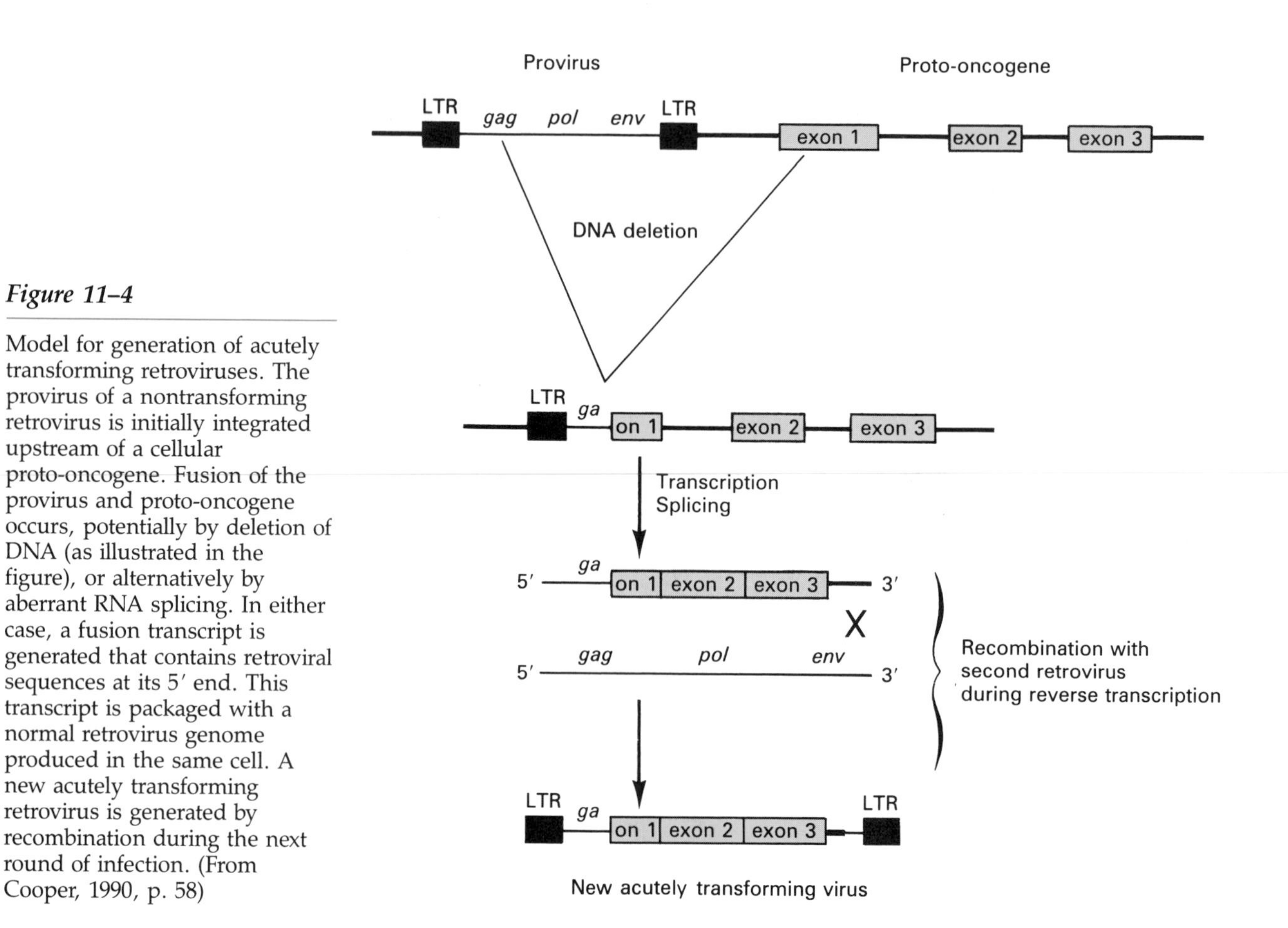

Figure 11–4

Model for generation of acutely transforming retroviruses. The provirus of a nontransforming retrovirus is initially integrated upstream of a cellular proto-oncogene. Fusion of the provirus and proto-oncogene occurs, potentially by deletion of DNA (as illustrated in the figure), or alternatively by aberrant RNA splicing. In either case, a fusion transcript is generated that contains retroviral sequences at its 5′ end. This transcript is packaged with a normal retrovirus genome produced in the same cell. A new acutely transforming retrovirus is generated by recombination during the next round of infection. (From Cooper, 1990, p. 58)

structure or expression of those genes, without the mediation of viruses. It therefore became important to examine human tumors for evidence of alterations in known proto-oncogenes. Several experimental approaches have been taken, and several types of alterations in proto-oncogenes have been discovered which convert the proto-oncogenes into *cellular oncogenes*. In the course of that work, many proto-oncogenes have been discovered that are not related to any known viral oncogene.

Gene Transfer Experiments The identification of retroviral oncogenes and their homologs in normal cells, the proto-oncogenes, was a seminal stage in the history of cancer genetics. Nevertheless, it gradually became clear that there was no reason to attribute the majority of human tumors to viral causation. Therefore, the possibility that most cancers were caused by somatic mutations became an urgent subject for investigation. The problem was how to identify a mutant gene with oncogenic potential among the tens of thousands of genes in the human genome.

The strategy that was adopted relied upon the assumption that an oncogene, if it is present in DNA extracted from a tumor, can be detected by transferring it to nontransformed cells in culture, which should then grow without restraint. The basic technique has been available since the classic experiments of Avery, McLeod, and McCarty in 1944, who demonstrated that DNA extracted from rough strains of *Pneumococcus* bacteria could convert smooth strains to rough strains in an inheritable manner. This was the first direct evidence that DNA is the genetic material. Subsequently, the ability of bacterial cells to take up purified DNA and integrate it into their genomes in such a way that the host cell was "transformed," has been widely exploited for the study of bacterial cell metabolism and for genetic mapping.

Around 1980, several cancer biologists independently began to explore the possibility that the presence of mutant, cancer-causing genes in DNA from tumors could be detected by a similar transformation assay. Not long before that, other investigators had surmounted a technical problem that stood in the way of applying the transformation assay to mammalian cells, which do not take up purified DNA as voraciously as bacteria nor integrate it as readily into their genomes. It was found that uptake of DNA by mammalian cells in culture was increased dramatically if the DNA was enmeshed in tiny crystals of calcium phosphate, which apparently stimulated endocytosis. With this method, uptake, integration, and expression of exogenous genes by mammalian cells took place at a usable frequency.

It thus became possible to ask whether DNA extracted from human cells could transform nontumor cells in culture, causing them to acquire the properties of cancer cells. What are those properties? Tumor cells in culture share two invariant properties with tumor cells in an organism: *immortality* (that is, no limit on the number of cell divisions that can occur if space and nutrients are available) and *uncontrolled growth* (which, in culture, is expressed as an escape from density-dependent inhibition of growth). Cells that exhibit both of those properties are said to be *transformed.*

The reader should note that the word "transformed" has been used with two distinct meanings in the last few paragraphs. In general, a transformed cell is one that expresses a gene acquired from exogenous DNA; this applies to any type of cell, whether it be bacterial, mammalian, or anything else. However, there is a special sense of the word "transformed" that applies to cell culture.

A cell that has acquired the properties of immortality and unrestrained growth, with or without the help of exogenous DNA, is also said to be transformed. The assay for oncogenes creates a cell that is transformed in both senses: it expresses exogenous DNA that has become incorporated into its genome and it grows without regard to its neighbors in culture. To avoid confusion, it has become common practice to use the term *transfection* to refer to the uptake and genomic integration of foreign DNA in mammalian cells, reserving *transformation* for unregulated growth in culture or in organisms into which the cells have been transplanted.

The first successful attempts to identify oncogenes from human tumors via transformation of cultured cells employed NIH 3T3 cells, which are mouse cells that have been immortalized, but which still exhibit contact inhibition; that is, they will form a monolayer in a culture vessel and then stop growing. It was a good choice of experimental systems, because, as it turned out, 3T3 cells need only one additional oncogene to become transformed; and cells that have incorporated such a gene will be easily detected, because they will grow as little foci (piles of cells) on top of the confluent monolayer of growth-arrested cells.

An assay for oncogenes that makes use of the preceding facts is really quite straightforward (Figure 11–5). DNA is extracted from a tumor and applied in the form of a calcium phosphate precipitate to a monolayer of 3T3 cells in a culture dish. Several weeks later, each transformed cell will have produced a visible focus of cells on top of the monolayer. A colony of transformed cells can be isolated and multiplied to any desired level; then DNA can be extracted and used to initiate another cycle of transformation. This is done so that the probability of there being more than one piece of tumor cell DNA integrated into the recipient cells will be very low.

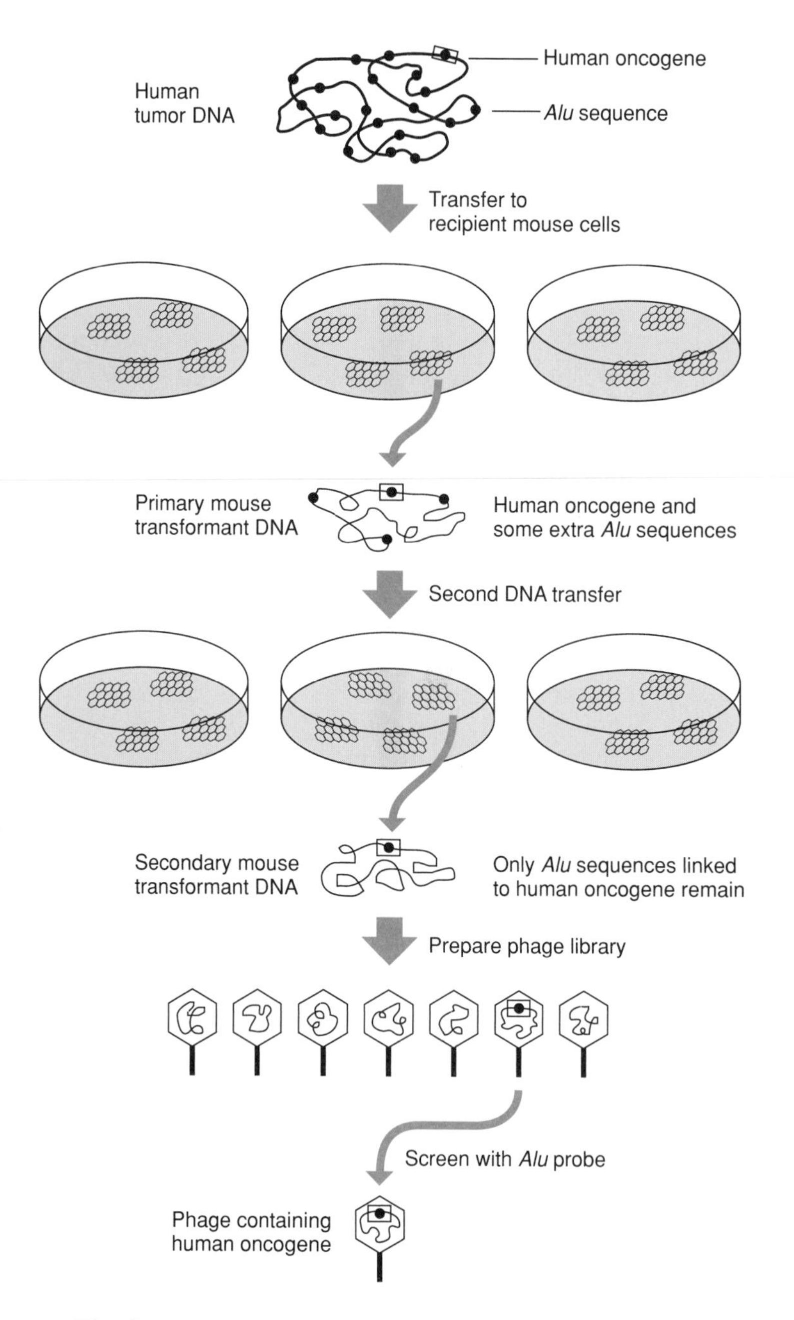

Figure 11–5

Molecular cloning of novel human oncogenes. Human DNA is used to transform mouse cells. The transformed cells are isolated and used as donors of DNA in a second cycle of gene transfer to dilute out human DNA sequences other than those that are closely linked to the human oncogene of interest. A phage library is then prepared from DNA of the secondary transformants and screened by hybridization with a probe for human *Alu* sequences. (From Cooper, 1990, p. 75)

The human gene that was responsible for the transformation of the mouse cells can then be isolated, after DNA from the transformed cells has been used to make a clone library, by identifying the clones that contain human-specific Alu sequences (see Chapter 2); these are so common that most pieces of human DNA of at least 20 kbp will contain at least one Alu sequence.

The first cellular oncogene isolated by the gene transfer strategy was derived from a human bladder carcinoma cell line called EJ. Subsequently, it became apparent that this oncogene was related to the *ras*H oncogene of Harvey sarcoma virus. Another cellular oncogene, initially isolated from a human lung carcinoma, proved to be homologous to *ras*K, from Kirsten sarcoma virus. A third member of the cellular *ras* gene family is *ras*N, so named because it was first isolated from a neuroblastoma.

Many such experiments have been done, using DNA from a variety of spontaneous tumors, such as carcinomas of the bladder, colon, and lung, as well as sarcomas, neuroblastomas, and leukemias. DNA from tumors produced by carcinogens in the laboratory has also been used. The general conclusion is that many tumors, regardless of origin within the body, have alterations in their DNA that will cause transformation of suitable fibroblasts in culture.

However, only about 20% of human tumors contain DNA that will transform 3T3 cells. This is believed to be a limitation of the assay, rather than evidence that no oncogene is present. The 3T3 cell transformation assay only detects mutant genes that are phenotypically dominant and are able to complement the effects of an endogenous mutation that has made the cells immortal. It turns out that the 3T3 assay is most effective in detecting the presence of *ras* oncogenes and their relatives. The function of *ras* proto-oncogenes and the mutations that convert them to cellular oncogenes will be discussed in a later section. Other naturally occurring oncogenes that have been identified by gene transfer include *neu*, *met*, and *trk*.

Another group of oncogenes has been identified as the result of DNA rearrangements during gene transfer. In the standard NIH 3T3 cell assay, DNA from normal cells does not cause efficient transformation; that is, DNA from tumor cells transforms 3T3 cells at least 100X more frequently than normal cell DNA. However, transformation by normal DNA is not zero. Investigation of the genes responsible for this initially surprising result revealed that they are oncogenes created during the process of gene transfer, by recombination of DNA fragments within a recipient cell.

Figure 11–6 diagrams two types of recombination that may convert a proto-oncogene into an oncogene. At least 10 human oncogenes have been identified as a result of DNA rearrangements during gene transfer into cultured cells. The normal functions of the corresponding proto-oncogenes are only partially understood.

Oncogenes Derived by Insertional Mutagenesis Integration of proviral DNA from retroviruses occurs at random sites throughout host cell genomes. Occasionally an oncogene is activated by retroviral integration, a

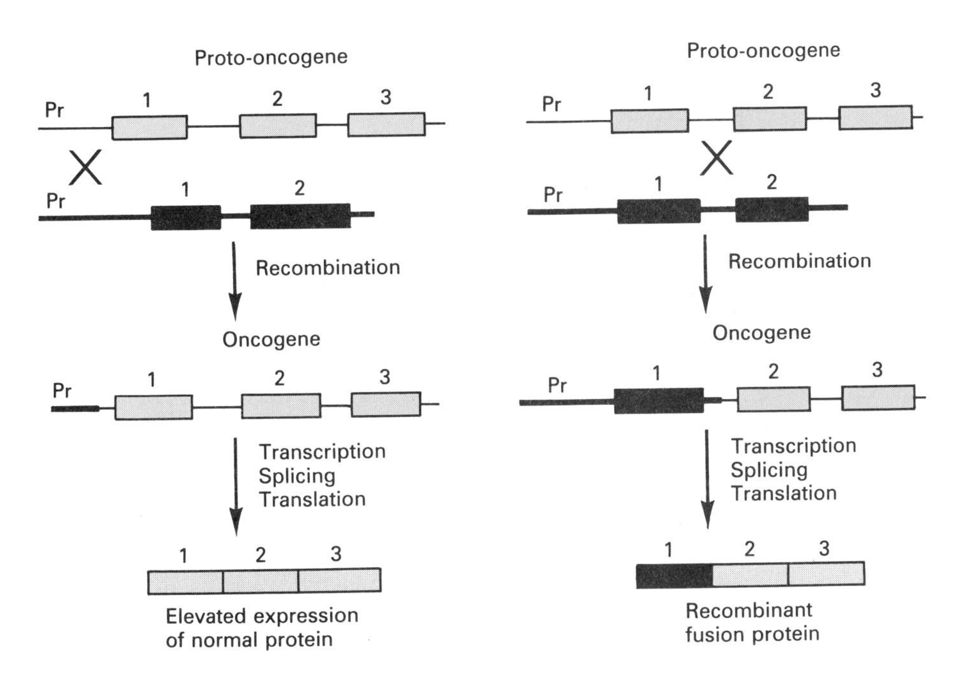

Figure 11–6

Oncogene activation by DNA rearrangements. Proto-oncogenes can become activated oncogenes by recombination with other DNA sequences, either during neoplasm development or during the process of gene transfer. In some cases (left side), these recombination events result only in the replacement of the regulatory sequences (indicated by Pr in the figure) of the proto-oncogene with regulatory sequences from another gene. Oncogene activation is then the result of abnormal expression of the normal proto-oncogene product. In other cases (right side), recombination events occur within coding regions, leading to the formation of recombinant fusion proteins. (From Cooper, 1990, p. 77)

process known as *insertional mutagenesis*. This appears to be the mechanism for tumor induction by weakly oncogenic transforming viruses. In the case of ALV, more than 80% of lymphomas have activated *myc* oncogenes. One way in which the ALV provirus can convert the *myc* proto-oncogene to c-*myc* is shown in Figure 11–7.

At least 15 oncogenes that can be activated by retroviral DNA integration have been found. Most of them have been studied primarily in mice, but the relevance of insertional mutagenesis as a potential cause of human tumors is obvious. It is also a factor that adds uncertainty to gene therapy protocols, in which retroviral vectors are used (Chapter 10).

Oncogenes Created by Chromosomal Translocations Chromosome instability is one of the most conspicuous characteristics of neoplastic cells. In most tumors, the karyotypic abnormalities are not predictable and they may be late consequences of transformation. However, some tumors appear to be the result of specific chromosomal rearrangements.

The original example involved chronic myelogenous leukemia (CML) and the "Philadelphia chromosome," a short version of chromosome 22 that is actually the product of a reciprocal translocation between small pieces at the ends of the long arms of chromosomes 9 and 22 (Figure 11–8). At least 90% of CML patients have the Philadelphia chromosome in their leukocytes (Rowley, 1973). Molecular analysis showed that most of the proto-oncogene *abl* from 9q is joined to a gene called *bcr* ("breakpoint cluster region") on 22q. The fusion protein expressed by this compound gene is a tyrosine kinase with much greater activity than normal Abl gene products. (Tyrosine kinases are discussed later in this chapter.)

Similarly, an 8:2, 8:14, or 8:22 reciprocal translocation is found in many Burkitt's lymphomas and plasmacytomas. The *myc* gene from chromosome 8 is modified by the loss of exon 1, which contains translational regulatory sequences, and the remainder of the gene is attached to part of an immunoglobulin gene (Figure 11–9). The immunoglobulin genes involved are the kappa light chain gene in the case of chromosome 2, a heavy chain gene

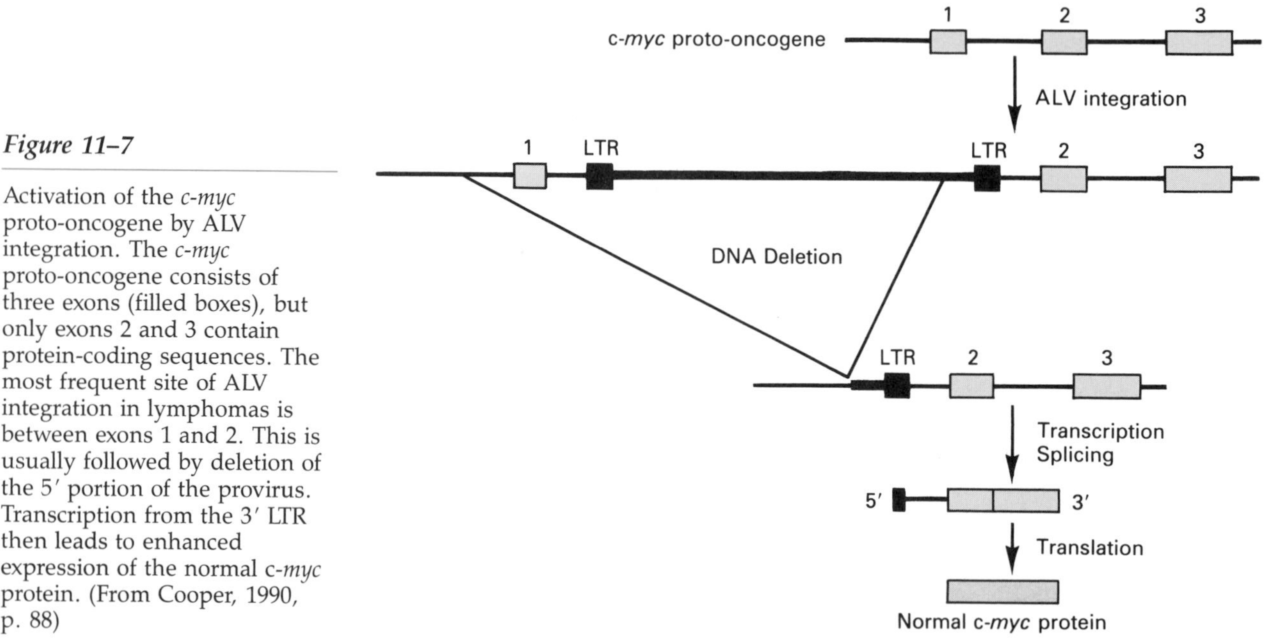

Figure 11–7

Activation of the *c-myc* proto-oncogene by ALV integration. The *c-myc* proto-oncogene consists of three exons (filled boxes), but only exons 2 and 3 contain protein-coding sequences. The most frequent site of ALV integration in lymphomas is between exons 1 and 2. This is usually followed by deletion of the 5′ portion of the provirus. Transcription from the 3′ LTR then leads to enhanced expression of the normal c-*myc* protein. (From Cooper, 1990, p. 88)

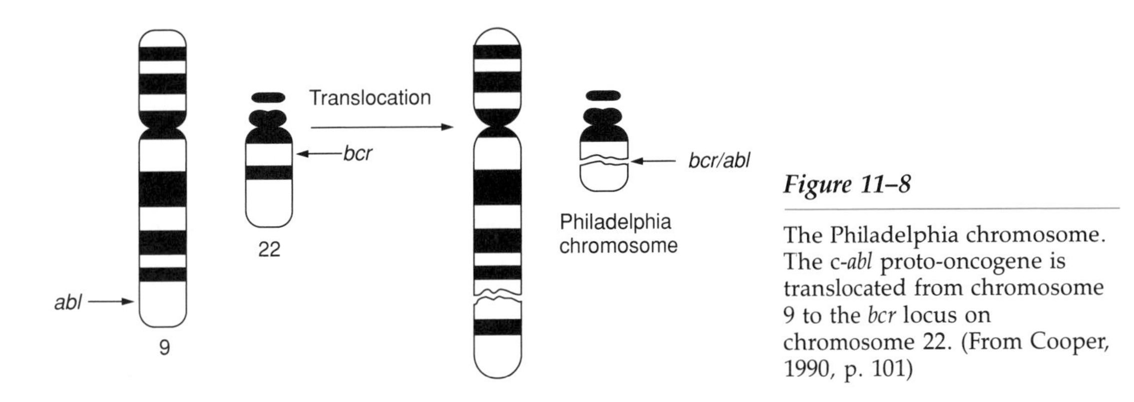

Figure 11–8

The Philadelphia chromosome. The c-*abl* proto-oncogene is translocated from chromosome 9 to the *bcr* locus on chromosome 22. (From Cooper, 1990, p. 101)

in the case of chromosome 14, and the lambda light chain gene in the case of chromosome 22 (see Chapter 12).

Although the details are not yet entirely clear, it appears that transcriptional control elements from the immunoglobulin gene lead to the synthesis of an unusually large amount of *myc* transcript. This, plus the fact that the absence of exon 1 eliminates translational control, results in synthesis of an exceptionally large amount of normal *myc* protein. The manner in which excess *myc* protein leads to unregulated growth is not known, but *myc* protein is a nuclear component and it is synthesized in response to growth-factor stimulation of quiescent normal cells in culture.

Oncogenes and Gene Amplification It is often observed that a tumor gradually becomes insensitive to a chemotherapeutic agent that was initially effective. Molecular analysis has shown that drug resistance is often the result of *gene amplification*. The classic example involves resistance to methotrexate, an inhibitor of dihydrofolate reductase (an enzyme essential for nucleotide synthesis). Tumors that become resistant to methotrexate can

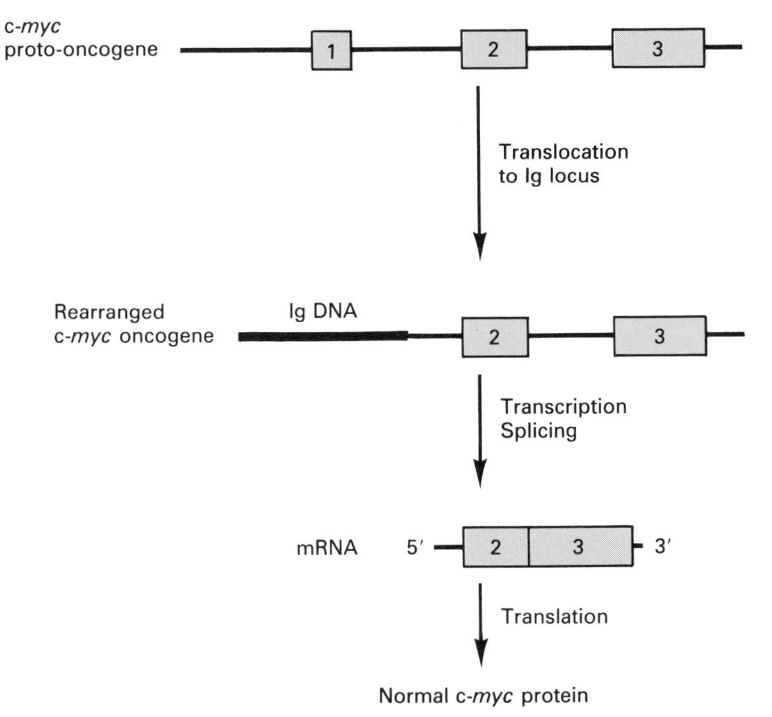

Figure 11–9

Activation of c-*myc* by translocation to an immunoglobulin locus in mouse plasmacytomas. Translocation frequently results in deletion of noncoding exon 1 and flanking sequences. Transcription of the rearranged gene is initiated from a cryptic promoter in the first intron, probably in response to enhancers of transcription in the immunoglobulin gene. This leads to constitutive, high-level expression of the normal C-*MYC* protein. (From Cooper, 1990, p. 100)

usually be shown to have multiple copies of the dihydrofolate reductase gene—sometimes as many as several hundred copies per cell.

Amplified genes are usually organized in tandem arrays, which may be detected intrachromosomally as *homogeneous staining regions* or as *double minute chromosomes* (Figure 11–10). The latter are independent minichromosomes that lack centromeres. They are distributed randomly at mitosis, and they are maintained in a population of cells only if they confer a growth advantage to those cells that possess them.

More than a dozen oncogenes have been found to be amplified in human tumors. Among the more frequently encountered amplified genes are the three members of the *myc* family, any one of which is often amplified in small cell lung carcinoma. Other examples include *erb*B, which tends to be amplified in glioblastomas and squamous cell carcinomas, and *erb*B-2, which is amplified in breast and ovarian carcinomas.

In most cases of amplified oncogenes, it is the proto-oncogene that is

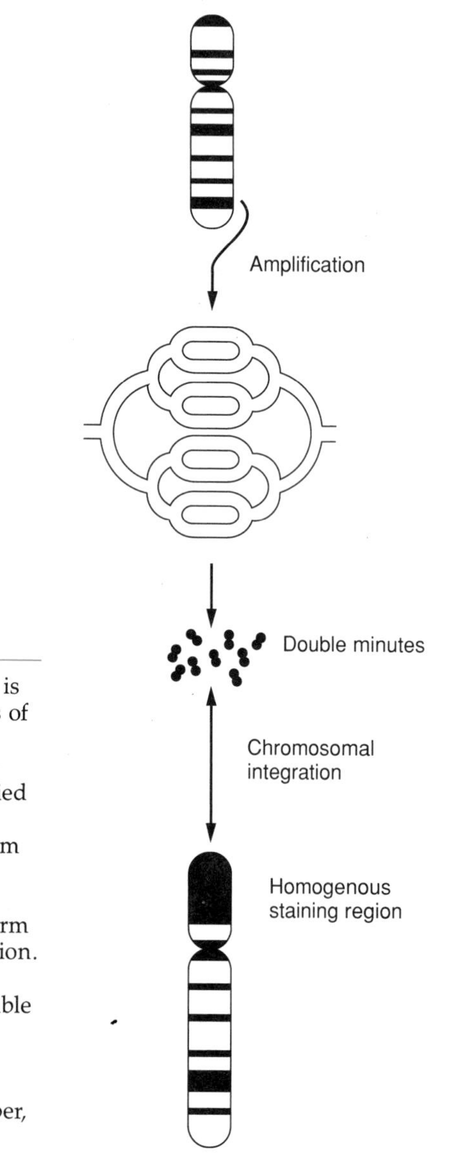

Figure 11–10

DNA amplification. A locus is amplified by repeated cycles of local DNA replication. Recombination can generate tandem arrays of the amplified DNA, which can be excised from the chromosome to form double minutes. Double minutes can integrate into another chromosome and form a homogeneous staining region. Because chromosome integration is reversible, double minutes and homogeneous staining regions are interchangeable forms of amplified DNA. (From Cooper, 1990, p. 110)

amplified, and the normal protein is produced. Presumably, an excessive amount of certain proteins deregulates growth control mechanisms, resulting in unrestrained cell growth. In some tumors, an amplified oncogene is found which has also undergone a mutation that may enhance its transforming potential. There is considerable uncertainty as to whether gene amplification is usually a primary or secondary event in tumorigenesis.

Tumor Suppressor Genes

In the preceding section, we discussed oncogenes that act in a dominant negative manner; that is, when introduced into a suitable recipient cell, they cause that cell to become transformed, to exhibit unconstrained growth in culture, despite the presence of two copies of the corresponding normal proto-oncogene. We shall now consider a different class of oncogenes—those whose normal function is to inhibit cell division. When both copies of one of these genes are lost or inactivated in a given cell, unrestrained growth ensues and a tumor may develop. These genes have been called anti-oncogenes, recessive oncogenes, tumor suppressor genes, and several other variations on the same theme.

The first evidence for the existence of genes that could suppress unrestrained cellular growth came from studies on cells that were hybrids between a tumor cell and a normal cell. Most of the combinations had normal growth control (Harris et al., 1969; Harris, 1988). Moreover, in the case of certain hybrid cells, the loss of specific chromosomes correlated with reversion to the transformed phenotype. Those observations implied that one way in which a tumor cell might arise would be the loss of normal function of a growth suppressor gene. Molecular studies have now confirmed that hypothesis.

Retinoblastoma

The best-known example of a tumor suppressor gene is the retinoblastoma gene, *RB*, which has a very interesting history. Retinoblastoma has long been known as a rare form of inheritable cancer, found in about one child in 20,000. Tumors in one or both eyes develop in the retina within the first few years of life, and if untreated, spread along the optic tract to the brain and become lethal. Currently, a tumor detected in its early stages can easily be removed or destroyed with a laser, with complete cure ensuing.

There are both inherited and sporadic (non-inherited) forms of retinoblastoma, with the latter representing somewhat more than half of the total cases. In inherited retinoblastoma, the disease is passed from generation to generation as an apparent autosomal dominant, with about 90% penetrance (see Chapter 8).

Children with inherited retinoblastoma usually develop more than one tumor in each eye, and the tumors appear earlier than in sporadic cases, where there is generally only one tumor in one eye. These and other data on the pattern of inheritance led Knudson in 1971 to suggest that two mutations are necessary for the appearance of a retinoblastoma. At that time, it was not possible to predict whether the two mutations would be in different genes or in the two alleles of one gene. We now know the latter to be true.

According to the Knudson hypothesis, inherited retinoblastoma occurs in individuals who have received one copy of a mutant *RB* gene from a parental germ line. A second mutation occurs in one or more retinoblasts during early childhood, as the retina develops its definitive structure. Individuals with sporadic retinoblastoma must have experienced two separate somatic cell mutations in a retinoblast or its precursors (Figure 11–11).

The first clue to the location of the *RB* gene came in 1978, when Yunis and Ramsay noticed that cells from a small percentage of retinoblastomas had a cytologically visible deletion in band 14 of the long arm of chromosome 13. Subsequently, it was observed that in some patients, the deletion was homozygous in the tumor, although it was heterozygous in the patient's normal tissues. Comparable observations were then made on ester-

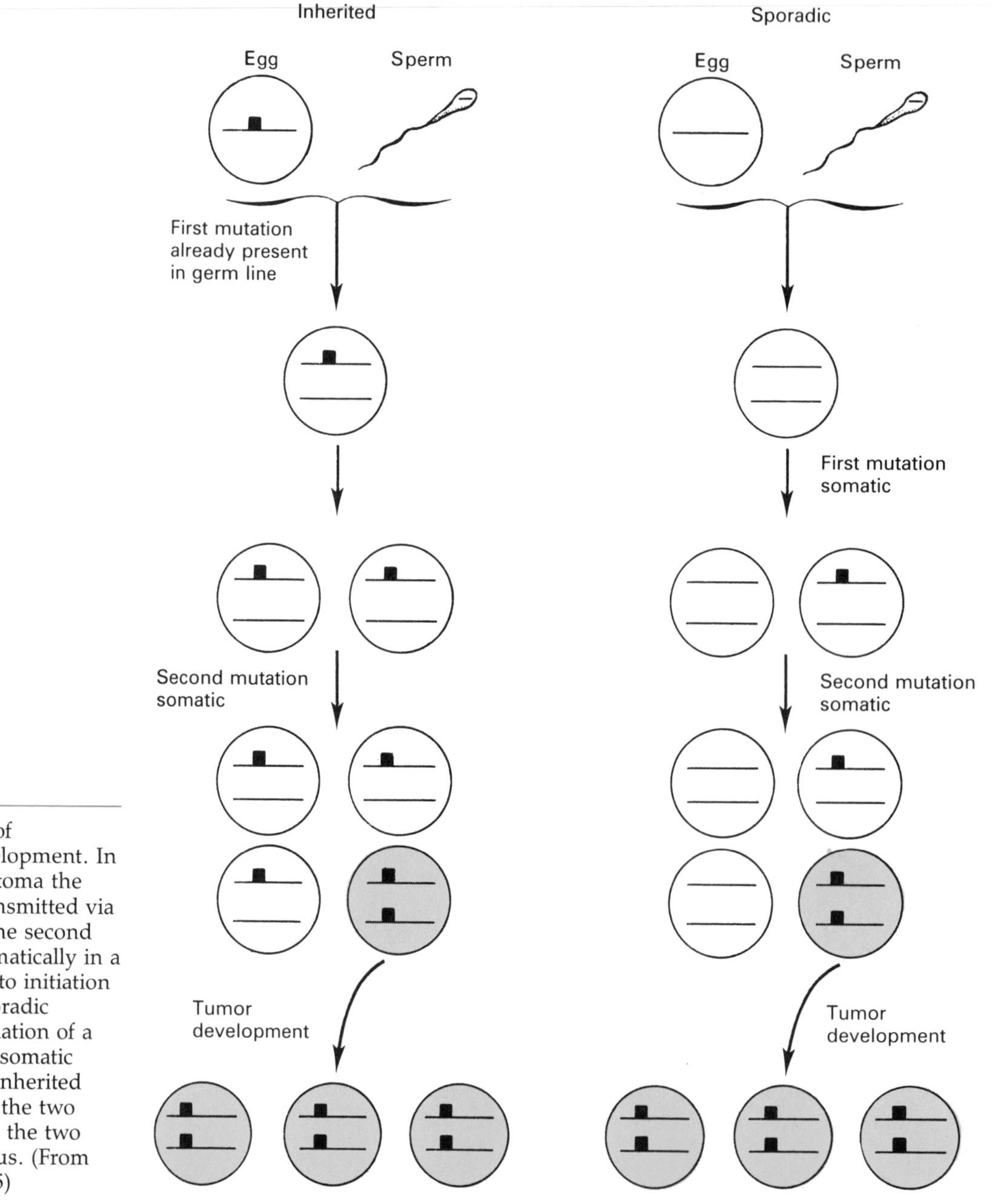

Figure 11–11

The two-hit model of retinoblastoma development. In inherited retinoblastoma the first mutation is transmitted via the germ line and the second mutation occurs somatically in a retinal cell, leading to initiation of the tumor. In sporadic retinoblastoma, initiation of a tumor requires two somatic mutations. In both inherited and sporadic cases, the two mutations inactivate the two alleles at the RB locus. (From Cooper, 1990, p. 125)

ase D, a protein encoded by a gene closely linked to *RB*, which also mapped to 13q14.

This information led to the hypothesis that both copies of the *RB* gene must be missing in order for a tumor to form. A child born with one mutant *RB* allele, such as the interstitial deletion at 13q14, need only lose the wild-type allele from one retinoblast during development of the retina, in order for a tumor to form (Figure 11–11). Inasmuch as there are several million retinal cells in each eye, the probability that at least one such loss would occur approached 100%, thus accounting for the high penetrance of the disease.

Several potential mechanisms for loss of the normal allele from a retinoblast are illustrated in Figure 11–12. The simplest is nondisjunction, the loss of the chromosome 13 that contains the normal allele at anaphase, resulting in one daughter cell that is trisomic for chromosome 13 and one that is monosomic, with the mutated allele. In the latter case, there would be no remaining functional *RB* allele in that cell, and it would initiate a tumor. It has also been suggested that duplication of the remaining chromosome 13 may occur after loss of the other chromosome 13, although a mechanism by which this might take place has not been documented.

Two other mechanisms that could lead to a cell with two nonfunctional *RB* alleles are mitotic recombination (Figure 11–13) and an independent second mutation.

In contrast, it is presumed that a child with sporadic retinoblastoma must have acquired two somatic mutations, one in each *RB* allele. This seems to require an exceptionally high mutation rate at the *RB* locus, given that there are only about 8 to 10 million retinal cells in one person, so it may well be that molecular analysis of DNA from patients with sporadic retinoblastoma will reveal some new mechanisms.

One possibility is that inactivation of one allele by somatic mutation early in retinal development is followed by mitotic recombination, yielding one cell with two copies of the mutant allele. Another possibility is that dominant negative mutations (where the mutant gene product interferes with the function of the normal gene product) sometimes occur. These would require only one mutational event to produce the tumor.

The need to abrogate the activity of both *RB* alleles in one retinal cell in

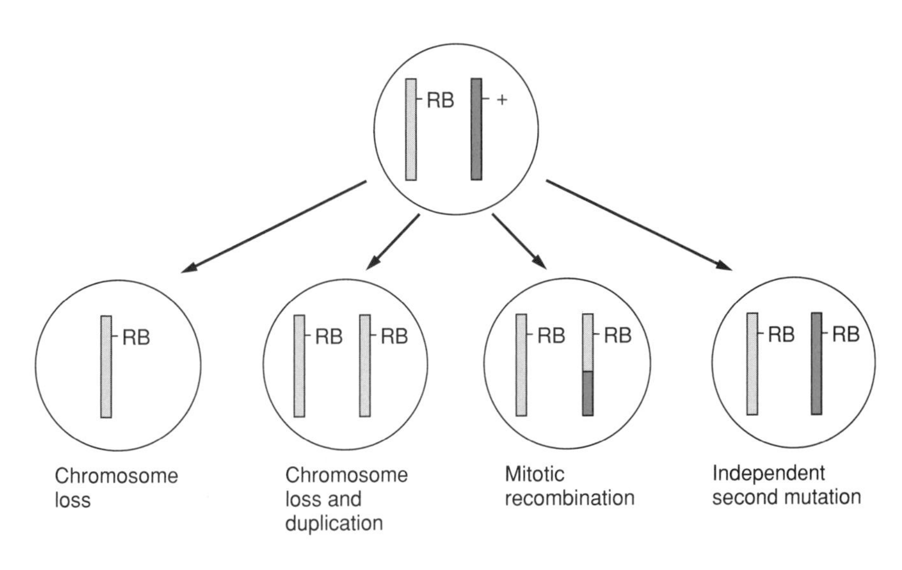

Figure 11–12

Mechanisms for inactivating the second copy of the retinoblastoma susceptibility gene. The first allele (designated RB) has been inactivated by a recessive mutation. Phenotypic expression of this mutation requires inactivation of the remaining wild-type allele (designated +). This can occur by chromosome loss with or without duplication of the remaining chromosome 13, by mitotic recombination, or by an independent second mutation. (From Cooper, 1990, p. 128)

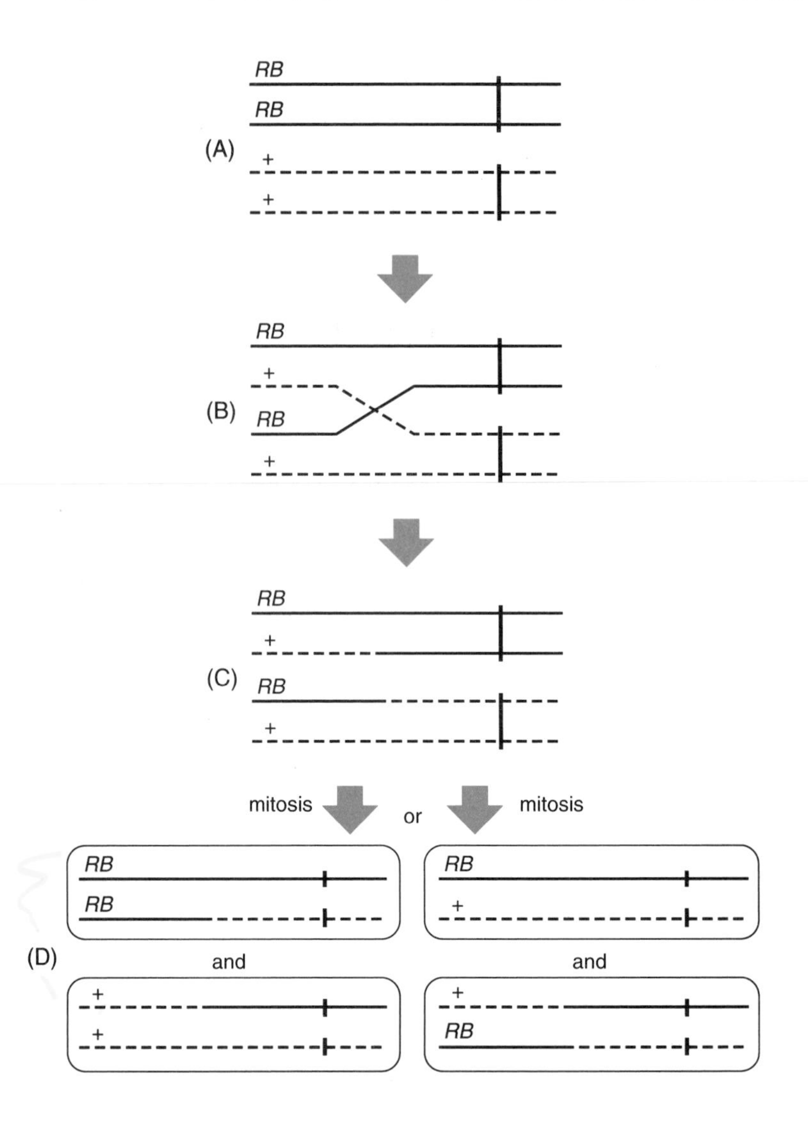

Figure 11–13

Mitotic recombination, which occurs in somatic cells at 0.1% or less of the frequency with which it occurs in meiotic cells, can lead to homozygosity at loci that were initially heterozygous. Panel (A) indicates a somatic cell in which the chromatids have been duplicated, prior to mitosis (centromeres are indicated by dark bars). One chromosome has a recessive mutation at the retinoblastoma locus (*RB*); the other chromosome is wild-type (+). A crossover between the retinoblastoma locus and the centromere (panel (B)) leads to the chromatids in panel (C). At metaphase, two orientations of the chromosomes are possible, which lead to the daughter cells shown in panel (D). On the left, one daughter cell is homozygous for the *RB* allele (and therefore, may produce a tumor) and the other daughter cell is homozygous wild-type. On the right, both daughter cells are heterozygous at the retinoblastoma locus (and therefore, phenotypically normal)

order for a tumor to develop presents us with a semantic conundrum: retinoblastoma is inherited as a Mendelian dominant condition at the organismal level, but, so it is claimed, the condition is recessive at the cellular level. Such statements refer to different aspects of genetics. Inheritance of the disease is a property of the organism and retinoblastoma is a dominant genetic disease; it does not require one mutant allele from each parent. Expression of the transformed phenotype is recessive at the cellular level; two mutant alleles are required in one cell in order for it to become a tumor cell. Retinoblasts do not develop retinoblastoma; when they are transformed, they *are* the disease.

Before leaving retinoblastoma, it is appropriate to point out some of the remaining mysteries. First, there is the indication that the only gene that needs to be altered in order to produce a retinal tumor in childhood is the *RB* gene. In fact, loss of both normal alleles at the *RB* locus may only be the initiating event. We don't yet know whether malignant retinoblastomas reproducibly have mutations in other genes, acquired during tumor progression.

Second, there is a problem involving the role of the *RB* gene in tumorigenesis in various tissues. The gene appears to be active almost ubiquitously, but mutations in *RB* are found mostly in retinoblastomas, osteosarcomas, small cell carcinomas of the lung, and bladder carcinomas. Is its role in growth control different, or perhaps redundant, in other tissues?

The polypeptide encoded by the *RB* gene is described in a latter section of this chapter.

Other Tumor Suppressor Genes

Wilms Tumor (WT) *Wilms tumor* is another childhood cancer, inherited as an autosomal dominant. It develops in one or both kidneys in about one in 10,000 children. Although there are probably two or three genes that can cause the tumor, the major locus is at 11p13, which was recognized because a small percentage of WT patients have a deletion there. Children with the deletion tend to have additional defects: aniridia, genitourinary abnormalities, and mental retardation (the WAGR syndrome), so several genes are probably missing.

Wilms tumor, like retinoblastoma, apparently develops in cells that have lost both copies of the *WT* gene. The evidence for this comes from studies on RFLPs, which have shown several cases in which the patient's tumor was found to be homozygous for RFLPs in the vicinity of 11p13, whereas the somatic cells were heterozygous.

The *WT1* gene encodes a 345-amino acid protein that appears to be a transcription factor (Call et al., 1990; Gessler et al., 1990). It has four zinc finger domains, which indicates sequence-specific DNA-binding capability. Indeed, both WT1 and a related protein, EGR1, bind strongly to CGCCCC-CGC. EGR1 is a transcriptional activator at promoters with that sequence, whereas if WT1 is bound to CGCCCCCGC, activation by EGR1 is prevented. Evidently, inactivation of both *WT1* alleles might allow unregulated expression of EGR1-activated genes in cells where WT1 protein is the normal transcriptional repressor. Understanding the effects of that change on cell growth will require identification of EGR1-activated genes and their functions.

The p53 Gene A nuclear phosphoprotein known as p53 is encoded by a gene that has recently been implicated as a potential anti-oncogene. It forms a stable complex with SV40 large T and with adenovirus E1B protein, (another oncoprotein); it also cooperates with the *ras* gene product to achieve full transformation of primary cells in culture. Although it can be argued that *p53* is therefore an oncogene, the fact that mutations throughout the gene give it immortalizing ability suggests that the effect of the mutations is actually to inactivate the normal function of the protein, rather than to acquire a specific abnormal function. Moreover, the p53 gene product is a polypeptide that oligomerizes; therefore, a mutant polypeptide may block the function of the normal polypeptide by inactivating the entire oligomer. Protein p53 may be a transcriptional regulator; it is a sequence-specific DNA-binding protein (Kern et al., 1991)

Loss of the region on chromosome 17p in which the *p53* gene resides has been observed in a variety of tumors, including those of the colon, lung, esophagus, breast, and other tissues. Many different mutations in the *p53* gene have been identified, and some correlations of specific types of mutations with tumors in specific tissues have been observed; for example, G:C and T:A transversions are the most frequent changes noted in cancers of the lung and liver, whereas transitions are more common in tumors of the colon and brain (Hollstein et al., 1991). It is hoped that these patterns of mutation can eventually be interpreted in terms of cell-specific causation.

Germ-line mutations in *p53* are responsible for the very rare Li-Fraumeni syndrome, which is characterized by high susceptibility to several malignancies, most commonly breast cancer (Malkin et al., 1990).

Studies on noninherited colorectal tumors have shown that the loss or inactivation of both wild-type alleles at the *p53* locus is often correlated with progression from the adenoma stage, which is not malignant, to carcinoma.

Other Colon Cancer Genes Approximately one person in 10,000 in American, British, and Japanese populations is affected with *familial adenomatous polyposis* (FAP), an autosomal dominant condition that first leads to the formation of hundreds of benign colorectal polyps, some which eventually progress to full malignancy. The major gene responsible for FAP, which was mapped by linkage analysis to chromosome 5q21, has now been identified (Kinzler et al., 1991). It is called *APC* (adenomatous polyposis coli). It encodes a very large protein (2843 amino acids), and the amino acid sequence in the N-terminal 25% of the protein indicates that it can form a coiled-coil, a structure that suggests the possibility of interactions with other proteins in multi-molecular assemblages.

Mutations in *APC* have been found, not only in tumors from sporadic colorectal cancer patients, but also in normal somatic cells of FAP patients (Nishisho et al., 1991). Thus, a mutant allele at the *APC* locus confers an inheritable predisposition to develop colorectal cancer. There are indications that the other allele of *APC* does not have to be inactivated by a second mutation in order for a tumor to form. Instead, mutations in several other genes appear to be necessary, including the *p53* gene described in the preceding section. There is also an *MCC* gene, located very close to *APC*, mutations in which are routinely associated with colon cancer. Germ-line mutations in *MCC* have not been found in FAP patients. Another candidate tumor suppressor gene, termed *DCC*, has been mapped to a region of chromosome 18q, which is lost in more than 70% of colorectal carcinomas (Fearon and Vogelstein, 1990).

Neurofibromatosis *Neurofibromatosis Type I* (NF1) may also be caused by malfunctions in a growth suppressor gene, although that hypothesis remains unproven at the moment. NF1 is a dominant genetic disorder that occurs in about one person in 3500 in all human races. It is characterized by the appearance of numerous cafe-au-lait spots and benign tumors on the skin known as neurofibromas. The latter are sometimes large enough to be disfiguring. Occasionally, malignant sarcomas develop. Some NF1 patients also have learning disorders, optic gliomas, or renal hypertension.

Roughly 50% of all NF1 patients have unaffected parents and are therefore presumed to be new mutants. This implies a mutation rate on the order of 1×10^{-4}, which is among the highest known for human genes. It raises the possibility that the *NF1* gene may be exceptionally large, like the Duchenne muscular dystrophy gene. Recent analysis at the molecular level supports that possibility.

The *NF1* gene was mapped to 17q11.2, using a set of translocations that were found in NF1 patients. Two groups cloned a candidate gene and published their results simultaneously in July of 1990 (Wallace et al.; Cawthon et al.). The gene has not yet been fully characterized, but it encodes a polypeptide of at least 2485 amino acids and consists of a series of small, widely spaced exons.

A variety of other associations between loss of a specific chromosomal region and certain malignancies are being reported. There may be a large set of growth-suppressing genes, most of which are tissue-specific and/or function only at specific stages of development.

Functions of Proto-Oncogenes and Oncogenes

In the preceding sections, we described the discovery of oncogenes and tumor suppressor genes and we discussed the general ways in which alterations in the expression of those genes can lead to neoplastic growth. In this section, we shall consider some specific biochemical properties of those genes and their functions in normal cells. The general theme underlying these examples is that proto-oncogenes and tumor suppressor genes all play controlling roles in the growth response of various cell types to external stimuli.

Protein-Tyrosine Kinases

The first oncogene to be described was the *src* oncogene of Rous sarcoma virus. It was also the first oncogene whose primary biochemical function was elucidated. The *src* gene product is a protein kinase (Collett and Erikson, 1978); that is, it transfers phosphates from ATP to amino acid residues in proteins. Unlike most protein kinases, which phosphorylate serine or threonine residues, the *src* kinase phosphorylates tyrosine residues.

There are now more than twenty known oncogenes that are protein kinases; collectively, they constitute the largest functional class of oncogenes. They can be classified into two major groups; growth factor receptors and nonreceptor kinases.

The prototype of the growth factor receptor class of oncogenes is *erb*B, whose proto-oncogene encodes the epidermal growth factor (EGF) receptor. Like several other growth factor receptors, the normal EGF receptor consists of an extracellular ligand-binding domain, a short transmembrane domain, and an intracellular domain that contains the kinase activity. Normally, the receptor does not act as a kinase unless EGF is bound to the extracellular domain, but when the protein loses the extracellular domain, as in the case of the *erb*B oncogene, it becomes constitutively active (Figure 11–14).

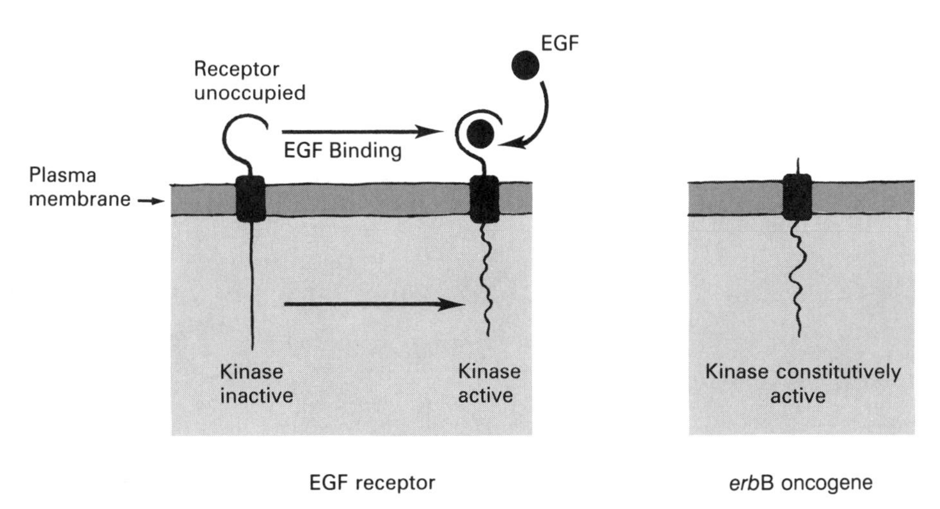

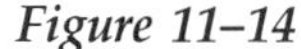

Figure 11–14

Mechanism of *erb*B oncogene activation. The tyrosine kinase activity of the EGF receptor (encoded by the *erb*B proto-oncogene) is controlled by EGF binding. In contrast, the kinase encoded by the *erb*B oncogene is constitutively active. (From Cooper, 1990, p. 183)

At least six other growth factor receptors with similar organization are known, but as yet, only one of them has been shown to be produced by a proto-oncogene. The gene that encodes the receptor for macrophage colony-stimulating factor 1 (CSF-1) can become the *fms* oncogene. There are also at least seven oncogenes that are protein-tyrosine kinases with amino acid sequences similar to the known growth factor receptors, for which the corresponding proto-oncogenes have not been identified.

The second group of protein-tyrosine kinases, the nonreceptor kinases, is encoded by *src* and half a dozen of its close relatives, plus a few other oncogenes. Considerable information about the regulation of SRC kinase activity is available. As shown in Figure 11–15, two sites of phosphorylation are important. Phosphorylation of tyrosine 416 is necessary for kinase activity, but phosphorylation of tyrosine 527 inhibits kinase activity. Tyrosine 416 can be autophosphorylated, but tyrosine 527 phosphorylation is mediated by another enzyme.

Several proteins that are substrates of the SRC kinase have been identified, but so far, none of them is clearly implicated in the control of cell division. In general, the pathways between activation of protein-tyrosine kinases and the commitment of a cell to undergo mitosis are still obscure.

Protein-Serine/Threonine Kinases

Most protein kinases do not phosphorylate tyrosine; instead, they phosphorylate serine and/or threonine residues. Several oncogenes encode protein Ser/Thr kinases. Among them are the three members of the *raf* family, which were first identified in chicken and mouse retroviruses. All three genes code for proteins of somewhat more than 600 amino acids. The N-terminal half of the molecule appears to have regulatory functions. When

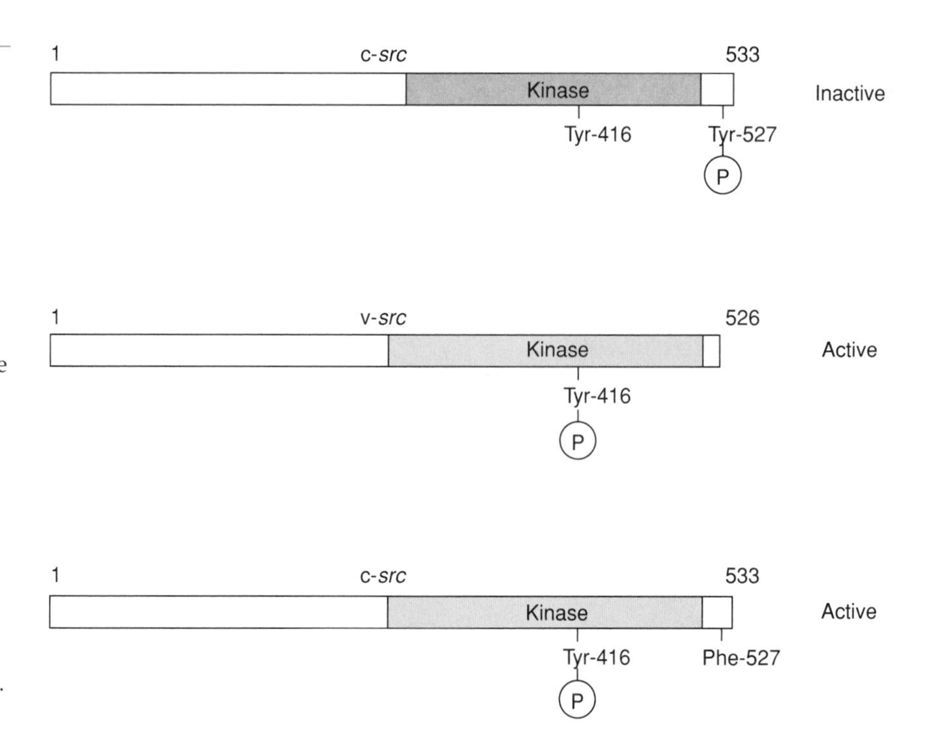

Figure 11–15

Regulation of SRC kinase activity by phosphorylation. The protein produced by the normal c-*src* proto-oncogene is phosphorylated *in vivo* at Tyr-527, which down-regulates kinase activity. The DNA encoding Tyr-527 and several other amino acids has been deleted from the viral v-*src* oncogene, which encodes a protein that is active as a kinase and is autophosphorylated at Tyr-416. The significance of Tyr-527 in regulation of SRC kinase activity is demonstrated by mutation of C-*SRC* codon 527 to encode phenylalanine instead of tyrosine. A C-*SRC* protein with Phe-527 substitution is, like the V-*SRC* protein, activated as a kinase and is autophosphorylated on Tyr-416. (From Cooper, 1990, p. 186)

most of it is deleted, the C-terminal portion has unregulated Ser/Thr kinase activity (Figure 11–16).

Protein kinase C is a family of at least seven Ser/Thr kinases, whose overall organization is similar to the products of the *raf* genes, although the amino acid sequences are distinctly different. Like *raf*, protein kinase C can become constitutively active if most of the amino-terminal domain is deleted. Inasmuch as protein kinase C plays a central role in signal transduction across the cell membrane, it is likely that the genes for these proteins can become oncogenes, but that has not yet been verified.

Oncogenes and Growth Factors

Among the stimuli necessary for the proliferation of normal cells is a group of polypeptide growth factors. It was logical to expect that abnormally large production of a growth factor by a cell capable of responding to the same factor (autocrine stimulation) would lead to unrestrained growth (Figure 11–17). That prediction was verified when it was discovered, via a computer search, that the amino acid sequence of the protein encoded by the oncogene *sis* (which had first been found in simian sarcoma virus) was closely related to the sequence of platelet-derived growth factor (PDGF). Active PDGF is a dimer that may consist of two A chains, two B chains, or one of each. The protein encoded by the human *sis* proto-oncogene corresponds to the B chain of PDGF.

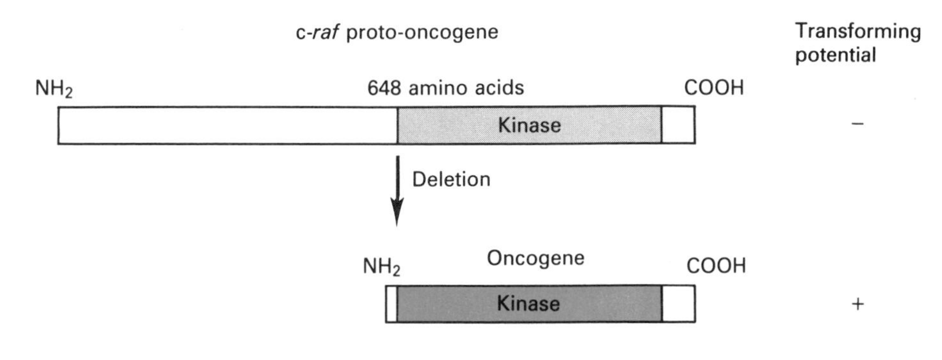

Figure 11–16

The c-*raf* proto-oncogene is converted to an active oncogene by deletion of the sequence that encodes the normal amino-terminal domain. (From Cooper, 1990, p. 216)

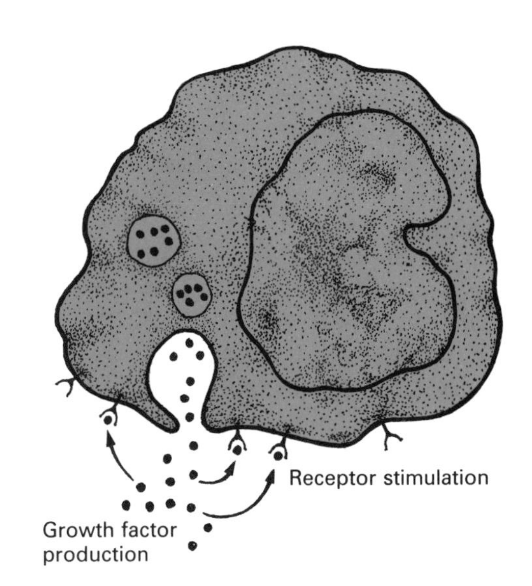

Figure 11–17

Autocrine stimulation of cell growth. A cell produces a growth factor to which it also responds, resulting in continuous cell proliferation. (From Cooper, 1990, p. 164)

Cells that express PDGF receptors are efficiently transformed by *sis* oncogenes; those that do not have the receptors are not susceptible to transformation. Therefore, interaction of the oncogene product with a cell surface receptor appears to be necessary for transformation.

Another family of growth factor genes encodes the fibroblast growth factors (FGFs) and their relatives. Three oncogenes (*int*-2, *hst*, and *fgf*-5) are members of the FGF family. They acquire transforming activity as a result of transcriptional overexpression, leading to synthesis and secretion of abnormally large amounts of their normal proteins. This is another instance of transformation as a result of autocrine stimulation.

Overexpression of EGF or its relative, transforming growth factor-alpha, both of which bind to the EGF receptor, can induce transformation of certain recipient cells. Several hematopoietic growth factors, such as interleukin-2, can also function as transforming agents via overexpression in susceptible cells.

Guanine Nucleotide-binding Proteins

A variety of cell surface receptors are coupled to intracellular metabolic pathways via guanine nucleotide-binding proteins, known more simply as G proteins. The archetype of these systems involves the activation of adenylate cyclase via the Gs protein (Figure 11–18). There are also receptors that associate with Gi, a G protein that inhibits adenylate cyclase. The response of a cell to a given hormone depends upon the presence of surface receptors for that hormone, and whether Gs or Gi is released from the complex when the hormone binds.

G proteins typically consist of three subunits. It is the alpha subunit that binds guanine nucleotide (Figure 11–19). When GDP is bound, the alpha subunit remains part of the ternary complex, which binds to the hormone receptor. Binding of the hormone to its receptor releases the alpha subunit, GDP is exchanged for GTP, and in this form, the G protein alpha subunit is able to interact with its target molecule (such as adenylate cyclase). However, alpha subunits have intrinsic GTPase activity, and when the bound GTP is hydrolyzed to GDP, the alpha subunit dissociates from the target molecule and reunites with the hormone receptor complex.

The three members of the *ras* oncogene family encode polypeptides that share substantial amino acid sequence similarities with portions of the alpha subunit of G proteins, although their size (about 21 kDa) is only half that of a typical alpha polypeptide. In addition, the *ras* proteins bind GTP and GDP, and have GTPase activity. When the polypeptides from c-*ras* oncogenes have been compared with the polypeptides from the corresponding proto-oncogenes, they usually differ in one amino acid, typically at position 12 or 61.

The single-base changes that activate the transforming potential of *ras* oncogenes have one of two effects: they either increase the rate at which GDP is exchanged for GTP, or they decrease the rate of hydrolysis of bound GTP. Both changes serve to increase the fraction of time that RAS proteins are bound to GTP, and therefore the time that they may be able to activate their putative target proteins.

Are RAS proteins signal transducers, like other G proteins? So far, we do not know the answer for mammalian cells, although the common finding of mutant *ras* genes in human tumors certainly suggests that normal RAS proteins have an important function in growth control. Fortunately, the *ras* genes have been maintained conservatively during evolution, and

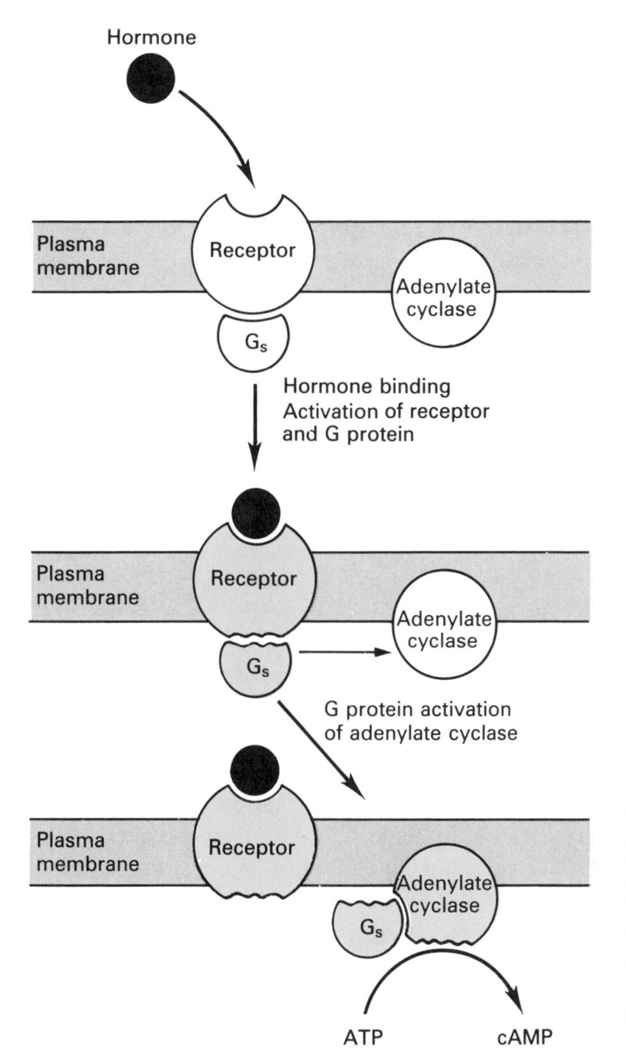

Figure 11–18

Hormonal activation of adenylate cyclase. Hormone binding to its receptor activates the G protein, Gs. Activated Gs stimulates adenylate cyclase, which catalyzes the conversion of ATP to cAMP. (From Cooper, 1990, p. 198)

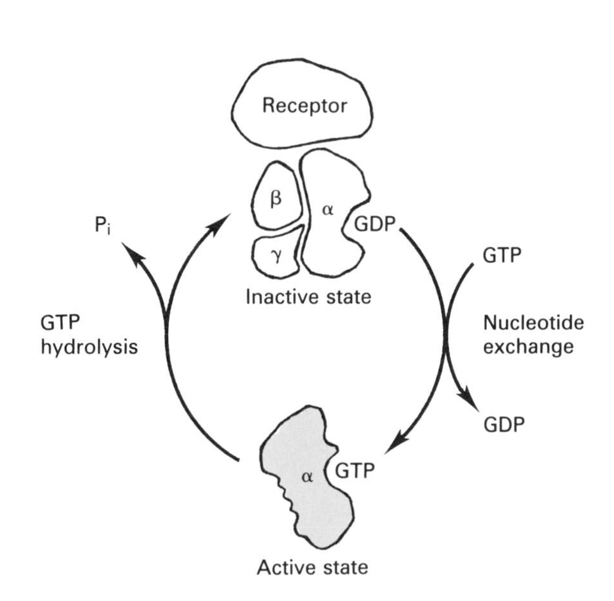

Figure 11–19

Regulation of G proteins by guanine nucleotide binding. In the inactive state, the G protein alpha subunit is bound in a complex with the beta and gamma subunits and a hormone receptor. Ligand binding to the receptor induces the exchange of GDP for GTP. The activated alpha subunit then dissociates from the complex and is able to interact with its target molecule (for example, adenylate cyclase). Hydrolysis of the bound GTP to GDP terminates the target-binding activity of the alpha subunit, and it reassociates with the receptor-beta subunit-gamma subunit complex. (From Cooper, 1990, p. 199)

two homologs of the mammalian genes exist in yeast, where they are known as *RAS1* and *RAS2*.

Mutational analysis of yeast has shown that normal proliferation can occur if either *RAS*1 or *RAS*2 is inactivated, but not if both *ras* genes are nonfunctional. Further studies showed that yeast RAS proteins are activators of adenylate cyclase, like the mammalian Gs protein. However, it is clear that mammalian RAS proteins do not activate adenylate cyclase. Their regulatory function must involve some other target molecule.

The mammalian RAS story has recently been complicated by the discovery of a large protein that appears to be a negative regulator of RAS activity; that is, it increases the GTPase activity of RAS proteins about 100 fold. This GTPase-activating protein (GAP) may be the product of one member of a family of genes, which includes the *NF1* gene mentioned in the section on tumor suppressor genes. Comparison of the amino acid sequence of the NF1 protein with known proteins revealed that a 360-amino acid segment of the NF1 protein is similar to the human GAP.

In yeast, there are two genes with homology to mammalian *GAP*, known as *IRA1* and *IRA2*. Both genes encode polypeptides of roughly 3000 amino acids, whose function is to stimulate GTPase activity of RAS proteins. Thus, yeast IRA proteins are functionally homologous to mammalian GAP proteins. The human NF1 protein is actually more similar to yeast IRA1 protein than it is to human GAP, and when a segment of the *NF1* gene was transfected into yeast, it complemented mutants in yeast *IRA1* and *IRA2* (Xu et al., 1990).

Currently, it is speculated that loss of normal function of the *NF1* gene in humans leads to the cutaneous abnormalities characteristic of the disease, via some consequence of deregulating *ras* gene function. It is not yet clear whether both copies of the *NF1* gene must be inactivated in order for a neurofibroma to form, because tests for homozygosity of markers in the *NF1* region would be futile in these nonclonal tumors. No doubt such studies will be soon be done on the less common neurofibrosarcomas found in some NF1 patients.

The products of the three *ras* genes mentioned at the beginning of this section share 30–50% amino acid sequence homology with about half a dozen RAS-related proteins that also have guanine nucleotide-binding capacity. The genes that encode these proteins have not yet been shown to be involved in neoplasia, but they do imply the existence of a complex network of signal-transducing elements involved in a growth control pathway that has yet to be defined. It would not be surprising if other members of this network have oncogenic potential in specific cell types.

Nuclear Oncogenes

Most of the oncogenes described so far in this chapter have been cytoplasmic components, involved in various aspects of signal transduction. Ultimately, signals generated at the cell surface must affect gene expression. The primary aspect of gene expression—RNA synthesis—is controlled by transcriptional regulatory proteins.

At least three groups of oncogenes have been shown to be transcriptional regulators. The first to be identified was *erb*A, whose polypeptide product belongs to a class of intracellular hormone receptors that interact with DNA. Specifically, the *erb*A proto-oncogene encodes the thyroid hormone receptor. This protein normally represses transcription of a set of genes when thyroid hormone is absent; but when the hormone binds to the DNA-receptor complex, the receptor dissociates from the chromosome and transcription of those genes is allowed (Figure 11–20).

The oncogenic form of the thyroid hormone receptor gene found in avian erythroblastosis virus (v-*erb*A) encodes a polypeptide that has lost the ability to bind thyroid hormone. In this form, it is a constitutive transcriptional repressor. It is likely that the *erb*A protein acts to repress genes in erythroblasts that would otherwise lead to differentiation into erythrocytes. This traps the erythroblasts in a continuously dividing state, and possibly potentiates the transforming activity of the other oncogene carried by the virus (*erb*B, derived from the EGF receptor gene).

Two other oncogenes, *jun* and *fos* (each of which is actually a small family of genes), also encode transcription factors. They are part of the complex known as AP1, a factor that interacts with regulatory elements of genes that are activated in response to some growth factors and to phorbol ester tumor promoters. The JUN and FOS proteins interact with each other to form dimers (Figure 11–21), which bind to the AP1 target sequence of certain genes. AP1 may include additional polypeptides, and the various members of the JUN and FOS families may form dimers that vary in their affinity for regulatory sequences of different genes in the set of phorbol ester-responsive genes.

The *myc* genes are often activated in human tumors. We have already mentioned examples involving chromosomal translocations, insertional mutagenesis by retroviruses, and gene amplification. MYC proteins have a nuclear location and bind DNA *in vitro*. Nevertheless, there is as yet no proof that MYC proteins function as transcription factors.

The retinoblastoma gene has recently been cloned, via the principles of positional analysis (as described in Chapter 5), and the gene product has been identified. The gene encompasses about 190 kbp; it codes for a phosphoprotein of molecular weight 105,000. Its precise function is not yet

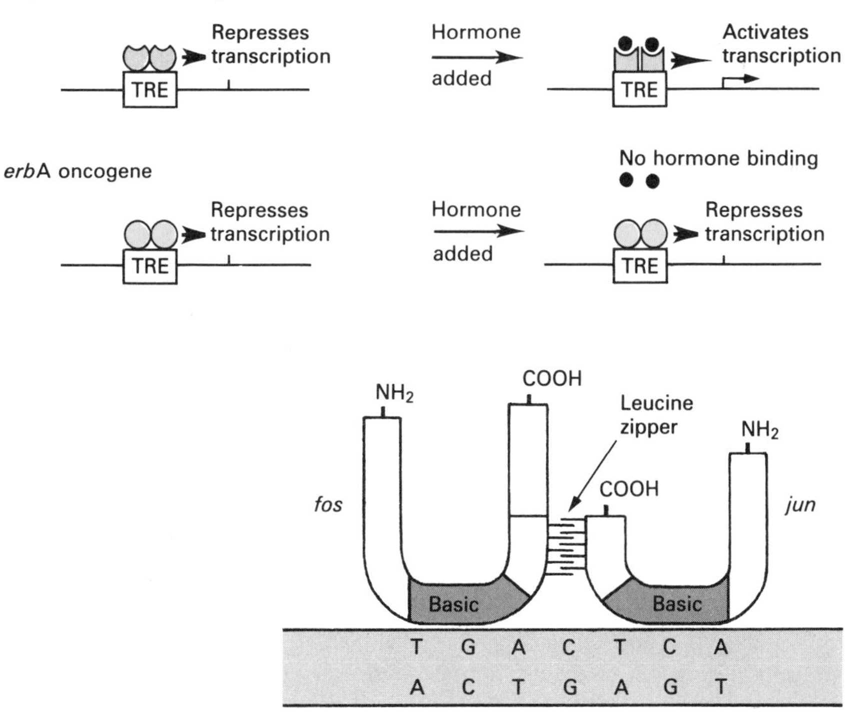

Figure 11–20

Regulation of transcription by the thyroid hormone receptor and the *erb*A oncogene protein. In the absence of hormone, the thyroid hormone receptor (encoded by the *erb*A proto-oncogene) binds to the thyroid hormone response element (TRE) and represses transcription of a set of genes. Hormone binding converts the normal receptor to a transcriptional activator, resulting in hormonal induction of target gene expression. The *erb*A oncogene protein, however, has lost the ability to bind thyroid hormone. Consequently, it acts as a constitutive repressor of thyroid hormone-responsive genes. (From Cooper, 1990, p. 229)

Figure 11–21

Binding of a JUN-FOS heterodimer to DNA. Dimerization involves interdigitation of leucine side chains to form a "leucine zipper." The basic regions of both JUN and FOS proteins then bind cooperatively to an AP-1 target site in DNA. (From Cooper, 1990, p. 234)

known. Perhaps the most suggestive evidence is that the RB polypeptide is one member of a complex that forms *in vivo* in association with the E1A protein, the product of the transforming gene of adenovirus. Similar complexes form between the RB polypeptide and SV40 large T antigen. In this case, a point mutation is known that simultaneously abolishes the transforming ability of the large T gene and the ability of large T antigen to bind the RB gene product, thus strengthening the argument that inactivation of RB is a necessary step in transformation.

The RB protein is a phosphoprotein in the S and G2 phases of the cell cycle, but it is not phosphorylated in G1 (nor in G0 in quiescent cells). T antigen binds to the non-phosphorylated or under-phosphorylated RB protein. This suggests that it is the non-phosphorylated form of the protein that acts as a growth suppressor, possibly by inhibiting the entry of cells into S phase.

The Role of Oncogenes in Tumor Formation

In the preceding sections of this chapter we have learned about the properties of oncogenes, the proto-oncogenes from which they are derived, and tumor suppressor genes. We have considered some of the ways in which activated oncogenes may pervert the functions of proto-oncogenes and the effect of loss of both normal alleles of tumor suppressor genes on growth restraints.

Most of the experimental work involved in the identification of oncogenes and tumor suppressor genes has been done with cultured cells. Transformed cells are immortal and have escaped the growth constraints of normal cells. Cells that display the transformed phenotype in culture may be tumorigenic *in vivo*, but that is not always the case.

The favorite cell culture test system uses NIH 3T3 cells, which are already immortalized, and only one oncogene is needed to confer density-independent growth. However, primary fibroblasts were not transformed by either *ras* or *myc* alone in early experiments, whereas transfection with both oncogenes did lead to transformation. Subsequently, it was found that primary cells could be transformed if either *ras* or *myc* alone was expressed at exceptionally high levels.

The two-hit model of Knudson (1971) for development of retinoblastoma was mentioned previously in this chapter. Here we have a system where inactivation of both alleles at one genetic locus appears to be the primary requirement for tumorigenesis. The cell-specific nature of this relationship is striking.

Transgenic mice have also provided evidence for the multi-step nature of tumorigenesis in certain cases, as well as exceptions to that rule. Mice carrying the *myc* oncogene attached to the MMTV (mouse mammary tumor virus) promoter develop mammary tumors. However, there is usually only one tumor, despite the fact that there are 10 mammary glands in a mouse and every cell has the transgene. Moreover, the tumor does not develop until the mouse is several months old, and the tumor is monoclonal. All of these facts imply that additional events, other than activation of the oncogene in mammary cells, are necessary for tumorigenesis.

Transgenic mice that carry both MMTV-*ras* and MMTV-*myc* develop tumors earlier than transgenic mice with only one oncogene. This result also emphasizes the contribution of multiple oncogenic events to tumor formation *in vivo*. However, there was still a latent period of about two months before tumors became detectable. Thus, more than two events must be necessary.

The induction of tumors by treatment of mouse skin with carcinogens also illustrates the multistep nature of tumor development. Although the carcinogens used in this assay are mutagens, they do not induce tumors by themselves. Treatment with a *promoter*, such as a phorbol ester, is also necessary. The promoter apparently gives a mitogenic stimulus to cells that have undergone an "initiating" event (presumably a mutation) in response to the carcinogen. Tumor promotion can occur up to a year after exposure to the carcinogen.

Most of the tumors that form on mouse skin after carcinogen and phorbol ester treatment are benign papillomas, but eventually some of them progress to malignancy. Activated *ras*H oncogenes can be found routinely in both the carcinomas and the premalignant papillomas. The other genetic lesions that are responsible for the malignant state have not yet been identified.

An interesting contrast is provided by transgenic mice carrying the *ras*H oncogene attached to the promoter of a gene (elastase) that is specifically expressed in pancreatic acinar cells. In this case, virtually all the cells of the fetal pancreas become tumor cells within days of pancreatic differentiation. Transformation thus appears to be a single-step event in this special situation.

The preceding examples, together with many other observations on a variety of experimental systems, support the generalization that development of the malignant phenotype is usually a multistep process, requiring several genetic changes, as was stated in the introduction to this chapter. However, in certain cells at specific stages of development, activation of a single oncogene or inactivation of both alleles at one tumor suppressor locus may rapidly induce tumor formation. This fact does not necessarily exclude the possibility that additional genetic changes occur in those systems with unusual speed and uniformity, but it is also compatible with the possibility that specific cell types may be convertible to the neoplastic phenotype with only one mutation. These and many other uncertainties must be resolved before we have a solid molecular understanding of cancer.

SUMMARY

Cancer is fundamentally a genetic disease, the end result of mutations that alter normal growth controls and lead to unregulated cell growth. Some mutations that predispose to development of tumors are inherited, but most of them occur in somatic cells.

Most tumors are clonal descendants of a single cell. There is strong evidence that at least two mutations are required for a tumor to form, and in some tissues, more than two mutations are necessary. Tumor development involves stages known as initiation, promotion, and progression.

There are two classes of cancer-causing genes. Oncogenes are altered versions of genes whose normal function is to promote cell growth in a regulated manner; oncogenes usually have phenotypically dominant activity. Tumor-suppressor genes, whose normal function is to restrain cell growth, can allow a tumor to form when both copies of a gene have been inactivated by mutations. Such mutations are usually phenotypically recessive at the cell level, although their effect may be dominant at the level of the organism.

Oncogenes were first described in viruses that have the ability to induce tumors. There are both DNA tumor viruses and RNA tumor viruses. Retroviruses are RNA viruses whose virion RNA is transcribed into DNA after infection of a host cell, followed by insertion of the DNA into the host genome. Transforming (oncogenic) retroviruses have acquired an oncogene, whose expression overwhelms the normal growth control mechanisms of the host cell.

Retroviral oncogenes all have homologues in

normal cells, which are called proto-oncogenes. In addition, there are many proto-oncogenes that are not known to have oncogenic equivalents in viruses, but which can become oncogenes by somatic mutations, producing cellular oncogenes. One important group of cellular oncogenes was identified by using DNA from human tumors to transform tissue culture cells. Transformed cells in culture display immortality and escape from density-dependent inhibition of growth.

Oncogenes can arise within cells by point mutations, by insertional mutagenesis, by chromosomal translocations, and by gene amplification.

The first tumor suppressor gene to be identified causes retinoblastoma (RB), a rare childhood tumor that develops in one or both eyes. The responsible gene has been identified; it is located on chromosome 13q14 and encodes a protein that appears to have a role in transcriptional control in retinoblasts and several other cell types. When both copies of the *RB* gene are inactivated in one cell, a tumor forms. Patients with inherited RB have a germ-line mutation that inactivates one *RB* gene copy; the second "hit" is a somatic mutation that inactivates the other copy. Non-inherited RB results from two somatic mutations.

A variety of other tumor suppressor genes have been described. They include a gene responsible for Wilms tumor (a childhood kidney cancer); the *p53* gene (loss of which is associated with tumor formation in several tissues); several genes whose inactivation predisposes to colon cancer; and the neurofibromatosis gene (mutations in which cause one variety of skin cancer).

The normal functions of proto-oncogenes are all related to some aspect of cellular growth control. They include protein kinases of several types, some polypeptide growth factors, guanine nucleotide-binding proteins, and transcription factors. We do not yet know the full biochemical effects of the normal function of any proto-oncogene, nor of any oncogenic derivative.

REFERENCES

Avery, O. T., MacLeod, C. M., and McCarty, M. 1944. Studies on the chemical nature of the substance inducing transformation of pneumococcal types. J. Exp. Med. 79:137–158.

Carney, D. and Sikora, K. 1990. Genes and cancer. Wiley-Liss, New York.

Cawthon, R. M., Weiss, R., Xu, G., et al. 1990. A major segment of the neurofibromatosis Type 1 gene: cDNA sequence, genomic structure and point mutations. Cell 62:193–201.

Collett, M. S. and Erikson, R. L. 1978. Protein kinase activity associated with the avian sarcoma virus *src* gene product. Proc. Natl. Acad. Sci. USA 75:2021–2024.

Cooper, G. M. 1990. Oncogenes. Jones and Bartlett, Boston.

Dix, D. 1989. The role of aging in cancer incidence: an epidemiological study. J. Gerontol. 44:10–18.

Fearon, E. R. and Vogelstein, B. 1990. A genetic model for colorectal tumorigenesis. Cell 61:759–767.

Harris, H. 1988. The analysis of malignancy by cell fusion: the position in 1988. Cancer Research 48:3302–3306.

Harris, H., Miller, O. J., Klein, G., Worst, P., and Tachibana, T. 1969. Suppression of malignancy by cell fusion. Nature 223:363–368.

Hoffman, R. M. 1990. Unbalanced transmethylation and the perturbation of the differentiated state leading to cancer. BioEssays 12:163–166.

Hollstein, M., Sidransky, D., Vogelstein, B., and Harris, C. C. 1991. p53 mutations in human cancers. Science 253:49–53.

Kern, S. E., Kinzler, K. W., Bruskin, A., et al. 1991. Identification of p53 as a sequence-specific DNA-binding protein. Science 252: 1708–1710.

Kinzler, K. W., Nilbert, M. C., Vogelstein, B., et al. 1991. Identification of a gene located at chromosome 5q21 that is mutated in colorectal cancers. Science 251:1366–1370.

Knudson, A. G. 1971. Mutation and cancer: statistical study of retinoblastoma. Proc. Natl. Acad. Sci. USA 85:1590–1594.

Knudson, A. G. 1985. Hereditary cancer, oncogenes, and anti-oncogenes. Cancer Res. 45:1437–1443.

Levine, A. J. 1990. Tumor suppressor genes. BioEssays 12:60–66.

Malkin, D., Li, F. P., Strong, L. C., et al. 1990. Germ line p53 mutations in a familial syndrome of breast cancer, sarcomas, and other neoplasms. Science 250:1233–1238.

Nishisho, I., Nakamura, Y., Miyoshi, Y., et al. 1991. Mutations of chromosome 5q21 genes in FAP and colorectal cancer patients. Science 251:665–669.

Peto, R. 1977. Epidemiology, multistage models, and short term mutagenesis tests. In H. H. Hiatt, J. D. Weston, and J. A. Winsten (eds.), Origins of human cancer, pp. 1403–1428. Cold Spring Harbor Laboratory, NY.

Prescott, D. M. 1988. Cells. Jones and Bartlett, Boston.

Rous, P. 1911. A sarcoma of the fowl transmissible by an agent separable from the tumor cells. J. Exp. Med. 13:397–411.

Rowley, J. D. 1973. A new consistent chromosomal abnormality in chronic myelogenous leukemia identified by quinacrine fluoresence and Giemsa staining. Nature 243:290–293.

Rowley, J. D. 1990. Molecular cytogenetics: Rosetta stone for understanding cancer. Cancer Res. 50:3816–3825.

Sager, R. 1989. Tumor suppressor genes: the puzzle and the promise. Science 246:1406–1412.

Steel, C. M. 1990. Genetic abnormalities in cancer. Curr. Opinion Biotech. 1:188–195.

Stehelin, D., Varmus, H. E., Bishop, J. M., and Vogt, P. K. 1976. DNA related to the transforming gene(s) of avian sarcoma viruses is present in normal avian DNA. Nature 260:170–173.

Vogt, P. K. 1971. Spontaneous segregation of nontransforming viruses from cloned sarcoma viruses. Virology 46:939–946.

Wallace, M. R., Marchuk, D. A., Andersen, L. B., et al. 1990. Type 1 neurofibromatosis gene: identification of a large transcript disrupted in three NF1 patients. Science 249:181–186.

Weinberg, R. A. 1989. Oncogenes, antioncogenes, and the molecular bases of multistep carcinogenesis. Cancer Res. 49:3713–3721.

Weinberg, R. A. 1990. Oncogenes and the molecular origins of cancer. Cold Spring Harbor Laboratory Press, Cold Spring Harbor, NY.

Xu, G., Lin, B., Tanaka, K., et al. 1990. The catalytic domain of the neurofibromatosis Type 1 gene product stimulates *ras* GTPase and complements *ira* mutants of *S. cerevisiae*. Cell 63:835–841.

Yunis, J. J. 1983. The chromosomal basis of human neoplasia. Science 221:227–236.

Yunis, J. J. and Ramsay, N. 1978. Retinoblastoma and subband deletion of chromosome 13. Am. J. Dis. Child. 132:161–163.

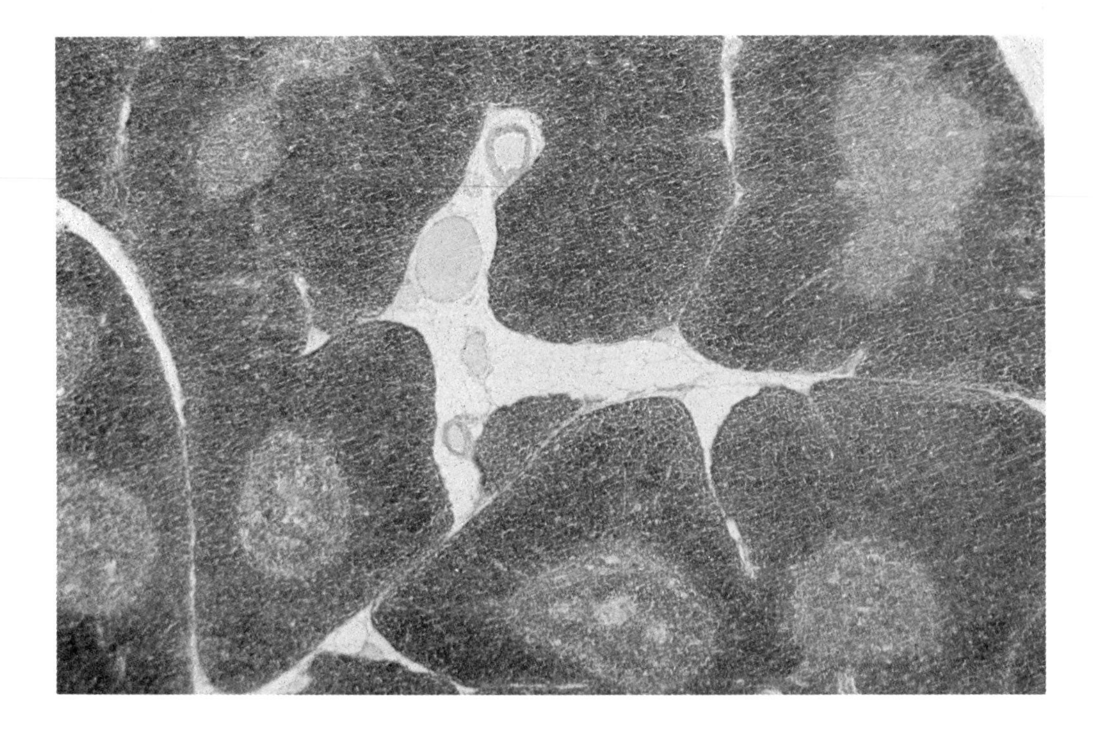

A section through the thymus gland, showing lobules where T-cells develop. (© 1989 Bruce Iverson, BSc.)

12 Genetics of the Immune System

The immune system constitutes the body's chief line of defense against foreign invaders, which may be viruses, bacteria, eukaryotic parasites, or cancer cells (which are internally generated "foreigners"). The two major components of the immune system are (1) B cells, which produce antibodies that are the primary defense against bacteria; and (2) T cells, which destroy virus-infected host cells, parasites, and cancer cells by direct cell-cell interactions, and which also play a key role in the activities of B cells.

A related system, known as the major histocompatibility complex (MHC), contains the principal determinants of self/nonself recognition and has an essential role in antigen presentation. The components of all three systems are characterized by great diversity, high specificity, and complex interactions with each other. In this chapter, we shall describe the essential features of the genetic organization of the immune system and the MHC. We shall also briefly survey the ways in which variations in immune system genes lead to specific diseases and disease susceptibilities.

B Cells and the Antibody Response

B cells are the most abundant class of lymphocytes, which in turn are a subset of leucocytes (white blood cells). In mammals, they are produced in the bone marrow, along with all the other white and red blood cells, from pluripotent stem cells (Figure 12–1). B cells also become functionally mature (i.e., ready to respond to antigens) in the bone marrow of mammals; but in birds, maturation of B cells takes place in a separate organ, the Bursa of Fabricius. The letter B is conveniently mnemonic for "bone," but it actually was used originally in reference to the Bursa.

B cells circulate continually through the lymph and blood. There are many billions of B cells in the human body and they are collectively capable of producing many millions of different antibodies during the lifetime of an individual. However, B cells are only *potential* antibody factories. They carry membrane-bound antibodies on the cell surface, and are stimulated by the binding of antigens to those surface antibodies, and by interaction with helper T cells, to undergo a series of cell divisions. In the process, they

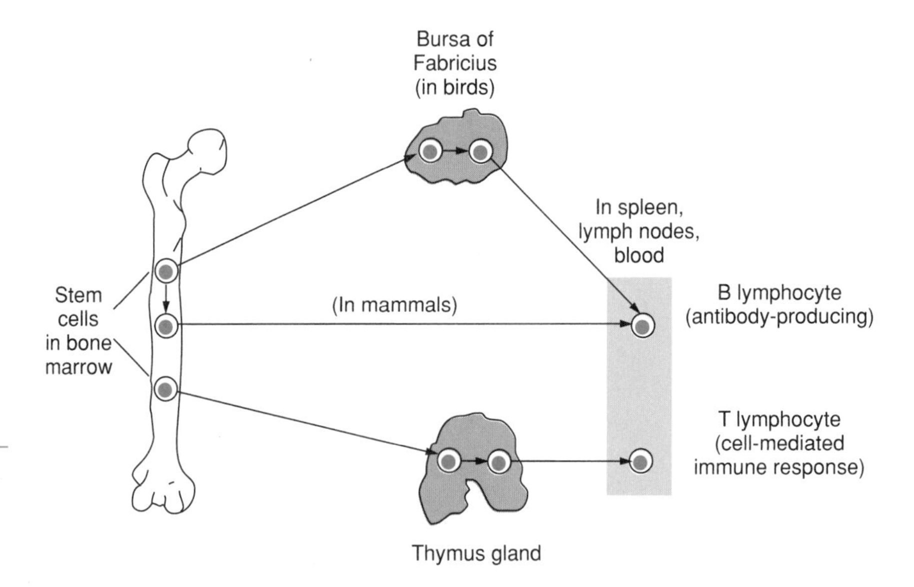

Figure 12–1

Origin of B and T lymphocytes during development in birds and mammals. (From Prescott, 1988, p. 529)

become specialized to produce large quantities of soluble antibodies, at which time they are known as *plasma cells* (Figure 12–2).

Most antibodies are not cell-associated; they move freely through the blood and lymph. The antibody response to infection is known as *humoral* immunity, in reference to "humors," an archaic term for body fluids.

Anything that elicits an antibody response can be considered to be an antigen. Most antigens are proteins or complexes of small molecules with proteins, although some are polysaccharides on the surface of bacteria and other cells. Antibodies are proteins that can combine specifically with portions of antigens (called *antigenic determinants* or *epitopes*), thereby marking those antigens and/or cells that display those antigens on their surface for destruction by other components of the self-defense system.

Antibody Proteins and Their Genes

Antibodies are oligomeric proteins; the basic unit is a tetramer, consisting of two *light chains* and two *heavy chains,* joined to one another by disulfide bonds and noncovalent interactions to form a Y-shaped structure (Figure

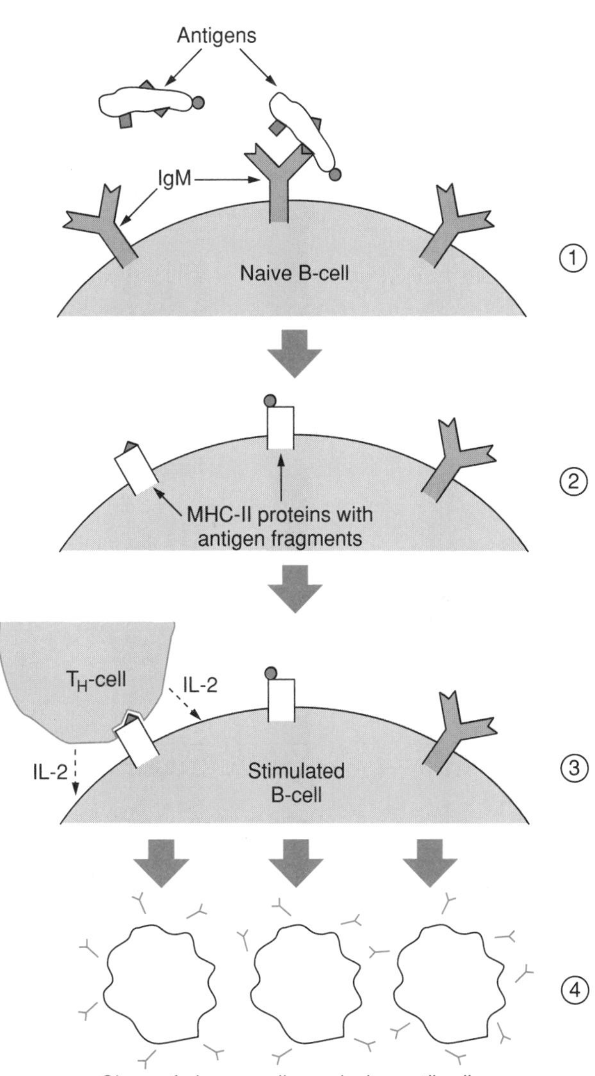

Figure 12–2

The recognition steps needed to activate a specific B cell for proliferation. In step 1, an antigen binds to an IgM antibody on the surface of a naive B cell. The antigen-antibody complex is taken into the cell by endocytosis, and the antigen is degraded by proteolysis. Fragments of the antigen bind to class II MHC proteins and are displayed on the surface of the B cell (step 2). A T_H cell with specificity for the same antigenic determinant binds to the antigen-MHC protein complex and secretes interleukin-2 and other polypeptides that stimulate the B cell to proliferate (step 3). The plasma cells that are the descendants of the stimulated B cell secrete large quantities of antibody that can bind to the same antigenic determinant, which may be circulating in body fluids and/or exposed on the surface of invading organisms (step 4).

12–3). The stem and the bases of the arms of the Y represent the *constant region*, where all antibodies of a given class have the same primary structure. The distal halves of the arms of the Y, consisting of about 110 amino acids at the N-terminal end of each polypeptide, constitute the *variable region*; it is here that the basis of antibody specificity is found.

The Y-arms of an *immunoglobulin* (an alternative name for an antibody) form two identical antigen-binding sites (Figure 12–3), into either of which an epitope on an antigen fits like a key into a lock. The number of possible antigenic determinants is astronomical, and fortunately for us, the number of possible antibodies is also huge. Our bodies are able to produce antibodies that combine specifically with almost any naturally occurring antigen. Moreover, synthetic antigens, made by scientists and which have never been seen by living systems before, also elicit the production of specific antibodies.

One of the great debates in the early days of immunology concerned the mechanism by which antibody diversity is generated. As soon as the basic facts of molecular genetics were known, it became obvious that there could not be a separate gene for every possible antibody; there simply was not enough DNA to code for tens of millions of polypeptides, and the total antibody repertoire of a human is at least that large. One group of immunologists favored an *instructional theory*; that is, an antigen must somehow tell B cells how to make antibodies that can combine with it.

The general idea was that the three-dimensional structure of newly synthesized antibodies must be very flexible, and that interaction of an antigen with a virgin antibody on the surface of a B cell caused the antibody to fold around the antigen. This molecular embrace was then supposed to stimulate the B cell to grow, divide, and produce large quantities of antibody. The hypothesis was rejected when it was shown that after separation from bound antigen, denatured antibodies could be renatured in the absence of antigen and still have precisely the same binding specificity as before.

It was the *clonal selection theory* of Macfarlane Burnet and Niels Jerne (Burnet, 1962) that provided the correct explanation for antibody diversity. Clonal selection claims that B cells capable of making every antibody to

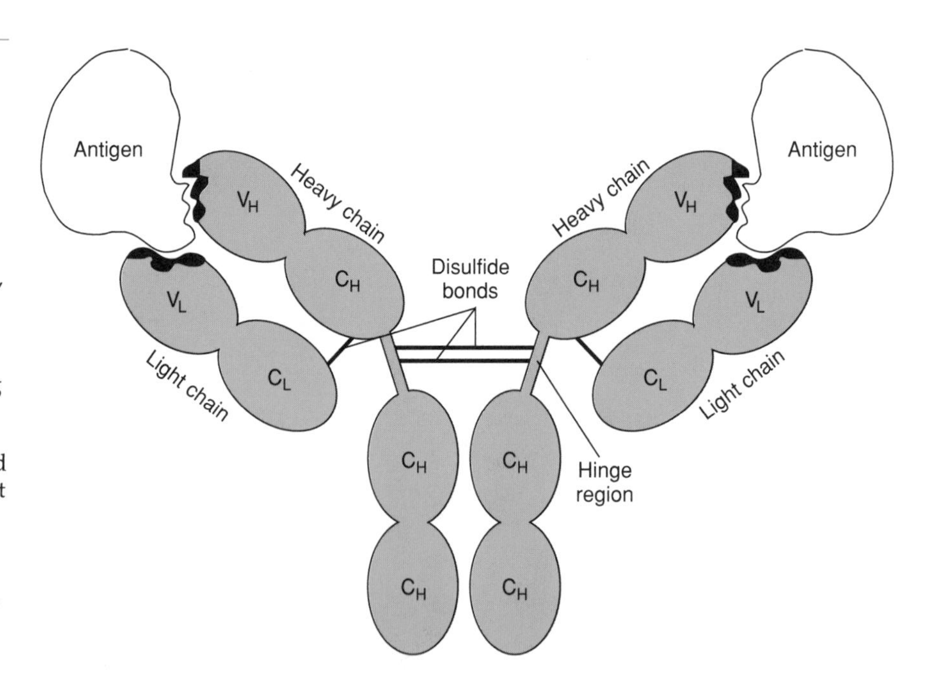

Figure 12–3

The Y-shaped IgG antibody. The left and right halves are identical. The heavy chains contain approximately 440 amino acids, while the light chains contain approximately 220. In each polypeptide, the N-terminal 110 amino acids constitute the variable domains, indicated by V_L and V_H respectively. The constant domains are indicated by C_L and C_H. The antigen-combining sites of the antibody are indicated diagrammatically in black, to emphasize the detailed complementarity that must exist between antigen and antibody, a complementarity that is three-dimensional in reality. (From Mange and Mange, 1990, p. 358; by permission of Sinauer Associates)

which an individual can respond already exist in a normal human, and that the arrival of a foreign antigen leads to the stimulation and multiplication of only those B cells that can make antibodies capable of binding to epitopes on that antigen (Figure 12–4). The paradox implied by the clonal selection theory was that the existence of a vast repertoire of antibody-specific B cells implied a corresponding diversity at the DNA level—a diversity too great to be encoded in ordinary genes. Where could the diversity come from?

A theoretical explanation was offered by Dreyer and Bennett (1965), who suggested that there are separate genes for variable and for constant portions of antibody molecules, and that functional antibody genes can be generated by a combinatorial process of joining any one variable unit to any one constant region unit. The problem was solved when Hozumi and Tonegawa (1976) demonstrated that antibody genes do something unprecedented; they are extensively rearranged when stem cells differentiate into B cells. This heretical conclusion violated the sacrosanct status of DNA as an epigenetically constant molecule, but it opened the door to a molecular explanation of antibody diversity, and the experimental evidence was indisputable (Figure 12–5).

It is now clear that antibody diversity results from a combinatorial process of gene shuffling in each of the three clusters of antibody genes in humans. Two of those clusters code for light chains: the lambda chains are encoded on chromosome 22 and the kappa chains are encoded on chromosome 2. An individual B cell makes either a kappa or a lambda light chain,

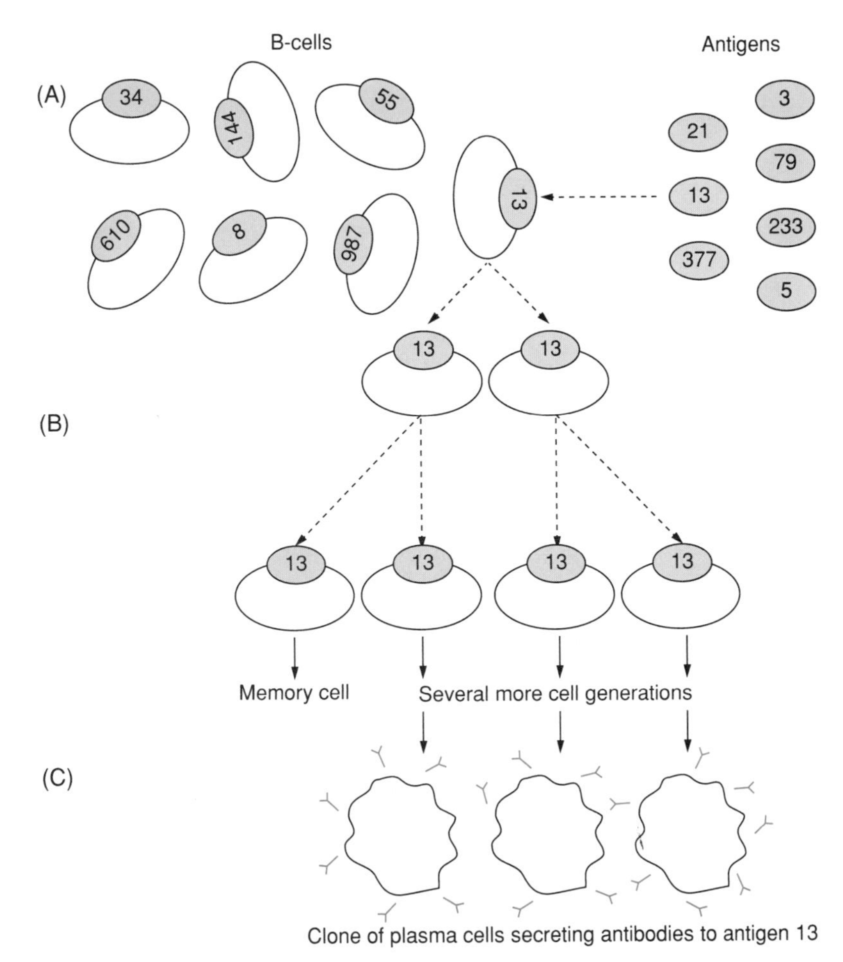

Figure 12–4

The clonal selection hypothesis. A specific B lymphocyte recognizes a foreign antigen (A), and proliferates (B), differentiating into a clone of antibody-producing plasma cells (C). A few immature plasma cells may stop differentiating and enter a long-lived resting state as memory cells, which can mount a rapid response to a subsequent challenge with the same antigen (B).

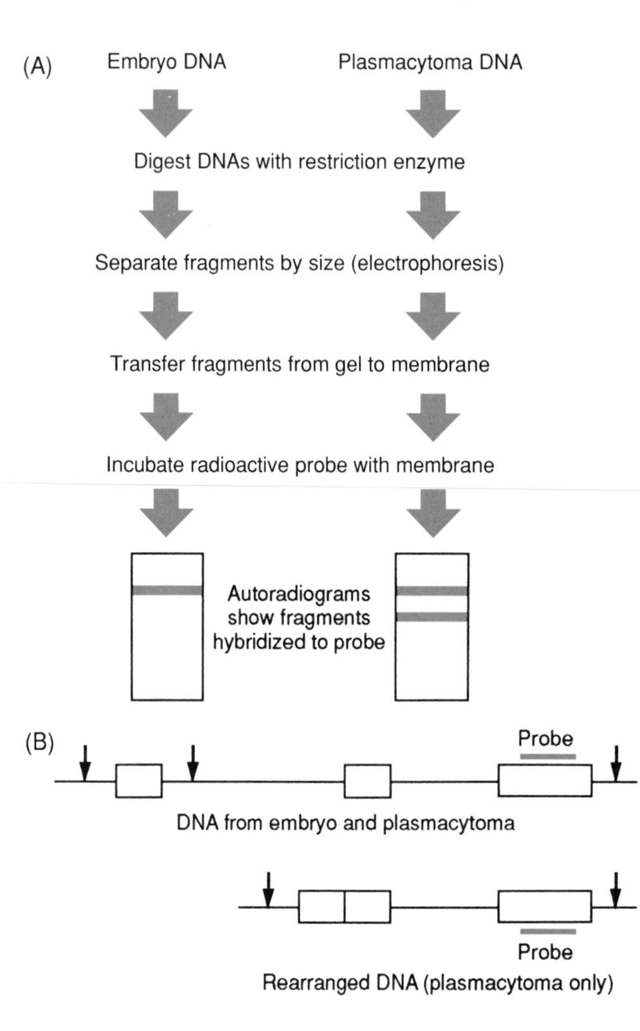

Figure 12–5

(A) An experimental protocol that demonstrates gene shuffling during development of the immune system. The basis idea is to compare the size of restriction fragments of DNA containing a constant-region gene (detected by the probe) in mouse embryos before differentiation of the immune system and in a B-cell tumor known as a plasmacytoma. DNAs from each source are separately digested with a restriction enzyme, separated according to size by electrophoresis, transferred to a nitrocellulose membrane, and incubated with a radioactive probe. Autoradiography reveals the bands to which the probe hybridizes. Embryonic DNA yields one band; plasmacytoma DNA yields two bands, one of which is the same size as the embryonic DNA band, while the other is smaller. Panel (B) interprets that result. One chromosome containing the antibody gene has undergone a deletion that brings two restriction enzyme sites closer together. Subsequent research has fully verified the hypothesis that the deletion is due to a special type of recombination of various units of the antibody gene complex and is a normal part of B-cell differentiation, by which they become able to produce specific antibodies.

not both. The third gene cluster, which specifies all the heavy chains, is on chromosome 14.

Each antibody gene cluster consists of three (in light chains) or four (in heavy chains) groups of units (Figure 12–6). During the maturation of a B cell, a functional light chain gene is formed when one unit from group V is randomly joined to a unit from group J. This joining takes place at the DNA level, with excision of all sequences that were between the joined segments. It is a type of recombination, but it is not reciprocal and it must be quite different in detail from meiotic recombination. The conjoined V-J unit becomes an exon of the mature light chain gene. After the gene is transcribed, RNA processing splices the V-J transcript to the constant region transcript to make the mature mRNA (Figure 12–7).

The number of possible light chains that can be produced is thus the product of the number of units in the V region times the number of units in the J region. The V regions of the human kappa and lambda clusters have not yet been completely sequenced, so there is some uncertainty about the total number of V units, but one estimate puts the number at about 300 for each cluster. In the kappa cluster, there are 5 J units and one C unit, so there are about 1500 possible kappa polypeptides. Organization of the lambda light chain gene cluster is somewhat different, but the total number of possible combinations is similar. Together, the kappa and lambda clusters have the potentiality to make about 3000 different light chains, via random combination of V and J units within each cluster.

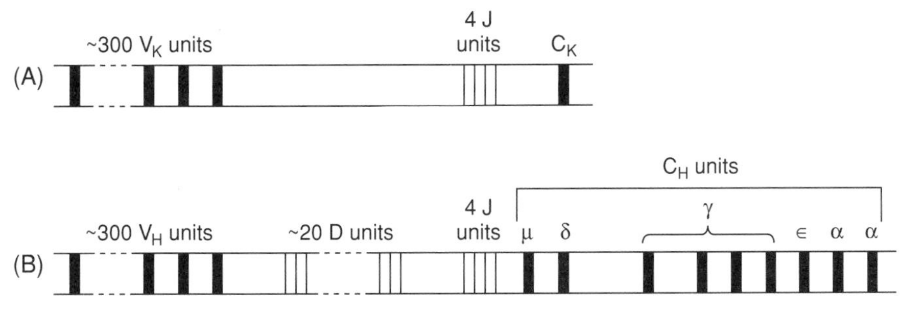

Figure 12–6

The human kappa light chain gene family (A) and the human heavy chain gene family (B). Considerable uncertainty exists about the number of V units in each case; the number of D units in the heavy chain complex is also a rough estimate. The various C_H units indicated in panel B are not involved with the generation of antibody specificity; they define several classes of antibodies via the interaction of other proteins with the constant region of the heavy chain (the stem of the Y diagrammed in Figure 12–3). The heavy chain gene family spans at least 300 kilobases on chromosome 14; the kappa family is somewhat smaller and is on chromosome 2; the lambda light chain gene family on chromosome 22 (not illustrated) differs in some details from the kappa gene family, but encodes approximately the same number of variable units.

Functional heavy chain genes are formed by a similar combinatorial process, but more possibilities exist, because there are roughly 20 units in a fourth cluster of gene segments, the D cluster, between the V and J clusters (Figure 12–6) There are two rearrangements at the DNA level in maturation of heavy chain genes. First, a D and a J segment are joined, then a V segment joins to the DJ unit. The VDJ sequence is not joined to the C sequence at the DNA level; after transcription of the gene, the VDJ and C sequences are brought together by RNA splicing.

The number of functional D units is somewhat uncertain, but if we take 20 as the nominal value, multiply that by 4 J units, and multiply that product by 300 V units, we see that there are 24,000 possible heavy chains.

The overall potential diversity of antibodies resulting from gene shuffling is then the product of the light chains times the heavy chains. Pursuing the examples given previously, this number is 3000 × 24,000 = 72 million possible antibodies. Thus, the combinatorial aspects of antibody gene rearrangements allow a few hundred coding sequences to form

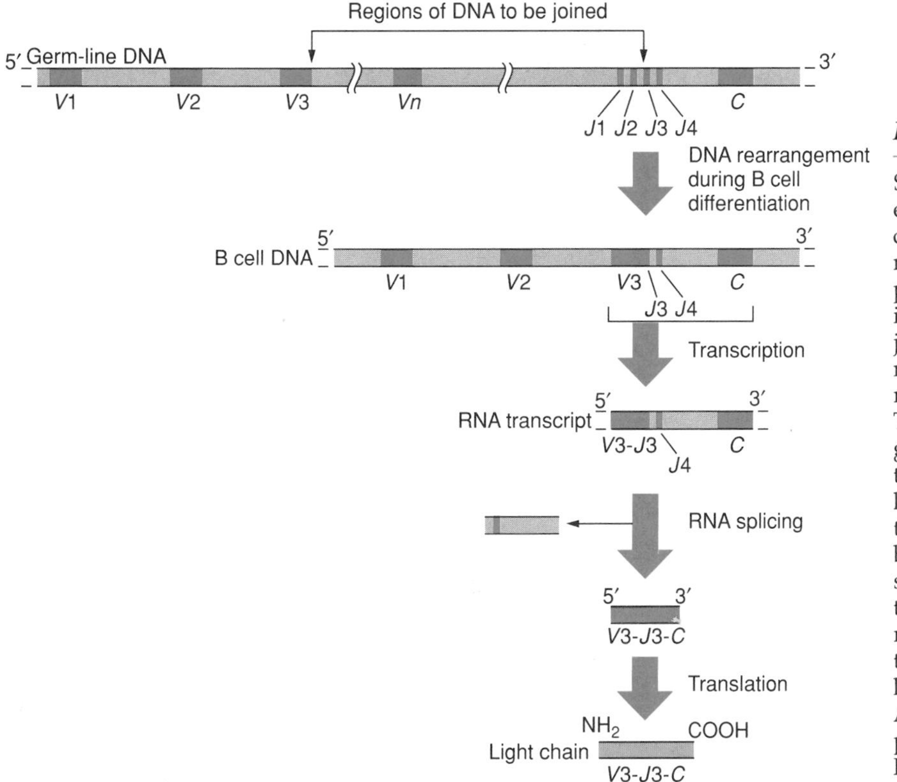

Figure 12–7

Stage in the maturation and expression of a kappa light chain gene. As an example of many combinatorial possibilities, the figure indicates that V3 becomes joined to J3 by a DNA rearrangement during maturation of a B cell. Transcription of the rearranged gene begins with the V unit that is now adjacent to a J unit, leading to the RNA shown in the third line. The sequences between J3 and the C unit are spliced out of the primary transcript, yielding the mature mRNA, which is then translated to yield the kappa light chain polypeptide. (From Alberts et al., 1989, p. 1025; by permission of Garland Publishing)

many millions of antigen-recognition sites at the ends of the Y-shaped antibody molecules.

Even more possible combining sites are generated by a variety of *mutational processes*. First, the joining reaction is not precise; often, one or a few nucleotides are omitted from one or more of the segments to be joined. In the case of heavy chain genes, base pairs may also be added at the D-J and V-D junctions, possibly via the activity of the enzyme terminal transferase. Additions and deletions involving a multiple of three base pairs will result in addition or deletion of amino acids from the corresponding polypeptide; additions or deletions of one or two base pairs or multiples thereof will shift the reading frame for mRNA translation (Figure 12–8).

Another source of antibody gene variability is nucleotide substitutions, which occur in the portion of the V unit that is upstream of the joining point. There may be several single base substitutions in one antibody light or heavy chain. Some of these will cause amino acid changes in the polypeptide; others will simply create synonymous codons. It is not yet clear what the molecular mechanism for this type of somatic mutation in immunoglobulin genes may be, but the trigger appears to be stimulation of a mature B cell by antigen. Thus, the binding specificity of the antibody produced by a given clone of B cells may change as plasma cells develop. It is believed that those descendants of the original B cell that produce antibodies with the strongest affinity for the stimulating antigen give rise to the greatest number of plasma cells. The implication is that precision of fit between antigen and antibody is sensed by the cell division control system, but no mechanism has been identified.

A price must be paid for these sources of diversity, because some of the additions or deletions change the reading frame in such a way that stop codons will be generated. When this happens, a non-productive rearrangement has occurred. It has been suggested that a B cell that fails in its first attempt to produce a functional light chain gene or heavy chain gene will then make a second attempt, using the homologous cluster (i.e., the other allele), but a mechanism by which the cell could know that an unproductive gene rearrangement had occurred has not been identified. In any case, if

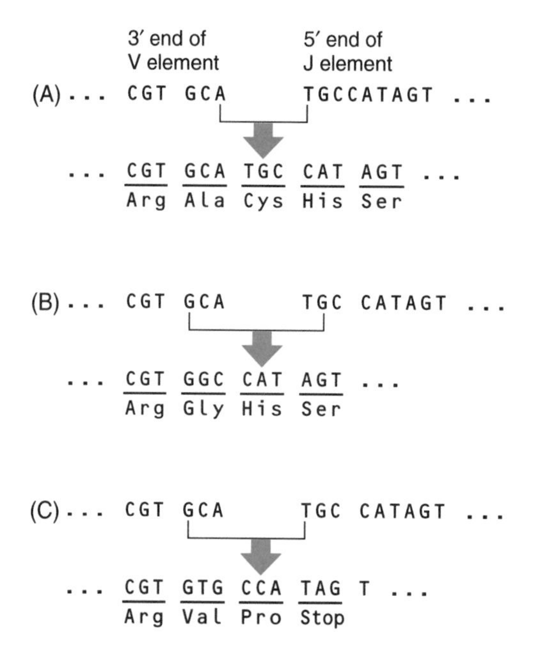

Figure 12–8

A hypothetical example of imprecise joining of a V element and a J element in a light chain gene complex. In (A) the last nucleotide of the V element has been joined to the first nucleotide of the J element, producing a DNA sequence that encodes the indicated amino acids in the immediate neighborhood of the joining point. In (B) the joining process has deleted two nucleotides from the V element and one nucleotide from the J element, producing a DNA sequence that preserves the same reading frame as in (A) but alters the amino acid sequence slightly. In (C), V–J joining has deleted two nucleotides from the V element, leading to a shift in the reading frame and creation of a stop codon.

one productive rearrangement has occurred, there will be no rearrangement of the other allele. That is, only one functional light chain gene (out of two lambda and two kappa clusters) and one functional heavy chain gene will be produced in a given B cell. We do not understand the basis for this *allelic exclusion*, but it appears to operate with great rigor.

It is also important to realize that a variety of B cells will respond to a single antigen. There are two reasons for this. First, most antigens will have more than one antigenic determinant (also called an epitope). That is, microorganisms and even large molecules such as proteins are likely to have several distinct surface features to which different antibodies can fit tightly enough to elicit the B cell response.

Second, most epitopes will fit with varying degrees of precision into the combining sites of different B cells. Stimulation of a naive B cell does not appear to be an all-or-none process; therefore, a group of antibodies with varying affinities to each epitope will usually be made. In other words, a typical antibody response to a single antigen will be *polyclonal*. If the plasma cells descended from a single B cell could be isolated, a *monoclonal antibody* could be obtained. In fact, this is possible in experimental systems, and it has provided a tool of great power, as explained in Box 12-1.

Cell Biology of the Antibody Response

As we have just seen, a B cell acquires the capacity to synthesize a specific antibody as a result of gene rearrangements during B cell maturation. We don't know whether all the combinatorial possibilities are realized in an individual human; if so, then there may be only one or a few B cells capable of making any particular antibody at a given time. In any case, it is clear that a normal human can respond to almost any antigenic challenge that Nature happens to provide.

B cells that have not been stimulated by antigen are said to be *naive*. A naive B cell does not secrete antibodies. Instead, it carries on its surface a special form of antibody, known as class M (or IgM, where Ig stands for immunoglobulin). Initially, M antibodies are bound to the plasma membrane via a hydrophobic peptide that is part of the heavy chains, with the combining sites facing the exterior (Figure 12-9(A)). When a suitable antigen binds to the membrane-bound IgMs, the B cell is stimulated to grow and divide into a clone of plasma cells, a process that takes approximately four to seven days. Participation of a T_H cell (see next section) is also required.

Subsequent to the initial stimulation by antigen, a series of interesting events occurs at the level of antibody expression. First, processing of the heavy chain gene transcript changes, so that the last several exons are deleted (Figure 12-10). This produces a protein without the segment that anchored the antibody to the cell membrane of the naive B cell; it also lacks a small cytoplasmic peptide that followed the membrane-spanning peptide. As a result, M antibody is now secreted. Moreover, secreted M antibody is an oligomer—a complex of five M proteins held together by disulfide bonds and a completely different molecule called the J protein (this has nothing to do with the J segments of antibody genes). The antigen-recognition sites all face the exterior of the pentamer (Figure 12-9(B)).

After a few days plasma cells undergo *class switching*. There are actually five classes of antibodies, defined by the constant regions of the heavy chains, as diagrammed in Figure 12-11. We will concentrate here on the most frequent switch, from class M to class G. Immunoglobulins of class G

BOX 12–1 Monoclonal Antibodies

Mature plasma cells do not divide, and they are relatively short-lived. If a single plasma cell could somehow be stimulated to grow and multiply, producing a clone of descendants, the antibody that was secreted would be *monoclonal*; every antibody molecule would have exactly the same specificity as every other antibody molecule secreted by those cells. If monoclonal antibodies were available in unlimited quantity, they would have enormous potential for diagnostic and therapeutic medicine. An ingenious solution to the problem of producing monoclonal antibodies was achieved by Georges Koehler and Cesar Milstein (1975), while working at the Medical Research Council Laboratory of Molecular Biology in Cambridge, England (Figure 12B1–1).

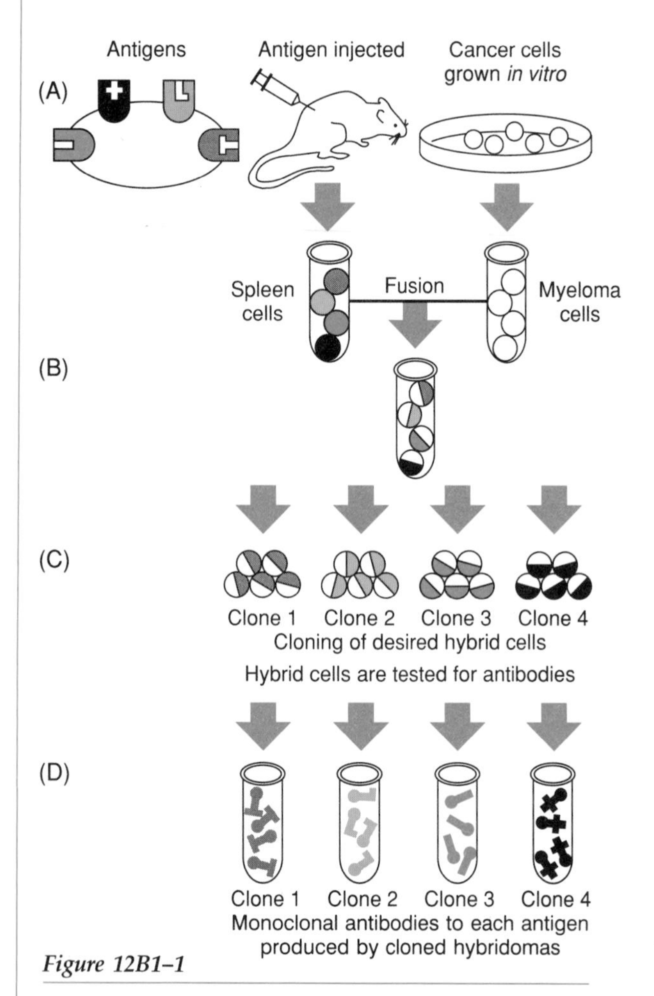

Figure 12B1–1

Construction of hybridomas that produce monoclonal antibodies. (A) An antigen is injected into a mouse one or more times and the animal's spleen is removed when the IgG antibody titer is maximal. (B) Spleen cells, which include antibody-producing plasma cells, are fused with cultured myeloma cells. (C) Clones of hybrid cells are expanded, and the culture medium is tested for presence of the desired antibody. (D) Clones that produce the desired antibody can be grown indefinitely and in virtually unlimited quantities in culture. (Modified from Edlin, 1990, p. 318)

Koehler and Milstein made *hybridomas* (a type of hybrid cell) by fusing plasma cells from mouse spleen with a special type of tumor cell known as a *myeloma*. Myelomas are naturally occurring tumor cells derived from B lymphocytes. Myelomas, like most tumor cells, are immortal; and like B cells, they synthesize antibodies, or at least parts of antibodies. Koehler and Milstein, however, chose a modified mouse myeloma clone that no longer synthesized antibody (so that the only antibodies made by the hybridoma cells would come from the spleen cell genes). The myeloma clone was also $HPRT^-$.

They injected a mouse with an antigenic protein, and some weeks later, when the immune response was expected to be maximal, they removed the animal's spleen, which is a good source of antibody-secreting plasma cells. The spleen cells were then fused to the modified myeloma cells, and the hybrids were grown in HAT medium (see Chapter 3). Myeloma cells that had not hybridized with spleen cells could not grow, because they were $HPRT^-$; spleen cells could not grow, because they were not transformed; therefore, only spleen-myeloma hybrids could survive.

Koehler and Milstein then cloned many of the hybrid cells and checked to see which ones were making antibodies to the antigen that had been injected into the spleen donor. They then showed that the antibody-producing clones could be grown indefinitely, and could be used for commercial scale production of monoclonal antibodies. Both men were awarded the Nobel Prize in 1985.

Monoclonal antibodies are currently in use for a wide variety of research and clinical purposes, and their applications continue to expand. In research, they are often given radioactive or fluorescent tags and used to locate a specific antigen, either in cells and tissues or in cell-free fractions. In medicine, monoclonal antibodies are used in pregnancy tests, where they make it possible to obtain an answer as early as 10 days after fertilization. They are used for the diagnosis of a multitude of viral and bacterial diseases, where their specificity and speed are considerably greater than previous methods of diagnosis. Red blood cell typing and HLA typing are now done with monoclonal antibodies.

In the future, it is hoped that modified monclonal antibodies can become the long-sought "magic bullets" that can be targeted to specific molecules on the surface of cancer cells, where they will deliver a toxin or some other cytocidal agent that destroys the tumor without harming normal cells or suppressing the entire immune system. That is still an elusive goal, but it is receiving intense effort, using the full armory of recombinant DNA technology and protein engineering (Winter and Milstein, 1991).

REFERENCES

Edlin, G. 1990. *Human Genetics: A Modern Synthesis*. Jones and Bartlett, Boston.

Koehler, G. and Milstein, C. 1975. Continuous cultures of fused cells secreting antibody of predefined specificity. Nature 256:495–497.

Winter, G. and Milstein, C. 1991. Man-made antibodies. Nature 349:293–299.

are the major circulating antibodies—about 80% of the total. Figure 12–3 is based on IgG antibodies. The switch from class M to class G antibody synthesis involves another gene rearrangement; the section of DNA coding for the constant region of the M heavy chain is deleted in such a way that transcription now proceeds directly from the J region to the C region of the G class (Figure 12–11).

Notice that class switching does not change the specificity of the antibody; the same antigen-recognition site that was present on the IgM antibodies made by a given plasma cell is also present when the switch is made

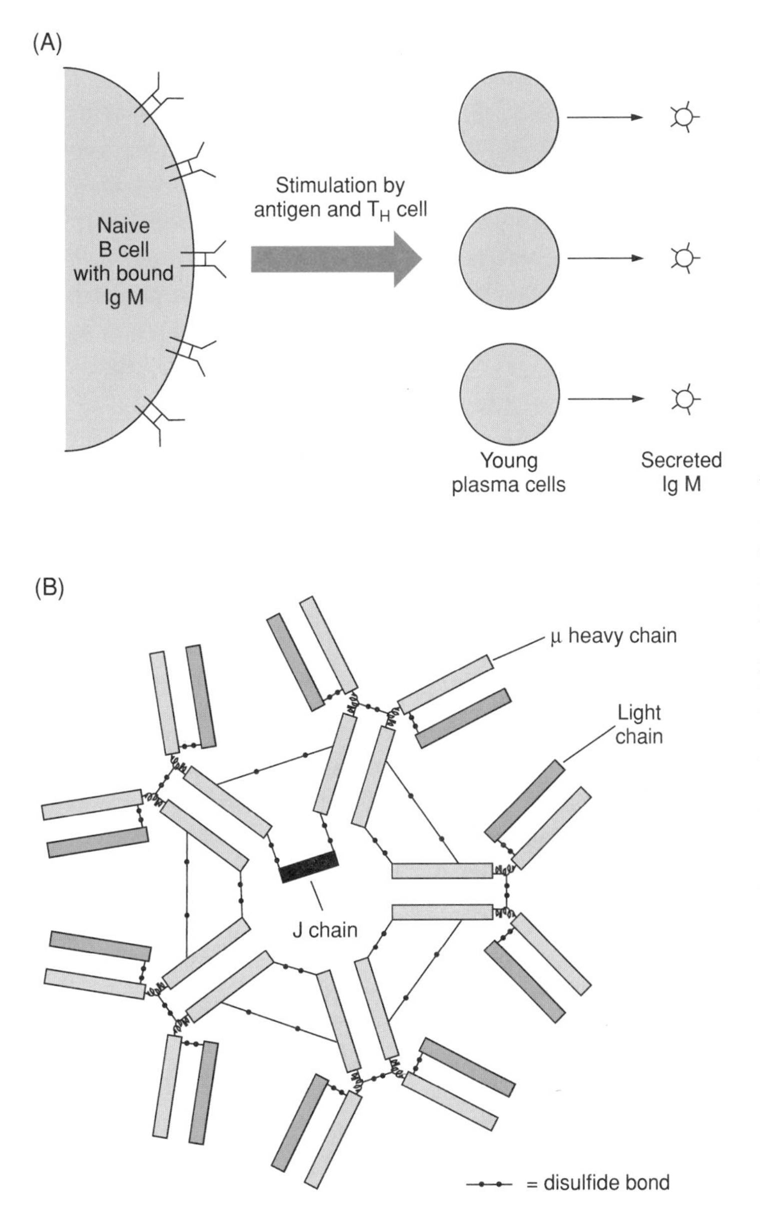

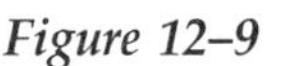

Figure 12–9

(A) A mature, but naive B cell displays IgM antibodies on its surface. The antibodies are anchored to the cell membrane by a hydrophobic series of amino acids. After the B cell has been stimulated by binding of an appropriate antigen to the surface IgMs and by interaction with a helper T cell, it begins to divide, forming a clone of plasma cells. During that time, the cells stop making membrane-bound IgM and begin to secrete a soluble variety of IgM that forms pentamers. (B) The pentameric form of secreted IgM antibodies. The five IgM units are held together by disulfide bonds. A single J chain, disulfide-bonded between two heavy chains, closes the ring structure. (From Alberts et al., 1989, p. 1015; by permission of Garland Publishing)

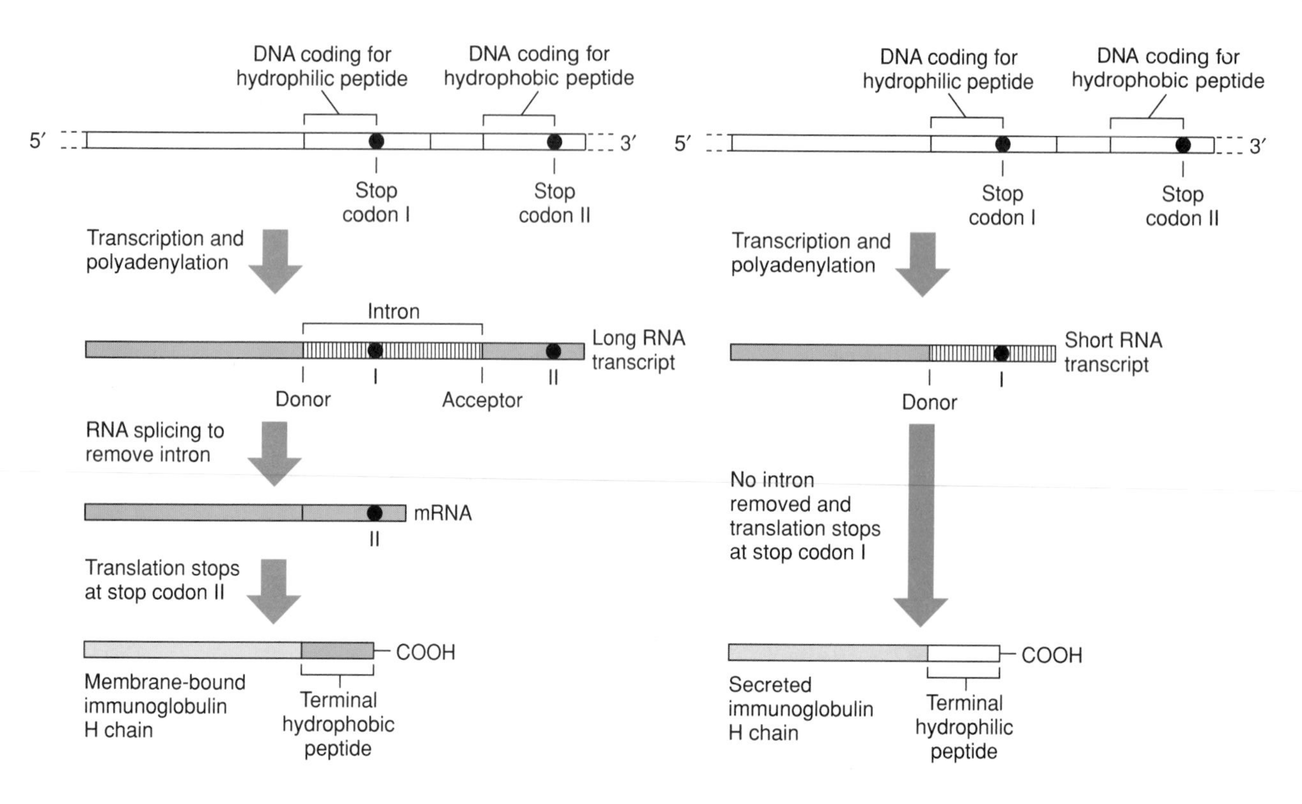

Figure 12–10

A B cell switches from making a plasma membrane-bound form to a secreted form of the same antibody molecule by altering the heavy chain mRNAs it produces when it is activated by antigen. This is thought to result from a change in the way the initial H-chain RNA transcripts are cleaved and polyadenylated at their 3′ ends. The two forms of H chain differ only in their carboxyl terminals: the membrane-bound form has a hydrophobic tail, which holds it in the membrane, whereas the secreted form has a hydrophilic tail, which enables it to escape from the cell. (From Alberts et al., 1989, p. 1028; by permission of Garland Publishing)

to IgG or any other class. Why does class switching occur? In the case of IgM and IgG, it has been suggested that the pentameric secreted form of IgM, with its ten antigen-combining sites, has the ability to form large aggregates with invading microorganisms, which will be attractive targets for phagocytosis by macrophages. If that does not eliminate the challenge, the switch to IgG occurs. When IgG is bound to antigen on the surface of a cell, the constant region is recognized by a group of enzymes, collectively known as *complement*. The ultimate consequence of binding complement is destruction of the target cell's membrane, and death of the cell. This is a more powerful defense than random phagocytosis.

The functions of the D, A, and E classes of antibody are not well understood. D antibodies are membrane-bound on naive B cells, but their role in the immune response is almost totally obscure. Class A antibodies are present in saliva, tears, and other secretions that come into direct contact with the outer world, and therefore it seems likely that they must be more effective in eliminating microorganisms before they enter the body than IgG antibodies would be; but how they achieve that functional specificity is not understood.

Class E antibodies bind to receptors on the surface of *mast cells*, via the constant region of the heavy chain. Mast cells are a type of granulocyte (a class of white blood cells) that contain histamine granules. When an antigen interacts with a complementary IgE antibody on a mast cell, the cell releases its histamine (Figure 12–12), which increases tissue permeability to fluids. Heparin, which inhibits blood clotting, is also released. If antigen is encountered in the respiratory tract, the familiar *allergic reaction* (sneezing, coughing, mucous secretion, and tissue swelling) may occur.

Cynics often ask why we need a special class of antibodies to make us miserable, but the answer must be that class E antibodies are actually part of our defenses against harmful invaders. Mast cells are not limited to the lungs, eyes, nose, and mouth; they also occur in lymph nodes, skin, and

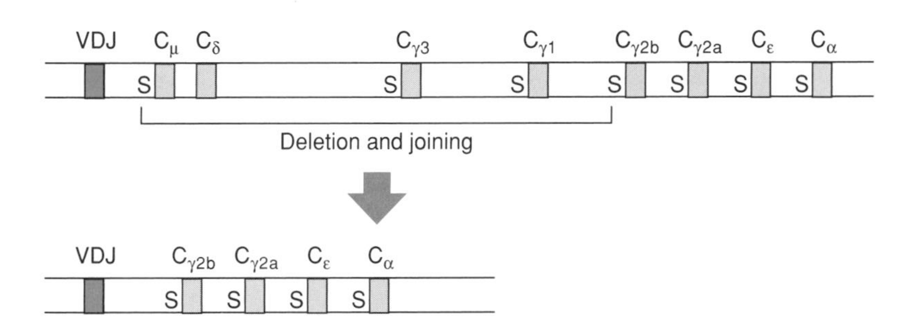

Figure 12–11

An example of the DNA rearrangement that occurs during "class switching" of heavy chain constant region gene elements. During maturation of an activated B cell into a plasma cell, production of IgM antibodies is replaced by production of IgG, IgE, or IgA antibodies. This is accomplished by a special type of recombination event that deletes the DNA between the VDJ sequence and one of the four C_γ sequences, the C_ϵ sequence, or the C_α sequence. Recombination occurs at specific switch sequences, indicated by the letter S. The diagram shows a class switch from IgM to IgG, using the $C_{\gamma2b}$ sequence.

the intestines (where they can be involved with allergic reactions to certain foods). It has been suggested that class E antibodies are directed against non-microbial parasites, but this has yet to be established.

In any case, allergy is generally considered to be a malfunction, involving the presence of abundant class E antibodies against foreign proteins that are not actually threatening. In extreme cases of hypersensitivity, when mast cells all over the body release their histamine simultaneously, the sudden reduction in blood flow to the heart, brain, and lungs can result in *anaphylactic shock*, which may be fatal if it is not quickly counteracted by an injection of epinephrine.

Finally, we must note that not all B cells capable of responding to a particular antigen become fully mature plasma cells. In most cases, a fraction of responding B cells stop dividing after they become capable of secreting IgG and remain in the circulation as *memory cells* (Figure 12–4), which may survive for years, or even for a human lifetime. Memory cells are what make a person "immune" to a pathogen, following an initial infection. They respond immediately to subsequent challenges by the same pathogen, quickly completing their differentiation into plasma cells, which produce a large amount of IgG antibody that overwhelms the invader before it can cause clinically evident symptoms. Memory cells are highly effective in providing immunity against common bacterial and viral diseases. They may be generated either through an overt infection (e.g., chicken pox) or through immunization (e.g., smallpox, polio, measles, diphtheria, and pertussis).

The T Cell Response: Cell-Mediated Immunity

Lymphocytes whose maturation requires a period of residence in the thymus gland are known as T cells. They have a variety of functions in the immune system, but none of them involves secreted antibody. Whatever T cells do, they do by cell-to-cell contact. There are two major classes of T cells: (1) cytotoxic T cells (T_C cells) recognize and lyse host cells with for-

Figure 12–12

Mast cells have cell surface receptors that bind IgE antibodies via the constant region of the class E heavy chain. An individual mast cell may bind a large number of IgE antibodies, with varying specificities for antigens. When a multivalent antigen crosslinks several antibodies, it triggers the mast cell to release histamine from stored granules by exocytosis. (From Alberts et al., 1989, p. 1017; by permission of Garland Publishing)

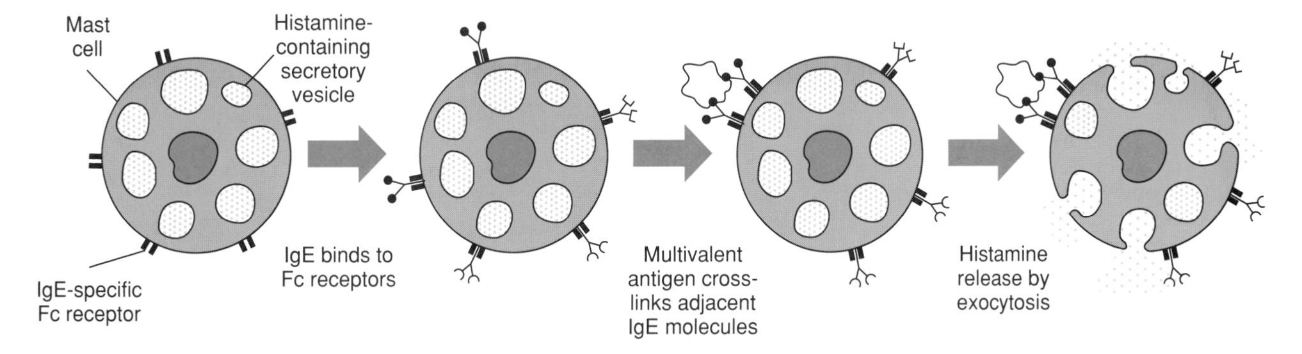

eign antigenic determinants held on their surface by MHC proteins (described later); and (2) helper T cells (T_H cells) play an essential role in stimulating B cells that have bound foreign antigens to multiply and secrete antibodies.

A third class of T cells is known as *suppressor T cells*. Their main function appears to be to modulate the B and T cell responses to antigens by inhibiting T_H cells, although the way in which they recognize specific T_H cells to be inhibited is not known. Possible uses of the inhibitory reaction are terminating an immune response after a threat has been conquered and preventing autoimmunity.

The crucial molecules for T cell function are the *T cell receptors* (TCRs), highly variable proteins on the surface of T cells that are capable of recognizing foreign antigens (Marrack and Kappler, 1987; Strominger, 1989). TCRs show some evolutionary relationship to antibodies, but unlike most antibodies, TCRs never leave the T cell. TCRs are the only known human proteins, besides antibodies, that owe their diversity to a series of genomic rearrangements.

There are two types of T cell receptors. The gamma/delta class is found on only about 5% of T cells in mice; it is made relatively early in development; its functions are not yet clear and we shall not consider it further. The alpha/beta class of TCRs, found on 95% of T cells, consists of two polypeptide chains, each having a molecular weight of about 40,000 (Figure 12–13). There are substantial amino acid homologies with immunoglobulin genes, and the mode of generating diversity is similar.

The TCR alpha cluster (Figure 12–14) consists of about 50 V units and about 100 J units, plus a single C unit. Diversity is generated by random combinations of V and J units. Nucleotides may also be added or deleted at the junctions between V and J. The TCR beta gene cluster contains at most 60 V units, any one of which can be joined at the DNA level to one of two D-J-C clusters (Figure 12–14). The only variable component of the D-J-C clusters is the J units: there are 6 in one cluster and 7 in another. Variability in the precise point of joining, plus the occasional addition of new nucleotides, adds to the potential diversity.

If we consider only the combinatorial possibilities, we see that there may be 5000 different alpha chains and at least 700 beta chains. Inasmuch as antigen-recognition sites in TCRs are formed jointly by the alpha and beta chains, there are potentially at least 3,500,000 such sites. The actual number must be considerably larger, because of the variability in the join-

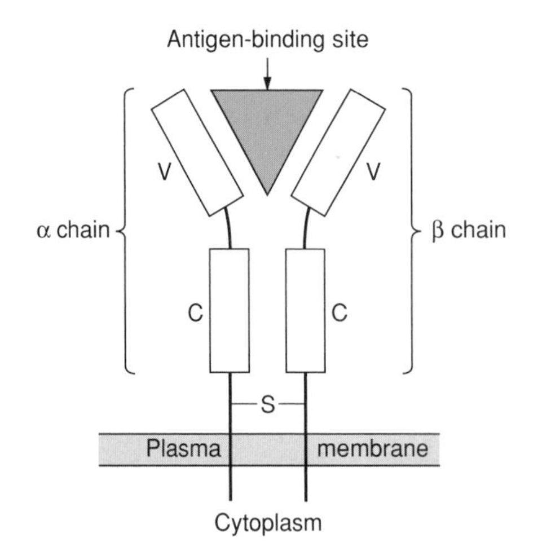

Figure 12–13

Diagram of a T cell receptor, which is composed of an alpha chain and a beta chain. Each polypeptide has a variable portion V and a constant portion C on the cell surface, as well as transmembrane and cytoplasmic portions. The two chains are held together by a disulfide bond S. In contrast to antibodies, T cell receptors have only one antigen-binding site. However, a typical T cell has tens of thousands of these receptors on its surface.

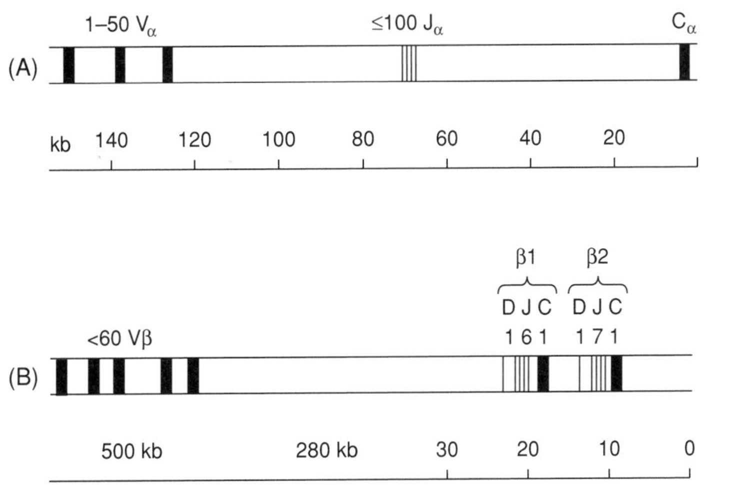

Figure 12–14

Organization of the TCR alpha gene cluster (panel A) and the TCR beta gene cluster (panel B). (From Lewin, 1990, p. 724; by permission of Cell Press)

ing reactions mentioned earlier. Note, however, that there does not seem to be somatic mutation within the V segments of TCR genes.

T cells become activated by interacting with fragments of antigenic macromolecules presented on the surface of macrophages or dendritic cells (Figure 12–15). The latter are many-branched cells found in the thymus, lymph nodes, skin, and elsewhere. Their relationship to macrophages is not yet clear, but they may be the major antigen-presenting cells in the thymus (King and Katz, 1990).

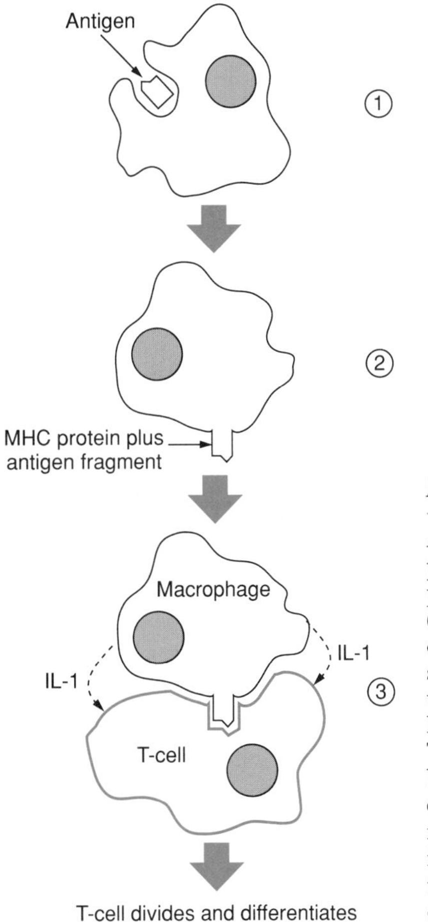

Figure 12–15

Activation of a specific T cell for proliferation. An antigen is phagocytized by a macrophage (step 1), which processes it and displays fragments of it on its surface (step 2) in conjunction with an MHC protein (class I or II). The receptor of an immature T_C cell or an immature T_H cell binds to the antigen-MHC complex (step 3). The macrophage then secretes interleukin-1, which stimulates the T cell to divide and differentiate.

How do T cell receptors recognize antigen? The process is fundamentally different from antigen recognition by B cells. T cells recognize antigen only when it is presented on the surface of another cell in association with one of the molecules of the *major histocompatibility complex* (MHC). We shall now consider the components of the MHC and their functions.

The Major Histocompatibility Complex

A family of genes located on human chromosome 6 codes for the highly polymorphic proteins of the major histocompatibility complex (MHC). The principal function of the MHC proteins is in *antigen presentation* on cellular surfaces (for a recent review, see Harding and Unanue, 1990). T cells interact with the MHC-antigen complexes. T cell responses are absolutely dependent upon recognition of antigens that are held by MHC proteins; soluble antigens are not recognized by T cells.

There are two major classes of MHC proteins. Class I molecules are found on the surface of nearly every cell type. They present antigens derived from intracellular proteins, including proteins normal for that cell type, proteins synthesized by viruses that may be replicating in the cell, and abnormal proteins that may be expressed by a tumor cell. Class II MHC proteins are found primarily on the surface of B cells, macrophages, and dendritic cells (many-branched, antigen-presenting cells that are a prominent component of lymphoid tissues). Antigens that have been taken into a cell via endocytosis are presented on the cell surface in association with class II molecules.

There are three genes in the MHC-I cluster in humans (Figure 12–16). They are called HLA-B, HLA-C, and HLA-A (HLA stands for Human Leucocyte Antigen, an old term that antedates the use of Major Histocompatibility Complex for these genes and their products). Each of them has many alleles; as a group, they are the most polymorphic gene loci known in humans, where there are at least 40 alleles at the A locus, 8 or more at the B locus, and at least 20 at the C locus.

The set of three MHC-I alleles on each chromosome 6 constitutes a haplotype, and because of the great diversity of alleles in the human population, most persons will have two different haplotypes. The probability that two unrelated individuals will have identical alleles at all three MHC-I loci is extremely small. This fact is the bane of organ transplantation programs, but it is one of our greatest bulwarks against infection and against cancer.

Each MHC-I molecule contains one 44,000 Da heavy chain and a 12,000 Da light chain known as beta-2-microglobulin (which is not polymorphic). The complex is anchored to the cell by a transmembrane segment in the heavy chain. The heavy chain has three extracellular sections, known as alpha-1, alpha-2, and alpha-3. The antigen-binding site is formed by the first two domains, while a supporting structure consisting of the alpha-3 domain and beta-2-microglobulin lies between the antigen-binding site and the cell membrane (Figure 12–17).

A cluster of three genes encodes the class II MHC molecules in humans; they are called HLA-DP, HLA-DQ and HLA-DR (Figure 12–16).

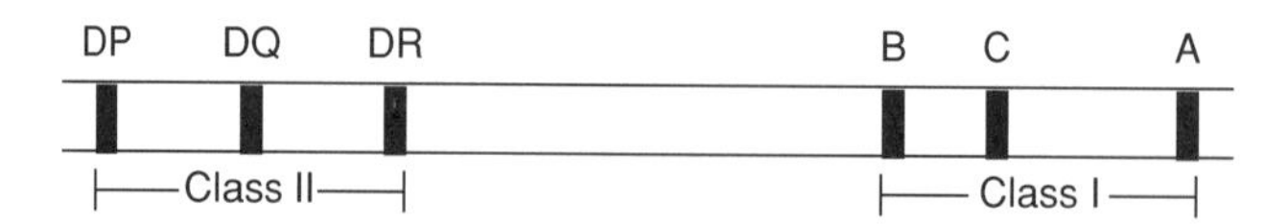

Figure 12–16

Diagram of the human major histocompatibility gene complex on chromosome 6. Several genes for components of complement and for some nonimmune functions are located between the class I and class II regions. The entire complex is at least 500 kbp long.

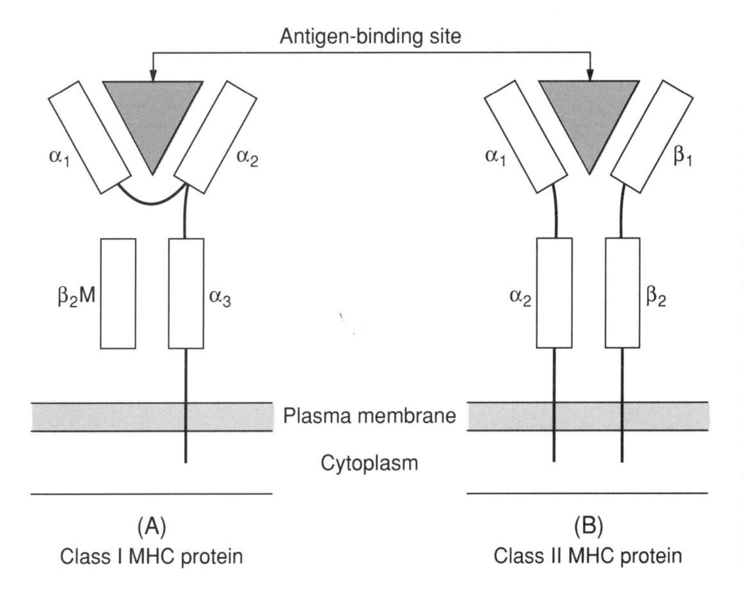

Figure 12–17

(A) Class I and (B) class II MHC molecules. The alpha chain of the class I molecule is noncovalently associated with a smaller polypeptide, beta-2 microglobulin (B_2M), which is not encoded within the MHC. The antigen-binding site of class I molecules is formed by the alpha-1 and alpha-2 domains of the alpha chain. The antigen-binding site of the class II molecules is formed by the alpha-1 and beta-1 domains of the alpha and beta chains, respectively.

MHC-II molecules do not contain beta-2-microglobulin. They consist of an alpha subunit (34,000 Da) and a beta subunit (28,000 Da), each of which contains two domains (Figure 12–17). The alpha-1 and beta-1 domains show homology with the alpha-1 and alpha-2 domains of class I MHC molecules. They are the polymorphic sections of the proteins and together they form the antigen-binding site. The alpha-2 and beta-2 domains of class II MHC proteins, like their functional counterparts in class I molecules, show some amino acid relationships with immunoglobulin molecules.

Both classes of MHC proteins contain antigen-binding sites that can accommodate peptides of 10–20 amino acids. An important difference between the MHC proteins and the antibodies of B cells or the T cell receptors is that diversity is entirely germ-line; there is no gene shuffling. Although there are several dozen alleles at some of the MHC loci, this is population diversity, not diversity within an individual. Every person has at most six alleles at the MHC-I loci and six at the MHC-II loci (only two alleles being possible at any one locus). Therefore, each MHC molecule must be able to bind a large variety of antigens.

The antigens that are held in the clefts of MHC class I and class II molecules are all derived from intact proteins by proteolysis, but there are important differences in the sites of proteolysis. For class I molecules, antigenic peptides are created primarily by the continual process of turnover of intracellular proteins. By a process that is not yet fully understood, fragments of proteins that have been degraded in lysosomes become available to nascent MHC-I molecules while they are being synthesized by ribosomes on the endoplasmic reticulum (Figure 12–18).

In fact, there is recent evidence implying that the heavy chain of MHC-I molecules *must* bind an acceptable peptide before it can bind beta-2-microglobulin and be exported to the cell surface. Not all proteolytic fragments are capable of being bound by a given MHC-I molecule, and although most humans will have six different varieties of MHC-I proteins (the products of two alleles at three loci), some protein fragments will not fit into any binding site in such a way that the signal to mature is given. Thus, there must be a large number of abortive bindings, and in some manner, those non-productive complexes must themselves become grist for the ly-

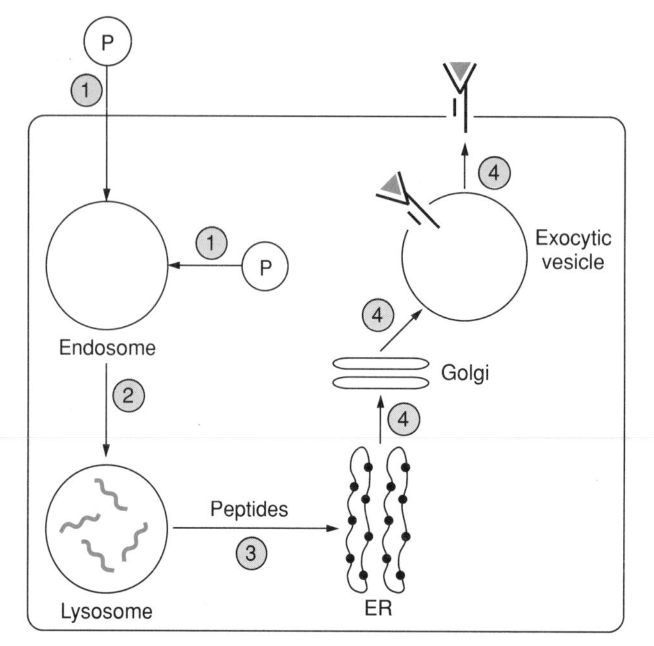

Figure 12–18

Proposed mechanism of antigen presentation by MHC-I molecules. The antigen is either synthesized endogenously or enters the cytosol from endocytic vesicles (1). It is degraded to peptides (2) by lysosomal enzymes. Some of the antigenic fragments bind to newly synthesized MHC-I polypeptides in the endoplasmic reticulum (3). The complex then migrates to the cell surface (4).

sosomal mills. In a limited sense, therefore, the antigen instructs the MHC-I protein how to fold, but of course, it cannot generate a binding site with specificity for only one antigen.

Antigens that are presented by class II MHC molecules are derived from external proteins that have been taken up by endocytosis into one of the three types of class II antigen-presenting cells (macrophages, B cells, and dendritic cells). Some proteolysis takes place in endosomes, and it is there that the antigenic fragments encounter class II MHC proteins (Figure 12–19). Prior to their arrival in endosomes, class II proteins are associated with a separate polypeptide, a molecule of 34,000 Da known as the "invariant" chain. A current hypothesis is that the invariant chain prevents class II proteins from binding intracellular peptides as they pass through the endoplasmic reticulum and the Golgi complex. Upon reaching an endosome, the MHC II-invariant chain complex dissociates and the MHC-II protein binds any peptide that will fit. Finally, the protein-antigen complex moves

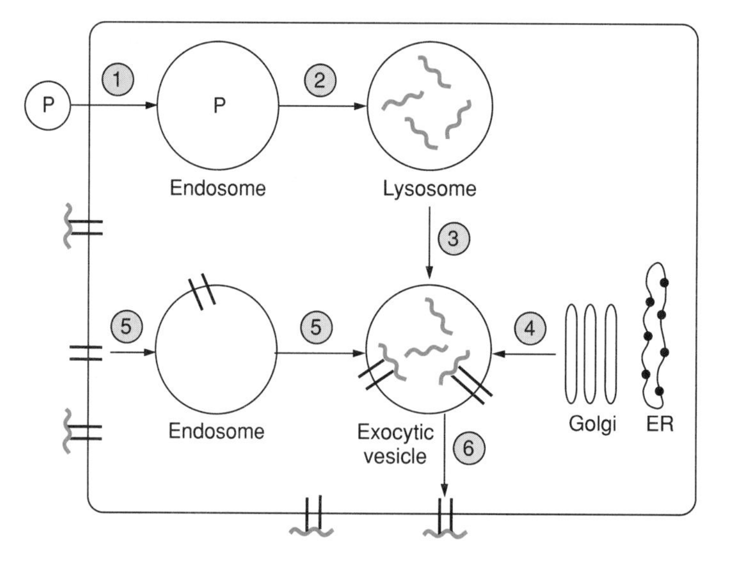

Figure 12–19

Proposed mechanism of antigen presentation by MHC-II molecules. (1) A protein is internalized and then (2) partially fragmented in lysosomes. (3) Some peptides are transported to vesicles containing MHC-II molecules, which (4) may be newly synthesized or (5) recycled from the plasma membrane. (6) Peptide-MHC-II complexes move to the plasma membrane.

to the cell surface, where the antigenic determinant becomes available for monitoring by T cells.

The two branches of the MHC complex have very different functions, which are achieved by interactions with two classes of T cells. Antigens bound to MHC-I proteins are sensed by cytotoxic T cells (T_C cells). If an antigen belongs to normal cells of the host, and if the system is functioning properly, all T_C cells will ignore it. However, if the antigen associated with an MHC-I molecule is caused by a viral infection, an intracellular parasite, or in some cases by a tumor cell protein, it will be recognized as an enemy and T_C cells will lyse the cell bearing that antigen.

Antigens presented by those cells that make class II MHC proteins are sensed by helper T cells (T_H cells). If the antigen fits a T cell's receptors satisfactorily, the T cells will secrete one or more lymphokines, polypeptides that affect the behavior of other cells, as well as themselves. In the case of a B cell, interaction with an appropriate T_H cell is required for maturation into a plasma cell and full-scale antibody production (see Figure 12–2). The helper T cell secretes several lymphokines that stimulate growth and differentiation of the B cell.

One aspect of the recognition of antigens by both cytotoxic and helper T cells depends upon cell surface proteins that associate with the TCRs and interact with constant regions of the MHC protein-antigenic determinant complex on the other cell. The protein that recognizes MHC-I proteins is found on T_C cells and is called CD8. (CD stands for "cell determinant," an immunological term applied to these molecules before anything was known about their function). T_H cells have CD4, which detects some constant feature of the MHC-II proteins on target cells.

Another protein that associates with the TCR-CD4 and TCR-CD8 complexes is CD3, which spans the cell membrane and interacts with a protein tyrosine kinase on the cytoplasmic side. It is likely that the kinase is part of the pathway by which signals generated via the TCRs at the surface of T cells are conveyed to the cell's interior, but the details are not yet known.

Tolerance and Autoimmunity

The ability of the antibody-mediated and cell-mediated aspects of the immune system to respond to virtually any foreign protein identifies a paradox. Why don't we attack our own cells? It would seem logical that, during the random combinations of V, D, and J units that occur as antibody genes and TCR genes mature, some B cells and T cells with specificity for self-antigens must be produced. However, as long as the immune system is functioning normally, there is *tolerance* to self (Nossal, 1989).

Tolerance: Clonal Deletion and Clonal Anergy

The basic mechanism of tolerance induction appears to be *clonal deletion* (also called clonal abortion by some investigators). T cells undergo a still-mysterious process of maturation in the thymus of young mammals, during which time they acquire functional T cell receptors. Studies on transgenic mice have shown that T cells capable of reacting with self-antigens do arise, but as a result of a totally obscure mechanism, they are mostly destroyed in the thymus.

In the case of B cells, which are not thymus-dependent for their maturation, a different mechanism for self-tolerance must exist. It has been shown that self-reactive B cells do exist in mice, but the vast majority of them make only IgM, while only a tiny fraction make IgG capable of binding a specific antigen. What prevents the former from being stimulated by self-antigens and forming a clone of antibody-producing plasma cells? The most likely explanation is that full activation of a B cell requires a T_H cell that also recognizes the same antigen, and will stimulate the B cell by binding to it and secreting interleukin-2. For self-antigens, the probability that both a responsive B cell and a responsive T cell will be simultaneously present is negligible.

B cells that have not been stimulated to produce antigen generally live only a few days (unless they are memory cells). Accordingly, if a self-reactive B cell has not been found by a helper T cell with specificity for the same antigen, the errant B cell will soon disappear. In effect, this constitutes clonal deletion of B cells. It is a passive form of clonal deletion, in contrast to T cell clonal deletion, which may be achieved by an active, but still unknown, process.

Clonal deletion is not 100% efficient, and it can be argued that it should not be, because this might compromise the organism's ability to react with foreign antigens. Recall that the response to any foreign antigen involves a spectrum of B and T cells: some bind epitopes from the antigen with great avidity, some bind with intermediate affinity, and some bind weakly. A T or B cell that binds weakly to a self-antigen might be useful in repelling an invading pathogen that carries a closely related antigen to which the cell can bind strongly.

The incompleteness of clonal deletion is supported by evidence that self-reactive T_C cells actually are present in normal mice at low frequency. However, they exhibit *clonal anergy*; that is, they don't attack the organism to which they belong. We do not yet know what turns them off, but it is likely that some second signal, analogous to the signal given to B cells by T_H cells, is missing.

Autoimmune Diseases

When the mechanisms of tolerance fail and an immune system attack against one's own tissues occurs, an autoimmune disease results. We do not yet understand the molecular details of any autoimmune disorder, but there is mounting evidence that autoimmunity is at least one factor in many common diseases.

Insulin-dependent diabetes mellitus (IDDM, or juvenile onset diabetes) arises when the beta cells of the islets of Langerhans in the pancreas are destroyed by autoantibodies. There is an impressive correlation with possession of class II histocompatibility alleles DR3 or DR4; 95% of patients with IDDM have one or the other. However, recent evidence suggests that the actual disease susceptibility locus is more likely to be the gene for the beta chain of the DQ protein, which is produced by another class II gene (Todd, 1990).

In any case, 40% of the general population also have either DR3 or DR4, so there must be other factors involved in the onset of IDDM. That conclusion is confirmed by studies on identical twins, which show 36% concurrence (presence of IDDM in both twins). It has often been suggested that individuals with DR3 or DR4 class II MHC genes are exceptionally susceptible to a virus, and that the viral infection leads directly or indirectly to the destruction of the insulin-producing beta cells, but there is no proof of that possibility at this time.

At least 10 serious human diseases show a significant association with possession of a particular HLA antigen, either class I or class II. In addition to IDDM, examples include systemic lupus erythematosus and celiac disease, both of which are associated with class I allele B8. The fact that individuals with one of those diseases do not ordinarily have the other disease emphasizes the role that additional, undefined factors must play.

The most striking correlation between an HLA antigen and a disease links antigen B27 and *ankylosing spondylitis*. Antigen B27 is present in 89% of persons with the disease (96% of Caucasians with the disease), but in only 9% of normal persons (7% of Caucasians). Ankylosing spondylitis is characterized by an inflammation of the joints, especially where tendons and ligaments attach to the spine and hip. It leads eventually to fusion of the vertebral and para-vertebral joints, resulting in the bent-over habitus that is a caricature of the "little old person." The precise molecular target of this apparently autoimmune reaction is not yet known. And, as is true for autoimmune diseases in general, more is required than presence of a given HLA antigen. The disease develops in about one Caucasian in 1000, and somewhat less in other races. Only about one person in 50–100 with the B27 antigen will develop ankylosing spondylitis (Benjamin and Parham, 1990).

Deficiency Disorders of the Immune System

In the first section of this chapter, we learned about somatic mutations that produce a constructive effect, by generating additional diversity in antibody genes. However, as with any genes, there can be mutations in the germ-line that destroy or diminish the function of components of the immune system. The most dramatic effects are produced by mutations that eliminate the entire B cell or T cell populations, or both.

Agammaglobulinemia is a X-linked recessive disorder characterized by total absence of B cells; an as-yet-unidentified step in the maturation of B cells from stem cells in the bone marrow is blocked. Infants are essentially normal for about one year, because of the presence of IgG antibodies from the maternal circulation. After that, they succumb quickly to bacterial infections, unless they receive regular injections of gamma globulin with antibodies to common bacterial diseases. The reciprocal defect—absence of T lymphocytes—occurs in patients with no thymus gland (DiGeorge syndrome).

Severe combined immunodeficiency disease (SCID) patients lack both B cells and T cells; they die in infancy unless kept in a germ-free enviornment. There are probably several genetic lesions that can produce SCID. The one that is currently receiving intense attention is the result of *adenosine deaminase deficiency*. For a reason that is still obscure, the buildup of adenosine and/or deoxyadenosine that occurs in body fluids of persons with this deficiency is selectively toxic to both B cells and T cells, or their precursors. Adenosine deaminase deficiency is the first target of gene therapy trials in humans (Chapter 10).

Another mutation that could, in principle, produce SCID is a defect in the recombinase that carries out V-(D)-J joining in B and T cells. At least two candidate genes are currently under investigation. If one turns out to be the catalytic polypeptide, the other may be an accessory factor. Indeed, the complete lymphoid-specific recombinase may be a complex protein; that would imply the possibility of SCID arising from mutations in any of the corresponding genes.

There are probably a large number of more subtle effects caused by mutations in the genes of the immune system. It seems safe to predict that mutations must occur in one or more of the V, D, and J combining segments of antibody and TCR genes, mutations which may eliminate an individual's ability to respond effectively to a particular pathogen. Someday, perhaps, we may find that there is a genetic reason why John never gets strep throat, while Mary is afflicted with it almost annually. We may also learn that some families who are prone to a particular cancer lack T cells that are able to attack the tumor with sufficient vigor to destroy it in an early stage. Such information will significantly enhance the power of preventive medicine.

SUMMARY

The two major components of the immune system are B cells, which produce antibodies, and T cells, which destroy a variety of targets by cell-cell interactions and also modulate the response of B cells. The major histocompatibility complex (MHC) contains the principal determinants of self-nonself recognition and has an essential role in antigen presentation to T cells.

B cells are lymphocytes that carry antibodies on their surface and respond to antigens by proliferating and differentiating into plasma cells, which secrete large amounts of antibodies. Each B cell is specialized for the production of only one specific antibody.

Antibodies are tetrameric proteins consisting of two light chains and two heavy chains, which form a Y-shaped structure. The distal portion of the arms of the Y forms the antigen-recognition site. Humans can make at least tens of millions of different antibodies.

Antibody diversity is not encoded by millions of genes. Instead, there are sets of gene subunits that are combined more-or-less randomly during differentiation of a B cell precursor. This is an important exception to the epigenetic constancy of DNA. There are three types of subunits that form light chains and four types of subunits that form heavy chains.

The combinatorial possibilities account for the human capacity to produce millions of different antibodies. Additional diversity is produced by variation in the precise point at which the different subunits of genes are joined and by somatic mutations that occur with high frequency during maturation of plasma cells.

Five classes of immunoglobulins (antibodies) are known. A naive B cell carries class M antibodies (IgM) on its surface. Following stimulation by an antigen and a helper T cell, the B cell begins to divide and differentiate. At the same time, IgMs are made in a secreted form. By the time fully differentiated plasma cells form, they usually switch from the synthesis of IgM to IgG, the most abundant type of antibody. Other types are IgD, IgA, and IgE, which function in special ways.

Lymphocytes whose maturation requires a period of residence in the thymus are called T cells. They are the agents of cell-mediated immunity. Cytotoxic T cells (T_C) lyse host cells with foreign antigenic determinants held on their surface by MHC proteins. Helper T cells (T_H) stimulate other cells of the immune system; for example, a full B cell response to an antigen requires interaction with a T_H cell that recognizes the same antigen. Suppressor T cells modulate responses of helper T cells.

T cells recognize foreign antigens via T cell receptors, variable proteins on the cell surface with some similarity to antibodies. T cell receptor diversity is also generated by a series of genomic rearrangements.

The major histocompatibility complex on chromosome 6 encodes a family of cell surface proteins that function in antigen presentation to T cells. Class I MHC proteins are found on nearly every cell type; they present antigens derived from proteins synthesized within a cell, whether normal or abnormal. Class II MHC proteins are found on B cells, macrophages, and some other lymphoid cells. They present antigens that have been taken into a cell by endocytosis.

Each of the three genes of the MHC-I cluster in humans has many alleles. As a result, the two haplotypes present in one person are almost never identical to the haplotypes in another person. This is the major reason for rejection of organs transplanted from one person to another. MHC-II genes also have many variants. There is no gene shuffling involved in MHC protein diversity.

Both classes of MHC proteins contain antigen-binding sites that can hold peptides of 10–20 amino acids, which are produced from larger molecules by proteolysis. Antigens bound to MHC-I proteins are recognized by cytotoxic T cells; if the antigen represents a foreign protein, the T_C cells will lyse the cell that carries it. T_H cells recognize antigens associated with MHC-II proteins; this stimulates the T cell, which in turn stimulates B cells.

Tolerance to self-antigens is not fully understood. During maturation of the thymus in young mammals, including humans, T cells capable of reacting with self-antigens are destroyed. This process of clonal deletion is not perfect, and mechanisms to suppress the activity of anti-self T cells must exist. Differentiation of anti-self B cells appears to be largely suppressed by the lack of corresponding helper T cells.

Autoimmune diseases result from failures of self-tolerance. Examples include juvenile onset diabetes and ankylosing spondylitis.

Inherited deficiencies of the immune system include agammaglobulinemia (absence of B cells) and DiGeorge syndrome (absence of T cells). Severe combined immunodeficiency disease (SCID) patients lack both B cells and T cells. SCID can arise in several ways, one of which is adenosine deaminase deficiency, a defect that is a candidate for gene therapy.

REFERENCES

Alberts, B., Bray, D., Lewis, J., Raff, M., Roberts, K., and Watson, J. D. 1989. Molecular biology of the cell, 2nd ed. Garland Publishing, New York.

Benjamin, R. and Parham, P. 1990. Guilt by association: HLA-B27 and ankylosing spondylitis. Immunol. Today 11:137–142.

Burnet, F. M. 1962. The integrity of the body. Harvard University Press, Cambridge, MA.

Dreyer, W. J. and Bennett, J. C. 1965. The molecular basis of antibody formation: a paradox. Proc. Natl. Acad. Sci. USA 54:864–869.

Hozumi, N. and Tonegawa, S. 1976. Evidence for somatic rearrangement of immunoglobulin genes coding for variable and constant regions. Proc. Natl. Acad. Sci. USA 73:3628–3632.

King, P. D. and Katz, D. R. 1990. Mechanisms of dendritic cell function. Immunol. Today 11:206–211.

Koehler, G. and Milstein, C. 1975. Continuous cultures of fused cells secreting antibody of predefined specificity. Nature 256:495–497.

Lewin, B. 1990. Genes IV. Oxford University Press, New York, and Cell Press, Cambridge, MA.

Mange, A. P. and Mange, E. J. 1990. Genetics: human aspects, 2nd ed. Sinauer Associates, Sunderland, MA.

Marrack, P. and Kappler, J. 1987. The T cell receptor. Science 238:1073–1079.

Nossal, G. J. V. 1989. Immunologic tolerance: collaboration between antigen and lymphokines. Science 245:147–153.

Prescott, D. M. 1988. Cells. Jones and Bartlett, Boston.

Strominger, J. L. 1989. Developmental biology of T cell receptors. Science 244:943–950.

Todd, J. A. 1990. Genetic control of autoimmunity in type 1 diabetes. Immunol. Today 11:122–128.

Tonegawa, S. 1983. Somatic generation of antibody diversity. Nature 302:575–581.

Winter, G. and Milstein, C. 1991. Man-made antibodies. Nature 349:293–299.

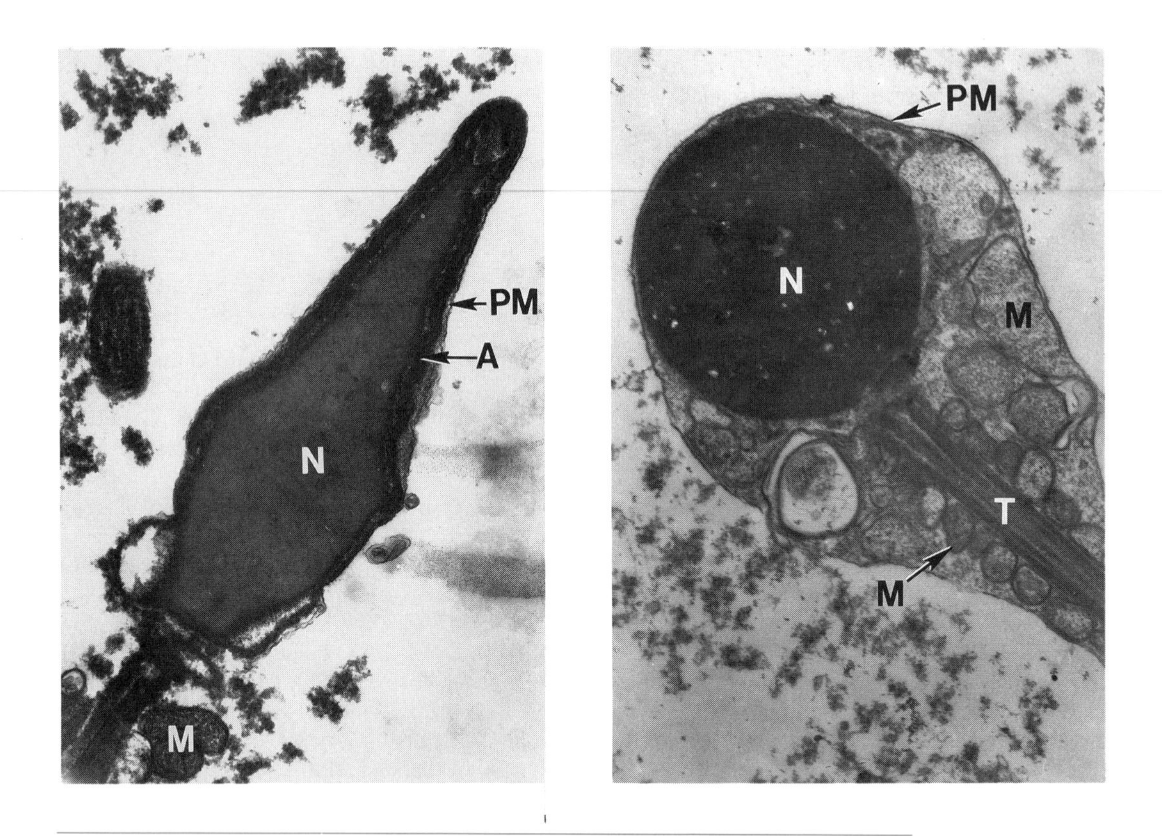

Left panel: a normal human sperm. Right panel: a sperm from an infertile man with acrosomal agenesis, a genetic defect. A = acrosome, M = mitochondria, N = nucleus, PM = plasma membrane, T = tail. (Courtesy of Jonathan Van Blerkom)

13 The X and Y Chromosomes

The fact that human sex is correlated with differences in chromosomal content was established early in the twentieth century, when the names X and Y were given to the sex-associated chromosomes by E. B. Wilson. Humans are typical mammals in this regard; females are the *homogametic* sex, with two X chromosomes in somatic cells, and males are the *heterogametic* sex, with one X and one Y chromosome. In other phylogenetic groups, sex determination can be quite different. Birds, for example, have homogametic males (with ZZ chromosomes) and heterogametic females (with ZW chromosomes). In some reptiles, sex is determined by the temperature at which the eggs develop.

Fruitflies also have XX females and XY males, but in this case, it is the ratio of X chromosomes to autosomes that determines sex; one X produces males, while two Xs produce a female in normal diploid flies. This relationship may have been partly responsible for the fact that the male-determining function of the human Y chromosome was not established until 1959, when improved methods for karyotype analysis demonstrated that XO individuals are phenotypic females and XXY individuals are phenotypic males (see Chapter 7)

In this chapter we shall first outline the main features of the molecular biology of the sex chromosomes (or *gonosomes*). A review of the principles of sex-linked inheritance, as seen in pedigrees, will then be given, followed by a few examples of X-linked diseases.

Molecular Biology of the X Chromosome

The human X chromosome contains approximately 150 Mbp of DNA, which constitutes about 5% of the haploid genome. In terms of genetic organization, the X appears to be quite similar to an average autosome, and genes on the X chromosome are no more likely to be connected with sexual differentiation than are autosomal genes. The density of mapped genes is currently greater for the X chromosome than for any autosome, because the unique pattern of inheritance of X-linked genes makes them easy to recognize. At this time, nearly 200 structural genes have been assigned to the X chromosome, together with more than 400 anonymous DNA segments. The recombination map is known in considerable detail; the total genetic length of the X chromosome exceeds 200 cM. A reference set of polymorphic marker loci has been mapped with precision; it is shown in Figure 13–1.

X chromosome genes have been less mobile during evolution than autosomal genes. As Ohno pointed out (1969), a gene found on the X chromosome in one mammalian species is highly likely to be found on the X chromosome in virtually all other mammals. Presumably this conservative aspect of genetic geography, often referred to as *Ohno's Law*, is related to the need for dosage compensation of X-linked genes, as explained next.

Random Inactivation of X Chromosomes in Females

Dosage compensation refers to mechanisms that lead to the same amount of expression for X-linked genes in females as in males, even though females have two X chromosomes, while males have only one X chromosome. Dosage compensation in placental mammals is achieved by random inactivation of most genes on one X chromosome during female embryogenesis. The first indication that one of the two X chromosomes in normal female mammals behaved differently from the other was the observation by Barr

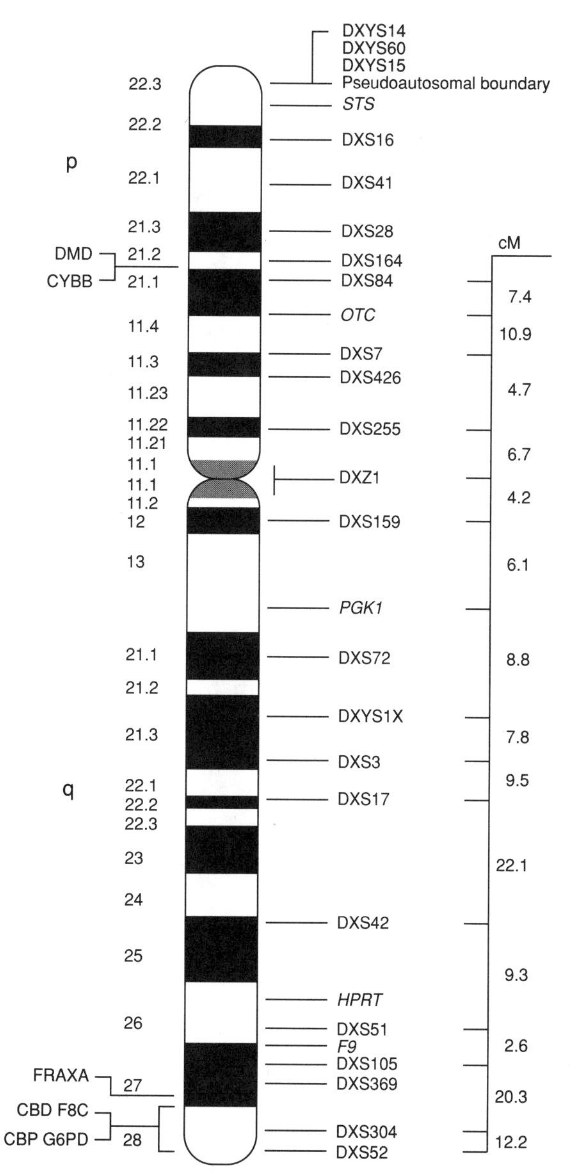

Figure 13–1

A map of the human X chromosome, with the locations of a set of well-defined polymorphic marker loci shown to the right of the chromosome. (From Davies et al., 1990, p. 259). Some of the loci are protein-coding genes (*STS* = steroid sulfatase, *OTC* = ornithine carbamoyltransferase, *PGK1* = phosphoglycerate kinase 1, *HPRT* = hypoxanthine-guanine phosphoribosyl transferase, *F9* = coagulation factor IX); the others, with names beginning with DXS, are single-copy anonymous DNA segments; and DXZ1 is a reiterated sequence at the centromere. The column on the far right gives the recombination fractions between some of the loci (From Keats et al., 1990, p. 394.) At the left of the chromosome are the numbers of the bands, and extending out to the left are the names and approximate locations of other genes discussed in this chapter and in Chapter 5 (DMD = Duchenne muscular dystrophy, CYBB = chronic granulomatous disease, FRAXA = site involved in FRA-X syndrome, CBD and CBP = genes for red and green color vision, F8C = coagulation factor VIII, and G6PD = glucose-6-phosphate dehydrogenase). No order is implied for the latter four genes, all of which are in Xq28.

and Bertram (1949) that interphase cells from females contain a heterochromatic spot (a dark-staining mass, now known as a *Barr body*) that is absent from males (Figure 13–2). They deduced that it represented one of the X chromosomes.

Subsequent studies on humans with sex chromosome aneuploidies showed that the number of Barr bodies per cell was always one less than the number of X chromosomes, and that the number of Y chromosomes did not affect the number of Barr bodies. For example, there are no Barr bodies in Turner syndrome patients (XO), there is one Barr body per cell in Klein-felter patients (XXY), and in XXX, XXXY, or XXXYY individuals there are two Barr bodies per cell. Another fact that emerged was that Barr body DNA, like heterochromatin in general, replicated later in the S phase of the cell cycle than euchromatic DNA. Late replication correlates positively with genetic inactivity.

These facts led Mary Lyon to propose (1961) the following, which has subsequently become known as the *Lyon hypothesis*.

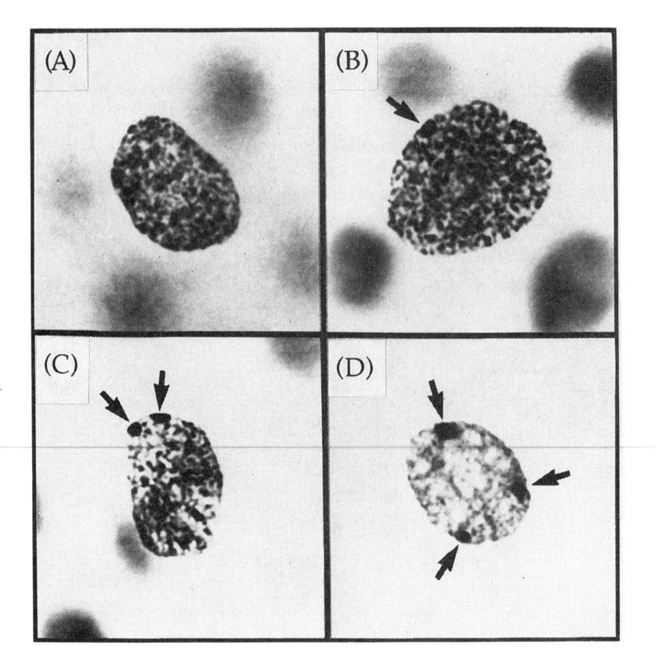

Figure 13–2

Barr bodies, as seen in interphase nuclei. (Panel A): A normal XY male with no Barr body. (Panel B): A normal XX female, with one Barr body (arrow). (Panel C): An XXX female, with two Barr bodies. (Panel D): An XXXX female, with three Barr bodies. (Courtesy of Arthur Robinson)

1. The heterochromatic X chromosome is inactive in terms of gene expression.

2. Inactivation, which occurs early in embryogenesis, is random; that is, either the paternal or maternal X in a given cell may be inactivated, and there is no preference for one or the other.

3. Once an X chromosome has been inactivated in a given cell, the same X chromosome will be inactive in all the somatic cell descendants of that cell. Inactivation is permanent.

The Lyon hypothesis has been extensively verified (Lyon, 1988). We now know that inactivation of one of the two X chromosomes in females occurs at the blastocyst stage, when there are 200–400 cells in a human embryo. Inactivation begins in the cells of the inner cell mass, from which the embryo arises, and spreads to the trophoblast somewhat later. Inactivation even involves the primordial germ cells, but when primary oocytes differentiate somewhat later in development, the inactive X is re-activated in the oocytes. Thus, only active X chromosomes are present in mature germ cells and during the earliest stages of embryonic development.

The most compelling evidence for the randomness of X chromosome inactivation came from studies of individuals who were heterozygous for X-linked genes with easily distinguished alleles. One of those genes codes for glucose-6-phosphate dehydrogenase (G6PD), which has relatively common alleles whose products can easily be separated by electrophoresis. Skin cells from individuals who were heterozygous for the *G6PD*A* and *G6PD*B* alleles were cloned; that is, single cells were allowed to multiply in culture until there were enough descendants to make enzyme assays possible. About half the clones expressed the **A* allele and the other half expressed the **B* allele; no clones expressed both alleles (Figure 13–3). This strongly implied that one of the two X chromosomes in each cell had been inactivated before the cell was cloned and that the inactivated state could persist through many cell generations in culture. Several other X-linked genes have given similar results.

Inactivation of a given chromosome begins at a point near the centro-

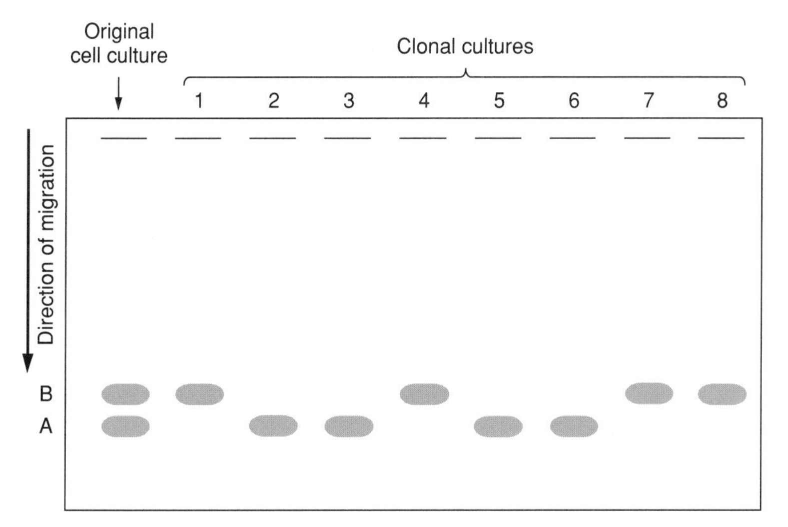

Figure 13–3

Electrophoretic pattern of G6PD isozymes of cell cultures from a woman heterozygous for the A and B alleles. The original cell culture, being non-clonal, shows both A and B bands. In clonal cultures 1–8, either the A band or the B band is found, but not both.

mere on the long arm (the inactivation center) and spreads in both directions toward the telomeres. We do not yet understand the molecular basis for X chromosome inactivation, although there is a large amount of evidence implicating DNA methylation as part of the process (methylation and gene inactivation were discussed in Chapter 8). Certainly, the inactive X chromosome is more heavily methylated than the active X, and demethylation of individual genes on an inactive X chromosome (either as an accident of nature or by experimental means) tends to be associated with activation of those genes. Nevertheless, it is not clear whether methylation is a primary or secondary event in the inactivation process. Probably the most mysterious aspect of the whole process is the mechanism by which the cell knows that all X chromosomes except one have been inactivated.

Another interesting aspect of X chromosome inactivation has been revealed by studies on *age-related reactivation* of X-linked genes. The general finding is that occasionally a gene on the inactive X may be reactivated, and that the frequency of such events increases as animals age. One clever experimental system involved a strain of mice that were heterozygous for a wild-type allele and a null allele in the tyrosinase gene (Cattanach, 1974). Tyrosinase is required for melanin production and is normally encoded by an autosomal locus; but in these mice, the autosomal fragment carrying the wild-type allele had been inserted into the X chromosome. As a consequence of random X inactivation, these mice had fur that was mottled black and white. But the older the mice grew, the more pigmented they became, thus reversing the usual effect of aging on hair color.

Some X-linked Genes Are Not Inactivated

If all the genes on the inactive X chromosome were inactivated, then it would be paradoxical that people with aneuploidies of the X chromosome are not fully normal; XO individuals should have all the X-encoded gene activity needed for normality and XXY individuals should have no unnecessary activity from the second X. In fact, it is clear that some genes remain active in both X chromosomes in normal females. Therefore, one can argue that persons with Turner syndrome suffer from a deficit of one or more gene products, while persons with Kleinfelter syndrome and other individuals with more than the usual number of X chromosomes owe their problems to a surfeit of those gene activities.

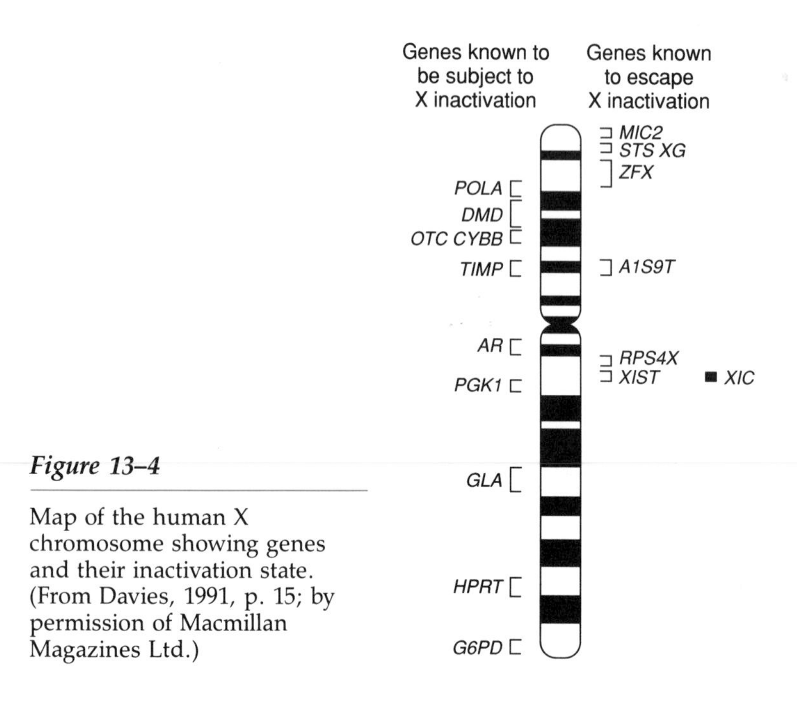

Figure 13–4

Map of the human X chromosome showing genes and their inactivation state. (From Davies, 1991, p. 15; by permission of Macmillan Magazines Ltd.)

The first genes that were found to remain active on the inactive X chromosome in normal females were all located near the tip of the short arm. They were: (a) the *MIC2* locus, which codes for a cell surface antigen; (b) the *Xg* locus, which codes for an antigen on the surface of red blood cells; and (c) the *STS* locus, which codes for the enzyme, steroid sulfatase. At that stage of knowledge, it was plausible to conclude that inactivation proceeded uniformly from the inactivation center, affecting all genes in its path until it terminated at a point proximal to the *STS* locus.

However, recent findings have identified additional non-inactivated genes (Davies, 1991), and it is clear that some of them are interspersed among otherwise inactive regions (Figure 13–4). *RPS4X* is unusual in that it is the only known gene on the long arm of the X chromosome that is active on both X chromosomes. The possibility that inactivation proceeds on a gene-by-gene basis is receiving careful consideration as a consequence of these findings.

Another newly described gene on Xq is *XIST* (Brown et al., 1991), which may be the locus corresponding to the X inactivation center (XIC); that is, the point at which inactivation of one X chromosome begins in each cell of the mammalian blastocyst. *XIST* has the unique property of being active on the inactive X chromosome, but inactive on the active X chromosome. An RNA product of the *XIST* locus has been found. Sequence analysis shows that it contains many stop codons, so it presumably does not code for a protein. The possibility that X chromosome inactivation is mediated by this RNA is under active consideration (Lyon, 1991).

Molecular Biology of the Y Chromosome

The human Y chromosome represents slightly more than 2% of the haploid DNA content; it contains about 65 Mbp. It is the third smallest human chromosome. The distal half of the long arm is heterochromatin, which can vary in amount from individual to individual without any apparent adverse consequences. A diagram of the human Y chromosome is presented in Figure 13–5.

The Pseudoautosomal Region (PAR)

This region is defined as those parts of the X and Y chromosomes that have a high degree of sequence homology and where crossing-over occurs during male meiosis. Pairing of the X and Y chromosomes actually involves most of the short arm of Y and an equal (but proportionately smaller) amount of X; but normally, crossing-over takes place only within the 2.6 Mbp at the tip of both chromosomes. Inasmuch as there are two copies of genes in the pseudoautosomal region, their location on the sex chromosomes cannot be deduced by ordinary pedigree analysis; the word "pseudoautosomal" refers to this fact (Burgoyne, 1982).

So far, only two structural genes are known to occur within the PAR in humans: *MIC2* and *CSF2R*. The function of the cell surface protein encoded by *MIC2* is not yet known. Granulocyte-macrophage colony stimulating factor (*CSF2R*), which is required for normal differentiation of some white blood cells, is also encoded by a gene in the pseudoautosomal region of the X and Y chromosomes (Gough et al., 1990). Several anonymous DNA segments that detect RFLPs have been assigned to the PAR. Their recombination frequencies (relative to maleness) are directly proportional to their physical distances. This implies that double crossovers do not occur within the PAR.

Occasionally, crossing-over is inaccurate and one or more genes proximal to the pseudoautosomal region may be transferred from X to Y or vice versa. As we shall see in the following discussion, such errors in sex chromosome recombination have been crucial in the search for the gene that determines maleness.

The Search for the Testis-determing Factor

We learned in the introduction to this chapter that the presence of one Y chromosome is sufficient to produce phenotypic maleness (at least male genitals), regardless of the number of X chromosomes in an individual genome. From this simple fact one can deduce that there must be at least one gene on the Y chromosome that controls the developmental switch into the pathway for male sexual differentiation. That gene has been generally referred to as *TDF*—the *testis-determining factor* gene. Since 1959, when the essential role of the Y chromosome was first established, the search for *TDF* has progressively narrowed the possibilities (Figure 13–6; McClaren, 1990; Cooke, 1990).

In 1966 the discovery that some XY females had lost the short arm of the Y chromosome, while retaining an isochromosome of Yq, showed that *TDF* must be on Yp (Jacobs and Ross, 1966). For a number of years, the

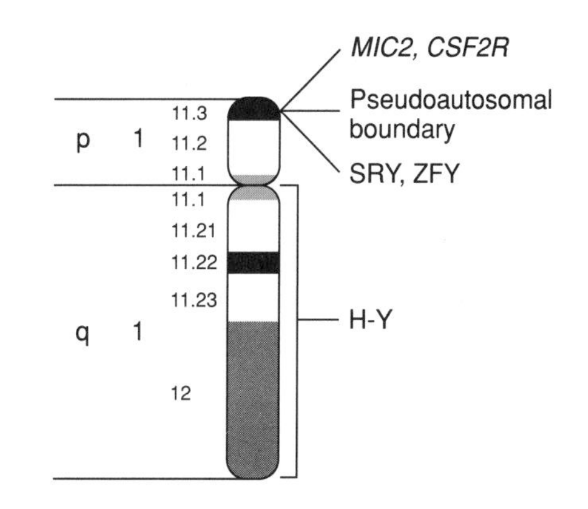

Figure 13–5

Diagram of the human Y chromosome. The line indicating the pseudoautosomal boundary is approximate. Location of the gene for H-Y antigen within the long arm is not precisely known. In addition to the genes indicated in the figure, several dozen anonymous DNA sequences have been assigned to the Y chromosome; a few of them come from the pseudoautosomal region. The gray region at the end of the long arm is heterochromatin, which is highly variable in amount.

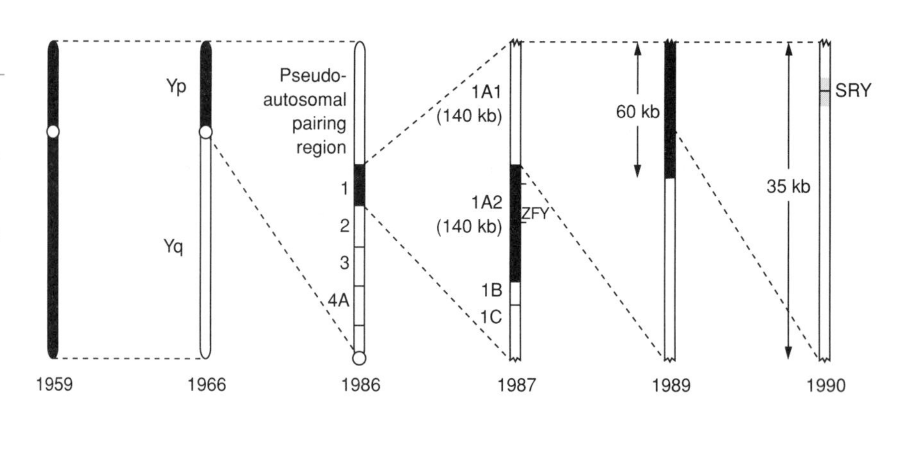

Figure 13–6

The search for the elusive testis-determining factor. The chromosomal region thought to include the factor (indicated in black) has shrunk from the entire Y chromosome in 1959 to less than 250 bases encoding the conserved 80-amino-acid motif of SRY in 1990. (From McLaren, 1990, p. 216; by permission of Macmillan Magazines Ltd.)

favorite candidate for *TDF* was the gene for H-Y, the male-specific histocompatibility antigen (see the next section). However, when mice with testes but no H-Y antigen were discovered, the case for H-Y as the human *TDF* became very weak; and the hypothesis was totally refuted when it was finally shown that the gene for H-Y maps to the long arm of human Y.

Identification of the pseudoautosomal region at the tip of Yp implied that *TDF* must lie proximal to PAR; that is, *TDF* could not be the male-determining gene if it were frequently exchanged with a homolog on the X chromosome. When Page et al. (1987) showed that XX phenotypic males often contained a short piece of Yp that originated just proximal to the PAR, and that XY females often lacked that piece of Yp, the probability that *TDF* was close to the PAR boundary (in region 1, Figure 13–6) rose dramatically.

Page et al. (1987) went on to study the PAR-proximal region in more detail and identified a gene (*ZFY*) that became an attractive candidate for *TDF*. It encoded a "zinc-finger" protein, that is, it belonged to a known class of transcription factors, and the idea that the TDF protein might act by controlling the expression of other genes was very appealing. However, analysis of more XX males identified several who lacked *ZFY*, and other data on expression of *ZFY* in mice indicated that it was unlikely to be *TDF* (Craig, 1990), because it was not active in the genital ridge, where testis development begins.

Finally, another gene was found within a few kilobases of the PAR boundary (Sinclair et al., 1990; Gubbay et al., 1990). It has been named *SRY* (sex-determining region of the Y chromosome). It is related to several known DNA-binding proteins, including the Mc protein, which is required for mating in fission yeast. More evidence that *SRY* is *TDF* comes from mice, where *SRY* is expressed in the urogenital ridge at a time in embryonic development just prior to overt sexual determination. The PAR-adjacent location of the *SRY* gene also suggests that no more than a small error in the position of X-Y recombination will move *SRY* from its proper location on the Y chromosome to the X chromosome, thus creating the possibility of an XX male and an XY female offspring.

Compelling evidence that *SRY* is *TDF* came from experiments with transgenic mice by Koopman et al. (1991). A 14 kbp DNA fragment containing *SRY* and some adjacent DNA that was expected to include regulatory sequences was injected into fertilized mouse eggs, which were then transferred to pseudopregnant females, where they developed normally. Among the progeny were several mice that were chromosomally female (XX), but phenotypically male, with fully developed genitalia. This dramatic result shows that *SRY* and its flanking sequences represent all the genetic information needed to switch sexual differentiation into the male pathway. Now the contest to unravel the regulatory cascade that the *SRY* gene product must initiate will begin.

Other Y-linked Genes

The *H-Y antigen* gene was the first gene correctly assigned to the Y chromosome. It was shown by Eichwald and Silmser (1955), using an isogenic strain of mice, that males would accept skin grafts from females, but females would reject skin grafts from males. This defined a histocompatibility antigen present only in males, which led to the name H-Y. Further research showed that males have the H-Y antigen on the surface of nearly all cells. The gene has been mapped to the long arm of the Y chromosome, but its function has not yet been clarified. There have been suggestions that it has some role in spermatogenesis, but other observations imply a role in growth control. The latter role would help to explain presence of the H-Y antigen on cells outside of the gonads (Heslop et al., 1989).

The gene for the H-Y antigen occurs in all vertebrates and is always characteristic of the heterogametic sex (e.g., male mammals but female birds). The gene has been closely conserved during evolution.

Very few other genes have been conclusively shown to reside on the Y chromosome. The genes for the surface antigen *MIC2* and the growth factor *CSF2R* were already mentioned in the discussion on the pseudoautosomal region. There is indirect evidence that one or more genes affecting stature are on the Y chromosome, and there may be another gene that affects tooth size. Overall, the Y chromosome is notable for its apparent lack of genes.

Patterns of Inheritance

Genes on the X or Y chromosomes, with the exception of those in the pseudoautosomal region, have unique patterns of inheritance which enable them to be readily distinguished from autosomal genes, provided that adequate pedigree information is available.

X-linked Genes

Recessive Alleles The inheritance of recessive mutations in X-linked genes has the following characteristics, which are also illustrated in Figure 13–7.

1. Males are more frequently affected than females. This is a direct consequence of the presence of a single X chromosome in male mammals; that is, males are *hemizygous* for X-linked genes. If a mutation is a reproductive lethal, only males will be affected by mutant alleles that are true recessives. If some reproduction by affected males occurs, then homozygous affected females are a possibility (Figure 13–7).

If a mutation has essentially no effect on reproduction (e.g. red/green color blindness), then the frequency of affected females in an equilibrium population will be the square of the frequency of affected males (Figure 13–8). This generalization is not rigorous, because the randomness of X chromosome inactivation makes every

Figure 13–7

Pedigree of color blindness. Note the affected female in generation II. Circles with central dots represent heterozygous females.

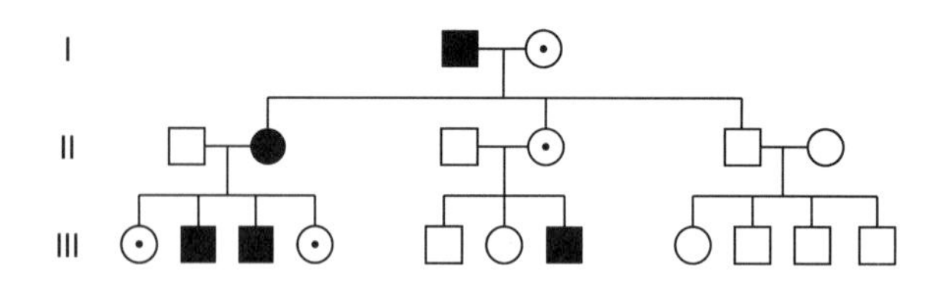

woman a mosaic in terms of X-linked gene expression. In females heterozygous for a deleterious allele at an X-linked locus, chance sometimes creates large areas of tissue without wild-type allele activity; those women may have clinically detectable symptoms. A classic example involves the distribution of sweat glands (Figure 13–9).

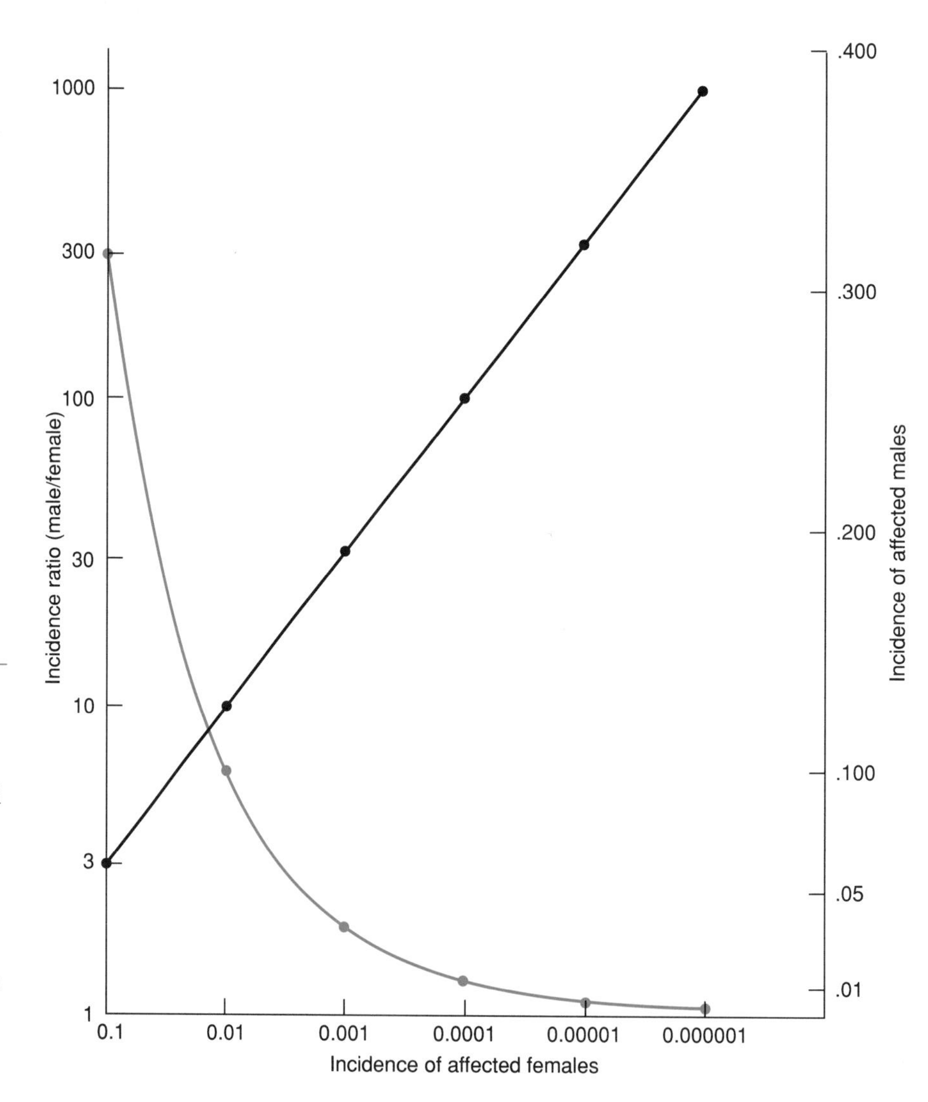

Figure 13–8

Relationship (curved line) between the frequency of a recessive X-linked allele, as indicated by the incidence of affected males (right scale), and the frequency of homozygous females (bottom scale). The rarer the allele, the higher the ratio of affected males to affected females (straight line). Note that the male/female ratio (left scale) is plotted against the incidence of affected females, not against the incidence of affected males.

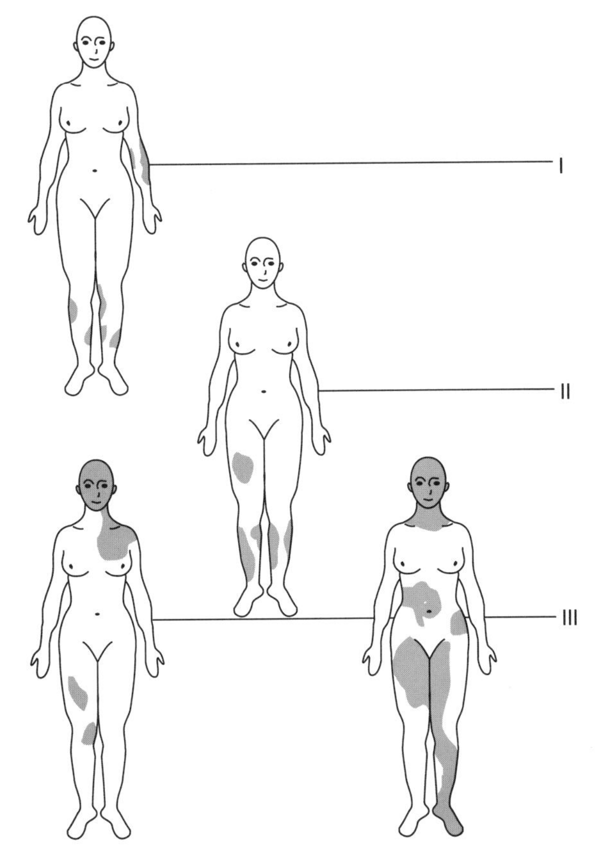

Figure 13–9

Mosaic expression of X-linked anhydrotic dysplasia in females, caused by random inactivation of X chromosomes in early development. Stippled areas indicate regions where sweat glands were absent or infrequent. The two females in generation III were identical twins. (From Kline et al., 1959)

The occurrence of partially affected females, resulting from mosiac inactivation of X-linked genes, has led some geneticists to question the applicability of the term "recessive." In fact, female mammalian cells are *functionally hemizygous* for most genes on the X chromosome. That, after all, is the goal of dosage compensation. Recessiveness and dominance may therefore not be applicable concepts at the cellular level, but for alleles of genes whose inactivity or abnormal activity in approximately half of the body's cells has no clinically evident effects on normal function, it still seems appropriate to use "recessive."

2. Affected males do not usually transmit their phenotype to their offspring. The sons of affected males do not receive an X chromosome from their fathers and therefore cannot be affected by their father's X chromosome. The daughters of affected males will usually be unaffected carriers (heterozygotes); but exceptions may occur (i) if the spouse of the affected male is a heterozygote, (ii) if the mutation is not completely recessive, or (iii) if Lyonization produces large enough regions of affected tissue to produce detectable abnormality.

3. The parents of affected males are usually phenotypically normal. This relationship occurs because most affected males either receive a mutant X chromosome from a heterozygous mother, or they are new mutations (see Chapter 6).

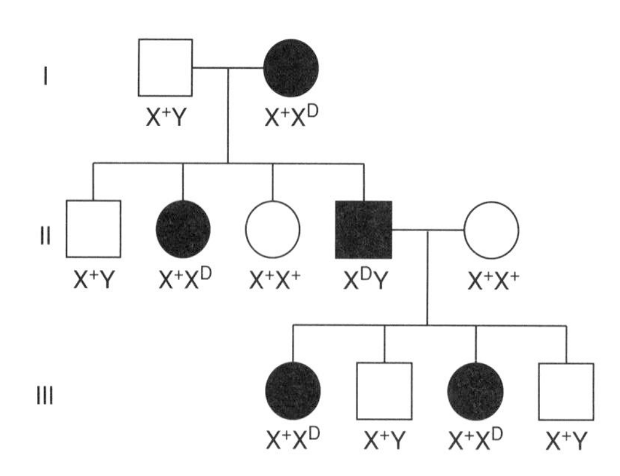

Figure 13–10

Hypothetical pedigree showing inheritance of an X-linked dominant trait. The affected female in generation I has equal numbers of affected and normal children, of both sexes. The affected male in generation II has only normal sons and only affected daughters.

Dominant Alleles A typical pattern of inheritance for an X-linked dominant allele is shown in Figure 13–10. Notice that an affected heterozygous female with a normal mate can have normal and affected sons, as well as normal and affected daughters. In contrast, an affected male with a normal mate will have no affected sons and no normal daughters.

Only a few diseases have been shown to be caused by X-linked dominant alleles. Among them are vitamin D-resistant rickets with hypophosphatemia and incontinentia pigmenti (a skin disease). The latter is genetically interesting because affected males apparently die before birth. This is the extreme example of a general tendency for males who have X-linked dominant alleles to be more seriously affected than their heterozygous female sibs. To a molecular biologist, this implies that the deleterious alleles, whatever they may be, usually do not totally prevent expression of the normal allele's product (see Chapter 8).

Y-linked Genes

The inheritance of mutations on the Y chromosome has the following characteristics (Figure 13–11).

1. Only males are affected.
2. All sons of affected males will also be affected.
3. Females do not transmit the trait. They cannot be heterozygotes for the Y chromosome.

The only potential exception to the preceding rules would involve mutations in genes in the pseudoautosomal region. If a recessive mutation were present in the Y chromosome allele of a gene in the PAR, males would not be affected unless a recessive allele were also present on the X chromosome. Such an affected male would probably not have affected sons, unless the mother supplied an X chromosome with another copy of the recessive allele.

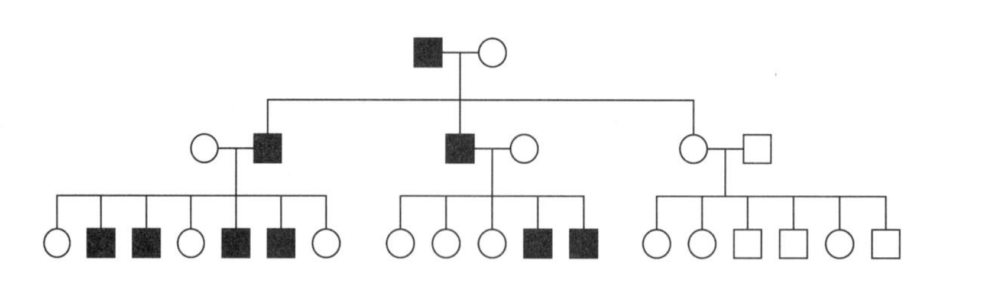

Figure 13–11

Hypothetical pedigree, showing inheritance of a trait caused by a Y-linked gene. Only males are affected.

Examples of X-linked Diseases

Several of the most familiar genetic diseases arise from mutations in X-linked loci. We shall not attempt a complete survey here, but we shall briefly summarize what is known about a few genetic abnormalities for which molecular analysis has been extensive. In Chapter 5 we discussed two of them: Duchenne muscular dystrophy and chronic granulomatous disease. Here we shall add color blindness, hemophilias A and B, and the fragile X syndrome.

Color Blindness

Human color vision depends upon three classes of cone cells in the retina, which have different spectral sensitivities. The visual pigments responsible for the light-absorbing power of cone cells are three related proteins, each of which binds a chromophore known as 11-cis-retinal. Small differences in the composition of the pigment proteins in the regions that bind 11-cis-retinal are responsible for the differences in spectral sensitivity of the chromophore. Only one pigment protein is present in a given cone cell; accordingly, there are blue-sensitive, green-sensitive, and red-sensitive cells. Because humans sense color as a mixture of stimuli from cone cells that respond to three primary colors, normal vision in humans is described as *trichromatic*.

All of the cone pigment proteins are related to rhodopsin, the protein in rod cells that responds to the entire visible spectrum. Evolutionary studies suggest that the gene for the blue-sensitive protein diverged from an ancestor of the red and green pigment genes long ago, and that the separation of the genes for red and green pigments occurred about 40 million years ago. The latter estimate is based on the fact that New World monkeys have dichromatic color vision, but African monkeys, who are more closely related to apes and humans, have trichromatic vision.

The gene for the blue-sensitive protein is autosomal (chromosome 7), so it is not surprising that defects in blue color vision are rare because of the requirement for two recessive alleles to produce an abnormal phenotype. In contrast, the genes for red and green pigment proteins are on the X chromosome, and males with abnormal red or green color vision are common, being about 8% of the Caucasian population. The term "color blindness" usually refers to persons with defects in the ability to detect red or green light, rather than to total inability to detect color. Color blindness is not an accurate term, but it is well established in the English language, and more precise alternatives tend to be too complicated.

The genes for the red and green pigment proteins have been cloned and analyzed in detail (Nathans, 1989). They lie adjacent to each other in band Xq28, with the green gene being downstream (in a transcriptional sense) from the red gene. Surprisingly, persons with normal color vision turn out to have variable numbers of the green pigment gene (one, two, or three copies in most cases) whereas the red pigment gene is ordinarily present in only one copy (Figure 13–12). At the amino acid level, the red and green pigment proteins are 98% alike.

The near-identity of the red and green genes, plus the presence of multiple green genes in some individuals, creates the potential for unequal crossing-over, which does occur, with the consequent creation of color vision abnormalities. As Figure 13–12 illustrates, misalignment of the red and green genes during meiosis can produce a chromosome that is green-minus

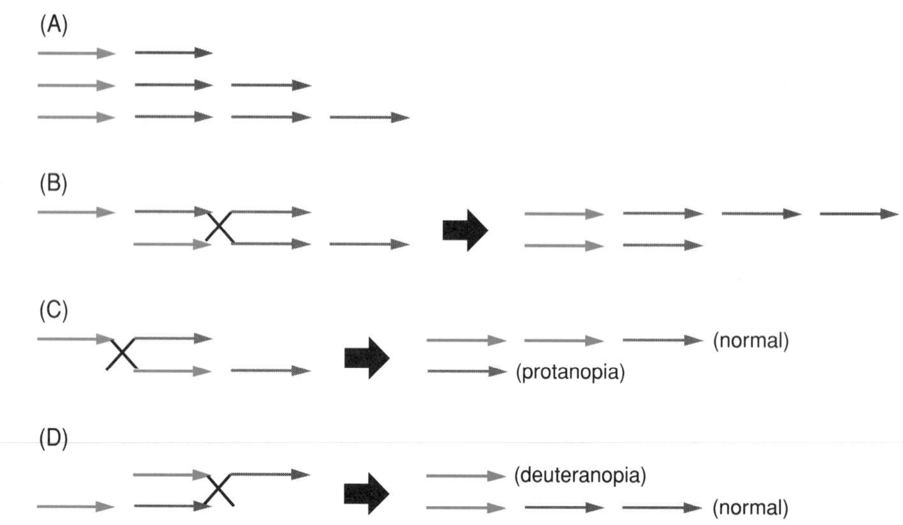

Figure 13–12

Normal X chromosomes have one red pigment gene and one, two, or three green pigment genes (panel A). Unequal crossing-over can alter the number of green pigment genes (panel B), or delete the red pigment gene producing protanopia (panel C), or delete the green pigment genes producing deuteranopia (panel D).

or a chromosome that is red-minus. Persons who have red-minus color vision are called *protanopes,* because red is regarded as the first primary color (the Greek word, *protos,* refers to the first item in a series; *anopia* refers to the lack of vision). Persons who have green-minus color vision are called *deuteranopes,* because green is the second primary color (Greek *deuteros* means second), and persons who are insensitive to blue light are *tritanopes* (Greek *tritos* means third).

Unequal crossing-over sometimes occurs within genes, creating hybrid genes that may be 5′red/3′green or 5′green/3′red. Individuals with a hybrid color vision gene often are more sensitive to red light than to green, or vice versa. They are known as red-anomalous trichromats or green-anomalous trichromats, respectively.

Molecular analysis of the genes for the red and green visual pigments has provided convincing proof of the role of unequal crossing-over in genetic variation, although it leaves unanswered questions about why some rearrangements are more common than others. The exceptionally high frequency of males with defects in red or green color vision raises questions in population genetics. For example, how frequent are new mutations resulting from unequal crossing-over, and was there no negative selection against persons with subnormal color vision in the past?

Hemophilias A and B

Blood clotting is a complex process that involves more than a dozen proteins, most of which are proteases or cofactors of proteases (Figure 13–13). Two of the genes are located on the X chromosome. Those genes encode factors VIII and IX, both of which have essential roles in the conversion of prothrombin to thrombin; thrombin then converts soluble fibrinogen to insoluble fibrin, the main component of blood clots.

In band Xq28, not far from the red and green visual pigment genes, is the exceptionally large gene for factor VIII, a protein of 2351 amino acids. Defects in the function of factor VIII are responsible for classical hemophilia, also known as hemophilia A (Antonarakis and Kazazian, 1988). The factor IX gene is in band Xq27; defective function of factor IX produces hemophilia B.

Hemophilia A affects one male in 10,000. Uncontrollable bleeding following even minor injuries formerly led to the death of most patients in

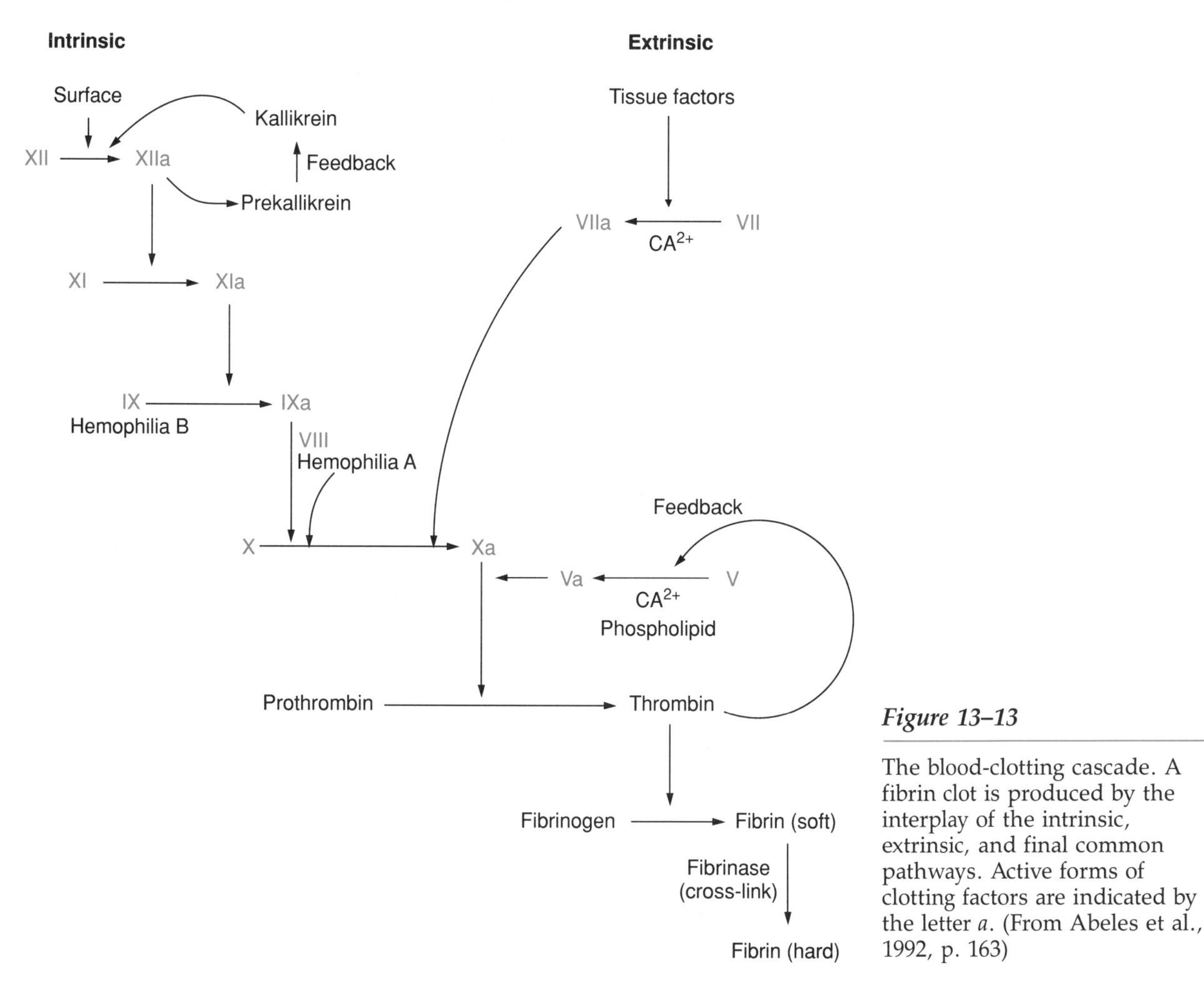

Figure 13–13

The blood-clotting cascade. A fibrin clot is produced by the interplay of the intrinsic, extrinsic, and final common pathways. Active forms of clotting factors are indicated by the letter *a*. (From Abeles et al., 1992, p. 163)

childhood or early adulthood, but the availability of blood plasma for transfusions in recent decades greatly ameliorated that situation. Tragically, many individuals with hemophilia were infected with AIDS in the 1980s, before the need to exclude infected blood bank donors was recognized. Inasmuch as the gene has been cloned, it is likely that reliably pure factor VIII will be produced in microorganisms in the near future.

History's most famous sufferer from hemophilia A was the Tsarevich Alexis, heir to the throne of Russia, who escaped the usual consequence of his disease when he was assassinated at the age of 14, along with the rest of his family. Alexis was a descendant of Queen Victoria of England, who apparently was a carrier of a new mutation in the factor VIII gene. Through her daughters, the hemophilia gene was disseminated to several royal houses. This human genetics book will spare the reader the necessity of studying Victoria's pedigree.

Hemophilia B affects one male in 30,000 to 50,000. It represents a defect in factor IX of the coagulation cascade. The gene and its encoded protein are of average size (see Chapter 2). The symptoms are similar to those of hemophilia A, as expected from the fact that the same process is inhibited. An alternative name for hemophilia B is "Christmas disease," for a rather odd reason—the surname of the family in which the disease was first recognized was Christmas!

Both of the hemophilia genes are providing molecular biologists with good material for the study of mutation rates, especially for measurement of the different rates in males and females. Like any X-linked recessive disorder, hemophilias A and B often represent new mutations, and the use of RFLPs and/or direct detection of a given mutation can identify the parent in which the new mutation arose (explained in Chapter 6). The gene for factor VIII mutates at a relatively high rate (3 to 4 $\times$ 10^{-5}), and analysis of the different types of mutation that occur within the gene is proving instructive. Of particular interest has been the discovery of two independent cases of insertion of an L1 repetitive sequence (see Chapter 2) into the gene.

The Fragile X Syndrome

One of the most common forms of mental retardation in humans is associated with the presence of a mutation at band q27.3 on the X chromosome. The mutation can be detected as a region of incomplete condensation in metaphase chromosomes (a secondary constriction, see Figure 13–14) in cultured cells, or under special culture conditions, as a tendency to break at that point. The latter characteristic is the basis for the name, fragile X syndrome (abbreviated FRA-X).

Affected males exhibit some degree of mental retardation, which varies from mild to severe. They may have language disabilities and some of the symptoms of autism, such as hand flapping and biting, and poor eye contact. They usually have exceptionally large testicles and large, protruding ears. Estimates of the frequency of affected males in various populations are mostly in the range of one in 1000 to 1500, which makes fragile X syndrome the second most frequent genetic cause of mental retardation (Down syndrome being somewhat more common).

Fragile X syndrome is exceptional among X-linked diseases in two ways. First, it is incompletely penetrant. Approximately 20% of males who carry the mutation do not express it at all, but are capable of transmitting it to their grandchildren through carrier daughters. Second, about one-third of heterozygous females express the mutation as mild mental retardation or learning disability. Mild and variable expression of X-linked recessive alleles in females is not unusual, because of Lyonization (random X inactivation in early development), but the frequency of clinically evident symptoms in females carrying the fragile X mutation is exceptionally high.

The site involved in FRA-X mutations has been found (Oberle et al., 1991; Yu et al., 1991; Verkerk et al., 1991). A gene, designated *FMR1*, has been identified that encodes a 4.8 kb mRNA that is expressed in the human brain. Near the 5′ end of the coding sequence is an uninterrupted series of approximately 40 arginine codons (CCG), which predicts a very basic re-

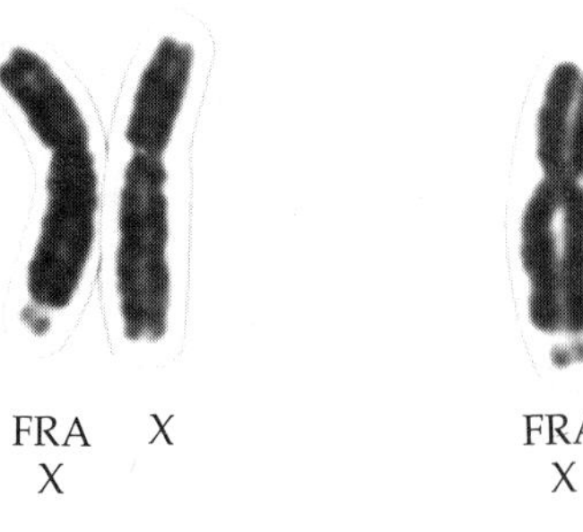

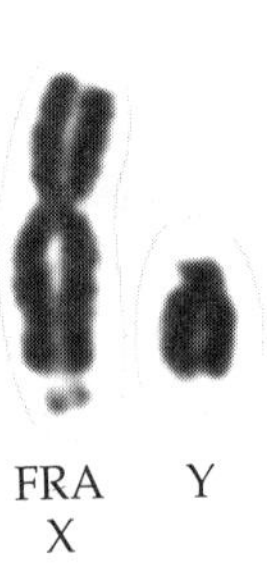

Figure 13–14

A fragile X chromosome, as seen in metaphase spreads, of (A) a heterozygous female and (B) a male, showing the small portion at the tip of the long arm that separates in suitable culture conditions (there is a connection, but it is not visible). (Courtesy of Grant Sutherland)

gion of the protein that might bind to DNA. This possibility is strengthened by the finding that the mRNA also encodes a consensus nuclear translocation sequence (i.e., a short sequence of amino acids needed to direct the protein from the cytoplasm, where it is synthesized, to the nucleus, where it presumably functions).

There is considerable polymorphism in the length of the $(CCG)_n$ series among normal persons, the range being at least 15–65 repeats (Kremer et al., 1991). However, the number of repeats appears to be stable within a family. In contrast, variants of this gene associated with the FRA-X syndrome have hundreds to thousands of tandemly repeated copies of the CCG codon, which apparently arise by a process of local amplification. Unaffected transmitting males tend to have less amplification than affected males, although they still have many more repeats than normal males have. It should not be long before we have some insight into the detailed biochemical effects of this unusual genetic variation.

Until recently, genetic counseling of families at risk for birth of a fragile X son was less than satisfactory (Brown, 1990). Polymorphic marker loci in the general region exist, but none is close enough to allow a nearly 100% prediction of the fetus's genotype. This problem may reflect an unusually high rate of recombination in the fragile X region. Another complicating factor is the variable expression of the mutation, including incomplete penetrance, mentioned earlier. It is likely that diagnosis and counseling will be greatly strengthened as a result of the detailed knowledge of the fragile X locus and its variants that is accumulating.

SUMMARY

The human X chromosome consists of about 150 Mbp of DNA and most of the genes it contains are unrelated to sexual differentiation. Many X chromosome genes have been identified because of their distinctive pattern of inheritance.

In females (XX) most of one X chromosome is inactivated in each cell in early embryogenesis. Inactivation is random; there is no preference for maternally derived or paternally derived X chromosomes. Inactivation is permanent in somatic cells, but the inactive X is reactivated during oocyte formation.

X chromosome inactivation achieves dosage compensation; that is, female cells have approximately the same number of active X chromosome genes as male cells, in which there is only one X chromosome. However, a few genes on the inactive X chromosome are not inactivated. The *XIST* gene is unique in being active on the inactive X and inactive on the active X; *XIST* is probably part or all of the X inactivation center, the locus from which inactivation spreads toward both termini.

The human Y chromosome consists of about 65 Mbp of DNA, most of which is reiterated sequences; few Y chromosome genes have been identified.

The pseudoautosomal region consists of about 2.6 Mbp at the tip of the short arm of both the X and the Y chromosomes. It is only in this region that extensive homology occurs. During male meiosis, there is exactly one crossover in the pseudoautosomal region.

Humans with one or more Y chromosomes are phenotypically male, regardless of the number of X chromosomes. The gene that determines maleness, which must be located on the Y chromosome, is known as *TDF* (testis-determining factor gene). Recent work has identified a gene, *SRY*, that encodes a DNA-binding protein that may be a transcription factor. It probably corresponds to *TDF*.

The pattern of inheritance of X-linked recessive alleles is characterized by the following: males are more frequently affected than females, affected males do not usually transmit their phenotype to their offspring, and the parents of affected males are usually phenotypically normal.

The pattern of inheritance of Y-linked traits is characterized by the following: only males are affected, all sons of affected males will also be affected, and females do not transmit the trait.

Several well-known diseases resulting from X-linked genes are described. Defects in red and/or green color vision are extremely common in human males. The relevant genes are closely related and lie adjacent to one another in band Xq28. Many mutations arise from unequal crossing-over in that gene cluster. Mutations in genes for two blood clotting factors are responsible for hemophilia A (factor VIII) and the less common hemophilia B (factor IX). The fragile X syndrome, the most common form of inherited mental retardation, is the result of an unusual type of variation in the *FMR1* gene in Xq27, involving local multiplication of a tandem series of arginine codons.

REFERENCES

Abeles, R. H., Frey, P. A., and Jencks, W. P. 1992. Biochemistry. Jones and Bartlett, Boston.

Antonarakis, S. E. and Kazazian, H. H., Jr. 1988. The molecular basis of hemophilia A in man. Trends in Genetics 4:233–237.

Barr, M. L. and Bertram, E. G. 1949. A morphological distinction between neurones of the male and female, and the behavior of the nucleolar satellite during accelerated nucleoprotein synthesis. Nature 163:676–677.

Brown, C. J., Ballabio, A., Rupert, J. L., et al. 1991. A gene from the region of the human X inactivation centre is expressed exclusively from the inactive X chromosome. Nature 349:38–44.

Brown C. J., Lafreniere, R. G., Powers, V. E., et al. 1991. Localization of the X inactivation centre on the human X chromosome in Xq13. Nature 349:82–84.

Brown, W. T. 1990. The fragile X: progress toward solving the puzzle. Am. J. Human Genet. 47:175–180.

Burgoyne, P. 1982. Genetic homology and crossing over in the X and Y chromosomes of mammals. Hum. Genet. 61:85–90.

Cattanach, B. M. 1974. Position effect variegation in the mouse. Genetical Res. 23:291–306.

Cooke, H. 1990. The continuing search for the mammalian sex-determining gene. Trends in Genetics 6:273–275.

Craig, I. 1990. Sex determination: zinc fingers point in the wrong direction. Trends in Genetics 6:135–137.

Davies, K. 1991. The essence of inactivity. Nature 349:15–16.

Davies, K. E., Mandel, J. L., Monaco, A. P., Nussbaum, R. L., and Willard, H. F. 1990. Report of the committee on the genetic constitution of the X chromosome. Cytogenet. Cell Genet. 55:254–313.

Eichwald, E. J. and Silmser, C. R. 1955. Communication. Transplant. Bull. 2:148–149.

Ellis, N. and Goodfellow, P. N. 1989. The mammalian pseudoautosomal region. Trends in Genetics 5:406–410.

Gough, N. M., Gearing, D, P., Nicola, N. A. et al. 1990. Localization of the human GM-CSF receptor gene to the X-Y pseudoautosomal region. Nature 345:734–736.

Gubbay, J., Collignon, J., Koopman, P., et al. 1990. A gene mapping to the sex-determining region of the mouse Y chromosome is a member of a novel family of embryonically expressed genes. Nature 346:245–250.

Heslop, B. F., Bradley, M. P., and Baird, M. A. 1989. A proposed growth regulatory function for the serologically detectable sex-specific antigen H-Ys. Hum. Genet. 81:99–104.

Jacobs, P. A. and Ross, A. 1966. Structural abnormalities of the Y chromosome in man. Nature 210:352–354.

Keats, B. J. B., Sherman, S. L., and Ott, J. 1990. Report of the committee on linkage and gene order. Cytogenet. Cell Genet. 55:387–394.

Kline, A. H., Sidbury, J. B. Jr., and Richter, C. P. 1959. The occurrence of ectodermal displasia and corneal dysplasia in one family: an inquiry into the mode of inheritance. J. Pediatrics 55:355–366.

Koopman, P., Gubbay, J., Vivian, N., Goodfellow, P. and Lovell-Badge, R. 1991. Male development of chromosomally female mice transgenic for *Sry*. Nature 351:117–121.

Kremer, E. J., Pritchard, M., Lynch, M., et al. 1991. Mapping of DNA instability at the fragile X to a trinucleotide repeat sequence $p(CCG)_n$. Science 252:1711–1714.

Laird, C. D. 1987. Proposed mechanism of inheritance and expression of the human fragile X syndrome of mental retardation. Genetics 117:587–599.

Lyon, M. 1961. Gene action in the X-chromosome of the mouse (Mus musculus L.). Nature 190:372–373.

Lyon, M. 1991. The quest for the X-inactivation centre. Trends in Genetics 7:69–70.

Lyon, M. F. 1988. X-chromosome inactivation and the location and expression of X-linked genes. Am. J. Human Genet. 42:8–16.

McLaren, A. 1990. What makes a man a man? Nature 346:216–217.

Nathans, J. 1989. The genes for color vision. Sci. Am. (Feb.) 42–49.

Oberle, I., Rousseau, F., Heitz, D., et al. 1991. Instability of a 550-base pair DNA segment and abnormal methylation in fragile X syndrome. Science 252:1097–1102.

Ohno, S. 1969. Evolution of sex chromosomes in mammals. Annu. Rev. Genet. 3:495–524.

Page, D. C. 1986. Sex reversal: deletion mapping the male-determining function of the human Y chromosome. Cold Spring Harbor Symp. Quant. Biol. 51:229–235.

Page, D. C., Mosher, R., Simpson, E. M., et al. 1987. The sex-determining region of the human Y chromosome encodes a finger protein. Cell 51:1091–1104.

Sinclair, A. H., Berta, P., Palmer, M. S., et al. 1990. A gene from the human sex-determining region encodes a protein with homology to a conserved DNA-binding motif. Nature 346:240–244.

Verkerk, A. J. M. H., Pieretti, M., Sutcliffe, J. S., et al. 1991. Identification of a gene (FMR-1) containing a CGG repeat coincident with a breakpoint cluster region exhibiting length variation in fragile X syndrome. Cell 65:905–914.

Yu, S., Pritchard, M. Kremer, E., et al. 1991. Fragile X genotype characterized by an unstable region of DNA. Science 252:1179–1181.

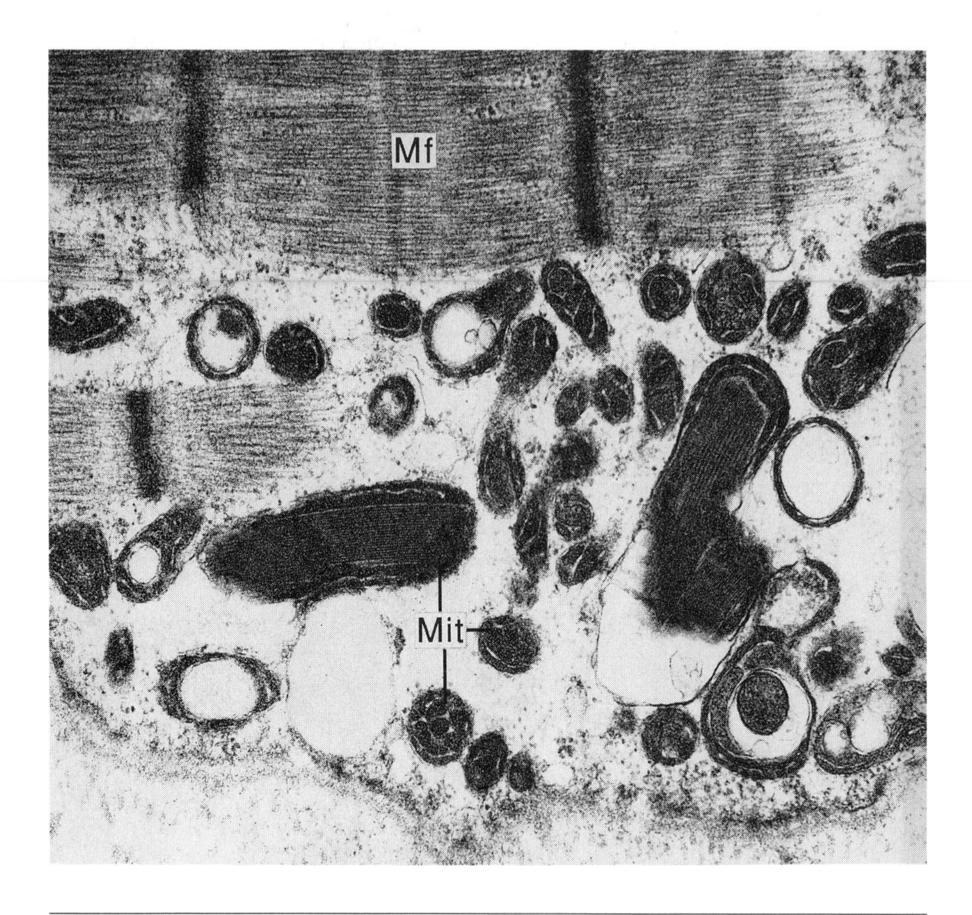

Electron micrograph of a skeletal muscle biopsy from a patient with MERRF disease. Most of the mitochondria have disorganized cristae and some show paracrystalline inclusions. Mf = myofibrils, Mit = mitochondria. (Photo courtesy of Marie Lott and Douglas Wallace)

14 The Mitochondrial Genome and Its Pathology

Mitochondria are complex organelles—probably the evolutionary descendants of prokaryotes that became intracellular symbionts of primitive eukaryotes in eons past, before plants and animals diverged from a common ancestor. There are significant differences in the structure, biochemistry, and genetics of mitochondria among major eukaryotic groups (plants, fungi, protozoa, animals), and it may be that several endosymbiotic relationships evolved independently into modern mitochondria. Mitochondria range widely in size; they may be ellipsoidal, as in mammalian liver (roughly 1 by 2 micrometers); or they may have a cylindrical form, about 1 micrometer in diameter and several to many micrometers in length. Their form is not static; mitochondria grow, bud, break off pieces, fuse with other mitochondria and separate again. Most mammalian cells contain several hundred mitochondria; some contain many more; and red blood cells have none.

Mitochondrial Structure and Function

The major structural features, as seen in the electron microscope (Figure 14-1), are the two *membranes* and the *cristae,* which are convolutions of the inner membrane that project into the *matrix,* or interior of the organelle.

Mitochondria are the principal sites of cellular energy production. Most of the ATP generated in eukaryotic cells is formed in mitochondria via a series of reactions termed *oxidative phosphorylation*. The molecular complexes that transfer electrons from NADH to oxygen, with the concomitant formation of ATP from ADP and inorganic phosphate, are collectively known as the *respiratory chain.*

In addition to oxidative phosphorylation, many other oxidative reactions take place within mitochondria. These include *fatty acid oxidation* and the *tricarboxylic acid cycle* (also known as the citric acid cycle or Krebs cycle). In all, several hundred proteins are involved in mitochondrial structure and contents, and nearly all of them are encoded by nuclear genes.

Much has been written about the mechanisms by which proteins synthesized on cytoplasmic ribosomes are targeted to mitochondria; these include specific sequences near the amino terminus which are required for recognition by components of the outer mitochondrial membrane, chaperone proteins that complex with and accompany newly synthesized polypeptides as they journey from cytoplasm to mitochondria, and special receptor proteins in the mitochondrial outer membrane (Hartl and Neupert, 1990).

Mitochondria contain their own genomes, which encode a small percentage of the macromolecules found within the organelle. Human mitochondrial DNA has been completely sequenced; it is circular and contains 16,569 bp (Anderson et al., 1981). Most mitochondria have 5 to 10 copies of mitochondrial DNA.

The genes encoded by human mitochondrial DNA are two ribosomal RNAs (12S and 16S), 22 tRNAs, and 13 polypeptides that form parts of the respiratory chain (Figure 14-2). Additional respiratory chain polypeptides are encoded by nuclear genes, as are all the other polypeptides that contribute to mitochondrial structure and function. Thus, mitochondria are the products of both nuclear and organelle genomes (Figure 14-3). The interactions between the genomes are complex, and so are the possibilities for malfunctions, as will be described presently.

The mammalian mitochondrial genome is noteworthy for its density of information; there are no introns and in several instances the first nucleotide of one gene is also the last nucleotide of the preceding gene. In lower euka-

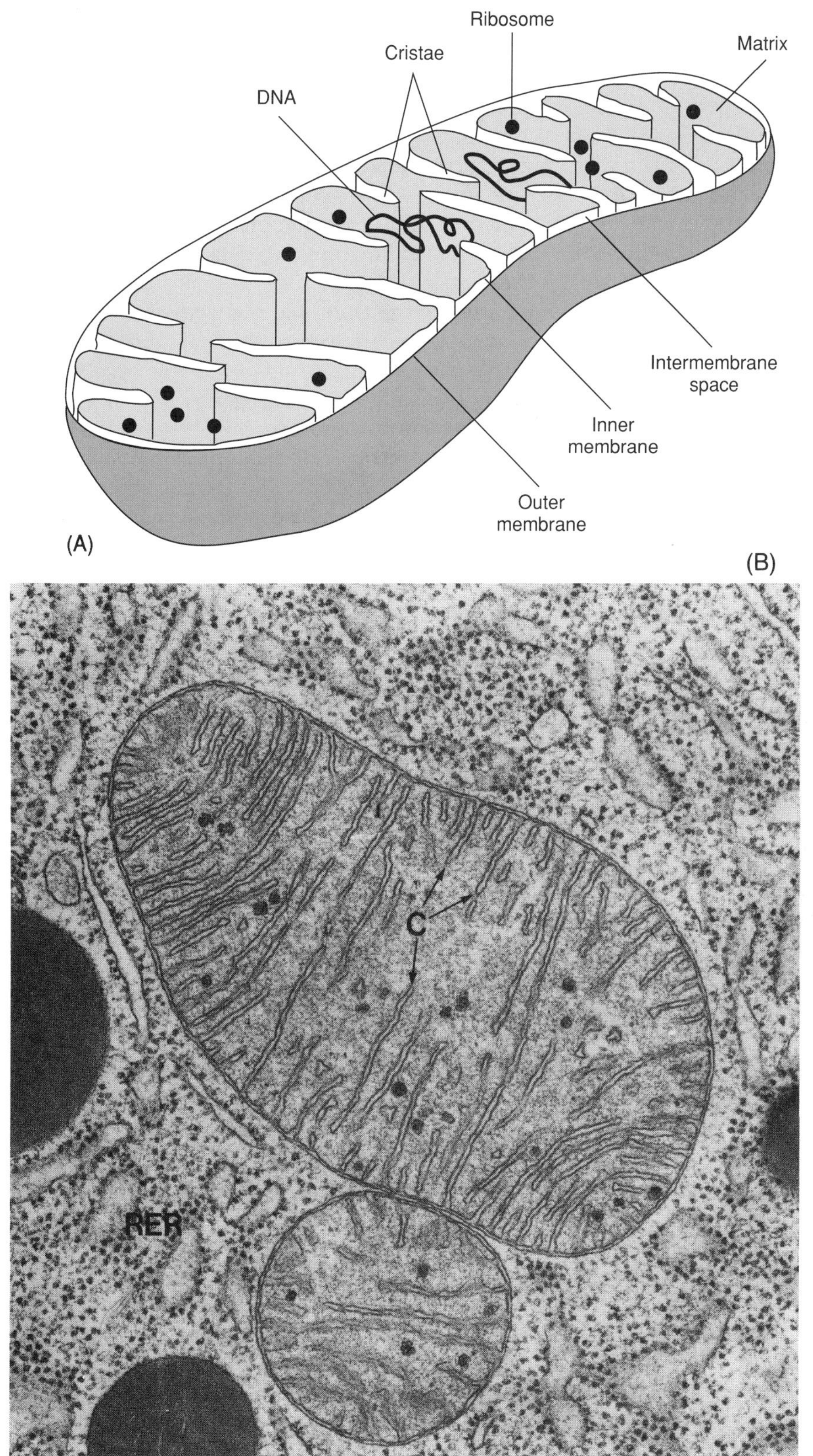

Figure 14–1

Structure of mitochondria. (A) A three-dimensional schematic diagram. (From Prescott, 1988, p. 137). (B) Electron micrograph of mitochondria in a bat pancreas cell. Note the sheetlike cristae (C). Rough endoplasmic reticulum (RER) surrounds the mitochondria in this cell. (From Prescott, 1988, p. 138)

ryotes, mitochondrial DNAs are much larger; yeast, for example, has a mitochondrial genome more than five times as large as that for mammals. Most of the difference is the result of introns and intergenic spaces, but there are also some differences in the encoded polypeptides. Another fundamental difference between mitochondrial and nuclear gene expression in general is that several codons do not have the same meaning as in the otherwise-universal genetic code; UGA, for example, is translated as tryptophan in mitochondria, instead of being a termination codon (Table 14–1).

DNA synthesis, RNA synthesis, and RNA processing within mitochondria all depend upon enzymes encoded by nuclear genes. Similarly, all of the protein components of mitochondrial ribosomes and the various accessory factors needed for protein synthesis are the products of nuclear genes in mammals; but they are all different from the genes that code for the cytoplasmic protein synthesis machinery.

It is an elaborate apparatus for the translation of a mere 13 messenger RNAs, and no one has yet offered a convincing explanation for the retention of those few genes within the mitochondrion, while the large majority of what must have originally been mitochondrial genes have been transferred to the nucleus during the several billion years of evolution that have elapsed since primitive eukaryotes incorporated the prokaryotic ancestors of mitochondria. Nowadays, the differences in the genetic code for mitochondrial protein synthesis versus cytoplasmic protein synthesis would

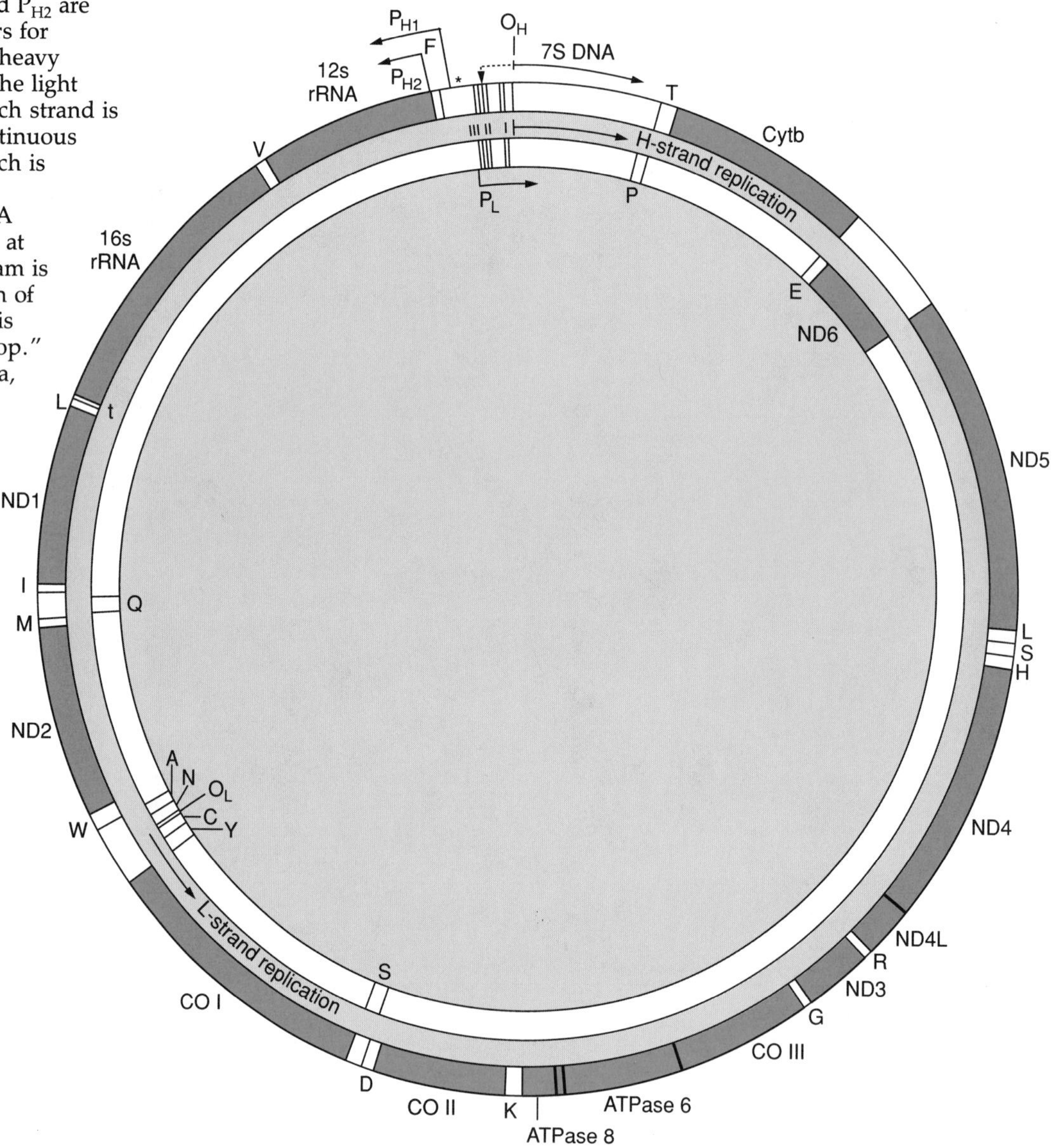

Figure 14–2

Mitochondrial DNA function map. The inner and outer circles are the C-rich light (L) strand and the G-rich heavy (H) strand, respectively. Dark shaded regions are the rRNA and polypeptide genes; tRNA genes are indicated by light blocks between the larger genes. CO I, II, and III are genes for cytochrome oxidase subunits, ND 1–6 encode subunits of NADH dehydrogenase, and tRNA genes are labeled according to the amino acid the tRNA carries, using the one-letter code. O_H and O_L are the origins of DNA replication for the heavy and light strands, respectively; P_{H1} and P_{H2} are alternative promoters for transcription of the heavy strand, while P_L is the light strand promoter. Each strand is transcribed as a continuous RNA molecule, which is then cleaved by a nuclease at the tRNA genes. The 7S DNA at the top of the diagram is involved in initiation of DNA replication; it is located at the "D-loop." (From Wallace, 1989a, p. 613)

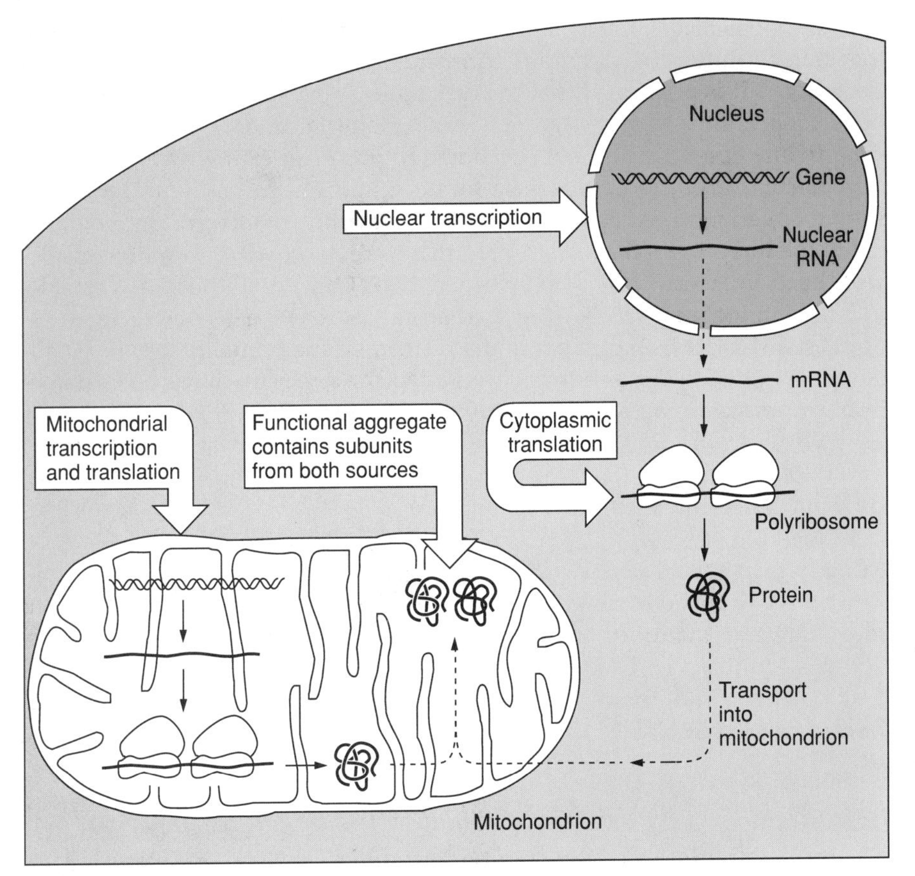

Figure 14–3

Complex mitochondrial proteins are assembled from the products of expression of nuclear genes and mitochondrial genes. (From Lewin, 1990, p. 521; by permission of Cell Press)

probably make transfer of the remaining mitochondrial genes to the nucleus impossible; but whether it was changes in the genetic code that prevented transfer or whether some other factor was responsible for retention of those few genes in a separate organelle, is a question to which no satisfactory answer exists at present.

One biochemical property that bolsters the hypothesis of the prokaryotic origin of mitochondria is the antibiotic sensitivity of mitochondrial protein synthesis. Several common antibiotics (such as erythromycin, streptomycin, and chloramphenicol), which exert their therapeutic effect by inhibiting bacterial protein synthesis, also inhibit mitochondrial protein synthesis, at least in cultured cells. It is not clear what effect prolonged use of those antibiotics might have on mitochondrial functions in a patient.

The existence of a separate genome within a cytoplasmic organelle im-

Table 14–1 Some Differences Between the "Universal" Code and Mitochondrial Genetic Codes*

		MITOCHONDRIAL CODES			
CODON	"UNIVERAL" CODE	MAMMALS	DROSOPHILA	YEASTS	PLANTS
UGA	STOP	*Trp*	*Trp*	*Trp*	STOP
AUA	Ile	*Met*	*Met*	*Met*	Ile
CUA	Leu	Leu	Leu	*Thr*	Leu
AGA, AGG	Arg	*STOP*	*Ser*	Arg	Arg

*Italics indicate that the code differs from the "universal" code.

Source: Alberts et al., 1989. p. 392.

plies the possibility of mutations in those genes. Indeed, it is well known that mitochondrial DNA tends to mutate about ten times faster than nuclear DNA. It is presumed that the difference is caused by lower fidelity of DNA replication and/or less accurate DNA repair mechanisms in mitochondria. In any case, the high rate of mutation has been exploited by students of molecular evolution, who have found mitochondrial DNA to be a rich source of variation. Analysis of the variants found in different populations led to the hypothesis that all living humans are descended from one female who lived in Africa some 200,000 years ago (Cann et al., 1987)—a female dubbed "mitochondrial Eve" by journalists. More extensive data and additional statistical tests have apparently confirmed the original estimate (Vigilant et al., 1991), but serious methodological criticisms have been made recently, and as of mid-1992, a lively controversy exists.

Mitochondrial Genetic Diseases

Mutations in mitochondrial DNA imply that diseases caused by the malfunction of mitochondrial genes must exist and, in fact, research in the last few years has identified several. Readers who would like to learn more about this subject can consult a variety of reviews, such as Capaldi, 1988; Wallace, 1989; and Shoffner and Wallace, 1990; or the editorial reports that have appeared in *Nature* (Grivell, 1989) and *Science* (Palca, 1990).

General Aspects of Mitochondrial Genetic Diseases

The most definitive characteristic of mitochondrial genetic diseases is *maternal inheritance*. Although sperm contain mitochondria and depend upon them for the energy needed to move up the female genital tract, those mitochondria are in the mid-piece, which does not accompany the sperm nucleus into the egg that it fertilizes. Therefore, all mitochondria are maternally derived. A male afflicted with a mitochondrial disease cannot pass the disease to his offspring, whereas all the offspring of an affected female may be affected (Figure 14–4).

However, a complicating factor in mitochondrial disease inheritance is the existence of *heteroplasmy*—genetic heterogeneity of mitochondrial populations within an individual. Heteroplasmy can arise because there are thousands of mitochondria in a human egg and several DNA molecules

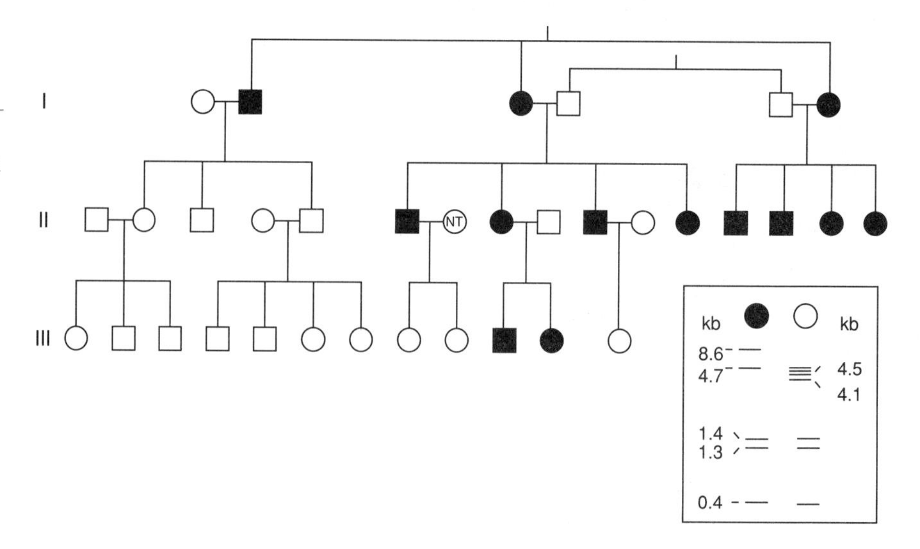

Figure 14–4

Maternal inheritance of the mitochondrial DNA in a human family. The lower right insert shows polymorphic *Hae*II restriction fragment patterns identified in blood platelet mitochondrial DNAs. Open symbols represent individuals with an Hae II site that produces 4.1 and 4.5 kbp fragments; solid symbols represent individuals without that Hae II site. NT indicates not tested. (From Wallace, 1989b, p. 10)

within each mitochondrion. Accordingly, a mutation may not be present in all the mitochondrial genomes that a person inherits. Moreover, the fraction of mutant mitochondrial DNA molecules may change differentially within tissues as the individual develops, matures, and ages. The result is that some mitochondrial diseases have highly variable symptoms, not only from family to family, but also within families. Thus, the clear-cut pattern of maternal inheritance exemplified by Figure 14–4 may not occur in all pedigrees where a genuine mitochondrial genetic disease exists.

In the sections that follow, an outline of diseases caused by mutations in the mitochondrial genome will be presented. We shall not attempt to survey diseases caused by mutations in nuclear genes which are expressed as mitochondrial malfunctions, but the reader should bear in mind that many such diseases must exist. The majority of polypeptides that comprise the respiratory chain (roughly 60 in total) are encoded by nuclear genes; therefore, there must be mutations in those genes that will have essentially the same phenotypic effects as some mutations in mitochondrially encoded respiratory chain polypeptides.

Moreover, we can confidently anticipate the discovery of mutations in nuclear genes that encode portions of the mitochondrial protein synthesis apparatus, which will mimic mutations in mitochondrial RNAs by having a general depressing effect on protein synthesis in that organelle. Finally, mutations in many of the nuclear genes for the multitude of enzymes and other proteins that reside in mitochondria must be able, in principle, to generate a plethora of syndromes that will ultimately be recognizable by physicians. The possibility of gene therapy for mitochondrial genetic diseases has been considered by several authors (reviewed by Lander and Lodish, 1990).

All diseases resulting from changes in mitochondrial DNA are fundamentally the result of malfunctions of the respiratory chain, whose major components are diagrammed in Figure 14–5. Mitochondrially encoded constituents of the respiratory chain are seven subunits (1,2,3,4L,4,5,6) of NADH dehydrogenase (complex I); one subunit of ubiquinol-cytochrome c oxidoreductase (complex III, also called cytochrome b); and three subunits of cytochrome c oxidase (complex IV). Two subunits of ATP synthase (complex V, not illustrated) are also encoded by mitochondrial DNA.

In general, the phenotypic effects of mitochondrial mutations reflect the extent to which a tissue relies on oxidative phosphorylation; the central nervous system is most sensitive, followed by skeletal muscle, heart muscle, kidney, and liver.

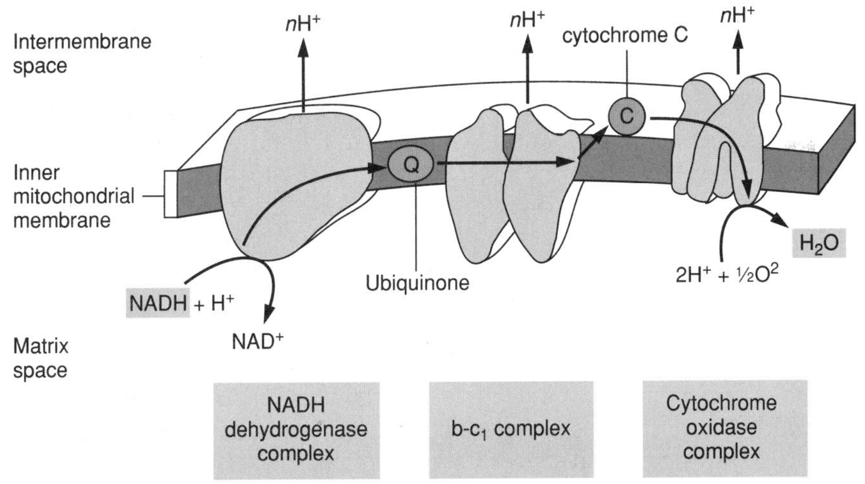

Figure 14–5

The flow of electrons through the three major respiratory enzyme complexes during the transfer of two electrons from NADH to oxygen. Ubiquinone and cytochrome c are mobile within the mitochondrial membrane; they serve as carriers of electrons between the complexes. (From Alberts et al., 1989, p. 363; by permission of Garland Publishing)

Two other sources of variability need to be borne in mind when considering the clinical expression of mitochondrial mutations. First, there are fluctuations in the extent of heteroplasmy. Chance plays a large role in the time and location of mitochondrial mutations, as well as the rate at which a mutation spreads through a population of dividing cells. Second, the components of the respiratory chain that are encoded by nuclear genes may not come from the same genes in all tissues. The existence of alternative, tissue-specific forms of many proteins (*isoforms*) is well established. Variation in isoforms of nuclear-encoded mitochondrial polypeptides could have a significant effect on the expression of a given mitochondrial mutation in different tissues or developmental stages.

Diseases Caused by Point Mutations in Genes for Polypeptides

Leber's hereditary optic neuropathy (LHON) is characterized by loss of vision, beginning centrally and proceeding peripherally. It is the result of decline of optic nerve function, usually beginning at ages 20–24. Cardiac dysrhythmia may also be present. Approximately 85% of patients are male—a sexual bias that is not understood.

One well-documented mutation associated with LHON is a missense mutation in a subunit of complex I (at nucleotide 11,778 of the DNA), which converts arginine to histidine. It was found in 9 of 11 LHON lineages tested in one study, and those 9 lineages appeared to be homoplasmic for the mutation. In another study, only 4 of 8 LHON lineages had the arg-to-his mutation, and a very small amount of heteroplasmy was detected (less than 5% wild-type mitochondrial DNA). It is not clear whether the small percentage of wild-type DNA can explain the variability in the age of onset and rate of progression of the disease seen in some patients.

The precise effect of the arg-to-his mutation in complex I is not yet known, but that specific arginine has been highly conserved in evolution, so it must play an important role in function of the complex. The shift to histidine may reduce the efficiency of electron flow enough to lead to cell death in the optic nerve in the third decade of life, on the average. Evidently, all cases of LHON cannot be ascribed to the same mutation, so we may expect to learn more about the genetic causes of this disease in the future.

Diseases Caused by Point Mutations in Mitochondrial RNAs

In principle, point mutations in mitochondrial rRNAs or tRNAs might have the effect of reducing the efficiency of mitochondrial protein synthesis, thus leading to a subnormal amount of respiratory chain enzymes. At least one such mutation has been identified. It is an A to G transition in the gene for mitochondrial $tRNA^{Lys}$, which is associated with MERRF syndrome.

Myoclonic epilepsy and ragged red fiber disease (MERRF) is characterized by central nervous system abnormalities (e.g. epilepsy, deafness, dementia) and deficiencies of skeletal and cardiac muscle function, associated with abnormal appearance of mitochondria in the electron microscope and "ragged" muscle fibers after staining with trichrome Gomori stain. Severity and type of symptoms vary substantially within a pedigree, which implies heteroplasmy of the mutant mitochondrial DNAs.

Biochemical analysis on muscle biopsies from some MERRF patients showed that there was deficient function of complexes I and IV. No mutations, other than the A to G transition mentioned previously, correlated reliably with MERRF symptoms in affected lineages. It is an easy mutation to assay, because it creates a cleavage site for a restriction enzyme (Cvi JI). Using that enzyme, researchers have shown a good correlation between the severity of symptoms and the proportion of mutant DNA in an individual's mitochondria.

Although the detailed consequences of the mutation in mitochondrial $tRNA^{Lys}$ have not yet been determined, it is plausible that a defect in one class of tRNA would have variable effects on different mitochondrial components, depending on the amount of lysine in the mitochondrially encoded polypeptide, the stability of each respiratory chain complex when one or more of its constituents is in short supply, and various other factors. Similarly, we may anticipate that mutations in mitochondrial rRNA genes will be found to underlie some human diseases, and that the extent of heteroplasmy will correlate with the severity of clinical symptoms. It is unlikely that many mutations in tRNA or rRNA genes will be found to be homoplasmic (100% mutant mitochondrial genomes), because they would probably be lethal.

Diseases Caused by Large Deletions in Mitochondrial DNA

Kearns-Sayre syndrome is characterized by a variety of neuromuscular symptoms, including paralysis of eye muscles, pigmentary retinopathy, dementia, and seizures. Numerous patients with Kearns-Sayre syndrome have been shown to have large deletions in a substantial fraction of their mitochondrial DNA molecules. The deletions all involve several mitochondrial genes, but the only consistent feature is that the origins of DNA replication are not deleted (Figure 14–6). Heteroplasmy is a necessary feature of these deletions; otherwise, cell death would ensue from total failure of essential mitochondrial functions.

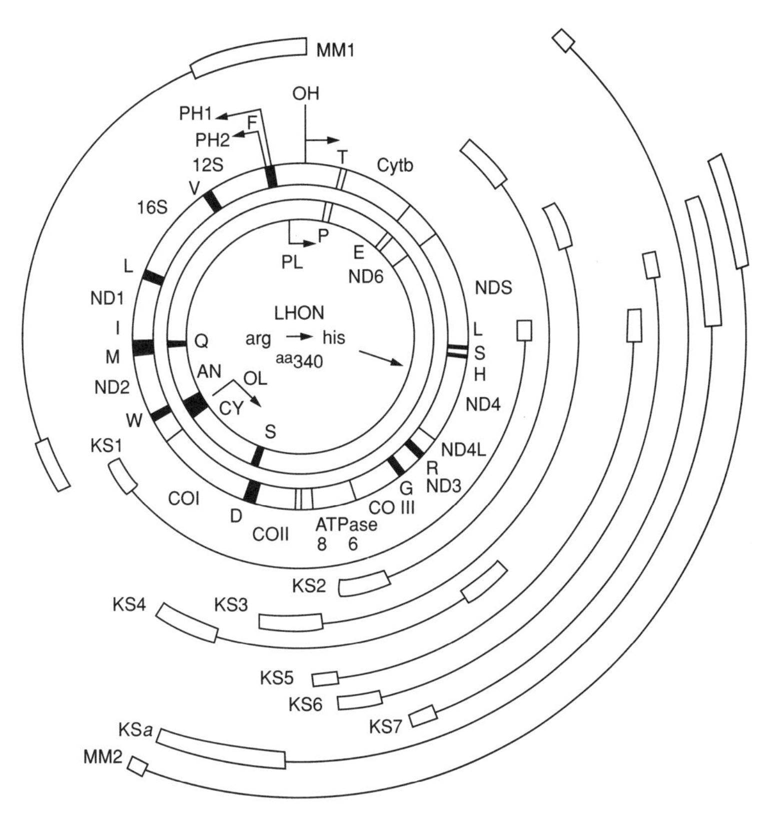

Figure 14–6

Clinical map of human mitochondrial DNA, showing deletions found in a series of patients with Kearns-Sayre syndrome (KS) and two patients with mitochondrial myopathy (MM1 and MM2). Bars at the ends of the arcs indicate that the ends of the deletions were not identified with certainty. The two circles and associated labels in the center of the diagram indicate the genes encoded by the two strands of circular mitochondrial DNA; see legend to Figure 14–2. (From Wallace, 1989b, p. 9)

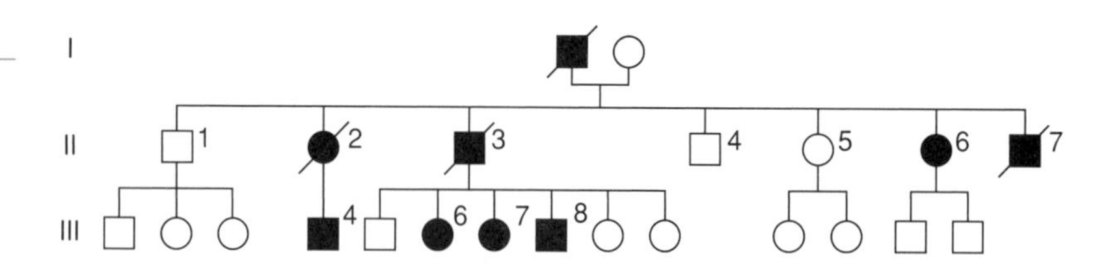

Figure 14–7

Pedigree of a family displaying late-onset mitochondrial myopathy, associated with deletions in mitochondrial DNA. The pattern is typical of an autosomal dominant condition. Solid symbols indicate affected individuals; a slash through a symbol represents a deceased person. (From Zeviani et al., 1989, p. 339; by permission of Macmillan Magazines Ltd.)

Within an individual, all deleted DNAs have the same deletion, which implies that all are descended from a single deletion event. It is presumed that smaller DNAs have a replicative advantage over larger DNAs within mitochondria; hence, clonal origin of a deletion seems logical. It is not yet clear what prevents the deletion-DNAs from becoming homoplasmic; there may be some point at which decline of mitochondrial function because of a deletion becomes self-limiting, or it may be that cells that become homoplasmic for the deletion simply die.

Most cases of human disease associated with large deletions of mitochondrial DNA are sporadic, rather than inherited. Presumably, deletion mutations occur at random times and places during development, with consequent variability in tissues affected, age when symptoms become evident, and severity of symptoms. Although deletion mutations could, in principle, be transmitted in heteroplasmic state from one generation to another, it may be that the large number of mitochondrial DNA replications that occur during development of a mature oocyte would allow most deletions to dominate the mitochondrial population, resulting in non-viability.

Deletions in mitochondrial DNA have also been found in the brains of patients who died of various causes, including Parkinson's disease. It has been suggested that the relatively high rate of mitchondrial DNA mutation is a major factor in the aging process in general (Linnane et al., 1989; Kadenbach and Mueller-Hoecker, 1990).

We close this chapter by referring to a remarkable case of mitochondrial DNA deletions apparently caused by a nuclear gene. Zeviani et al. (1989) described four members of a large Italian family who were afflicted with late-onset mitochondrial myopathy, many of the symptoms being the same as those in Kearns-Sayre patients. Direct biochemical analysis of mitochondrial DNA from muscle biopsies showed that all four patients had large deletions. All the deletions began in a region known as the D-loop (see Figure 14–2), where both DNA and RNA synthesis begin, but the 3′ ends of the deletions were quite variable. Nevertheless, the pattern of inheritance was clearly autosomal dominant (Figure 14–7). In all probability, a protein encoded by a mutant nuclear gene interacts with the mitochondrial DNA in such a way that the probability of deletions occurring is increased. This vividly illustrates the complexity of the factors influencing mitochondrial inheritance and function.

SUMMARY

Mitochondria are the principal sites of ATP production, which occurs by the process of oxidative phosphorylation. The main structural features of mitochondria are the two membranes, the matrix or interior of the organelle, and the cristae, which are projections of the inner membrane into the matrix.

Most human cells contain several hundred mitochondria. Mitochondria contain DNA, which consists of 16,569 bp in humans; it encodes 13 polypeptides, 22 tRNAs, and 2 rRNAs. There are usually 5 to 10 copies of mitochondrial DNA in each mitochondrion. Most of the several hundred polypeptides that are found in mitochondria are encoded by nuclear genes.

It is believed that mitochondria are the descendants of prokaryotes that became endosymbionts in primitive eukaryotes long ago, and that most of the original genetic information has been transferred to the eukaryotic nucleus. Some evidence supporting this hypothesis is that several aspects of protein synthesis in mitochondria are more similar to protein synthesis in present day prokaryotes than in eukaryotes.

Mammalian embryos derive all of their mitochondria from the oocyte; sperm mitochondria do not enter the fertilized egg. This maternal inheritance produces a distinctive pattern in pedigrees, which helps to identify mitochondrial genetic diseases.

Mitochondrial mutations often display genetic heterogeneity within an individual (heteroplasmy). The percentage of mutant mitochondrial DNAs may vary from cell to cell, from tissue to tissue, and may change within a tissue as an individual ages.

Mitochondrial DNA mutates about 10X faster than nuclear DNA. Diseases resulting from mitochondrial mutations are caused by malfunctions of the respiratory chain; they are basically the result of reduced ATP production. The phenotypic effects of mitochondrial mutations primarily reflect the extent to which a tissue relies on oxidative phosphorylation.

Leber's hereditary optic neuropathy (LHON) is characterized by loss of vision, beginning in the third decade of life. One common mutation that causes LHON is a missense mutation in one polypeptide of NADH dehydrogenase.

Myoclonic epilepsy and ragged red fiber disease (MERRF) is characterized by various abnormalities in the central nervous system and in skeletal and cardiac muscle function. MERRF has been shown to be caused by a mutation in one of the mitochondrial tRNAs.

Kearns-Sayre syndrome is characterized by several abnormalities in the eyes, the nervous system, and the muscles. It is associated with large deletions within the mitochondrial genome, which must have severe effects on mitochondrial gene expression. Kearns-Sayre mutations are always heteroplasmic.

REFERENCES

Alberts, B., Bray, D., Lewis, J., Raff, M., Roberts, K., and Watson, J. D. 1989. Molecular biology of the cell, 2nd ed. Garland Publishing, New York.

Anderson, S., Bankier, A. T., Barrell, B. G., et al. 1981. Sequence and organization of the human mitochondrial genome. Nature 290:457–465.

Attardi, G. and Schatz, G. 1988. Biogenesis of mitochondria. Ann. Rev. Cell. Biol. 4:289–333.

Cann, R. L., Stoneking, M., and Wilson, A. C. 1987. Mitochondrial DNA and human evolution. Nature 325:31–36.

Capaldi, R. A. 1988. Mitochondrial myopathies and respiratory chain proteins. Trends in Biochemical Sciences 13:144–148.

Grivell, L. A. 1989. Mitochondrial DNA: small, beautiful and essential. Nature 341:569–571.

Hartl, F-U. and Neupert, W. 1990. Protein sorting to mitochondria: evolutionary conservations of folding and assembly. Science 247:930–938.

Kadenbach, B. and Mueller-Hoecker, J. 1990. Mutations of mitochondrial DNA and human death. Naturwissenschaften 77:221–225.

Lander, E. S. and Lodish, H. 1990. Mitochondrial diseases: gene mapping and gene therapy. Cell 61:925–926.

Lewin, B. 1990. Genes IV. Oxford University Press, Oxford, England.

Linnane, A. W., Marzuki, S., Ozawa, T., and Tanaka, M. 1989. Mitochondrial DNA mutations as an important contributor to aging and degenerative diseases. Lancet, March 25,1989:642–645.

Palca, J. 1990. The other human genome. Science 249:1104–1105.

Prescott, D. M. 1988. Cells. Jones and Bartlett, Boston.

Shoffner, J. M. and Wallace, D. C. 1990. Oxidative phosphorylation diseases: disorders of two genomes. Adv. Hum. Genet. 19:267–330.

Vigilant, L., Stoneking, M., Harpending, H., Hawkes, K., and Wilson, A. C. 1991. African populations and the evolution of human mitochondrial DNA. Science 253:1503–1507.

Wallace, D. C. 1989a. Report of the committee on human mitochondrial DNA. Cytogenet. Cell Genet. 51:612–621.

Wallace, D. C. 1989b. Mitochondrial DNA mutations and neuromuscular disease. Trends in Genetics 5:9–13.

Zeviani, M., Servidei, S., Gellera, C., et al. 1989. An autosomal dominant disorder with multiple deletions of mitochondrial DNA starting at the D-loop region. Nature 339:309–311.

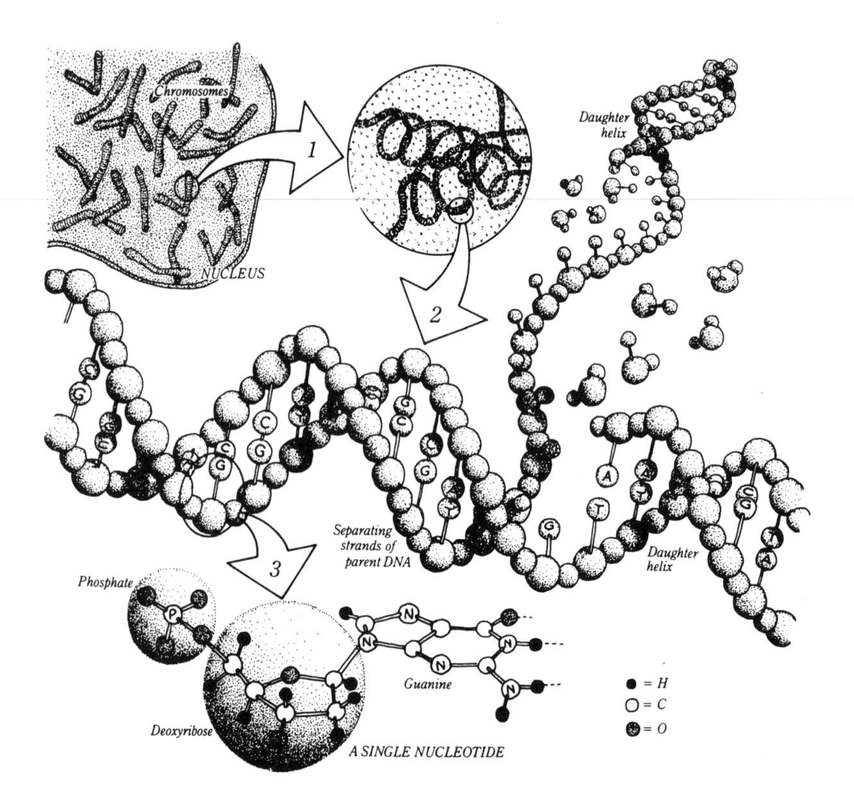

The human genome at four levels of detail. Apart from reproductive cells (gametes) and mature red blood cells, every cell in the human body contains 23 pairs of chromosomes, each a packet of compressed and entwined DNA (1, 2). Each strand of DNA consists of repeating nucleotide units composed of a phosphate group, a sugar (deoxyribose), and a base (guanine, cytosine, thymine, or adenine) (3). Ordinarily, DNA takes the form of a highly regular double-strand helix, the strands of which are linked by hydrogen bonds between guanine and cytosine and between thymine and adenine. Each such linkage is a base pair (bp); some 3 billion bp constitute the human genome. The specificity of these base-pair linkages underlies the mechanism of DNA replication illustrated here. Each strand of the double helix serves as a template for the synthesis of a new strand; the nucleotide sequence (i.e., linear order of bases) of each strand is strictly determined. Each new double helix is a twin, an exact replica, of its parent. (Figure and caption text provided by the LBL Human Genome Center. Reproduced from Human Genome: 1991–1992 Program Report)

15 The Human Genome Project

The seeds of the Human Genome Project were sown in 1977, when simple and efficient methods for sequencing DNA were described. Before long, the possibility of sequencing the entire human genome must have germinated in dozens of minds independently. At first it was no more than wishful thinking, a fantasy so extreme that one would mention it only to one's closest friends, and even then in a bantering tone, lest one be suspected of having lost one's grasp on reality. But as the years passed, and the DNA sequences accumulated in the data banks at an ever-increasing rate, the ultimate extrapolation began to seem feasible.

The first serious group discussion of the possibility of sequencing the entire human genome took place at a meeting in Alta, Utah, in December, 1984. The meeting was sponsored by the Department of Energy (DOE) for the purpose of solving the problem of detecting extremely low levels of mutations in humans exposed to radiation and other environmental hazards. During the meeting, participants realized that the mutation detection problem required technologies and a level of effort that would be similar to the requirements for sequencing the human genome.

Further discussion of the feasibility of producing a complete human genome sequence occurred at workshops held in Santa Cruz, California, in 1985 and at Santa Fe, New Mexico, in March, 1986, and at a larger meeting in the summer of 1986 on "The molecular biology of *Homo sapiens*," held at Cold Spring Harbor, New York.

It is impossible to credit any one person with bringing the idea forward initially, but the first formal proposal appears to have been Renato Dulbecco's letter to *Science* magazine, published March 7, 1986. Dulbecco focused on the potential benefits to cancer research, but one can readily argue that any aspect of human biology in which detailed knowledge of gene structure is important would benefit from the availability of the complete genomic sequence. One of the most vocal early proponents of the Human Genome Project was Walter Gilbert, who had already won a Nobel Prize for developing a method to sequence DNA (Gilbert, 1987).

Initial Concerns

As soon as public discussion of the possibility of completely analyzing the human genome began, questions about feasibility, impact of the project on the rest of biological research, definition of goals, and potential misuses of the information began to pour in from all directions.

Cost

The human genome contains *three billion base pairs!* In 1986, it was commonly estimated (by optimists) that one skilled person could sequence 100,000 base pairs (100 kbp) per year at an average cost of $1/base pair. The money was one problem: there had never been a biological project that cost anywhere near $3,000,000,000, but that amount was well within the range of precedent set by high-energy physics. The prospect of biology entering the same financial league as physics and space exploration was not daunting to everyone; to some, it was exhilarating.

Use of trained personnel was another serious problem. At 100 kbp/year, it would take at least 30,000 person-years to sequence the human genome once. Of course, the full task would be much larger, both in terms of cost and effort, because of the need to confirm the initial sequences and to do comparative studies on other organisms. Nevertheless, the proposal to sequence the human genome gained momentum.

The Threat of "Big Science"

By the middle of 1986, there was a substantial group of influential scientists who were convinced that the human genome *could and should* be sequenced. They focused their efforts on questions of *when and how*. The Department of Energy had already expressed enthusiastic interest. They had a long tradition of organizing big science, and the human genome sequence would be a big project—no doubt about it. Moreover, DOE had been involved in biology for 40 years, mostly in the area of radiation effects, but more recently in molecular genetics. DOE co-sponsored GenBank, one of the world's two major data storage facilities for DNA sequences. DOE also ran the National Laboratory Gene Library Project, which prepares clone libraries of DNA from specific human chromosomes (Chapter 3).

A potential major role for DOE in the Human Genome Project was appealing for another reason—avoiding dilution of resources. Several scientists were quick to point out that a multi-billion dollar project could have a devastating effect on the rest of biomedical research, unless it was separately funded. They were concerned that if the National Institutes of Health—the usual funding agency for such research—were given primary responsibility for the Human Genome Project, there might not be enough money provided to maintain the current level of support in other areas. The possibility that DOE could get a special appropriation more easily than NIH seemed worth considering.

On the other hand, some people worried that DOE's tradition of big, centralized science projects was not in the best interests of biology in general. Do we want a Human Genome Institute that will soak up a large fraction of the skilled people in the profession for the next 10–20 years, they asked? Most scientists found that possibility unappealing.

The Role of Sequencing

One of the major questions of experimental strategy concerned the importance of determining the complete human genomic DNA sequence. Should the sequence be done randomly, in the expectation that all the pieces would fit together eventually, or should we first sequence portions of the genome that we already know to be biochemically interesting, such as those in the vicinity of genes that are responsible for common genetic diseases? Should we do one chromosome at a time? Should we do any massive sequencing at all with present technology, or should we wait until more rapid, automated techniques are available?

Others asked whether the detailed nucleotide sequence was really the primary goal. After all, the coding fraction of the human genome is probably less than 5% of the total DNA, so why bother sequencing all the introns and intergenic DNA? Some argued that a physical map was more important, at least at first. There are various types of physical maps, but in the present context, a physical map is an ordered collection of overlapping cloned DNA segments that together span a given region of the genome.

Physical maps, together with linkage maps, are helpful in identifying genes responsible for diseases where the primary biochemical defect is unknown (Chapter 5), but they do not by themselves identify genes. Physical maps are also a necessity for sequencing. One cannot sequence large pieces of DNA. Sequences in the multi-kilobase range must be assembled from many small sequences, and the existence of ordered, overlapping clones is essential for that process (Chapter 3).

Uses of the Data

It is so easy to assume that knowing the complete human genome sequence is a desirable goal, that it almost seems unnecessary to ask, "What will we do with the information?" Let's break that general question down into more specific questions.

Interpretation of Sequence Data First, when we know the sequence of a long stretch of DNA, will we know where the genes are? In most cases, yes. Sequences that code for most exons will be recognizable because they are ORFs (open reading frames, which are series of nucleotide triplets that contain no stop codons), although many spurious ORFs occur by chance. Consensus sequences for RNA splicing, RNA termination, and so on, are also identifiable in many cases. Undoubtedly, some small exons and possibly even whole genes that code for small proteins and other DNA sequences that code for RNA molecules that are not translated will be missed at first, but these should be a small fraction of the total. Sequences that control the expression of specific genes will probably be the hardest to identify.

Several computerized analysis systems for gene recognition in long blocks of DNA sequence are under development. One of these is already available online. It is called GRAIL (Gene Recognition and Analysis Internet Link). In a recent trial run on 19 human genes that had already been sequenced, GRAIL located 90% of the coding exons that were at least 100 bp long; however, it missed about half of the shorter exons. Another weakness of the program in its present state is that the ends of the exons cannot be precisely identified. However, GRAIL and other gene identification programs are in their infancy; they will surely become more powerful in the near future. Therefore, one of the consequences of knowing the complete human genomic sequence is that we will have a very good count of the total number of human genes (a major problem discussed in Chapter 2).

If we know where a gene begins and ends, and which sections are open reading frames, we can deduce the amino acid sequence of the encoded protein. Does that mean that we know the function of the protein? Not necessarily. It is often possible to find out what known proteins are related to a protein encoded by a newly sequenced gene, using computerized comparison methods; but if the protein in question has no relatives in the data bank, its function may be unguessable from its amino acid sequence. Therefore, the complete human genomic sequence will tell us something about gene function, but many thousands of person-years of biochemistry will remain to be done before we know what every gene product does and how expression of that gene is controlled.

However, it is widely expected that significant advances in our understanding of how protein structure and function are derived from the primary amino acid sequence will be made in the next few years. By the time that we have sequenced most of the human genome, it should be possible to deduce the function of many proteins from the amino acid sequence, which in turn can be read from the nucleotide sequence of the gene. Will large families of genes be discovered that have no relatives in currently available sequences? One fascinating possibility is that we might find whole tribes of proteins that carry out as-yet-unknown neural functions.

Indirect Consequences of Increased Genomic Knowledge Last, but surely not least, there are questions about possible negative consequences of knowing the complete sequence of the human genome and the functions mediated by each of our genes. To be sure, the sequence produced by the Human Genome Project will only be a reference standard. It may come mostly from one individual, or more likely, it will be a composite of DNA from various persons. The little variations that make each of us unique will not be revealed by the reference sequence, but in the course of determining that sequence, thousands of cloned DNAs will be produced which could be used as probes with anyone's DNA.

The Human Genome Project will accelerate the acquisition of probes for genes that determine an individual's susceptibility to heart disease, to certain types of cancer, to diabetes, and to some types of mental illness. It would be naive to assume that at some future date we will all know how long each of us will live (barring accident) and what we will die from, based on our individual genotypes. Nevertheless, we will be able to know a lot about our own metabolic weaknesses; and although this sounds good at first, it does carry the potential for limiting individual freedom. It also raises serious problems about the protection of privacy. Should insurance rates be based on genotype? Should people be denied employment in certain industries because their genotypes make them more likely to suffer occupation-related illness than other people?

In a more general sense, exhaustive knowledge of the human genome and its functions will be viewed by some people as de-humanizing. It cannot fail to make us more aware of the extent to which we are chemical machines. Probably we shall never know everything about the human brain, but we may well discover the neurochemical basis for most of our behavior, including emotions that seem sacred to us now. Most of us would have few objections if hunger and sexual desire could be explained as patterns of gene expression, but there are many who would deem it a pity if love, faith, and poetic inspiration could be reduced to the interactions of a few dozen proteins! We may find that our concept of human nature changes as our knowledge of human genetics expands.

These and many other ethical, legal, and social problems will not be the result of the Human Genome Project by itself. The Human Genome Project has accelerated the rate of acquisition of genetic knowledge, but even without the HGP, gradually increasing understanding of the human genome and its functions would have brought the same issues to our attention eventually. We may blame the HGP for forcing us to deal with more problems sooner than would otherwise have occurred, but we cannot accuse it of creating problems that are unique consequences of the project itself.

Organization of the Human Genome Project

Early Stages

Despite the evident uncertainties of many types, interest in the Human Genome Project grew rapidly, and a major genome analysis program is now a reality. Indeed, for proponents of the project, its inevitability was obvious from the beginning, as exemplified by a comment written in the May 1, 1987, issue of *Science* by the magazine's editor, Daniel Koshland, ". . . the obvious answer as to whether the human genome should be sequenced is, 'Yes. Why do you ask?' "

Nevertheless, decisions on projects of this magnitude are not made capriciously. The pros and cons of the Human Genome Project have been studied carefully by at least two government agencies that are not associated with DOE or NIH, which were obviously the potential prime contractors. The National Research Council, a branch of the National Academy of Sciences, established a committee, chaired by Bruce Alberts, while the Office of Technology Assessment, a congressional agency, carried out an independent evaluation with a different advisory committee.

The National Research Council and Office of Technology Assessment committees published their reports in 1988 (see references). Both groups concluded that mapping and sequencing the human genome was a goal worthy of support at the national level, with government funds that should be specifically designated for the project, in order to avoid reduction of funding for other areas of research. The NRC committee made several recommendations, one of which was to defer massive sequencing projects until technological innovations had made it possible to sequence DNA at least 10X faster than can be done with present machines. In the first few years of the project, emphasis should be given to a high-resolution genetic map and to physical maps at various levels of resolution.

Another major recommendation of the NRC report was that the Human Genome Project include analysis of several non-human DNAs, such as the laboratory mouse (*Mus musculus*), the fruitfly (*Drosophila melanogaster*), a nematode worm (*Caenorhabditis elegans*), yeast (*Saccharomyces cerevisiae*), and *E. coli*. There are two main reasons for comparative studies. First, it is unethical to use humans as experimental organisms; therefore, the function of a human gene must often be learned by studying its homologs in simpler creatures. Second, evolution of protein-coding DNA sequences has been much more conservative than evolution of the rest of the genome. Accordingly, identification of genes will often be facilitated by comparing human sequences with sequences from other organisms, especially the mouse, looking for genomic segments that have very similar sequences in both species. These comparative studies are essential to the goal of understanding how the human genome functions, but it must be realized that they significantly increase the scope and the cost of the overall project.

In regard to organization, the NRC group felt that a lead agency should be designated. Preference for NIH was based partly on its historical role as major supporter of biomedical research in individual laboratories, and partly on its tradition of peer review of applications. The OTA report concluded that it might not be necessary to name a lead agency, but if one were chosen, NIH would be the logical choice. However, NIH moved slowly. Apparently, the administrators were concerned about possible disruptions of their normal policy of supporting diversified research by several thousand small-to-medium laboratories throughout the nation.

By the spring of 1988, however, a strategy had been formulated, and NIH Director James Wyngaarden announced that a separate Office of Human Genome Research, reporting directly to him rather than to one of NIH's existing institutes, would be established. James Watson, who shared the Nobel Prize in 1962 for deducing the double-helical structure of DNA, was invited to be the director. Watson wanted the job, but he was reluctant to sever his connection with the Cold Spring Harbor Laboratory, where he has been director for many years.

A compromise was reached; Watson would run the Human Genome Project for NIH on a part-time basis, spending two days per week in Bethesda; he would continue as director of the Cold Spring Harbor Laboratory. He formally accepted the NIH position in September, 1988. On October 1, 1989, the status of Watson's office was upgraded to that of an independent funding unit within NIH, with authority to award grants and contracts. It was renamed the National Center for Human Genome Research (NCHGR). Watson resigned the directorship of NCHGR in May, 1992.

At about the same time, the Department of Energy reorganized its genetic research program. Ongoing research in human genetics at Los Alamos and Lawrence Livermore National Laboratories was expanded and both programs were elevated to the status of Genome Centers. A new Human Genome Center was established at the Lawrence Berkeley Laboratory (LBL). Charles Cantor became Principal Scientist for the entire DOE program in human genetics. Cantor was one of the principal developers of the pulsed-field gel electrophoresis technique for separation of very large DNA molecules, which will be of major significance in genome mapping projects (Chapter 3).

An important organizational step took place in early fall, 1988, with the signing of a "Memorandum of Understanding" (MOU) between DOE and NIH. The MOU formally establishes "interagency cooperation that will enhance the human genome research capabilities of both agencies." It sets forth joint goals of the Human Genome Project and establishes a joint advisory committee to advise and review the relevant activities of NIH and DOE. An interagency working group has also been formed to deal with more specific decisions relevant to "the need for and the feasibility of initiating a variety of cooperative and complementary programs and projects in order to advance knowledge in human genome research." The organizational relationships between NIH and DOE in connection with the Human Genome Project are summarized in Figure 15-1.

The MOU also specifies a variety of measures for information exchange which will ensure that scientists working under the auspices of DOE or NIH will be able to avoid unnecessary duplication of effort and will be frequently updated on each others' progress. The need for information exchange with other agencies active in human genome research, both domestically and internationally, is also an explicit element of policy.

Current Activities of the Human Genome Project

Activities related to the Human Genome Project have proceeded at a rapid rate since the latter part of 1988. The 15-year project formally began in the fall of 1990 (fiscal year 1991), and perspectives on its future were published in October of that year by the two principal scientists (Cantor, 1990; Watson, 1990). In addition, an important report outlining the plans for the first five years, prepared jointly by DOE and NIH, appeared in March, 1990, with the title, "Understanding Our Genetic Inheritance. The U. S. Human Genome Project: the First Five Years, FY 1991–1995."

Among the goals set forth in that document are: (1) completion of a human genetic map with markers spaced an average of 2 centimorgans apart, with no gaps larger than 5 cM; (2) assembly of a physical map of the human genome with markers spaced at approximately 100,000 bp intervals; (3) improvement of DNA sequencing technology and sequencing of 10 million bp of human DNA in a few large, continuous stretches; (4) preparation of a genetic map of the mouse; (5) development of effective software and database designs to support the mapping and sequencing of large genomes; and (6) identification of the major ethical, legal, and social implications of the Human Genome Project and development of policy options to address them.

Funding for the Human Genome Project was begun by Congress at

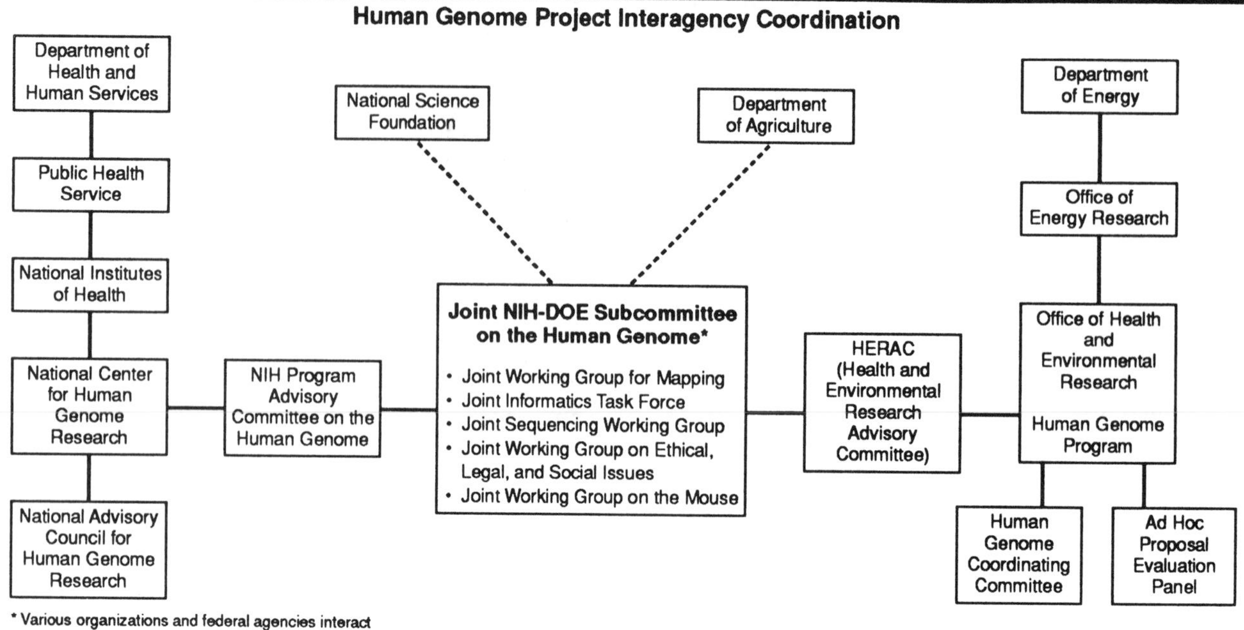

Figure 15–1

Human Genome Project Interagency Coordination. (From Human Genome: 1991–92 Program Report. U.S. Department of Energy, Washington, D.C., p. 65)

modest levels in fiscal year 1988 ($17.2 million for NIH and $10.7 million for DOE) and is expected to grow to a total of about $200 million per year in the near future, although current federal budgetary problems may postpone the attainment of that level of support. Present plans call for about two-thirds of the total to go to the National Center for Human Genome Research at NIH. The money will be used in several ways. One will be to award grants to individual scientists, following established procedures for peer review of investigator-initiated proposals. Another use of the funds will support a training program for pre-doctoral, post-doctoral, and senior investigators.

A major initiative by NCHGR will be the establishment of 14 or 15 centers for human genome research throughout the country. These centers will support several senior investigators and all of their associates. Each will have a specific goal. The first four centers were established in late 1990. Their locations and goals are: Washington University at St. Louis, mapping of human chromosomes 7 and X; University of California at San Francisco, mapping of human chromosome 4; Massachusetts Institute of Technology, mapping of the mouse genome, with emphasis on chromosomes 1,11, and X initially; and University of Michigan, mapping of genes responsible for human diseases. As of December, 1991, additional center grants have been awarded to Baylor College of Medicine, the University of Utah, and Children's Hospital of Philadelphia.

In response to widespread concern about possible negative consequences of increased knowledge of the human genome, NCHGR has set aside 3% of its budget for a program on the Ethical, Legal and Social Implications (ELSI) of genomic knowledge. The program sponsors interdisciplinary conferences, where scientists, legal scholars, ethicists, sociologists, medical professionals, and others are brought together to consider the implications of our expanding knowledge of the human genome. Research projects, such as a study on quality control in genetic testing and interpretation, are also supported.

Most of DOE's contribution to the Human Genome Project was initially supposed to be focused on technology development and informatics, and most of that work will be done at the National Laboratories, where there will also be an in-house traineeship program. DOE also awards some grants for extramural research. However, the roles of NIH and DOE in the Human Genome Project are becoming less distinct than was originally envisioned, and it now seems likely that each agency will do whatever the administrators decide needs to be done. Because of the high level of inter-agency communication on the Human Genome Project, there is very little potential for wasteful duplication of effort.

At Lawrence Livermore National Laboratory (LLNL), chromosome 19 is under intensive study, and contigs covering more than 70% of the chromosome have been created. Los Alamos National Laboratory (LANL) is concentrating on chromosome 16, where about 90% of the chromosome has been assigned to contigs. In addition, both laboratories are continuing to develop overlapping clone libraries of individual human chromosomes, using lambda phage and cosmids. LANL is preparing libraries of chromosomes 4, 5, 6, 8, 10, 13, 14, 15, 16, 17, 20, and X; LLNL is preparing libraries of chromosomes 1, 2, 3, 7, 9, 11, 12, 18, 19, 21, 22, and Y. Preliminary work on construction of YAC libraries from individual human chromosomes is underway.

The STS Proposal

As planning for the Human Genome Project progressed, a major logistical problem became apparent. At first, it was assumed that there would be a need for a national clone bank; that is, a central facility where cloned segments of human DNA in recombinant organisms would be collected, maintained, and distributed to the research community upon request. However, the project soon began to outgrow the bounds of feasibility.

The scale of the problem is illustrated by the following calculations. If a physical map of an average chromosome (about 120 Mbp) is constructed with cosmids that carry an average insert of 40 kbp, then an end-to-end set of cosmids to cover the whole chromosome requires 3000 cosmids. Of course, contigs always require overlapping clones, so allowing for 50% overlap, we can estimate that roughly 6000 cosmids would be needed for each chromosome. Multiplying by 24 types of chromosomes implies 144,000 cosmids for the entire genome.

Should each of those cosmids be kept in perpetuity, to be distributed to any interested investigator upon demand? If so, what would it cost? The answer to the cost question depends upon the level of quality control that the clone bank would be required to provide, but leaving aside that uncertainty, all estimates of the cost of maintaining such a collection involve many millions of dollars per year, and the cost would continue indefinitely.

The clone bank problem was aggravated by the virtual certainty of technical obsolescence; that is, new cloning vectors are constantly being devised. Would a researcher 10 years from now want to work with clones in today's vectors? Probably not. On the other hand, an investigator who was trying to identify a gene responsible for a specific disease, and who had narrowed the search to one or two Mbp of DNA, would certainly be spared a lot of effort and expense if a set of clones spanning the region of interest could be ordered from a central repository.

This logistical problem was significantly simplified when four distinguished geneticists devised the *STS proposal* (Olson et al., 1989). An STS is a *sequence tagged site*, a region of a few hundred base pairs that has been sequenced, and that contains unique sequences of about 20 bp at each end, which can be used as PCR primers (see Box 4–1). The use of PCR technology allows an investigator anywhere in the world to replicate the DNA defined by an STS cheaply and quickly. That DNA can then be used as a probe for further analysis of neighboring regions; for example, to initiate a chromosome walk or to search for polymorphic sites.

The STS proposal specifies that any lab that reports a contig from a known genomic location should also publish the sequence of a small region from that contig, which can be used as an STS. It is desirable to have STSs at least every 100 kbp. Quite recently, DOE has announced plans to clone and sequence as many human cDNAs as possible, which will thereby give everyone access to expressed genes. Those cDNA sequences will also define STSs, although there will be some problems caused by adjacent nucleotides in cDNA sometimes representing exons separated by thousands of bp in introns in genomic DNA. The British genome program will also emphasize cDNAs, and other groups are expressing interest, recognizing that a physical map of expressed genes is inherently more interesting than a map based largely on anonymous sequences. STSs from cDNAs are currently referred to as Expressed Sequence Tagged Sites.

The use of STSs will probably eliminate the need for clone banks containing hundreds of thousands of clones with inserts of 40 kbp or less. At this time, it not clear whether a national clone bank with large-insert vectors, such as YACs, will be economically justifiable. At 200 kbp per clone, and with an overlap of 50%, the entire human genome could be covered with 30,000 clones. The larger the inserts, the smaller the storage problem will be.

The International Scene

Although serious discussion of a large-scale human genome analysis project first developed in the United States, other countries have expressed interest. As each year passes, it becomes increasingly clear that many nations will be involved. From the beginning, most scientists felt that mapping and sequencing the human genome should be an international effort. After all, there is only one human species, and the results of analyzing our genome will be useful for all of us.

One of the first countries to announce its interest in the Human Genome Project was Japan. In the laboratory of A. Wada, a project was undertaken in 1981 to develop super-sequencing machines. Initially, the goal was a machine that could do one million bases per day, but that goal has been scaled back to 100,000 bases per day. A fully automated system for DNA analysis was announced in the summer of 1991. It is a series of machines that carry out the various steps, from initial purification of DNA to computerized analysis of the sequence. Its output is estimated at approximately 100,000 bp/day. Current commercial machines, both in the United States and Japan, can produce 10,000–15,000 bp/day. Otherwise, the extent to which Japan will become involved in human genome analysis is not yet clear. There is much discussion among scientists, government officials, and industrial leaders, but no consensus has been reached on long-term strategy. A DNA analysis center is planned, with an initial budget of less than

$10,000,000 per year. It will concentrate on mapping until advanced sequencing technology has been adequately developed.

On the other side of the world, the European Community has established a program to sequence the genome of the yeast, *S. cerevisiae*. A consortium of 35 laboratories collaborated to produce the sequence of yeast chromosome III (Oliver et al., 1992), the first complete sequence of any eukaryotic chromosome. This chromosome contains 315 kbp of DNA; 182 open reading frames (ORFs), each potentially encoding a protein of at least 100 amino acids, were identified. Surprisingly, more than half of the ORF sequences were unrelated to any known proteins. This indicates that our knowledge of the genetics of even this relatively well-known and relatively simple organism is far from complete. As sequence information about the rest of the yeast genome accumulates, it will provide an informative model for the interpretation of data on higher organisms, including humans.

Whether there will be a multi-national project on the human genome in Europe is not yet clear. In Germany there is widespread opposition to learning a lot about human genetics before regulations have been established to prevent misuse of the information. The ghost of Nazi eugenics haunts the dreams of present-day Germans.

There is a project on human chromosome X underway in Italy, and the Russians have begun a human genome project, although its specific goals are not yet clear. Britain and the United States have established a collaborative effort to complete the sequence of the worm, *C. elegans*, which already has a fairly detailed physical map. The initial goal is to sequence 3 Mbp by the end of 1993. This work is being done at the Medical Research Council Laboratory of Molecular Biology in Cambridge, England, and at Washington University in St. Louis. Significant support for other human genome projects has now become a standard item in the Medical Research Council budget in Britain; sequencing and mapping expressed sequences will be emphasized.

A substantial program of research on the human genome has recently gotten underway in France. Formally announced in January, 1991, it is budgeted for $50 million in 1992. The French program will emphasize development of automated mapping technologies, including a series of robots that will become commercially available. Biological research will be concentrated on regions of the genome that contain coding sequences.

It is likely that more countries will be heard from, as the work in the United States gathers momentum. At one time, James Watson spoke out in favor of carving up the human genome on a national basis, one chromosome to country *a*, another to country *b*, and so on; but the idea didn't find much support. Whatever happens, it is clear that there is widespread desire for international cooperation among scientists. Whether chauvinistic reasoning by politicians will ultimately lead to unnecessary duplication of effort remains to be seen.

There is also an international Human Genome Organization (HUGO) which was organized in April, 1988, by an independent group of scientists for the purpose of assisting with coordination of national efforts, facilitating exchange of research resources, encouraging public debate, and providing information on the implications of human genome research. The 42 founding members were increased to more than 200 in 1990. The organization has headquarters in England (McKusick, 1989), with branch offices in the United States, Japan, and Moscow. HUGO now has primary responsibility for organizing workshops on individual human chromosomes; it will also coordinate projects on sequencing and mapping cDNAs.

In summary, a worldwide effort to analyze the human genome is underway, with the United States taking the leading role so far. It is likely that the goal of having virtually all of the sequence by the year 2005 is attainable, assuming that efforts to devise more rapid and cheaper sequencing technologies are successful, and that general economic conditions do not necessitate major reductions in funding. However, we cannot expect to understand the normal functions of more than a minor fraction of all human genes by that time, nor the regulatory mechanisms that govern the expression of most genes, nor the effects of more than a few of the nearly infinite possibilities for mutations that alter gene expression and lead to disease. Completion of the Human Genome Project, as it is presently defined, will provide the foundation upon which many decades of additional research will be based.

SUMMARY

The United States Human Genome Project is a long-term effort to develop detailed genetic and physical maps of the human genome, to locate and identify virtually all human genes, and to determine the DNA sequence of most or all of the human genome. It is organized and supported primarily by the Department of Energy (DOE) and the National Institutes of Health (NIH). Related programs exist in several other countries and there is extensive international cooperation.

Detailed analysis of the human genome became possible as an outgrowth of the development of efficient DNA sequencing methods. The Department of Energy's program on the biological effects of radiation created a need to identify very rare changes in DNA (mutations), which required automation of DNA analysis techniques that would also make it possible to analyze the entire human genome.

Early discussions of the feasibility of undertaking full analysis of the human genome dealt with problems of cost, optimal use of trained personnel, whether the full DNA sequence should be the primary goal, and uses of the information to be obtained.

The official 15-year Human Genome Project began in October, 1991. Somewhat more than half of the funds provided by Congress (which are now between $100 and $200 million per year) go to the National Center for Human Genome Research at NIH, an independent funding unit that awards grants and contracts to investigators throughout the nation. The rest of the money goes to DOE, which conducts research on human genetics at three national laboratories and also awards grants to outside investigators.

The Human Genome Project also supports research on the mouse and several non-vertebrate organisms. Information on simpler genomes is helpful in analyzing the human genome. Moreover, the ability to transfer human genes to mice and to create models of human disease in mice is invaluable for learning the functions of human genes and their variants.

Approximately 3% of the funds for the Human Genome Project are reserved for studies of the ethical, legal, and social implications of genomic information.

The Human Genome Organization (HUGO) was formed by an international group of scientists to help coordinate national programs of research on the human genome. They are now responsible for organizing international workshops on individual human chromosomes. HUGO's other activities include helping to coordinate projects whose goal is to map and sequence expressed genes (via cDNAs).

REFERENCES

Cantor, C. R. 1990. Orchestrating the Human Genome Project. Science 248:49–51.

Department of Energy. 1990. Human Genome. 1989–90 Program Report. DOE Office of Health and Environmental Research, Washington, DC.

Dulbecco, R. 1986. A turning point in cancer research:

sequencing the human genome. Science 231:1055–1056.

Gilbert, W. 1987. Genome sequencing: creating a new biology for the twenty-first century. Issues in Science and Technology, Spring, 1987, pp. 26–35.

Human Genome Project Interagency Coordination. From Human Genome: 1991–92 Program Report. U.S. Department of Energy, Washington, D.C., p. 65.

Koshland, D. E., Jr. 1987. Sequencing the human genome. Science 236:505.

McKusick, V. A. 1989. The Human Genome Organization: history, purposes, and membership. Genomics 5:385–387.

National Institutes of Health—Department of Energy. 1990. Understanding our genetic inheritance. The U.S. Human Genome Project: the first five years, FY 1991–1995. U.S. Gov't Printing Office.

National Research Council. 1988. Mapping and sequencing the human genome. National Academy Press, Washington, D.C.

Office of Technology Assessment. 1988. Mapping our genes. Genome projects: how big, how fast? Johns Hopkins Press, Baltimore.

Oliver, S. G., van der Aart, Q. J. M., Agostini-Carbone, M. L., et al. 1992. The complete DNA sequence of yeast chromosome III. Nature 357:38–46.

Olson, M., Hood, L., Cantor, C., and Botstein, D. 1989. A common language for physical mapping of the human genome. Science 245:1434–1435.

U.S. Department of Energy. 1992. Human Genome: 1991–92 Program Report. U.S. Department of Energy, Office of Energy Research, Office of Health and Environmental Research, Washington, D.C.

Watson, J. D. 1990. The Human Genome Project: past, present, and future. Science 248:44–49.

Index